Particle Deposition and Aggregation

Series Editor
Professor R. A. Williams
University of Exeter, UK

Series Advisors
Professor J. Gregory
University College, London, UK

Dr. K. Nishinari
National Food Research Laboratory, Japan

Professor D. C. Prieve
Carnegie-Mellon University, USA

Professor T. Ring
University of Utah, USA

Professor H. N. Stein
Technical University of Eindhoven, The Netherlands

Dr. D. J. Wedlock
Shell Research Ltd, UK

Series Titles
Colloid and Surface Engineering: Applications in the process industries
Controlled Particle, Droplet and Bubble Formation

Particle Deposition and Aggregation

Measurement, Modelling and Simulation

M. Elimelech
J. Gregory
X. Jia
R. A. Williams

BUTTERWORTH
HEINEMANN

Butterworth-Heinemann Ltd
Linacre House, Jordan Hill, Oxford OX2 8DP

A member of the Reed Elsevier plc group

OXFORD LONDON BOSTON
MUNICH NEW DELHI SINGAPORE SYDNEY
TOKYO TORONTO WELLINGTON

First published 1995

British Library Cataloguing in Publication Data
Elimelech, M.
Particle Deposition and Aggregation:
Measurement, Modelling and Stimulation. –
(Colloid & Surface Engineering Series)
I. Title II. Series
620.44

ISBN 0 7506 0743 2

Library of Congress Cataloguing in Publication Data
Particle deposition and aggregation: measurement, modelling, and
simulation/M. Elimelech . . . [et al.]
p. cm. — (Colloid and surface engineering series)
Includes bibliographical references and index.
ISBN 0 7506 0743 2
1. Particles. 2. Agglomeration. I. Elimelech, M. (Menachem)
II. Series.
TP156.P3P35 94–41069
660'.29485–dc20 CIP

Composition by Genesis Typesetting, Laser Quay, Rochester, Kent
Printed and bound in Great Britain by
Hartnolls Limited, Bodmin, Cornwall

Contents

Preface

Deposition and aggregation of small solid particles from a liquid dispersion are encountered in a variety of situations in manufacturing and process technologies and in natural aquatic environments. Deposition may occur onto a surface immersed into the suspension, that is of macroscopic dimensions compared with the particle, or solids may settle due to gravity to form a sediment at the bottom of the vessel containing the mixture. Aggregation involves the formation of assemblies of particles brought together by collisions, held in contact by surface forces. From a mechanistic point of view, deposition onto a macroscopic surface can be considered to be a limiting case of particle–particle aggregation in which one of the particles has an infinite dimension.

These phenomena are of significant industrial importance in the chemical, environmental, electronics, mineral and biological sectors. For instance, deposition processes are used in preparing magnetic tapes, in water purification using packed-bed filters, and in selective capture of solids, cells and macromolecular species. Aggregation to form larger particle assemblies is also an important step in enhancing the rate of gravitational settling to allow improved dewatering of suspensions, and in selective flocculation processes. It is often desirable to control the rate and extent of aggregation or deposition to control the morphology of the aggregates or deposits that are being formed. Many industrial operations involve both aggregation and deposition simultaneously.

Particle deposition and aggregation also play an important role in the transport of pollutants in aquatic environments and in the geochemical cycling of trace elements in the ocean. The transport of viruses, bacteria and particle-bound reactive pollutants in soils and ground waters is controlled by the rate of particle deposition. Aggregation in rivers, lakes and oceans is an important mechanism of modifying the size of particles, and, subsequently, in controlling their settling rates.

This book describes how such processes can be measured, modelled and simulated in a systematic manner. It brings together the necessary disciplines of colloid and surface chemistry, hydrodynamics, experimental methods, and computational methods to present a unified approach to this problem. Advances in instrumental and computational techniques are enabling the engineering of these complex colloid systems to be undertaken with greater confidence, although many problems still need to be addressed. For example, the advantages and limitations of employing computational simulation are discussed. The three aspects of *experimental measurement, modelling and simulation* are intended to provide an introduction to the key concepts involved augmented by bibliographic recommendations.

A reasonable level of mathematical knowledge is assumed within the text, although basic concepts of physical and surface chemistry and micro-hydrodynamics are introduced at an appropriate level. The book should be useful to graduates working in process and environmental engineering research or industrial development at a postgraduate level, and to scientists who wish to extend their knowledge into more realistic process conditions in which the fluid hydrodynamics and other complicating factors must be accommodated.

The book is divided into four parts. Part I presents the theoretical principles governing deposition and aggregation phenomena, including a discussion of the forces that exist between particles and the hydrodynamic factors that control the movement of the particles and suspending fluid. Part II introduces methods for modelling the processes, first at a simple level (e.g. single particle–surface, single particle–single particle interactions in model flow conditions) and then describes the simulation protocols and computation tools which may be employed to describe more complex (multiple-particle interaction) systems. Part III summarizes the experimental methods of quantifying aggregating and depositing systems and concludes with a comparison of experimental results with those predicted using simple theoretical predictions. Part IV is largely based on illustrative examples to demonstrate the application of simulation and modelling methods to particle filtration, aggregation and transport processes.

A separate nomenclature is provided for each chapter. Wherever possible the use of notation is standardized throughout the text, with the same symbol being used consistently for important variables. Inevitably some conflicts are bound to arise in a multidisciplinary work and, therefore, we have chosen to adopt the nomenclature that is most commonly used within the subject area under consideration. The meaning of the symbols employed should be obvious from the context and can be confirmed by consulting the nomenclature at the front of each chapter.

The book is of multiple authorship. The principal areas of interest of the authors were characterization of particle–particle interactions (JG, RAW), hydrodynamics (XJ), experimental and theoretical behaviour of particle deposition on model collectors (ME), experimental and theoretical aspects of aggregation (JG), and simulation and computation methods (RAW, XJ). For this reason the individual chapters may differ slightly in style and approach, nevertheless we have sought to deliver a coherent text. We are particularly grateful to several colleagues who have commented on the different portions of the text, undertaken proof-reading and provided additional experimental data. These include Dr Tom Dyakowski, Professor Graham A. Davies, Dr Alec E. James, Dr Geoff Lyman and Dr Lianfa Song.

Professor Menachem Elimelech
Department of Civil and Environmental Engineering
University of California
Los Angeles, USA

Professor John Gregory
Department of Civil and Environmental Engineering
University College
London, UK

Dr Xiaodong Jia
Camborne School of Mines
University of Exeter
Redruth, UK

Professor Richard A Williams
Camborne School of Mines
University of Exeter
Redruth, UK

Part I

Theoretical analysis of deposition and aggregation phenomena

Chapter 1

Introduction

Dispersions of particles in liquids are commonly encountered in a wide range of process industries. Dispersed particles can have sizes ranging from fractions of a millimetre down to macromolecular dimensions (a few nanometres). Figure 1.1 gives an overview of typical particle size ranges, together with some related information. Traditionally, a distinction has been drawn between *colloidal* and *suspended* particles, with the former taken as having at least one dimension with a size less than about 1 μm. This boundary is rather arbitrary, but can be broadly justified for a number of reasons:

(a) dispersed particles are very often examined by optical microscopes and it is difficult to resolve particles smaller than the wavelength of ordinary light (say, less than 0.5 μm). Thus colloidal particles, as defined above, would be largely 'invisible' to the optical microscopist. Suspended particles are more easily observed

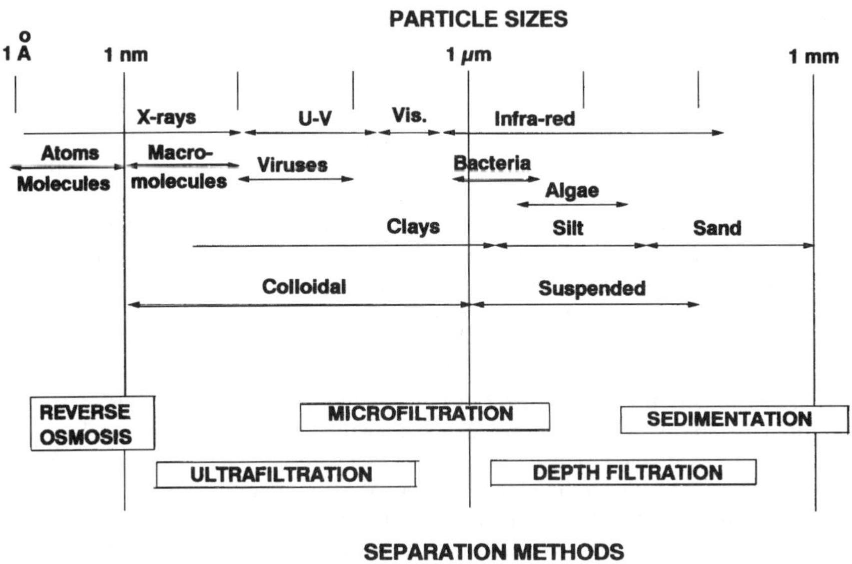

Figure 1.1 *Typical size ranges of particulate material and the wavelengths of different forms of electromagnetic radiation. The conventional boundary between colloidal and suspended matter is at 1 μm. Appropriate solid–liquid separation processes for different particle sizes are also shown*

(b) sedimentation rate is dependent on the square of particle size, and particles smaller than about 1 μm tend to settle very slowly. Thus colloidal particles remain in a dispersed state for long times, whereas suspended particles are fairly rapidly removed by sedimentation under normal gravity
(c) the specific surface of particles (i.e. the surface area per unit mass or volume) depends inversely on particle size. Spherical particles with a diameter of 1 μm have a surface area of about $6\,m^2\,cm^{-3}$ or $3\,m^2\,g^{-1}$ for a material with a specific gravity of 2. Processes occurring at surfaces, notably adsorption, become more significant with smaller particles, and 1 μm is a size below which surface effects begin to predominate
(d) certain interactions between particles, often referred to as colloid interactions or surface forces (to be discussed in Chapter 3), are proportional to the first power of particle size. 'External' forces such as those arising from gravitational and hydrodynamic effects depend more strongly on particle size. For instance, the gravitational attraction force is proportional to the mass of a particle and hence on the cube of its size. Hydrodynamic forces, as a result of flow, depend roughly on the square of particle size. It follows that, as particle size decreases, colloid interactions become increasingly significant relative to external forces. Again, 1 μm is a convenient dividing line.

In natural waters, an operational distinction between 'suspended' and 'dissolved' impurities is often drawn on the basis of particle size. Impurities which are retained on a standard membrane filter (0.45 μm) are defined as 'suspended', whereas those which are small enough to pass through the membrane are 'dissolved'. Clearly, much of the 'dissolved' fraction would be classed as 'colloidal' on the basis of the previous definition.

However, it must not be supposed that there is a sudden change from one type of behaviour to another at a given particle size. In most cases there is a gradual transition, and for this reason the terminology that we employ will not usually imply anything about particle size. We shall refer to particles, colloids, colloidal dispersions and colloidal suspensions, rather indiscriminately.

In this book we shall be concerned with two related aspects of the behaviour of suspended particles – *aggregation* and *deposition*. These processes have many features in common and some important differences. Aggregation involves the association of particles to form clusters. Deposition is the process whereby particles are transported to a surface, where they become attached. These processes are illustrated schematically in Figure 1.2.

An extremely important aspect of the behaviour of dispersed particles is their colloidal stability, which determines whether they remain in a dispersed state or form larger aggregates. The process by which small particles aggregate is known by various terms depending on the type of system involved and on specific conventions. These include coagulation, flocculation, agglomeration and agglutination. In this book, the generic term aggregation is used for the most part, but there will be cases where more specific terminology is needed.

Aggregation depends essentially on two distinct influences:

(a) particles must move in such a way that collisions occur. This may be achieved by Brownian diffusion, fluid motion or sedimentation

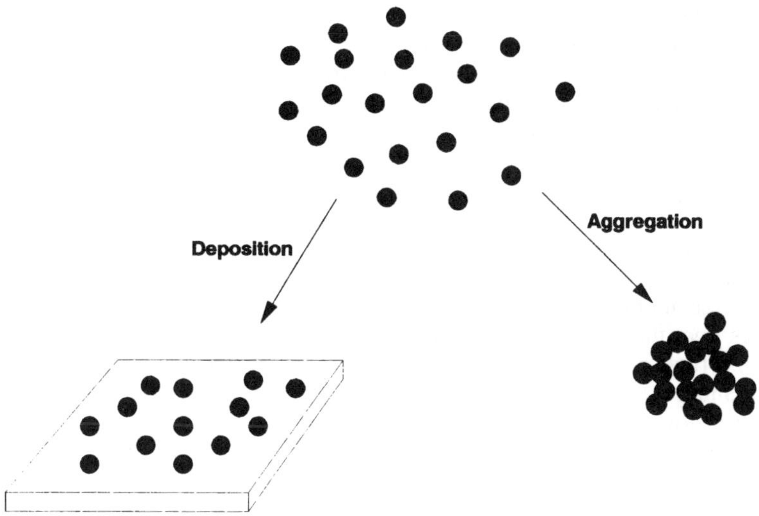

Figure 1.2 *Schematic illustration of the processes of aggregation and deposition of particles*

(b) interactions between colliding particles must be such that permanent contacts can be formed. Particles which repel each other are said to be stable, since they do not form aggregates.

These two processes, transport and attachment, can, to a good approximation, be treated independently in most practical cases. The reason is that colloid interactions (Chapter 3) are of very short range, usually much less than the particle size, so that particles have to approach very close to each other before any significant interaction is felt. The interaction may be attractive (e.g. van der Waals or hydrophobic attraction) or repulsive (e.g. electrical or steric repulsion). The nature of the interaction is greatly dependent on the nature of the particle surfaces and on the solution chemistry. The transport step (Chapters 4 and 6) has to bring particles together from comparatively large distances, over most of which colloid interactions play no role.

Deposition also involves transport and attachment steps, which, again, can be treated separately to a large extent. In fact, deposition has several features in common with aggregation and could be regarded as an extreme case of *heteroaggregation* (which normally involves particles of different types). An important feature of deposition is that particles which are colloidally stable, by virtue of, say, electrical repulsion, may readily deposit on a surface of a different type, with which there is no repulsion.

Aggregation and deposition are extremely important in the process industries and in the natural environment. The aggregation of aquatic colloids is of major importance in determining the transport and ultimate fate of many pollutants in natural waters. Particulate impurities in lakes and oceans settle at rates which depend greatly on the nature of aggregates formed.

Particle separation processes such as sedimentation, flotation and filtration are significantly influenced by particle size, and the process chosen can depend greatly on the size of the particles to be removed (see Figure 1.1). For most operations there is an optimum particle size and in very many practical cases the actual particles are in a smaller size range. Consequently, separation can be greatly enhanced by increasing the particle size and this is achieved by some form of aggregation process. An obvious example is sedimentation, where colloidal particles would have too low a settling rate for practical purposes. Aggregation (flocculation) of such particles is often essential in order to give more rapid settling.

The process of *microfiltration* involves filtration of suspensions through media with pores finer than the particle size (e.g. membranes). The separated particles may form a filter cake, which plays an important role in the removal of further particles. Small particles, especially those in the colloidal size range, give filter cakes of very low permeability and hence low filtration rates. Again, aggregation can give a marked improvement.

With *depth filtration* (or *packed-bed filtration*) particles are removed during passage through beds of granular or fibrous filter media, where the pores are considerably larger than the particle size. Removal depends on particle size and flow rate in a complicated manner (see Chapter 12), but there is usually a minimum efficiency for particles a few micrometres in size. Aggregation of small particles can greatly enhance the capture efficiency.

The deposition of particles from suspension onto solid surfaces is a subject of great significance in many industries. In some cases, deposition is desirable, as in filtration and various processes in which a coating of deposited particles needs to be formed. In many other instances deposition needs to be prevented as far as possible. These include detergency, mineral processing (where 'slime coating' of colloidal particles onto larger grains can be a serious problem) and the fouling of heat exchanger surfaces. A number of biological examples could be mentioned, such as the deposition of bacteria on the tooth surface and thrombus formation associated with artificial organs.

There is considerable interest in the transport of colloidal contaminants (including bacteria and viruses) in ground water and this depends greatly on the deposition behaviour of such particles in soils and aquifers. This topic is the subject of Chapter 13.

Particle deposition is of great importance in depth filtration since particles must be deposited on collectors, such as filter grains or fibres, in order to be captured in the filter. In the early stages of a filtration process particles deposit on essentially clean collector surfaces. This aspect has been studied most extensively in model investigations (see Chapter 5). Particles may also become attached to previously deposited particles and extensive deposits may build up. This would depend greatly on the nature of the particle–particle interactions and should only occur when the particles are colloidally unstable. In such cases aggregation of particles may also occur *before* deposition, depending on kinetic factors.

Fundamental studies of deposition and aggregation have been undertaken with different perspectives and with various objectives. Hydrodynamic factors and colloidal interactions play very important roles, but in many studies only one aspect has been emphasized. For instance, deposition on the rotating disc (Chapter 5) has been investigated primarily because the hydrodynamic conditions are easily controlled and are well understood. Deposition rates are then used to gain insight into the colloid interactions between the disc surface and suspended particles. Similarly, many investigations of filtration through packed beds (Chapter 10) have been concerned primarily with colloidal aspects.

In such cases, the effect of colloid interactions is often expressed as a collision efficiency factor, which is the ratio of the actual rate of aggregation or deposition to the transport-limited rate. A collision efficiency of unity implies that every collision leads to permanent attachment. This condition should be achieved when surface-chemical conditions are adjusted to eliminate repulsion. Repulsive forces can reduce the collision efficiency by many orders of magnitude, such that practically no aggregation or deposition occurs. Attractive interactions can, in principle, give a rate higher than the transport-limited value, but the increase is usually quite small, as a result of the short-range nature of colloid interactions. Interaction between particles, especially electrical interaction, can be changed by adjustments to solution chemistry (ionic strength, pH) and the collision efficiency can be controlled in this way. Comparison between measured and computed collision efficiencies can give very useful information on colloid interactions. There are still some major discrepancies between theoretical predictions and measured collision efficiencies (Chapter 11), which are the subject of continuing debate.

Because of the central role played by colloid interactions in aggregation and deposition processes, Chapters 2 and 3 are devoted to a discussion of charged interfaces and surface interaction potentials.

Most 'real world' problems involving aggregation and deposition are very complex and the systems cannot be described in sufficient detail to enable a complete analysis to be undertaken. The emphasis in this book is on 'ideal' systems, which may be somewhat remote from practical processes.

Modelling and simulation studies inevitably have to make simplifying assumptions, such as uniform spherical particles and smooth collector surfaces, constant surface-chemical conditions and laminar flow. Without such simplification, computational and theoretical difficulties would be insurmountable with present techniques. Experimental studies have very often used ideal systems, such as monodisperse latex particles and spherical collectors, so that comparisons with theoretical predictions can be made.

Nevertheless, model studies, providing systematic information on well-defined systems, can usefully contribute to greater practical understanding of real processes. The influence of certain parameters, such as particle size and flow rate, can be thoroughly investigated and the results will often enable, at least, a qualitative interpretation to be made of similar phenomena in real systems.

Although the distinction is not entirely clear-cut, it is worth pointing out the essential differences between modelling and simulation approaches to problems of particle aggregation and deposition, before detailed discussion of these topics in later chapters.

Modelling is an essentially *macroscopic* process, which seeks to provide functional relationships between important variables. It deals with suspensions in terms of overall properties such as particle concentrations, size distributions and shear rates, using either analytical or numerical methods of analysis. Examples of modelling of colloid transport, deposition and aggregation are given in Chapters 4–6. Depending on the nature of the assumed model and on the level of approximation used, quite simple analytical results can be derived in some cases, which give an immediate indication of the effect of key variables on the process being considered. In many instances, however, the results can only be obtained by numerical techniques. The motivation for most modelling studies is to predict the response of a process, such as packed-bed filtration (Chapter 12), to changes in certain variables, such as flow rate and particle size.

By contrast, *simulation* is performed at a *microscopic* level, dealing with individual particles and based on a stochastic approach. A typical simulation process involves a detailed consideration of particle interactions and hydrodynamics and starts from basic principles. Details are given in Chapters 7 and 8 and applications to filtration processes and colloidal dispersions are discussed in Chapters 14 and 15. One of the major advantages of simulation is that information on the *structure* of aggregates or deposits may be derived. Some of the earliest applications were in the field of particle aggregation, from which fundamental ideas on the fractal, self-similar nature of aggregates emerged (Chapter 6). Conventional analytical models of aggregation and deposition give no idea of structure.

While the major emphasis in this work is on theoretical aspects, relevant experimental techniques are described briefly in Chapters 9 and 10. The methods covered are mainly appropriate to fundamental studies, based on ideal systems. Results from such studies are especially useful for comparison with theoretical predictions, e.g. of particle deposition kinetics (Chapter 11).

Chapter 2

Electrical properties of interfaces

Nomenclature

a	radius of sphere
a_i	activity of ion i
C_δ	capacitance of Stern layer
c_i	molar concentration of ion i
e	electron charge
E	electric field strength
E_S	streaming potential
F	Faraday constant
i	current
I_S	streaming current
k	Boltzmann constant
n_+, n_-	local number concentrations of cations and anions
n_0	number concentration of ions in bulk solution
n_i	local number concentration of ion i
n_{i0}	number concentration of ion i in bulk solution
N_S	surface density of adsorption sites
q	total charge on sphere
Q	flow rate
r	radial distance from centre of sphere
R	gas constant
T	absolute temperature
U	electrophoretic mobility
u_i	specific adsorption potential of ion i
v	velocity
v_{eo}	electro-osmotic velocity
x	distance from surface
X	$= \kappa x$
x_i	mole fraction of counterions in solution
y	$= ze\psi/kT$
y_0	$= ze\psi_0/kT$
z	valence of ions in symmetrical (z–z) electrolyte.
z_i	valence of ion i (including sign of charge)
γ	$= \tanh(y_0/4)$
δ	thickness of Stern layer
ΔP	pressure difference

ϵ	permittivity
ζ	zeta potential
κ	Debye–Hückel parameter, eqn (2.7)
λ_0	specific conductivity of solution
λ_S	surface conductivity
μ	viscosity
ρ	volume charge density in solution
σ_δ	charge density in the Stern layer
σ_0	surface charge density
σ_1	diffuse layer charge per unit area of surface
ψ	electrical potential
ψ_0	surface potential

2.1 INTRODUCTION

Many of the important properties of colloidal systems are determined directly or indirectly by the interaction forces between particles. These colloidal forces consist of several components, principally the electrical double layer, van der Waals, Born, hydration and steric forces, which act over relatively short distances and are greatly dependent on the surface properties of particles or collectors. Consequently, a detailed understanding of the solid–liquid interfacial region is important in any attempt to model and simulate aggregation and deposition phenomena. The nature of the forces will be discussed in further detail in Chapter 3, while this chapter will focus on the nature of charged interfaces.

Electrical interaction between colloidal particles (especially in aqueous systems) is one of the most important influences on particle stability, aggregation and deposition. Most particles in aqueous media are charged, for reasons which will be considered briefly in Section 2.2.1. In an electrolyte solution, the distribution of ions around a charged particle is not uniform and gives rise to an *electrical double layer.* This topic has been the subject of several reviews, including the comprehensive texts of Hunter (1981,1987) and Lyklema (1991) and others cited in the bibliography. The essential point is that the charge on, say, a particle surface is balanced by an equivalent number of oppositely charged *counterions* in solution. These counterions are subject to two opposing influences: electrostatic attraction tending to localize the counterions close to the particles and the tendency of ions to diffuse randomly throughout the solution, owing to their thermal energy. The surface charge on a particle and the associated counterion charge together constitute the electrical double layer. The surface charge and the distribution of counterions near to the surface depend greatly on the nature and concentration of salts in solution, giving a relatively simple means of modifying colloid interactions. In most cases, the only practical means of manipulating the stability of particles and their tendency to deposit on surfaces is by changing the electrical interaction through changes in the solution chemistry.

For this reason, it is worthwhile to outline the important properties of the electrical double layer, and to show how physical models can be applied. It will become apparent that although modern experimental techniques and computational tools have given a better insight into charged interfaces and the various colloidal forces, a complete analysis is not usually possible in real systems. It is nearly always necessary to resort to approximate expressions, which prove to be convenient and reasonably accurate for many practical purposes. It is helpful to appreciate the limitations of these expressions which may form the basis for later modelling and simulation strategies.

2.2 The electrical double layer

2.2.1 Origin of surface charge

When two dissimilar phases are in contact it is likely that there will be a separation of charge and a difference in electrical potential between them. If we restrict attention to the case of a solid in contact with an aqueous electrolyte solution, there are several possible mechanisms for the separation of charge:

(a) a difference in affinity of ions for the two phases
(b) ionization of surface groups
(c) physical restriction of certain ions to one phase.

In the important case of ionic solids, one or other of the constituent ions may have a greater tendency to dissolve in the aqueous phase. Examples include the extensively studied cases of silver halide suspensions (Bijsterbosch and Lyklema, 1978), the best known example being *silver iodide*. In contact with pure water, silver ions have a greater tendency to escape from the crystal lattice, leaving a crystal with an excess negative charge. The solubility product of silver iodide is about $10^{-16}\,(\text{mol dm}^{-3})^2$. If the concentration (strictly, *activity*) of silver ions in solution is increased (and hence the iodide concentration decreased), a point is reached where the greater escaping tendency of the silver ions is just balanced by their higher concentration in solution, and the solid then has no net charge. This is the *point of zero charge* (*PZC*) and occurs when the silver concentration is about $3.26 \times 10^{-6}\,\text{mol dm}^{-3}$ or pAg = 5.5 (and hence pI = 10.5).

Because the surface electrical properties of ionic solids such as silver iodide are dependent on the concentrations of the constituent ions, these are known as *potential-determining ions* (PDI). The relationship between the surface potential, ψ_0, and the concentration of PDIs can be derived by straightforward thermodynamic reasoning and the result is known as the *Nernst equation*:

$$\psi_0 = \text{constant} + \left(\frac{RT}{z_i F}\right) \ln a_i \tag{2.1}$$

where R is the gas constant, T the absolute temperature and F the Faraday constant. z_i and a_i are the valence and the activity of the PDI, i.

At low electrolyte concentrations the ion activity can be approximated by the concentration, c_i and in this case eqn (2.1) can be differentiated to give:

$$\frac{d\psi_0}{d(\log_{10} c_i)} = 2.303 \frac{RT}{z_i F} \left(= \frac{59.2}{z_i} \text{ mV at } 25°\text{C} \right) \tag{2.2}$$

Thus, for a singly charged PDI, the potential of the solid should change by about 59 mV for a ten-fold variation in concentration. Note that the term z_i in these expressions must include the sign of ion charge, so that for a positive PDI the potential will *increase* for an increase in concentration and vice versa.

Other ionic solids which show similar behaviour include calcium carbonate, where the PDIs are Ca^{2+} and CO_3^{2-}. However, in this case there are complications caused by the reaction of the carbonate ion to form HCO_3^- and H_2CO_3, depending on the pH (Thompson and Pownall, 1989). This gives an apparent pH dependence of surface charge, but H^+ is not a PDI in the conventional sense.

Another very important example of pH-dependent surface charge is the case of metal oxides, such as Al_2O_3, Fe_2O_3 and TiO_2. In contact with water, the oxide surfaces become hydroxylated, giving the possibility of surface ionization to yield either positive or negative groups. In a simplified manner, the ionization of such groups can be written as (Stumm, 1992):

$$S{-}OH_2^+ \rightleftharpoons S{-}OH \rightleftharpoons S{-}O^-$$

where S denotes a solid surface.

The process occurring from left to right involves the progressive loss of two protons, and the steps can be characterized by appropriate equilibrium constants K_1 and K_2. The extent of deprotonation depends on the values of these constants and on the pH of the solution. As the pH is increased there is a greater tendency to lose protons and the oxide surface becomes more negative. There is a characteristic pH value at which the surface has no net charge and this can also be termed the PZC, by analogy with the case of ionic solids discussed above. Because of the nature of the oxide surface charge, experimental determination of surface charge densities and PZC values can be carried out fairly simply by acid–base titration.

According to the simplest model of the oxide–water interface, the PZC is just the mean of pK_1 and pK_2. These depend very much on the chemical nature of the oxide. Acidic materials such as silica (PZC about 2) have a great tendency to lose protons and are negatively charged over most of the pH range. Basic oxides such as MgO (PZC about 12) acquire protons very readily and are positively charged at most pH values. Intermediate cases include Fe_2O_3 (PZC 6.8) and TiO_2 (PZC 6.2), where the surface charge can change sign as a result of quite small pH changes in the neutral region. Such considerations are of great practical importance in the aggregation and deposition of oxide particles.

The description given above is highly simplified and more refined models are available (Hiemstra and van Riemsdijk, 1991). In particular, it should be noted that PZC values for oxides are dependent on such factors as crystalline form and mode of preparation. Literature values should be treated with some caution. Although H^+ and OH^- are not strictly constituent ions of oxides, the change of surface potential with pH can show Nernstian behaviour, especially in the region of the PZC.

Some surfaces have only monofunctional ionic groups, such as latex particles with carboxylic and sulphate groups. In these cases, the degree of ionization, and hence the surface charge density, may be dependent on pH, but there is no PZC. The surface charge approaches zero as the pH is changed, but does not reverse.

Preferential adsorption of ions on surfaces may occur and modify the surface charge. In the case of ionic surfactants, hydrophobic interaction can give strong adsorption and the surface acquires a charge characteristic of the surfactant ion. However, in practice, the more usual case is where anions or cations from solution adsorb on charged surfaces. This will be discussed in the context of electrical double layer models in Section 2.2.3.

The examples given so far fall into categories (a) and (b) listed at the beginning of this section. In many important cases, such as oxides, it is difficult to distinguish the concepts of surface ionization and different affinities of ions for the solid surface.

The third category is rather different and includes materials with an 'inherent' excess charge as a result of isomorphous substitution. For instance in the tetrahedral silica layer of kaolinite, substitution of Si(IV) by Al(III) gives a net negative lattice charge. For similar reasons, many clay minerals in water are negatively charged.

2.2.2 The Gouy–Chapman double layer model

Whatever the origin of the surface charge, it must be exactly balanced by an equal and opposite charge in solution. This balancing charge is accounted for by an excess number of oppositely charged ions or *counterions* in the solution adjacent to the charged surface and a deficit of similarly charged ions or *co-ions*. The question now is how is this charge distributed in the solution?

Considering, for simplicity, a planar interface of infinite extent, with a surface potential ψ_0 (V) and a surface charge density σ_0 ($C\,m^{-2}$), it is possible to describe the distribution of charge and potential in the solution as a function of distance from the surface. An approach due, independently, to Gouy and Chapman (see, for example Hunter, 1981) is based on a number of simplifying assumptions:

(a) an infinite, flat impenetrable interface
(b) ions in solution are point charges, able to approach right up to the interface
(c) the surface charge and potential are uniformly 'smeared out' over the surface
(d) the solvent is a uniform medium with properties (especially permittivity) that are independent of distance from the surface.

Assumptions (a)–(d) are all questionable and can be modified in refined versions of the theory. The assumption of a 'smeared-out' surface charge may seem especially suspect in view of known charging mechanisms which involve discrete ionic sites. However, the rate of exchange of protons and other ions between surface sites and the adjacent solution is usually very rapid, compared to, say, the time of approach of colloidal particles. For this reason, many surfaces may be regarded as having an effectively uniform surface charge density.

A fundamental result from electrostatics gives the relationship between charge density, ρ ($\mathrm{C\,m^{-3}}$), and potential, ψ, at any point. This is the *Poisson equation*:

$$\nabla^2\psi = -\frac{\rho}{\epsilon} \tag{2.3}$$

where ϵ is the permittivity of the medium.

In an electrolyte solution the charge density depends on the local concentrations of anions and cations and these depend on the local potential through the *Boltzmann expression*:

$$n_i = n_{i0} \exp\left(-\frac{z_t e\psi}{kT}\right) \tag{2.4}$$

where n_i is the number concentration of ion i at a point where the potential is ψ, n_{i0} is the corresponding concentration in the bulk solution (where the potential is zero), e is the electron charge and k is the Boltzmann constant.

The exponential term represents the electrical energy of an ion with a charge $z_i e$, relative to its thermal energy kT. If the ion has a charge opposite to the sign of the potential, then the local concentration will be higher than the bulk value and vice versa. Thus, when the potential has a negative sign, there will be a local excess of cations and a deficit of anions. It is convenient to restrict attention to symmetrical, z–z, electrolytes and in this case the charge density is given simply by:

$$\rho = ze(n_+ - n_-) = -2zen_0 \sinh\left(\frac{ze\psi}{kT}\right) \tag{2.5}$$

where n_+ and n_- are the local number concentrations of anions and cations and n_0 is the number concentration of each ion in the bulk solution. In this case z is the charge number of the ions and does not include the sign of charge.

Since we are considering a flat interface and are only concerned with the variation of potential normal to it, the one-dimensional form of the Poisson equation can be used, which, together with eqn (2.3), gives the *Poisson-Boltzmann expression*:

$$\frac{\mathrm{d}^2\psi}{\mathrm{d}x^2} = \frac{2zen_0}{\epsilon} \sinh\left(\frac{ze\psi}{kT}\right) \tag{2.6}$$

It is convenient to introduce the dimensionless parameters:

$$y = ze\psi/kT \qquad \text{and} \qquad X = \kappa x$$

where κ is given, for z–z electrolytes, by:

$$\kappa^2 = \frac{2e^2 n_0 z^2}{\epsilon kT} \tag{2.7}$$

The *Debye–Hückel parameter,* κ, has the dimensions of reciprocal length and plays a very important part in electrical interaction between colloidal particles.

Substituting y and X in eqn (2.6):

$$\frac{d^2y}{dX^2} = \sinh y \tag{2.8}$$

This form of the Poisson–Boltzmann expression can easily be solved, with the following boundary conditions (since the potential takes the value ψ_0 at the surface and zero in the bulk solution, far from the surface):

$$y = y_0\ (= ze\psi_0/kT) \quad \text{when } X = 0;\ y = 0 \text{ and } dy/dX = 0 \text{ when } X = \infty$$

The solution of eqn (2.8) is then:

$$y = 2\ln\left[\frac{1 + \gamma\exp(-X)}{1 - \gamma\exp(-X)}\right] \tag{2.9}$$

where γ is a function of the dimensionless surface potential, y_0:

$$\gamma = \frac{\exp(y_0/2) - 1}{\exp(y_0/2) + 1} = \tanh\left(\frac{y_0}{4}\right) \tag{2.10}$$

There are two useful approximations to eqn (2.9). If the surface potential is quite low ($y_0 < 1$ or ψ_0 less than about 25 mV in 1–1 electrolytes), then a simple exponential form is appropriate:

$$y = y_0\exp(-X);\ \text{or}\ \psi = \psi_0\exp(-\kappa x) \tag{2.11}$$

This can be also be obtained directly by solving the linear form of eqn (2.8), i.e. assuming $\sinh y \approx y$, and is often known as the *Debye-Hückel approximation.* Actually, eqn (2.11) is an adequate approximation to eqn (2.9) for values of y_0 up to about 2.

For any value of surface potential, at points not too close to the surface ($X > 1$), another exponential form is suitable:

$$y = 4\gamma\exp(-\kappa x) \tag{2.12}$$

Since, for $y_0 < 1$, $\gamma \approx y_0/4$, eqn (2.12) becomes equivalent to eqn (2.11) for low values of surface potential. Eqn (2.12) is the basis of the 'linear superposition' approximation for the interaction of charged particles (see Section 3.2.1).

The approximate exponential fall of potential from a charged surface means that the Debye–Hückel parameter, κ, defined in eqn (2.7) for z–z electrolytes, determines the extent of the counterion charge in solution (or the *diffuse layer*). At a distance $1/\kappa$ from the surface the potential has fallen to a value $1/e$ (1/2.72) of the surface potential. For this reason, the length $1/\kappa$ is usually known as the 'thickness' of the diffuse layer. For the general case of electrolyte solutions containing a number of dissolved salts, κ is defined by:

$$\kappa = \sqrt{\left(\frac{e^2 \sum n_{i0} z_i^2}{\epsilon kT}\right)} \tag{2.13}$$

where the sum is taken over all ions present in the solution.

Inserting numerical values appropriate to aqueous solutions at 25°C and converting the ion concentrations into molar terms gives:

$$\kappa = 2.32 \times 10^9 \left(\sum c_i z_i^2\right)^{1/2} \tag{2.14}$$

where the ion concentrations c_i are in mol dm^{-3} and κ is in m^{-1}.

For typical solutions, values of the diffuse layer thickness, $1/\kappa$, are in the range 1–100 nm. For instance, in 10^{-4} M NaCl the value is about 30 nm and in 10^{-4} M $CaSO_4$ about 15 nm. These values are indicative of the range of electrical interaction between colloidal particles.

Another important result from Gouy–Chapman theory relates the surface charge density, σ_0, to the surface potential, ψ_0. The former is easily derived, since the surface charge must be balanced by the excess charge in the diffuse layer. Considering unit area of surface, the total diffuse layer charge is derived simply by integration of the charge density given by eqn (2.5):

$$\sigma_0 = -\int_0^\infty \rho \, dx$$

It can be shown that, for symmetrical electrolytes, the result is given by (Hiemenz, 1985):

$$\sigma_0 = (8n_0\epsilon kT)^{1/2} \sinh\left(\frac{y_0}{2}\right) \tag{2.16}$$

For small potentials ($y < 1$), this simplifies to:

$$\sigma_0 = \epsilon\kappa\psi_0 \tag{2.17}$$

If the double layer system is regarded as a parallel-plate capacitor, then the capacitance is simply $\epsilon\kappa$ and $1/\kappa$ is the effective distance between the plates. This is another reason why the length $1/\kappa$ can be thought of as the 'thickness' of the diffuse layer.

The Gouy–Chapman theory, as described briefly above, has several shortcomings. For instance, measured capacitances at certain interfaces can be much lower than those predicted by the theory. Also, counterion concentrations close to charged interfaces, calculated from eqn (2.4), can become unreasonably high, even for only moderate values of surface potential. Quite simple modifications to the theory can be made, the most important of which are to allow for *finite ion size* and *specific adsorption of ions*, which lead to the *Stern* and *Stern–Grahame* models of the double layer.

2.2.3 The Stern model

Ions in solution are not point charges, but have a finite size. Hydrated ions have diameters typically of the order of 0.5 nm or less and allowance for the size is probably the most important correction to the Gouy–Chapman theory. If we think of the ion charge as being located at the centre of a sphere of finite size, then the charge cannot touch the interface, but must be limited to a distance of approach equal to the sphere radius. This distance is usually given the symbol δ and it defines a region which is inaccessible to counterion charge. This concept, originally due to Stern, accounts for many of the observed discrepancies with Gouy–Chapman theory. The plane at distance δ from the surface is usually known as the *Stern plane*, and the region between this plane and the surface is known as the *Stern layer.* Outside the Stern layer there is a diffuse layer of counterions, just as in the Gouy–Chapman model.

In the simplest Stern model, counterions are entirely mobile and not adsorbed on the surface. In this case there is no charge in the Stern layer and the diffuse layer charge just balances the surface charge. The Stern plane can be regarded as a capacitor, with a capacitance which depends on its thickness, δ, and the permittivity of the medium close to the surface. In aqueous systems the permittivity of water close to a charged interface may not be the same as the bulk value, but could be significantly lower as a result of the high field strength. The relative permittivity (dielectric constant) of bulk water is quite high (about 78) because of the highly polar nature of the water molecule. In high field strengths, most of the water dipoles become orientated in the field and the dielectric constant can be reduced to a value as low as 6 for the case of electric saturation (Hiemstra and van Riemsdijk, 1991). It is likely that the bulk value would be reached at a distance of only one or two hydrated ion diameters from the surface. If we assume $\delta = 0.3$ nm and a value of 10 for the dielectric constant of water in the Stern layer, then the inner layer capacitance turns out to be about $0.3\,\mathrm{F\,m^{-2}}$, which is comparable to values found for silver iodide.

In the case of low surface potential, the capacitance of the diffuse layer can be calculated from eqn (2.17) and becomes greater at higher ionic strengths (higher κ). For instance, in 10^{-1} M 1–1 salt solution, the diffuse layer capacitance would be about $0.7\,\mathrm{F\,m^{-2}}$, which is rather higher than the Stern layer capacitance calculated above. Two capacitors in series have a total capacitance, C_t given by:

$$\frac{1}{C_t} = \frac{1}{C_1} + \frac{1}{C_2} \qquad (2.18)$$

The total capacitance cannot exceed the smaller of the two individual values, which means that in the example above the total capacitance would be mainly determined by the Stern layer value. In dilute salt solutions, where the diffuse layer capacitance is much lower, this determines the overall value. The surface potential, ψ_0, falls to zero in the bulk solution and this drop is divided between the Stern and diffuse layers according to their capacitances. The largest potential drop occurs over the layer with the smallest capacitance, so that at high ionic strength the potential will fall sharply through the Stern layer, to the value ψ_δ (the *Stern potential*). In very dilute salt solutions, where the diffuse layer capacitance is smaller than that of the Stern layer (which is largely independent of ionic strength), most of the potential drop occurs over the diffuse layer and the Stern potential has a value close to the surface potential ψ_0.

If the Stern layer has no adsorbed ions (see below), then the diffuse layer charge is equal and opposite to the surface charge and is given by an expression analogous to eqn (2.16), but with the Stern layer potential in place of the surface potential. The potential drop across the Stern layer can then be calculated from the charge, if the Stern layer capacitance C_δ is known. With numerical values appropriate to aqueous solutions at 25°C, the result is:

$$\psi_0 - \psi_\delta = \frac{0.117}{C_\delta} c^{1/2} \sinh(19.4 z \psi_\delta) \tag{2.19}$$

where c is the concentration of z–z electrolyte (mol dm^{-3}).

Results from this expression are shown in Figure 2.1, where a surface potential of 250 mV and a Stern layer capacitance of 0.2 F m^{-2} have been assumed. The Stern potential ψ_δ is plotted against concentration of 1–1 and 2–2 salts. The Stern potential falls with increasing salt concentration and has lower values in the case of more highly charged ions.

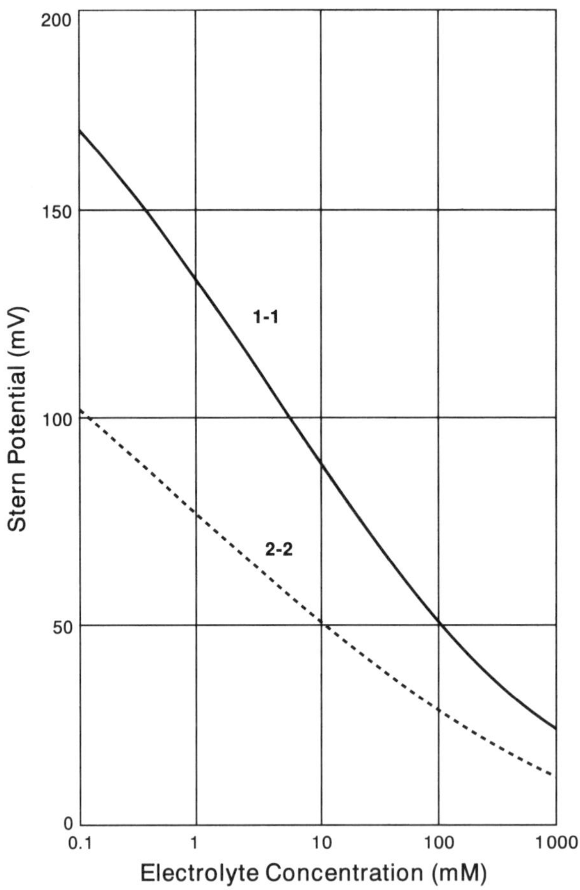

Figure 2.1 *The variation of Stern potential, ψ_δ, with the concentration of symmetrical 1–1 and 2–2 electrolytes. The surface potential, ψ_0, is assumed to be 250 mV and the Stern layer capacitance is 0.2 F m^{-2}*

The fact that the Stern potential is often much lower than the true surface potential has several important implications. One of these is that the diffuse layer can often be modelled quite well with the linear Poisson–Boltzmann (low-potential) approximation, where this would not be suitable if the surface potential were used.

The counterions may not be entirely mobile, but could be bound to the surface, to some extent. The electrical energy of an ion in the Stern plane, relative to the thermal energy, is $ze\psi_\delta/kT$. For univalent ions and Stern potentials greater than about 50 mV, this term becomes significant and may be responsible for some counterion binding to the surface. This is analogous to ion pair formation in electrolyte solutions (Davies, 1962).

There may also be non-electrostatic (or 'chemical') reasons for binding of ions, and the effect is then often known as *specific adsorption*. Although there is some confusion over this term, it draws attention to an important distinction. If a metal ion is to undergo specific binding to a surface, it is usually necessary for some water of hydration to be lost. There can then be direct bond formation between the metal and electron-donating groups on the surface (e.g. oxygen groups at an oxide surface). Such an interaction is said to lead to an *inner sphere complex*, as opposed to an *outer sphere complex* formed between a hydrated metal ion and a charged surface (Stumm, 1992). It is reasonable to suppose that outer sphere complexes are formed entirely as a result of electrostatic attraction. However, even ions forming outer sphere complexes can show 'specific' differences as a result of different sizes of the hydrated ions.

Strictly, the division of an interaction energy into 'electrical' and 'chemical' terms is not justified thermodynamically, but it is conceptually appealing and double layer models are often formulated in this way. The energy appearing in the Boltzmann expression then takes the form $(ze\psi_\delta/kT+u)$, where u is the *specific adsorption potential*. When this term exceeds a few kT units, it can dominate the electrical term. Using an approach similar to that of the Langmuir adsorption isotherm, it is possible to calculate the surface concentration of adsorbed counterions, which, when converted to an equivalent charge density in the Stern layer, σ_δ, becomes (Hunter, 1993):

$$\sigma_\delta = z_i e N_s x_i \exp\left[-\frac{(z_i e\psi_\delta + u_i)}{kT}\right] \tag{2.20}$$

where N_s is the density of adsorption sites and x_i is the mole fraction of counterions in solution. The counterion valence z_i includes the sign of charge, so that the electrical interaction energy is negative (attraction), provided ψ_δ is of opposite sign.

The presence of 'fixed' charge in the Stern layer means that it is no longer possible to equate the surface and diffuse layer charges. Instead we have:

$$\sigma_0 + \sigma_\delta + \sigma_1 = 0 \tag{2.21}$$

where σ_1 is the diffuse layer charge per unit area of surface.

If the specific adsorption potential is large and favourable (i.e. negative), adsorption of counterions can continue until the Stern layer charge more than

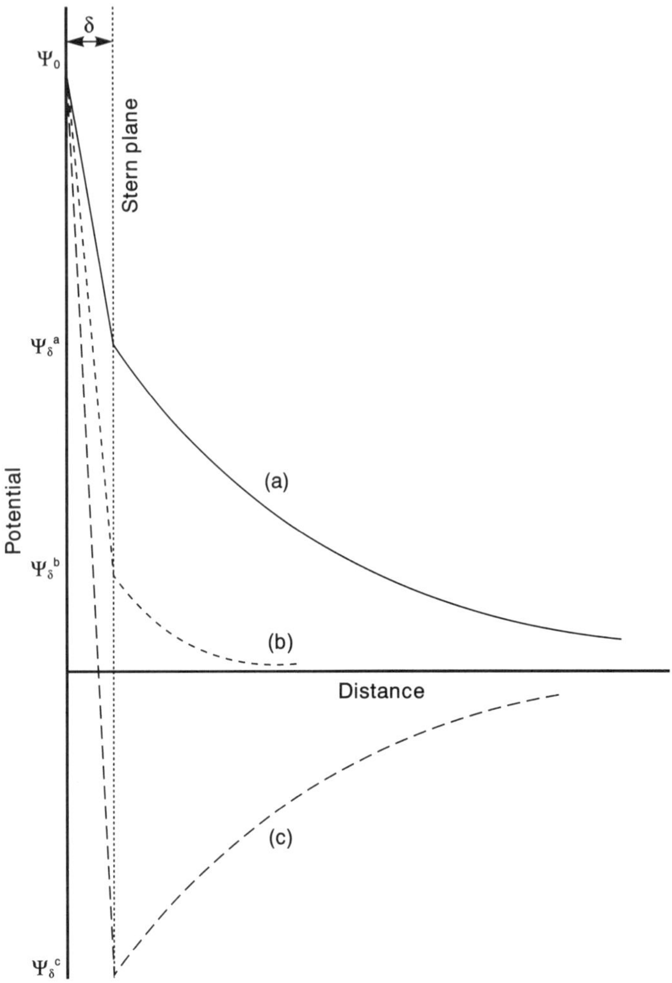

Figure 2.2 *The variation of potential close to a charged interface:* (a) *in the case of a low concentration of indifferent electrolyte,* (b) *in the case of a higher concentration of indifferent electrolyte and* (c) *in the presence of a salt with specifically adsorbing counterions, giving charge reversal*

compensates the surface charge and *reversal of charge* occurs. This provides unequivocal evidence that some specific, non-electrical interaction is involved. Charge reversal can be brought about by many specifically adsorbing counterions, such as hydrolysing metal ions and polyelectrolytes, and is of great practical importance in particle aggregation and deposition.

Figure 2.2 shows schematically the effect of electrolytes on the distribution of potential in the double layer, for the case of an 'indifferent' electrolyte (i.e. with no specifically adsorbing ions) at two concentrations, and a salt with a specifically adsorbing counterion, giving charge reversal.

2.2.4 The Stern–Grahame model

Although, for many purposes, the Stern model of the double layer is quite adequate, a number of further refinements can be made. One of these, due to Grahame (1947), subdivides the Stern layer into two regions: an inner layer occupied by specifically adsorbed, unhydrated ions (e.g. inner sphere complexes) and a second layer where hydrated counterions (outer sphere complexes) are located. The boundaries of these regions are often known as the *inner Helmholtz plane* (*IHP*) and *outer Helmholtz plane* (*OHP*). The OHP is equivalent to the Stern plane. Outside this plane is the diffuse layer, adequately described by Gouy–Chapman theory.

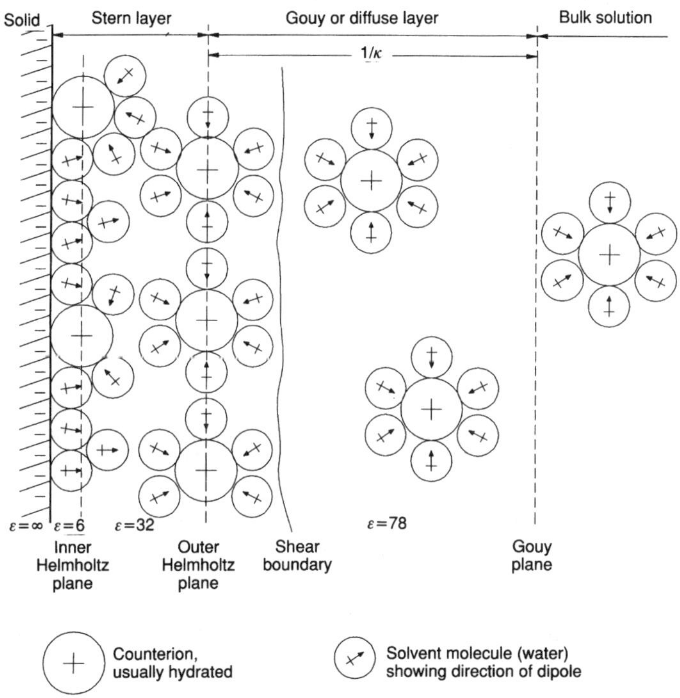

Figure 2.3 *The Stern–Grahame model of the electrical double layer, showing the adsorption of unhydrated ions at the inner Helmholtz plane and of hydrated ions at the outer Helmholtz plane, together with a diffuse layer which extends outwards into the bulk solution. Likely values of the dielectric constant, ε, are shown in the different regions. The electrokinetic shear boundary is thought to lie just outside the outer Helmholtz plane*

One important feature of this model is that it admits the possibility of three different values of solvent permittivity. In aqueous systems the dielectric constant in the inner region will be quite low and is often assumed to be about 6 (representing dielectric saturation). In the outer region, where the field strength is rather lower, a value of about 30 might be appropriate. In the diffuse layer, the value for bulk water (78) is assumed. This approach has been quite often used, especially in the modelling of the oxide–water interface (Stumm, 1992), giving what are known as *triple layer models*.

A schematic diagram showing the essential features of the Gouy–Stern–Grahame double layer model is given in Figure 2.3.

2.2.5 Further refinements to the model

It has been recognized for some time (see, for example Hunter (1981)) that there is a fundamental inconsistency inherent in the Poisson–Boltzmann approach and a similar problem arises in the Debye–Hückel theory of electrolytes (Robinson and Stokes, 1965). The basic difficulty is that the expression for charge density, eqn (2.5), violates the electrostatic principle of linear superposition, since it does not predict that the potential at any point due to a collection of charged species will be the sum of potentials due to the individual charges. The expected linear dependence is only found for the low-potential approximation, which leads to eqn (2.11). The problem lies in the fact that the potential used in the Boltzmann expression, eqn (2.4), should be the *potential of mean force*, rather than the mean (time-averaged) potential. The former is calculated from the work required to bring the ion from bulk solution up to a particular point, a process which may involve some redistribution of charge.

This inconsistency has prompted a number of alternative, statistical-mechanical approaches (Buff and Stillinger, 1963; Martynov, 1966; Henderson *et al.*, 1983). The arguments between supporters of the Poisson–Boltzmann approach and its alternatives still continue, although their impact on practical aspects of colloid science has not been of great significance. This is partly due to the fact that electrical effects are less important in concentrated electrolyte solutions (where deviations from the Poisson–Boltzmann equation become more serious). In reality, the difficulties of implementing the alternative models under practical conditions (e.g. involving surface roughness) are too great.

Another noteworthy approach to this problem has involved the use of computer simulation. For instance, Torrie and Valleau (1979, 1980, 1982) have performed extensive Monte Carlo simulations of the double layer using a *primitive* model in which the electrolyte is considered to be an assembly of charged 'hard spheres' in a uniform dielectric medium. Their results show very good agreement with the Gouy–Stern–Grahame theory for the case where there is no specific adsorption and for relatively dilute electrolyte solutions (<0.1 M).

In practice, there is probably little error involved in the use of the Poisson–Boltzmann expression, eqn (2.6), and this approach is adopted in nearly all cases.

2.2.6 Spherical double layers

So far, only planar interfaces have been considered and in this case the Poisson–Boltzmann equation can be solved analytically. However, for spherical particles an analytical solution is only possible for the low-potential (linear approximation) case.

The Poisson–Boltzmann equation for z–z electrolytes, eqn (2.6), transposed into polar coordinates becomes:

$$\frac{1}{r^2}\frac{\mathrm{d}}{\mathrm{d}r}\left(r^2\frac{\mathrm{d}\psi}{\mathrm{d}r}\right) = \frac{2zen_0}{\epsilon}\sinh\left(\frac{ze\psi}{kT}\right) \tag{2.22}$$

where r is the radial distance from the centre of the sphere.

For low potentials, such that $\sinh y \approx y$, the solution is simply:

$$\psi = \frac{\psi_0 a}{r}\exp\left[-\kappa(r-a)\right] \tag{2.23}$$

where a is the radius of the sphere and ψ_0 is the surface potential. (Strictly, eqn (2.23) applies only to a diffuse double layer. In the case of a Stern layer around the particle, ψ_0 can be replaced by the Stern potential, ψ_δ.)

It can be shown, by considering the total counterion charge, that the charge on a sphere is given by:

$$Q = 4\pi\epsilon\psi_0 a(1+\kappa a) \tag{2.24}$$

The surface charge density follows immediately from this expression:

$$\sigma_0 = \frac{\epsilon\psi_0}{a}(1+\kappa a) \tag{2.25}$$

As might be expected, in the case of diffuse layers which are thin compared to the particle radius (i.e. $\kappa a \gg 1$), eqns (2.23) and (2.25) reduce to the corresponding flat plate expressions, eqns (2.11) and (2.17).

Because of the linear approximation, these expressions will become inaccurate for surface (or Stern) potentials greater than a few tens of millivolts in 1–1 electrolytes. In many cases of interest in particle aggregation and deposition, we are concerned with fairly low potentials and this restriction is not serious. Exact, numerical approaches to the problem of spherical double layers have been reviewed by Hunter (1981), together with some useful approximations for higher potentials. These will not be considered further here.

2.3 Electrokinetic phenomena

2.3.1 General

The most widely used experimental approach to the study of charged surfaces is through the use of electrokinetic techniques. Electrokinetic phenomena arise whenever there is relative movement between a charged interface and the adjacent electrolyte solution, so that part of the double layer charge moves with the liquid. Fundamental to these phenomena is the concept of a plane of shear, which separates the 'fixed' from the 'mobile' parts of the electrical double layer. The electrical potential at the shear plane is the *electrokinetic potential*, or, more usually, the *zeta potential*, after the symbol ζ, which is conventionally used to denote this quantity.

It is usually assumed that the shear plane lies outside but fairly close to the Stern plane (see Figure 2.3), so that essentially all of the diffuse layer charge is 'mobile' and the counterions in the Stern layer are 'fixed'. This means that values of 'surface charge' determined by electrokinetic techniques can be considerably less than the 'true' surface charge, measured by a titration method. Another important difference is between the PZC, obtained from a titration technique (see Section 2.2.1), and the *isoelectric point* (*IEP*), which is the condition (usually pH value) at which the zeta potential becomes zero (see, for example, Lyklema (1978)). These two points coincide only in cases where there is no specific adsorption of ions. Where specific adsorption of counterions occurs, these become part of the 'fixed' charge and the point of zero electrokinetic potential will occur under conditions where the surface potential, ψ_0, determined by the concentration of potential-determining ions, has a non-zero value.

The relationship between the measured zeta potential and double layer structure is still a subject of debate and the shear plane cannot be located with any certainty. However, electrokinetic measurements are still of immense practical importance in colloid science, since they give information, however imperfect, which is not obtainable in any other way. There is a clear correlation, for instance, between zeta potential and colloid stability in very many cases. Colloidal particles are generally found to be stable if the zeta potential exceeds a certain value and to coagulate at lower values. Similarly, deposition of particles on surfaces is very often controlled by the zeta potentials of particles and substrate. Use of zeta potentials in quantitative expressions for electrical interaction (see Section 3.2) may not be strictly justified, but there is usually no real alternative.

We now briefly consider the main electrokinetic techniques and their interpretation. The available methods are as follows:

(a) a *streaming potential* is established when solution is forced to flow through a tube, or porous plug, of material which acquires charge in contact with the solution
(b) a *sedimentation potential* arises when charged particles settle through a solution
(c) *electro-osmosis* is the flow of liquid through a tube or porous plug under the influence of an applied electric field
(d) *particle electrophoresis* is the movement of charged particles in an electric field

Of these, the most commonly used techniques are streaming potential (sometimes streaming *current* is measured) to determine the zeta potential of granular or fibrous materials, and particle electrophoresis, to derive information on colloidal particles. However, it is convenient to start with an analysis of electro-osmosis, which provides a basis for a discussion of the other techniques. A more detailed account can be found in Hunter (1981).

2.3.2 Electro-osmosis

It is easiest to consider a flat interface, although the treatment is also applicable to electro-osmosis in cylindrical tubes, provided that the tube radius, r, is very much larger than the double layer thickness (i.e. $\kappa r \gg 1$). An electric field, E, is assumed to be applied in a direction parallel to the surface and an element of moving fluid at some distance from the charged interface is subject to both electrostatic and viscous forces. If the fluid element has a thickness dx and a charge density ρ and the fluid velocity changes by an amount dv across the element, then it follows that:

$$E\rho\, dx = \frac{d}{dx}\left(\mu \frac{dv}{dx}\right) dx = \mu \frac{d^2v}{dx^2}\, dx \tag{2.26}$$

Substituting for ρ from the linear version of the Poisson equation, eqn (2.3), gives:

$$-E\epsilon \frac{d^2\psi}{dx^2} = \mu \frac{d^2v}{dx^2} \tag{2.27}$$

where ϵ and μ are the (local) values of the permittivity and viscosity, respectively.

Integration of eqn (2.27) can be carried out easily if it is assumed: (a) that the fluid velocity is zero at a plane where the potential $\psi = \zeta$ (i.e the shear plane) and attains a constant value, v_{eo}, far from the interface, where both ψ and $d\psi/dx = 0$, and (b) that ϵ and μ are constant (i.e. have their bulk solution values) over this entire region. The latter condition is more likely to apply if the mobile charge is entirely within the diffuse part of the double layer, although no assumption about the detailed structure of the double layer is implied. Integrating eqn (2.27) twice gives a very simple result for the electro-osmotic velocity, v_{eo}:

$$v_{eo} = \frac{E\epsilon\zeta}{\mu} \tag{2.28}$$

For aqueous solutions, this expression should apply to electro-osmotic flow in tubes with diameter greater than a few micrometres. In such cases the fluid velocity only varies over a region very close to the tube wall and is constant over most of the tube width. It is thus possible to calculate the volume flow rate through the tube, Q, by electro-osmosis; it is simply $\pi r^2 v_{eo}$. The tube radius can be eliminated by introducing the current, i, which, by Ohm's law, can be written:

$$i = E\pi r^2 \lambda_0 \tag{2.29}$$

where λ_0 is the specific conductivity of the electrolyte solution. This gives:

$$\frac{Q}{i} = \frac{\epsilon\zeta}{\mu\lambda_o} \tag{2.30}$$

Thus, measuring the flow rate of a solution through a capillary tube under the influence of an applied electric field should, in principle, allow the zeta potential of the tube wall to be derived. It is only necessary to measure the flow rate and the electric current, and to have values of the conductivity, permittivity and viscosity of the solution.

The advantage of eliminating the tube dimension is that eqn (2.30) should then apply to porous media as well as tubes, which would provide a means of determining the zeta potential of granular and fibrous media. In practice, this approach is complicated by *surface conductance*. The accumulation of counterions close to a charged interface can lead to a higher current than expected from the bulk conductivity of the solution, especially at low salt concentrations. If the surface conductivity is λ_S, then the result for tube flow is:

$$\frac{Q}{i} = \frac{\epsilon\zeta}{\mu(\lambda_0 + 2\lambda_S/r)} \tag{2.31}$$

It is difficult to account for surface conductance in porous media, where the pore geometry is uncertain. One approach is to determine the effective conductivity of the solution in the porous medium, by measuring the resistance *in situ*, and this can give adequate results.

Electro-osmosis is not often employed as a practical means of deriving zeta potentials, but it has an important effect in particle electrophoresis (see Section 2.3.4). It may also be used in practice as a means of dewatering (Sunderland, 1987) and remediation of contaminated soil (Shapiro and Probstein, 1993).

2.3.3 Streaming potential

When a solution flows through a cylindrical tube, with a charged inner wall, then part of the double layer charge is carried with the flowing solution and establishes a *streaming current*, I_S. This causes an electrical potential difference between the ends of the tube, which is the *streaming potential*. At steady state the streaming potential causes a conduction current, I_C, which just balances the streaming current. When the double layer is very thin compared to the tube radius, so that most of the charge is close to the tube wall, it is not difficult to show, using Poiseuille's equation for laminar tube flow and the Poisson equation for the solution charge density, that the streaming current is given by:

$$I_S = -\frac{\epsilon\zeta}{\mu}\pi r^2 \Delta P \tag{2.32}$$

where ΔP is the pressure difference between the ends of the tube.

The streaming potential, E_S, at steady state is sufficient to produce a conduction current which just balances I_S and hence depends on the resistance presented by the tube. In general, the resistance depends on the tube dimensions, the solution conductivity, λ_0, and the surface conductivity, λ_S. The result is:

$$\frac{E_S}{\Delta P} = \frac{\epsilon\zeta}{\mu(\lambda_0 + 2\lambda_S/r)} \tag{2.33}$$

The similarity between this expression and eqn (2.31) for electro-osmotic flow is noteworthy. Comparison of the two shows that:

$$\left(\frac{E_S}{\Delta P}\right)_{i=0} = \left(\frac{Q}{i}\right)_{\Delta p=0} \tag{2.34}$$

This fundamental result is a direct consequence of the Onsager reciprocity principle in non-equilibrium thermodynamics.

If the surface conductance is negligible, then eqn (2.33) can be applied to porous media, since there is then no term for the dimensions of the tube. Measuring the streaming potential as a function of applied pressure should give a linear relationship and the zeta potential can be calculated from the slope. Where surface conductance is significant (as is often the case), the procedure becomes more complicated, although the measurement of the *in situ* conductivity, as in the case of electro-osmosis, can give satisfactory results.

To avoid problems of surface conductance, it is possible to measure the streaming *current*, I_S, directly and there are commercial devices available for this purpose (Dentel and Kingery, 1989). However, for studies of particle deposition on granular media, the zeta potential of the grains is usually determined by a streaming potential technique.

2.3.4 Particle electrophoresis

This is the most widely used electrokinetic technique and has very often been applied in studies of aggregation and deposition phenomena. In its simplest form, the motion of particles in an electric field is observed microscopically and the velocity is determined. The *electrophoretic mobility*, U, is defined as the velocity per unit field strength.

In practice electrophoretic mobilities are measured in either cylindrical or rectangular cells and modern instruments use laser Doppler techniques to derive the velocity, rather than direct microscopic observation. There is a practical problem caused by electro-osmotic flow at the cell walls. In a closed cell this results in a return flow through the centre of the cell and a characteristic velocity profile through the depth of the cell. Since the liquid flow is superimposed on the electrophoretic movement of the particles, the observed particle velocity is the sum of two terms. Provided that the liquid velocity is known, then it is, in principle, possible to calculate the true electrophoretic velocity. The more usual approach is to conduct the measurements at the *stationary levels*, where the liquid velocity is zero (the electro-osmotic flow is just balanced by the return flow). In this way the electrophoretic velocity may be observed directly.

For particles large compared with the double layer thickness ($\kappa a \gg 1$), the velocity can be derived in exactly the same way as the electro-osmotic velocity at a planar interface, eqn (2.28), since we are only concerned with the *relative* velocity between particle and liquid and it makes no difference whether one or the other is stationary. In this case the electrophoretic mobility has a very simple form, known as the *Smoluchowski equation*:

$$U = \frac{\epsilon\zeta}{\mu} \tag{2.35}$$

This expression applies to fairly large particles in electrolyte solutions which are not too dilute. For particles of 1 μm diameter in a 10 mM solution of 1–1 electrolyte, κa is about 160 and eqn (2.35) could be used with very little error. In such cases, for aqueous solutions at 25°C, inserting numerical values gives the following result for the zeta potential:

$$\zeta\,(\mathrm{mV}) = 12.8U\ (\mu\mathrm{m\,s^{-1}/V\,cm^{-1}}) \tag{2.36}$$

where the units of ζ and U indicated are those conventionally used.

At the other extreme, where particles are very much smaller than the double layer extent ($\kappa a \ll 1$), a different approach can be used. The electric force on the particle ($=QE$, where Q is the particle charge and E the field strength) and the Stokes drag ($=6\pi\mu av$, where v is the particle velocity) must exactly balance under steady-state conditions. The particle charge relevant to electrokinetics is given by eqn (2.24), with ψ_0 replaced by ζ, provided that the zeta potential is not too large. With the condition $\kappa a \ll 1$, this approach gives an expression known as the *Hückel equation*:

$$U = \frac{2\epsilon\zeta}{3\mu} \tag{2.37}$$

This differs from eqn (2.35) only by a factor of 2/3. The Hückel equation is not usually applicable to particle electrophoresis, since the condition $\kappa \ll 1$ is only satisfied for very small particles in dilute electrolyte solutions. For particles of 10 nm diameter, the concentration of 1–1 electrolyte would have to be less than 50 μM. The expression is more often applicable to electrophoresis of macromolecules, such as proteins (although, strictly, the result applies only to spherical particles).

For particles of colloidal size, in solutions of moderate ionic strength, κa often takes intermediate values and neither the Smoluchowski nor Hückel equation is appropriate. In such cases corrections have to be applied, which are discussed in some detail by Hunter (1981). Essentially there are two types of correction:

(a) An *electrophoretic retardation effect* arises from the opposite movement of charged particles and counterions in an electric field, and acts, together with frictional resistance, to reduce the mobility. The derivation of the Smoluchowski expression, eqn (2.35), assumes that the electro-osmotic flow is equal and opposite to the particle motion and does not consider the frictional resistance on the particle. Both electrophoretic retardation and frictional drag are correctly incorporated in the Hückel equation and this accounts for the 2/3 difference between the two expressions. However, for $\kappa \ll 1$, the electrophoretic retardation effect is small, since very little of the force is transmitted to the particle. In the general case both electrophoretic and frictional retardation have to be incorporated and the result can be expressed as a correction factor to be applied to the Hückel expression:

$$U = \frac{2\epsilon\zeta}{3\mu} f_1(\kappa a) \tag{2.38}$$

The function $f_1(\kappa a)$ depends on the particle shape and for spherical particles the value shows a smooth transition from 1 to 1.5 as κa varies from 0 to ∞.

(b) A *relaxation effect* arises because the motion of a charged particle causes the diffuse layer to become distorted. As the particle moves, it is not possible for the ions in the diffuse layer to respond instantly and so they lag behind the particle, giving an asymmetrical charge distribution, which causes the particles to move more slowly This effect depends on κa, the zeta potential and the limiting conductances of the ions. Modern treatments (e.g. O'Brien and White, 1978) include the electrophoretic retardation effect and so should give a complete description of the variation of mobility with zeta potential, particle size and ionic strength. Full computation of electrophoretic mobility using the numerical approach of O'Brien and White (1978) is quite straightforward and there are also simple analytical approximations which are in error by no more than a few per cent (Hunter, 1981).

The most important consequence of the relaxation correction is that the mobility can go through a maximum as the zeta potential is increased. However, the maximum does not occur until the zeta potential reaches 100 mV or more and such values are not often encountered in practice, especially in studies of particle aggregation.

Because of the uncertainties in calculating zeta potentials, electrophoretic data are often quoted as mobilities, which are unambiguous experimental quantities.

A more recent devevelopment is the field of *electroacoustics* (Hunter, 1993). When a suspension is subjected to high-frequency sound waves, the particles vibrate at the same frequency. Because of the different relaxation times of particles and ions in the diffuse layer, the latter respond more quickly to the ultrasonic waves, giving a small dipole. The cumulative effect of many such dipoles gives a measurable electric field which is known as the *colloid vibration potential*. The magnitude of this potential is proportional to the electrophoretic mobility of the particles. This is very useful for concentrated suspensions, where conventional electrophoresis is difficult. For essentially the same reason, application of an alternating electric field to a suspension of charged particles generates sound waves and gives rise to the *electrokinetic sonic amplitude* or *ESA effect*.

2.3.5 Interpretation of zeta potentials

Much of the above discussion has been concerned with the problem of deriving electrokinetic (zeta) potentials from electrokinetic measurements. Even when a completely satisfactory method is available, leading to reliable values of zeta potential, there remains the problem of relating zeta potentials to double layer properties. Essentially, the question is that of the location of the shear plane separating the 'fixed' from the 'mobile' part of the double layer.

The relationship between the zeta potential and the Stern potential, ψ_δ, is of great interest, but cannot be established with any certainty, since there is no independent, direct method of measuring ψ_δ. There are several cases, notably with latex particles, where the measured zeta potential passes through a maximum as ionic strength is increased. Such behaviour is not expected for the Stern potential, which should decrease as the concentration of (indifferent) electrolyte is increased. Several explanations have been proposed to explain this anomaly, such as the 'hairy' nature of the polymer latex surface as a result of protruding polymer chains (van den Hoven and Bijsterbosch, 1987). In such cases the approach of counterions to the surface may be restricted and the plane of shear may be displaced outwards. Similar effects might be expected with adsorbed polymers. Another explanation is that there may be increasing adsorption of co-ions on hydrophobic areas of the latex surface as the salt concentration is increased (Elimelech and O'Melia, 1990). Since latex particles are very often used in model studies of aggregation and deposition, it is important to understand their electrokinetic behaviour.

In the case of oxide particles it is possible to derive the surface charge density by acid-base titration and there have been many attempts to relate these measurements with electrokinetic data determined under similar conditions. Generally, the surface charge determined by titration can be much less than the 'electrokinetic charge', implying that much of the counterion charge lies inside the plane of shear. Several models of the oxide–water interface have been developed, including triple layer models (Yates *et al.*, 1974; Davis and Leckie, 1978). However, in order to obtain good fits with experimental data, rather arbitrary choices of affinity constants of protons and other ions to surface sites have to be made. An alternative approach of Hiemstra and van Riemsdijk (1991), which recognizes that different faces of oxide crystals contribute to the surface charge in different ways, seems to provide a more rational explanation of observed behaviour.

From the standpoint of colloid stability, a very important point is the distance of the shear plane from the 'true' particle surface. In calculating the interaction between particles as a function of separation distance, it is usual to assume that the attractive, van der Waals forces operate between the particles and so the correct distance to use is the separation between the surfaces. However, if zeta potentials are used to calculate electrical interaction, it is the separation between shear planes which is appropriate. If the shear plane is located 0.5 nm from the particle surface, then, effectively, the electrical interaction between two particles operates over a distance 1 nm less than the van der Waals interaction. Since both interactions decrease strongly with separation distance (see Chapter 3), this effect can give a substantially increased particle stability (because of the greater influence of the electrical repulsion term). This point has been discussed by Duckworth and Lips (1978) and can significantly influence interpretation of colloid stability data (Penners and Koopal, 1987).

Bibliography

Everett, D. H. (1992) *Basic Principles of Colloid Science*, 2nd edn, Royal Society of Chemistry, London

Hunter, R. J. (1981) *Zeta Potential in Colloid Science*, Academic Press, London

Hunter. R. J. (1987) *Foundations of Colloid Science*, Vol. 1, Oxford University Press, Oxford

Israelachvili, J. N. (1991) *Intermolecular and Surface Forces*, 2nd edn, Academic Press, London

Lyklema, J. (1991) *Fundamentals of Interface and Colloid Science*, Vol. 1: *Fundamentals*, Academic Press, London

Shaw, D. J. (1992) *Introduction to Colloid and Surface Chemistry*, 4th edn, Butterworth-Heinemann, Oxford

Stumm, W. (1992) *Chemistry of the Solid-Water Interface*, Wiley Interscience, New York

Williams, R. A. (1992) *Colloid and Surface Engineering: Applications in the Process Industries*, (ed.), Butterworth-Heinemann, Oxford

References

Bijsterbosch, B. H. and Lyklema, J. (1978) Interfacial electrochemistry of silver iodide. *Adv. Colloid Interface Sci.*, **9**, 147–251

Buff, F. P. and Stillinger, F. H. Jr (1963) Statistical mechanical theory of double-layer structure and properties. *J. Chem. Phys.*, **39**, 1911–1922

Davies, C. W. (1962) *Ion Association*, Butterworths, London

Davis, J. A. and Leckie, J. O. (1978) Surface ionization and complexation at the oxide/water interface. II Surface properties of amorphous iron oxyhydroxide and adsorption of metal ions. *J. Colloid Interface Sci.*, **67**, 90–107

Dentel, S. K. and Kingery, K. M. (1989) Using streaming current detectors in water treatment. *J. Am. Water Wks Assn*, **81**, 85–94

Duckworth, R. M. and Lips, A. (1978) The influence of Stern layers on the calculation of Hamaker functions from coagulation data. *J. Colloid Interface Sci.*, **64**, 311–319

Elimelech, M. and O'Melia, C. R. (1990) Effect of electrolyte type on the electrophoretic mobility of polystyrene latex colloids. *Colloids Surfaces*, **44**, 165–178

Grahame, D. C. (1947) The electrical double layer and the theory of electrocapillarity. *Chem. Rev.*, **41**, 441–501

Henderson, D., Blum, L. and Lozada-Cassou, M. (1983) The statistical mechanics of the electric double layer. *J. Electroanal. Chem.*, **150**, 291–303

Hiemenz, P. C. (1985) *Principles of Colloid and Surface Chemistry*, Marcel Dekker, New York

Hiemstra, T. and van Riemsdijk, W. H. (1991) Physical chemical interpretation of primary charging behaviour of metal (hydr)oxides. *Colloids Surfaces*, **59**, 7–25

Hunter, R. J. (1981) *Zeta Potential in Colloid Science*, Academic Press, London

Hunter, R. J. (1987) *Foundations of Colloid Science*, Vol. 1, Oxford University Press, Oxford

Hunter, R. J. (1993) *Introduction to Modern Colloid Science*, Oxford University Press, Oxford

Lyklema, J. (1978) Surface chemistry of colloids in connection with stability. In *The Scientific Basis of Flocculation* (ed. Ives, K. J.) Sijthoff and Noordhoff, Alphen aan den Rijn, 3–36

Lyklema, J. (1991) *Fundamentals of Interface and Colloid Science*, Vol. 1: *Fundamentals*, Academic Press, London

Martynov, G. A. (1966) *Research in Surface Forces*, Vol. 2 (ed. Derjaguin, B. V.), Consultants Bureau, New York, 75–102

O'Brien, R. W. and White, L. R. (1978) Electrophoretic mobility of a spherical colloidal particle. *J. Chem. Soc. Faraday II*, **74**, 1607–1626

Penners, N. H. G. and Koopal, L. K. (1987) The effect of particle size on the stability of haematite

hydrosols. *Colloids Surfaces*, **28**, 67–83
Robinson, R. A. and Stokes, R. H. (1965) *Electrolyte Solutions*, Butterworths, London
Shapiro, A. P. and Probstein, R. F. (1993) Removal of contaminants from saturated clay by electroosmosis. *Env. Sci. Tech.*, **27**, 283–291
Stumm, W. (1992) *Chemistry of the Solid–Water Interface*, John Wiley, New York
Sunderland, J. G. (1987) Electrokinetic dewatering and thickening. *J. Appl. Electrochem.*, **17**, 889–898
Thompson, D. W. and Pownall, P. G. (1989) Surface electrical properties of calcite. *J. Colloid Interface Sci.*, **131**, 74–82
Torrie, G. M. and Valleau, J. P. (1979) A Monte Carlo study of an electrical double layer. *Chem. Phys. Lett.*, **65**, 343–346
Torrie, G. M. and Valleau, J. P. (1980) Electrical double layers: I. Monte Carlo study of a uniformly charged surface. *J. Chem. Phys.*, **73**, 5807–5816
Torrie, G. M. and Valleau, J. P. (1982) Electrical double layers: IV. Limitations of the Gouy–Chapman theory. *J. Phys. Chem.*, **86**, 3251–3257
van den Hoven, Th. J. J. and Bijsterbosch, B. H. (1987) Streaming currents, streaming potentials and conductances of concentrated dispersions of negatively-charged, monodisperse polystyrene particles. Effect of adsorbed tetraalkylammonium ions. *Colloids Surfaces*, **22**, 187–205
Yates, D. E., Levine, S. and Healy, T. W. (1974) Site-binding model of the electrical double layer at the oxide/water interface. *J. Chem. Soc. Faraday Trans.*, **70**, 1807–1818

Chapter 3

Surface interaction potentials

Nomenclature

a	particle radius
A	Hamaker constant
C_s	electrolyte molar concentration
e	electron charge
f_R	double layer force per unit area
F_R	double layer force
$\hbar$	Planck's constant divided by 2π
h	surface-to-surface separation distance
h_{max}	energy-barrier distance
h_{min}	primary-minimum distance
H	distance between outermost peaks of surface irregularities
k	Boltzmann constant
M	molar concentration
n_0	limiting refractive index
n_∞	bulk number density of ions
N_A	Avogadro's constant
r	centre-to-centre or centre-to-surface separation distance
S	mean separation distance between two rough surfaces
T	absolute temperature
V_A	van der Waals interaction potential energy
V_B	Born interaction potential energy
V_R	electrical double layer interaction potential energy
V_T	total colloidal interaction potential energy
V_{LJ}	Lennard-Jones potential energy
x	distance
z	valence
α	radius ratio ($=a_2/a_1$)
δ_p	thickness of plate
ε	dielectric constant ($=\varepsilon_0\varepsilon_r$)
ε_E	depth of primary minimum in Lennard–Jones potential
$\varepsilon_i(0)$	contribution of species i to zero frequency dielectric constant
γ	reduced surface potential ($=\tanh(ze\varphi/4kT)$)
Φ	reduced potential ($=ze\varphi/kT$)
φ	electrical potential
κ	Debye–Huckel reciprocal length

λ	characteristic wavelength (usually 100 nm)
ρ	reduced distance ($=r/a_1$)
σ	surface charge density
σ_c	collision diameter
$\nu(r)$	potential energy per unit area between two plates
ω_v	characteristic dispersion frequency
ζ	zeta potential

CCA	constant-charge approximation
CPA	constant-potential approximation
DIM	Derjaguin integration method
ELM	extended McCartney and Levine formula
IPM	ion-penetrable membrane model
LPB	linear Poisson–Boltzmann equation solution
LSA	linear superposition approximation
NLPB	non-linear Poisson–Boltzmann equation solution

3.1 Introduction

The two most familiar kinds of colloidal interaction are van der Waals attraction and electrical double layer repulsion, which form the basis of the well-known DLVO theory of colloid stability, developed independently by Derjaguin and Landau (1941) and Verwey and Overbeek (1948). While these enable a large amount of experimental aggregation and deposition data to be explained, at least in a semiquantitative manner, there are many cases where other types of interaction have to be invoked. Initially, these were lumped together under some general heading, such as 'structural forces', but recent advances have given considerable insight into these 'extra' or 'non-DLVO' forces and it is now possible to give a more detailed classification. In aqueous systems, various kinds of hydration effects can be important, especially at close approach of surfaces. These are often associated with the hydration of ions at the particle or deposition surfaces and usually give an extra repulsion. It is now known that hydrophobic effects can also be important, mostly giving an extra attraction between particles. Other important effects arise from the presence of adsorbed polymers, giving either a repulsion ('steric' interaction) or an attraction ('polymer bridging').

These interactions will be discussed below, paying special attention to those aspects which have a direct bearing on particle aggregation and deposition processes and where some degree of control may be achieved. Quantitative discussion is focused on interaction of spherical particles with similar and dissimilar particles and macroscopic surfaces.

3.2 Double layer interaction between macroscopic bodies

When two charged particles approach each other in an electrolyte solution, their diffuse double layers overlap and, in the case of similarly charged particles, a repulsion is experienced between them. The precise way in which the double layers respond to each other depends on a number of factors. One distinction, which was recognized early in colloid stability theory (Verwey and Overbeek, 1948), is between interaction at *constant surface potential* or *constant surface charge*. The former case corresponds to the maintenance of surface-chemical equilibrium during approach, which, because of the very short time of an encounter between colloidal particles (typically, about 10 μs for a Brownian collision), may not be a realistic assumption (Frens and Overbeek, 1972). Constant-charge interaction might be expected when the particles have a fixed surface charge density (as with latex particles with bound ionic groups or clays with a certain ion exchange capacity). However, these considerations apply to the potential and charge at the particle surface, whereas double layer interaction depends on the potential at the Stern plane, which may respond quite differently to the approach of another surface. In particular, the Stern layer equilibrium may be much more rapidly established than that at the particle surface. Lyklema (1980) has shown that relaxation of the diffuse part of the double layer will always be very rapid on the timescale of a particle encounter. The link between double layer dynamics and surface conductivity has been pointed out by Dukhin and Lyklema (1987), who also showed that the rate of approach of two particles could be significantly slowed by such effects. Despite its significance, their model is a little too cumbersome for practical application and thus it will not be considered in the following discussion.

The interaction energy may be calculated in two ways. One is to solve the Poisson–Boltzmann equation directly for the system under consideration, but this is usually unable to give a simple analytical solution. The other method is to construct the formula from known expressions for each of the surfaces involved in the absence of the others. The approximations obtained in this manner are often of more practical importance because they are simple, yet fairly accurate. Although the expressions for sphere–sphere and sphere–plate interactions are of more interest to particle deposition and aggregation processes, many of them are actually derived from the corresponding plate–plate interactions. Therefore, the plate–plate interaction will be dealt with first.

3.2.1 Plate–plate interactions

A general formula for the force per unit area (i.e. pressure) between two approaching flat double layers in a symmetrical (z–z) electrolyte was given by Langmuir (1938) and Bell and Peterson (1972):

$$f_R = n_\infty kT \left[2\cosh\Phi - \left(\frac{d\Phi}{\kappa dx}\right)^2 - 2 \right] \tag{3.1}$$

where n_∞ is the bulk number density of ions, $\Phi = ze\varphi/kT$ and φ denotes the potential at a distance x from the plate. Integrating the force over distance then gives the potential energy per unit area υ_R:

$$\upsilon_R = -\int_\infty^h f_R \, dx \tag{3.2}$$

where h is the separation of the two surfaces. At equilibrium, f_R should be equal everywhere; therefore the potential and the potential gradient from eqn (3.1) may be evaluated, in principle, at any point in order to calculate the force. Some of the approximate expressions obtained using such integration approaches under various conditions are summarized in Table 3.1. Note that the expressions in Table 3.1 (and Table 3.2) are given in terms of the bulk number density of ions n_∞, i.e. the number of ions per cubic metre. It is related to the more familar salt concentration, or electrolyte molar concentration, C_s (in units of mol dm^{-3}) by $n_\infty = 1000 N_A C_s$, where N_A is Avogadro's constant.

Although a great deal of effort has been expended in the development of theories of double layer interaction, the results are of limited use in practical problems, owing to the lack of sufficient information on the systems. For instance, as discussed in Chapter 2, the relevant surface potential is not usually known with any certainty and is generally assumed to be the zeta potential, which can be determined experimentally. Also, dynamic aspects of double layer interaction are still not fully understood and calculations are often still made on the basis of 'constant potential' or 'constant-charge' assumptions, although neither of these extreme cases is likely to apply in practice. A simple expression, which represents a useful compromise between the two extremes, is the so-called *linear superposition approximation* (LSA). The LSA assumes that a region exists between the two interacting surfaces where the potential is sufficiently small and obeys the linearized Poisson–Boltzmann equation, so that contributions from each surface can be added to give the overall potential. In the vicinity of each surface the potential is assumed to be due to that surface alone. Clearly, the LSA holds only if the particle separation is relatively large (i.e. $\kappa h \gg 1$).

Gregory (1975) has made a comparison of his 'compression' model, eqn (3.6), with some of the expressions listed above, i.e. eqns (3.3) to (3.5), based on exact solutions for the constant-potential (Devereux and de Bruyn, 1963) and the constant-charge (Bell and Peterson, 1972) cases. As may be expected, the differences between the three approaches (i.e. CCA, LSA and CPA, Table 3.1) become most pronounced at short separations (e.g. $\kappa h < 3$) and the LSA always gives intermediate values. Since in reality neither the CCA nor the CPA is likely to be correct, the LSA may be the more reasonable solution to use. Gregory's expression, eqn (3.6), gives a better estimation for similar surfaces but becomes poorer for dissimilar surfaces than the LPB approximation, eqn (3.4). Therefore, these two expressions may be regarded as complementary in application. The model of Ohshima and Kondo (1988b) was devised specially for the ion-penetrable membrane surfaces and in the limiting case of small potentials, eqn (3.8), it gives exactly the same results as the LSA expression, eqn (3.7).

It may be noted that all the expressions in Table 3.1 are for symmetrical (z–z) electrolytes. The explicit incorporation of asymmetrical electrolytes is possible. For instance, Wilemski (1982) has derived an expression for the weakly interacting equal plates of dissimilar double layers in a mixture of symmetrical and asymmetrical electrolyte, within the framework of the Gouy–Chapman theory. However, the resulting expression is not convenient for practical uses since numerical integration procedures are required.

Table 3.1 *Plate–plate double layer interaction energies*

Expression		*Method*	*Validity*	*Source*
$\frac{64n_\infty kT}{\kappa}\gamma_1\gamma_2\exp(-\kappa h)$	(3.3)	LSA	Small Φ_1 and Φ_2 Symmetrical	Gregory (1975)
$\frac{n_\infty kT}{\kappa}\left[(\Phi_1^2+\Phi_2^2)(\coth\kappa h-1)+2\Phi_1\Phi_2/\sinh\kappa h\right]$	(3.4)	CCA LPB	Small Φ_1 and Φ_2 Symmetrical	Usui (1973)
$\frac{n_\infty kT}{\kappa}\left[(\Phi_1^2+\Phi_2^2)(1-\coth\kappa h)+2\Phi_1\Phi_2/\sinh\kappa h\right]$	(3.5)	CPA LPB	Small Φ_1 and Φ_2 Symmetrical	Hogg *et al.* (1966)
$\frac{2n_\infty kT}{\kappa}\ln\left[2\overline{\Phi}\left(\frac{Z+\overline{\Phi}\coth(\kappa h/2)}{1+\overline{\Phi}}\right)-\ln(\overline{\Phi}^2+\cosh\kappa h+Z\sinh\kappa h)+\kappa h\right]$ $\overline{\Phi}\equiv(\Phi_1+\Phi_2)/2 \qquad Z\equiv\{1+\overline{\Phi}^2/\sinh^2(\kappa h/2)\}^{1/2}$	(3.6)	CCA NLPB	$\lvert\Phi_1\rvert<2$ and $\lvert\Phi_2\rvert<2$ Symmetrical	Gregory (1975)
$\frac{32n_\infty kT}{\kappa}\gamma^2(1-\tanh\kappa h)$	(3.7)	CPA	Small Φ Equal plates Symmetrical	Verwey and Overbeek (1948)
$2\varepsilon\kappa^2\zeta^2\exp(-\kappa h)$	(3.8)	IPM	Small Φ Equal plates Symmetrical	Ohshima and Kondo (1988b)
$\frac{32n_\infty kT}{\kappa}\Big\{\gamma^2(1+\gamma^2+\gamma^4)[1-\tanh(\kappa h/2)]$ $-\gamma^4\left[(1+\gamma^2)\frac{\sinh(\kappa h/2)}{2\cosh^3(\kappa h/2)}+\frac{\kappa h/2}{2\cosh^4(\kappa h/2)}\right]$ $-\gamma^6\left[\frac{\sinh(\kappa h/2)}{4\cosh^5(\kappa h/2)}+\frac{5\kappa h/2}{4\cosh^6(\kappa h/2)}-\frac{(\kappa h/2)^2\sinh(\kappa h/2)}{\cosh^7(\kappa h/2)}\right]\Big\}$	(3.9)	CPA NLPB	All separations for Φ up to 5 Equal plates Symmetrical	Ohshima and Kondo (1988a)
$\sigma\sec h(\kappa h)+\left(\frac{2\sigma^2}{\varepsilon\kappa}-\frac{\varepsilon\kappa\varphi^2}{2}\right)(\tanh\kappa h-1)$	(3.10)	Mixed CPA and CCA LPB	Weak interaction Symmetrical	Kar *et al.* (1973)

3.2.2 Sphere–sphere interactions

In calculating the sphere–sphere double layer interactions, perhaps the most commonly used is the integration method of Derjaguin (1934). This is based on the corresponding plate–plate expression which is then used to construct the interaction energy between two closely spaced spheres:

$$V_R = \frac{2\pi a_1 a_2}{a_1 + a_2} \int_h^\infty \upsilon_R \, dx \tag{3.11}$$

or in the case of the force of interaction, F_R:

$$F_R = \frac{2\pi a_1 a_2}{a_1 + a_2} \upsilon_R(h) \tag{3.12}$$

where h denotes the (minimum) surface-to-surface separation between the spheres. The major disadvantage of this method is that it may be applied only when both the conditions $\kappa a_i > 5$ and $h \ll a_i$ $(i = 1, 2)$ are satisfied. Generally, $\kappa a_i > 5$ is considered to be adequate (Verwey and Overbeek, 1948; McCartney and Levine, 1969). In recognition of this limitation, McCartney and Levine (1969) developed a method in which the second limitation (i.e. $h \ll a_i$) has been lifted. Besides these two approaches, the simple LSA principle can also be used, as mentioned above in discussing plate–plate interactions. Some expressions for sphere–sphere electrical double layer interaction potential energy are listed in Table 3.2. The group of parameters

$$\frac{n_\infty kT}{\kappa^2}$$

in the given expressions in Table 3.2 may be replaced by

$$\frac{\varepsilon}{2}\left(\frac{kT}{ze}\right)^2$$

and both forms are in common use in the literature.

In general, the LSA method yields the correct interaction at large separations for *all* surface potentials and particle sizes; Derjaguin's integration gives accurate results for large particles at short distances; and the McCartney and Levine formulation is a good approximation at all separations but small potentials. It should be noted that although the first two methods themselves place no restriction on the potentials, the resulting expressions often do because of the difficulty in solving the Poisson–Boltzmann equation. Therefore, care must be taken in choosing the right expression from Table 3.2. Figure 3.1 compares the potential energy of interaction given by three equations taken from Table 3.2, using a LSA expression, eqn (3.13), a CPA expression, eqn (3.15), and a CCA expression, eqn (3.16). This serves to emphasize again that the use of the CCA approximation is inaccurate at small separation distances.

For sphere–sphere interactions, subject to the limiting conditions shown along with eqn (3.13) in Table 3.2, application of the LSA yields the following equation, which is in common use:

$$V_R = 64\pi \frac{a_1 a_2}{a_1 + a_2} \left(\frac{kT}{ze}\right)^2 \gamma_1 \gamma_2 \exp(-\kappa h) \tag{3.20}$$

where ε is the permittivity of the medium, z the valency of the ions (a symmetric z–z electrolyte is assumed) and e is the elementary charge. The terms γ_1 and γ_2 are dimensionless functions of the surface potentials, defined in eqn (2.10).

The pre-exponential term in eqn (3.20) depends on the zeta potentials of the particles and is always positive (repulsion) if the potentials are of like sign and negative (attraction) if they are of opposite sign. This is in accord with intuitive expectations, but not with results when either constant potential or constant charge are assumed. In the former case attraction may arise between particles with surface potentials of like sign but unequal magnitude, and, for constant charge, particles of opposite charge may repel each other (Gregory, 1975). For the interaction of different particles (heteroaggregation) or for deposition of particles on a substrate, these differences could be significant and considerable attention has been paid to this point (e.g. Rajagopalan and Kim, 1981). However, there is no convincing experimental evidence for these 'counterintuitive' effects and most results on heteroaggregation (James *et al.*, 1977) or deposition (Gregory and Wishart, 1980) indicate attraction for oppositely charged surfaces and repulsion for like signs of charge, which gives at least some support for simple expressions such as eqn (3.20).

For identical particles, eqn (3.20) reduces to:

$$V_R = 32\pi\varepsilon a (kT/ze)^2 \gamma^2 \exp(-\kappa h) \tag{3.21}$$

For small values of surface or zeta potential, this simplifies still further to:

$$V_R = 2\pi\varepsilon a \zeta^2 \exp(-\kappa h) \tag{3.22}$$

which shows the influence of ζ more clearly.

The effect of added salt on the repulsion will be two-fold: a decrease in the zeta potential (especially in the case of specifically-adsorbing counterions) and an increase in κ. Both of these will cause a decrease in the repulsion at a given separation distance. Another conclusion from eqns (3.21) and (3.22) is that the repulsion should be directly proportional to particle size, as will be shown later for the case of van der Waals attraction. The predicted exponential decay of the interaction with distance, and a decay length of $1/\kappa$, have been confirmed experimentally from direct force measurements between mica sheets in various electrolyte solutions (Pashley and Israelachvili, 1984), for separation distances greater than about 4 nm. At closer approach, hydration forces become important. The predicted dependence on zeta potential could not be directly checked, but values calculated from the measured force were in the expected range.

Double layer interaction between surfaces of high charge density and in solutions of quite high ionic strength can give behaviour which is not expected on the basis of the Poisson–Boltzmann approach (Kjellander and Marcelja, 1988). Effects due to ion size and ion–ion correlations may give, respectively, repulsive and attractive forces and the latter can be strong, especially with

Table 3.2 *Sphere–sphere double layer interaction energies*

Expression		*Method*	*Validity*	*Source*
$\frac{128\pi a_1 a_2 n_\infty kT}{(a_1 + a_2)\kappa^2} \gamma_1 \gamma_2 \exp(-\kappa h)$	(3.13)	LSA DIM	$h \ll a_i$ $\kappa a_i > 5$ Symmetrical	Gregory (1975)
$\frac{128\pi a_1 a_2 n_\infty kT}{r\kappa^2} \gamma_1 \gamma_2 \exp(-\kappa h)$	(3.14)	CPA LPB LSA	$\kappa r \gg 1$ $\kappa a_i \geq 10$ Symmetrical	Bell *et al.* (1970)
$\frac{2\pi a_1 a_2 n_\infty kT}{(a_1 + a_2)\kappa^2} (\Phi_1^2 + \Phi_2^2) \left[\frac{2\Phi_1 \Phi_2}{\Phi_1^2 + \Phi_2^2} \ln \left(\frac{1 + e^{-\kappa h}}{1 - e^{-\kappa h}} \right) + \ln (1 - e^{-2\kappa h}) \right]$	(3.15)	CPA LPB DIM	Small Φ_i $h \ll a_i$ $\kappa a_i \gg 1$	Hogg *et al.* (1966)
$\frac{2\pi a_1 a_2 n_\infty kT}{(a_1 + a_2)\kappa^2} (\Phi_1^2 + \Phi_2^2) \left[\frac{2\Phi_1 \Phi_2}{\Phi_1^2 + \Phi_2^2} \ln \left(\frac{1 + e^{-\kappa h}}{1 - e^{-\kappa h}} \right) - \ln (1 - e^{-2\kappa h}) \right]$	(3.16)	CCA LPB DIM	Small Φ_i $h \ll a_i$ $\kappa a_i \gg 1$	Wiese and Healy (1970) Usui (1973)
$\frac{\pi a_1 a_2}{a_1 + a_2} \left[\left(\frac{2\varphi\sigma}{\kappa} \right) \left(\frac{\pi}{2} - \tan^{-1} \sinh(\kappa h) \right) - \left(\frac{\sigma^2}{\varepsilon\kappa^2} - \varepsilon\varphi^2 \right) \ln (1 + e^{-2\kappa h}) \right]$	(3.17)	Mixed CPA and CCA LPB DIM	Weak interaction Symmetrical $h \ll a_i$ $\kappa a_i \gg 1$	Kar *et al.* (1973)

$\frac{2\pi n_\infty kT}{\kappa^2}\frac{a_1 a_2(r-a_1)(r-a_2)}{r[(a_1+a_2)r-a_1^2-a_2^2]}[\Omega_1 \ln(1+\Gamma)+\Omega_2 \ln(1-\Gamma)]$ $\Omega_1=\Phi_1^2+\Phi_2^2+\Lambda\Phi_1\Phi_2 \qquad \Omega_2=\Phi_1^2+\Phi_2^2-\Lambda\Phi_1\Phi_2$ $\Lambda=\sqrt{\frac{a_2(r-a_2)}{a_1(r-a_1)}}+\sqrt{\frac{a_1(r-a_1)}{a_2(r-a_2)}}$ $\Gamma=\sqrt{\frac{a_1 a_2}{(r-a_1)(r-a_2)}\exp[\kappa(a_1+a_2-r)]}$	(3.18)	LPB EML	Small Φ_i $\kappa a_i \geq 5$ Symmetrical All distances	Bell *et al.* (1970)
$\frac{64\pi a n_\infty kT}{\kappa^2}\left\{\gamma^2\left(1+\frac{2}{3}\gamma^2+\frac{23}{45}\gamma^4\right)\ln(1+e^{-\kappa h})\right.$ $-\frac{1}{3}\gamma^4\left(1+\frac{22}{15}\gamma^2\right)\left(\frac{\kappa h}{2}\right)\left[1-\tanh(\kappa h/2)-\frac{\tanh(\kappa h/2)}{2\cosh^2(\kappa h/2)}\right]$ $\left.-\frac{\gamma^4(1+23\gamma^2/30)}{6\cosh^2(\kappa h/2)}-\frac{\gamma^6[1-11(\kappa h/2)\tanh(\kappa h/2)]}{60\cosh^4(\kappa h/2)}+\frac{\gamma^6(\kappa h/2)}{6\cosh^6(\kappa h/2)}\right\}$	(3.19)	CPA NLPB DIM	$h \ll a_i$ $\kappa a_i \gg 1$ Medium Φ_1 Equal spheres Symmetrical	Ohshima and Kondo (1988a)

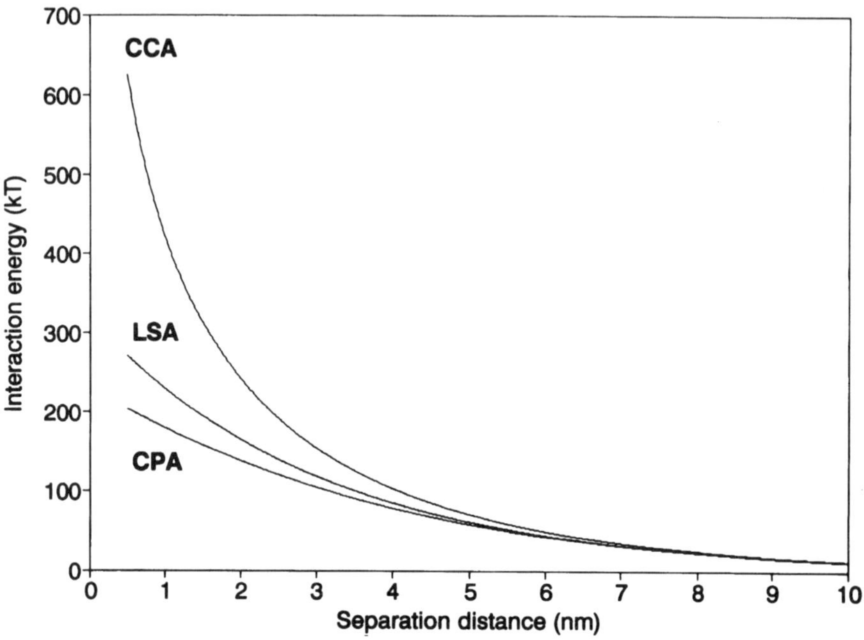

Figure 3.1 *Comparison of linear superposition (LSA), controlled-potential (CPA) and controlled-charge (CCA) assumptions in calculating the potential energy of interaction between two identical spheres using eqns (3.13), (3.15) and (3.16) respectively. Curves correspond to 0.5 μm spheres, $\zeta = 25\,mV$, C = 0.1 M, in 1–1 electrolyte at 298 K*

divalent electrolytes. Measurements with mica sheets in 0.15 M $CaCl_2$ have shown strong attraction and there is evidence of similar effects in biological systems (Wennerstrom and Jonsson, 1988).

3.2.3 Sphere–plate interactions

The three approximations (LSA, CPA, CCA) apply equally well for plate–sphere interactions, which can be estimated by modifying the sphere–sphere expressions given in Table 3.2 by allowing one of the radii to approach infinity.

3.3 Van der Waals interaction

3.3.1 Overview

The fact that an attractive force always exists between colloidal particles of the same material has long been recognized but a detailed understanding of these forces took a long time to emerge (Israelachvili, 1992). This attraction between two closely separated surfaces is generally called the London–van der Waals force. The force arises from spontaneous electrical and magnetic polarizations, giving a fluctuating electromagnetic field within the media and in the gap between them.

Essentially there are two theoretical approaches to the evaluation of London–van der Waals attraction. In the classical (or *microscopic*) approach, due largely to Hamaker (1937), the interaction between two macroscopic bodies is obtained by the pairwise summation of all the relevant intermolecular interactions. All expressions obtained in this manner (Table 3.3) may be split into a purely *geometrical* part and a constant *A*, the Hamaker constant, which is related only to the properties of the interacting macroscopic bodies and the medium. Typically, *A* lies between 10^{-21} J and 10^{-19} J. Hamaker's method can be easily applied to different geometries. However, the assumption of complete additivity is a rather serious deficiency and the resulting expressions always overestimate the interaction.

The additivity assumption is overcome through use of an alternative (*macroscopic*) approach suggested by Liftshitz (1956), in which the interaction is derived entirely from considerations of the macroscopic electromagnetic properties of the medium. In spite of its rigour, extensive application of the theory has so far been hampered because of:

- the difficulty of obtaining adequate experimental data on the frequency-dependent dielectric permittivity of the colloidal particles and of the solvent medium, and
- the prohibitive difficulty in its formulation for geometries other than the simplest plate–plate arrangement.

Certain simplifications may be possible as described clearly by Hough and White (1980) and Prieve and Russel (1988), but in general the computation required and lack of appropriate dielectric data limits practical use of the macroscopic approach. Several reviews and comparisons of the two approaches are available (e.g.Gregory, 1969, 1989; Visser, 1972; Israelachvili, 1992). Suffice it to say that the difference between the results obtained from the two approaches is not too great (usually less than 60%), and the results have been borne out by direct measurements of van der Waals forces between mica sheets (Israelachvili and Tabor, 1972). Hence, for practical purposes, it is more convenient to rely on the Hamaker-type expressions with corrections to account for the retardation effect (described in Section 3.3.4).

Having summarized the principal features of van der Waals interactions above, the following two sections describe further details associated with the use of Hamaker-type equations for sphere–sphere interactions, with a particular emphasis on the practical implications of the formulae.

3.3.2 Hamaker expressions for interacting spheres

The Hamaker expressions are based on the assumption of pairwise additivity of intermolecular forces. The interaction between two particles is calculated simply

Table 3.3 *Unretarded van der Waals interaction energies for sphere–sphere and sphere–plate geometries*

Expression		*Validity*
$-\frac{A}{6}\Bigg(\frac{2a_1a_2}{h^2+2a_1h+2a_2h}+\frac{2a_1a_2}{h^2+2a_1h+2a_2h+4a_1a_2}+\ln\left(\frac{h^2+2a_1h+2a_2h}{h^2+2a_1h+2a_2h+4a_1a_2}\right)\Bigg]$	(3.23)	Sphere–sphere
$-\frac{Aa_1a_2}{6h(a_1+a_2)}$	(3.24)	Sphere–sphere, $h \ll a_1$
$-\frac{A}{6}\left[\frac{a}{h}+\frac{a}{h+2a}+\ln\left(\frac{h}{h+2a}\right)\right]$	(3.25)	Sphere–plate
$-\frac{A}{12\pi}\left[\frac{1}{h^2}+\frac{1}{(h+2\delta_p)^2}-\frac{2}{(h+\delta_p)^2}\right]$	(3.26)	Plate–plate, finite thickness
$-\frac{A}{12\pi h^2}$	(3.27)	Plate–plate, infinite thickness

by summing the interactions of all molecules in one particle with all of the molecules in the other. Hamaker (1937) replaced the summation by a double integration procedure, which leads to very simple expressions, especially when the separation distance is small. For two spheres of radii a_1 and a_2, separated by a distance h, the *interaction energy* at close approach ($h \ll a$) is given by eqn (3.24) (see Table 3.3). In this expression the Hamaker constant is often written as A_{12}, which is the Hamaker constant for the media 1 and 2, of which the spheres are composed, and assuming, for the moment, that they are immersed in a vacuum. This depends entirely on the properties of the interacting media (see below). For equal spheres of radius a_1, the result is simply given by:

$$V_A = -A_{11}a_1/12h \tag{3.28}$$

These expressions only apply at close approach and become quite inaccurate at separations greater than about 10% of the particle radius. However, in many cases the interaction energy is insignificant at larger distances so that eqns (3.24) and (3.28) are acceptable for most practical purposes. The most important features of the Hamaker expressions include the inverse dependence of the interaction energy on the separation distance, and the direct dependence on particle size.

In principle, the attraction should become very strong on close approach and infinite on contact. However, short-range repulsion forces, such as Born repulsion (see Section 3.5) and other effects to be considered later, come into play at close approach and keep the attraction finite.

The results given above apply to the interaction of media across a vacuum. For interaction through a liquid (as in problems of colloid stability and deposition), the same expressions can be used, but with a modified Hamaker constant, which, for media 1 and 2 separated by medium 3, can be written:

$$A_{132} = A_{12} + A_{33} - A_{13} - A_{23} \tag{3.29}$$

where A_{13} is the Hamaker constant for materials 1 and 3, etc. Depending on the relative magnitudes of the individual constants, the presence of a third medium can significantly reduce the interaction.

A useful approximation for Hamaker constants of different media is the geometrical mean assumption:

$$A_{12} \approx (A_{11}A_{22})^{1/2} \tag{3.30}$$

With this and the corresponding expressions for A_{13} and A_{23}, eqn (3.29) becomes:

$$A_{132} \approx (A_{12}^{1/2} - A_{33}^{1/2})(A_{22}^{1/2} - A_{33}^{1/2}) \tag{3.31}$$

For similar materials, 1, interacting across medium 3:

$$A_{131} = (A_{11}^{1/2} - A_{33}^{1/2})^2 \tag{3.32}$$

Equation (3.32) led Hamaker to the conclusion that the van der Waals interaction between similar materials in a liquid would always be attractive (positive Hamaker constant), whatever the values of A_{11} and A_{33}. However, for different materials, the possibility of negative Hamaker constants and van der

Waals repulsion arises, for instance when $A_{11} > A_{33} > A_{22}$. Some practical consequences of negative Hamaker constants have been considered by van Oss *et al.* (1979) and the subject has been reviewed by Visser (1981). Although van der Waals repulsion may be important in wetting phenomena (Hough and White, 1980) and in phase separation from polymer solutions (van Oss et al, 1979), there is very little evidence of the effect in colloidal dispersions. Examples may be the carbon–water–Teflon (Visser, 1981) and carbon–butanol–Teflon (Cooper and Marsden, 1978) systems.

3.3.3 Calculation of Hamaker constants

It was shown by Gregory (1969) that an approximate Hamaker constant for a material can be calculated entirely in terms of its optical properties – the limiting refractive index n_0 and a characteristic dispersion frequency, ω_v, both of which can be derived from optical dispersion data. In fact, the results will be expressed in terms of the limiting dielectric constant (relative permittivity), ε_0, which is linked to n_0 by the Maxwell relation, $\varepsilon_0 = n_0^2$. For convenience, the subscripts 0 and v are omitted, so that for substance 1 the relevant terms are ε_1 and ω_1. For different media, 1 and 2, the Hamaker constant is then given by:

$$A_{12} = \frac{27}{32} \frac{\hbar\omega_1\omega_2}{\omega_1 + \omega_2} \frac{(\varepsilon_1 - 1)(\varepsilon_2 - 1)}{(\varepsilon_1 + 2)(\varepsilon_2 + 2)} \tag{3.33}$$

where $\hbar$ is Planck's constant divided by 2π. For similar media, the result is:

$$A_{11} = \frac{27}{64} \hbar\omega_1 \left(\frac{\varepsilon_1 - 1}{\varepsilon_1 + 2}\right)^2 \tag{3.34}$$

An equivalent expression was derived by Tabor and Winterton (1969).

Comparison of eqns (3.32) and (3.34) shows that the 'geometrical mean' assumption, eqn (3.30), will only be valid if the dispersion frequencies for the media are not very different. Equations (3.33) and (3.34) are based on the assumptions of pairwise additivity and that the major contribution to the interaction comes from a closely spaced group of frequencies in the ultraviolet (so that the optical dispersion behaviour can be represented by a single 'dispersion frequency'). Lifshitz computations should include contributions from a wide range of frequencies, although Hough and White (1980) have shown that, in many cases, ultraviolet relaxations provide most of the interaction and that infrared contributions are much less significant than had previously been supposed. Furthermore, they found that accurate information on the ultraviolet relaxation can be derived from optical refractive index data, an assumption which was implicit in deriving eqns (3.33) and (3.34).

Because of the very polar nature of liquid water (as evidenced by its very high dielectric constant), there is an important 'zero frequency' or 'static' term in its Hamaker constant, which is not included in the value calculated from eqn (3.34), since optical dispersion data can give no information on lower frequency terms.

This term makes a contribution to the 'composite' Hamaker constant, A_{132}, which depends on the static dielectric constants of the interacting media, $\varepsilon(0)$. The terms ε_1, etc., in eqn (3.33) are the limiting values at visible wavelengths and will only be similar to the static values in the case of non-polar materials, such as hydrocarbons. The zero frequency contribution to the Hamaker constant is given approximately by:

$$A_0 = \frac{3kT[\varepsilon_1(0) - \varepsilon_3(0)][\varepsilon_2(0) - \varepsilon_3(0)]}{4[\varepsilon_1(0) + \varepsilon_3(0)][\varepsilon_2(0) + \varepsilon_3(0)]} \tag{3.35}$$

The zero frequency term is the only temperature-dependent contribution to van der Waals interaction. Because water has a dielectric constant of about 80 and non-polar media have values in the range 2–3 (just the square of the refractive index), the zero frequency term for many aqueous colloids is about $3kT/4$ or around 3×10^{-21} J. For some systems this may be the dominant contribution to the interaction. A further complication is that this term is affected by the presence of dissolved salts and is considerably reduced at high ionic strengths, essentially by a 'damping' effect (Davies and Ninham, 1972). The relaxation of ions in solution is sufficiently slow to respond to microwave oscillations, but has no effect on the higher frequency contributions to the Hamaker constant.

Although reliable data are scarce, experimental values of Hamaker constants are in reasonable agreement with calculated values. Direct measurements of van der Waals attraction between mica surfaces in air (Tabor and Winterton, 1969) gives a Hamaker constant of $13.5 \pm 1.5 \times 10^{-20}$ J and the value calculated (Gregory, 1977) from eqn (3.34) is 11.3×10^{-20} J. Direct measurements for mica in water (Israelachvili and Adams, 1978) give a value of approximately 2.2×10^{-20} J and the value calculated from eqns (3.32) and (3.20) is 2.0×10^{-20} J. A small zero frequency term would bring the two values into very good agreement.

The effect of ionic strength on the zero frequency term can sometimes be seen experimentally. For instance, the Hamaker constant for polystyrene particles in 0.6 M NaCl, determined from the rate of aggregation in shear flow (Feke and Schowalter, 1985), is about 0.48×10^{-20} J, which is rather lower than values determined in more dilute salt solutions and the calculated Lifshitz (0.91×10^{-20} J) and Hamaker (0.79×10^{-20} J) results. Marra (1986) has found that the Hamaker constant for lipid bilayers in aqueous solutions is reduced by the expected amount in 0.2 M NaCl.

For practically all aqueous dispersions, Hamaker constants lie in the range $0.3–10 \times 10^{-20}$ J. Dense mineral particles have values towards the upper end of this range, whereas low-density, especially biological, materials have quite low values. In the latter case, the 'zero frequency' term may be the dominant contribution and so the effective Hamaker constant may be significantly reduced by increasing ionic strength. For materials with Hamaker constants greater than about 10^{-20} J, the van der Waals interaction can be assumed to be essentially independent of ionic strength.

In many studies the Hamaker constant has been treated as an adjustable parameter, with a value chosen to fit experimental data.

3.3.4 Retardation

Since dispersion forces are electromagnetic in character, they are subject to a retardation effect. The finite time of propagation causes a reduced correlation between oscillations in the interacting bodies and a smaller interaction. Retardation is implicitly included in the full Lifshitz treatment, but it is fairly easy to modify the simple Hamaker approach to take account of the effect.

It is convenient to think of a 'characteristic wavelength' of the interaction, λ (given by $\lambda = 2\pi c/\omega_v$, where c is the velocity of light and ω_v is the dispersion frequency), which has a value of around 100 nm for most materials. It is often assumed that retardation only becomes significant when the separation distance between particles is of the same order as the characteristic wavelength. However, for spherical particles, it can be shown that retardation is important at much closer approach (Gregory, 1981a). For instance, with spheres of radius 1 μm, separated by a distance of 10 nm ($\lambda/10$), the attraction energy is reduced by a factor of about 2. At a distance of 100 nm, retardation causes about a 10-fold reduction in the attraction.

As shown in Table 3.4, the retardation effect can be introduced into the Hamaker expressions by a simple empirical factor (Gregory, 1981a), and for spheres a suitable factor is $1/(1 + 14h/\lambda)$, so that instead of eqn (3.28), the retarded attraction between two identical spheres is given by:

$$V_A = -\frac{A_{11}a_1}{12h(1 + 14h/\lambda)} \tag{3.44}$$

This simple expression gives good agreement with the much more elaborate computations of Clayfield *et al.* (1971) up to distances of about 10% of the particle radius. At greater separations, the simple Hamaker expressions become rather inaccurate (for geometrical reasons rather than retardation), but at such distances the interaction is usually insignificant. If there are contributions to van der Waals interaction from lower frequency terms, these should be less subject to retardation and simple expressions such as eqn (3.44) would not then apply. Measurements of retarded interaction between mica surfaces in air can be fitted by assuming just one characteristic wavelength, but such an approach would not be appropriate in systems with a significant zero frequency contribution.

A comparison of the expressions listed in Table 3.4 (except eqn (3.43)) with the exact formula of Clayfield *et al.* (1971) has been made by Gregory (1981a). Based on this comparison it may be concluded that, within their range of validity, eqn (3.40) yields better results than eqn (3.39) in the case of sphere–sphere interactions, but for sphere–plate interactions eqn (3.42) is even better. Furthermore, it was shown that eqn (3.44) is an adequate approximation for all separation distances up to about 10% of the particle radius. The geometrical restriction of $h \ll a_i$, imposed by the use of the Derjaguin procedure (and thus having nothing to do with retardation), may not be a serious limitation in practice.

The effect of surface roughness on van der Waals interaction has been considered by Czarnecki and Dabros (1980) and Czarnecki and Itschenskij (1984). They carried out a statistical analysis of surface irregularities, giving a

Table 3.4 *Retarded van der Waals interactions for sphere–sphere and sphere–plate geometries*

Expression		*Validity*	*Source*
$-\frac{A\lambda}{237h^3}$	(3.36)	$2\pi h/\lambda \gg 1$ Semi-infinite spaces	Overbeek (1952)
$-\frac{Aa}{12\pi h}\left[\frac{2.45}{10}\left(\frac{\lambda}{h}\right)-\frac{2.17}{60\pi}\left(\frac{\lambda}{h}\right)^2+\frac{0.59}{280\pi^2}\left(\frac{\lambda}{h}\right)^3\right]$	(3.37)	$h \gg \lambda/6$ Equal spheres	Schenkel and Kitchener (1960)
$-\frac{Aa_1a_2}{6(a_1+a_2)h}\left[\frac{2.45}{10\pi}\left(\frac{\lambda}{h}\right)-\frac{2.17}{60\pi^2}\left(\frac{\lambda}{h}\right)^2+\frac{0.59}{280\pi^3}\left(\frac{\lambda}{h}\right)^3\right]$	(3.38)	$h > \lambda/4\pi$, $h \ll a_i$ Sphere–sphere	Ho and Higuchi (1968)
$-\frac{Aa_1a_2}{6h(a_1+a_2)}\left[\frac{1}{1+11.12h/\lambda}\right]$	(3.39)	$h < \lambda/\pi$, $h \ll a_i$ Sphere–sphere	Schenkel and Kitchener (1960)
$-\frac{Aa_1a_2}{6h(a_1+a_2)}\left[1-\frac{bh}{\lambda}\ln\left(1+\frac{\lambda}{bh}\right)\right]$	(3.40)	$h \ll a_i$, $b = 5.32$ Sphere–sphere	Gregory (1981a)
$-\frac{Aa}{6h}\left[1-\frac{bh}{\lambda}\ln\left(1+\frac{\lambda}{bh}\right)\right]$	(3.41)	$h \ll a$, $b = 5.32$ Sphere–plate	Gregory (1981a)
$-A\left[\frac{2.45\lambda}{60\pi}\left(\frac{h-a}{h^2}-\frac{h+3a}{(h+2a)^2}\right)-\frac{2.17\lambda^2}{720\pi^2}\left(\frac{h-2a}{h^3}-\frac{h+4a}{(h+2a)^3}\right)+\frac{0.59\lambda^3}{5040\pi^3}\left(\frac{h-3a}{h^4}-\frac{h+5a}{(h+2a)^4}\right)\right]$	(3.42)	$h > \lambda/4\pi$ Sphere–plate	Czarnecki (1979)
$-\frac{Aa_1a_2}{6H(a_1+a_2)}\left[1-\frac{bH}{\lambda}\ln\left(1+\frac{\lambda}{bH}\right)\right]\frac{H}{S}$	(3.43)	$b = 5.32$ Rough spheres	Czarnecki and Itschenskij (1984)

radial mass distribution function, which was then used in Hamaker-type integrations to give the interaction of a particle with a plane surface. The results show that, when the roughness layer is thick in comparison with the separation distance, the unretarded interaction can be very substantially reduced. When retardation is allowed for, the effect of roughness is not so pronounced.

3.4 Non-DLVO forces

The importance of the double layer and van der Waals interactions has been well demonstrated by the classical DLVO theory (Verwey and Overbeek, 1948). However, there are situations, notably the spontaneous 'repeptization' or reversible aggregation observed by many investigators, where combination of the two principal forces alone does not give satisfactory agreement with experimental results. In such cases, and assuming that the DLVO theory (see Section 3.5) itself is reliable, some kind of additional short-range forces may have to be included as well. For colloidal particles carrying adsorbed polymers, the forces are known as *steric* or *osmotic forces*. *Structural* or *hydration forces* are the terms associated with the forces that develop as particles with adsorbed fluid layers interact. For the cases where the precise details of the spatial variation of the short-range forces are not important, it is often more convenient to resort to the microscopically averaged *Born repulsion.*

'Structural force' is a generic term used to describe the interactions arising specifically from the adsorption of solvent, surfactant or macromolecules at the interface. It covers, broadly, three kinds of interactions, namely, the repulsive hydration interactions (due to orientated water molecules adsorbed at each interface), the attractive hydrophobic interactions (because of the very strong interaction between water molecules compared to that between water molecules and organic groups), and the repulsive steric interactions (from deformation or penetration of adsorbed macromolecules).

The presence of an adsorbed layer can sometimes have a significant influence on the stability of colloidal dispersions through one or more of the following mechanisms:

(1) by changing the electrical double layer force either directly, in the case of polyelectrolytes or by causing a displacement of the Stern surface
(2) by altering the interparticle van der Waals attraction by modifying the effective Hamaker constant
(3) by generating additional interactions either due to desorption of adsorbed molecules, or compression and interpenetration of adsorbed layers.

The existence of these structural forces has long been recognized, but their exact nature has yet to be firmly established. However, it is clear that in some colloidal dispersions, they can be unexpectedly strong and long-ranged. The structural forces and their consequences have been a subject of extensive studies and have been well-documented elsewhere (e.g. Langbein, 1969; Hesselink *et al.*, 1971; Tanford, 1980; Napper, 1983; Churaev and Derjaguin, 1985; deGennes, 1987; Xu and Yoon, 1989; Eriksson *et al.*, 1989; Gregory, 1989; Israelachvili, 1992).

3.4.1 Born repulsion

This short-range repulsion originates from the strong repulsive forces between atoms as their electron shells interpenetrate each other. Precise description of the interatomic potential must be based on quantum mechanical considerations. However, a number of simplified approximate analytical forms have been proposed (Fitts, 1966), of which the most widely used is the Lennard–Jones *m*–*n* potential:

$$V_{LJ} = \varepsilon_E \frac{n}{n-m}\left(\frac{n}{m}\right)^{\frac{m}{n-m}}\left[\left(\frac{\sigma_c}{r}\right)^n - \left(\frac{\sigma_c}{r}\right)^m\right] \tag{3.45}$$

where r is the interatomic separation, σ_c the collision diameter and ε_E the depth of the primary energy well. The attractive part is due to van der Waals interaction and the repulsive contribution is known as Born repulsion. A value of 6 for m is generally accepted, but there is much less agreement on the value of n (integers ranging from 8 to 20 have been suggested for n but 12 is typically used).

In the past, Born repulsion has been incorporated into the interparticle potential by directly taking the repulsive part of the interatomic potential and rewriting it for macroscopic spheres. Although it has found some utility in various applications (Nieuenhuis *et al.*, 1981), such an *ad hoc* treatment has little or no theoretical justification. A more realistic formulation, at least at the approximation level of DLVO theory, would be the Hamaker-type integration of all molecules in the system. Ruckenstein and Prieve (1976) first published a formula obtained in this manner for the sphere–plate Born repulsion:

$$V_B = \frac{A\sigma_c^6}{7560}\left[\frac{8a+h}{(2a+7)^7} + \frac{6a-h}{h^7}\right] \tag{3.46}$$

where the collision diameter σ_c is typically of the order of 0.5 nm. A more general representation has been given by Feke *et al.* (1984):

$$\begin{aligned} V_B = {} & 4A\left(\frac{\sigma_c}{a_1}\right)^{n-6}\frac{(n-8)!}{(n-2)!}\frac{1}{\rho} \\ & \times \left\{\frac{-\rho^2-(n-5)(\alpha-1)\rho-(n-6)[\alpha^2-(n-5)\alpha+1]}{(\rho-1+\alpha)^{n-5}}\right. \\ & + \frac{-\rho^2+(n-5)(\alpha-1)\rho-(n-6)[\alpha^2-(n-5)\alpha+1]}{(\rho+1-\alpha)^{n-5}} \\ & + \frac{\rho^2+(n-5)(\alpha+1)\rho+(n-6)[\alpha^2+(n-5)\alpha+1]}{(\rho+1+\alpha)^{n-5}} \\ & \left. + \frac{\rho^2-(n-5)(\alpha+1)\rho+(n-6)[\alpha^2+(n-5)\alpha+1]}{(\rho-1-\alpha)^{n-5}}\right\} \end{aligned} \tag{3.47}$$

where $\rho = r/a_1$ and $\alpha = a_2/a_1$, n may be any of the integers between 8 to 20, but 12 is more commonly used. In the cases where no better knowledge is available or necessary, these expressions are useful for estimating the depth of the primary minimum of the total interaction.

The effect of Born interaction may not be of great significance in aqueous systems since the presence of any hydrated ions, which are likely to be present, will prevent surface–surface separation distances approaching 0.3 nm.

3.4.2 Hydration effects

The nature of water close to a surface can be very different from that of bulk water, for a variety of reasons. Since most particles and surfaces carry a surface charge and hence ionic surface groups, some hydration of these groups would be expected, by analogy with ions in solution. Some particles, especially those of biological origin, have various types of hydrophilic material at their surface, such as proteins and polysaccharides. These can have quite large amounts of 'bound water', which plays a part in the interaction of such particles.

The approach of two particles with hydrated surfaces will generally be hindered by an extra repulsive interaction, distinct from electrical double layer repulsion. This hydration repulsion arises essentially from the need for the surfaces to become dehydrated if true contact between particles is to occur. This involves work and hence an increase in free energy of the system.

The most direct evidence for hydration effects comes from measurements of the force between mica sheets separated by different salt solutions (Pashley and Israelachvili, 1984). At low ionic strengths, the repulsion follows the expected exponential form for double layer interaction, but at salt concentrations above about 1 mM an extra, monotonic, short-range force is apparent, owing to adsorbed hydrated cations. This extra force increases with the degree of hydration of the counterions ($Li^+ \approx Na^+ > K^+ > Cs^+$) and decreases nearly exponentially over the range 1.5–4 nm, with a decay length of the order of 1 nm. Experiments on thin surfactant films have also provided evidence for hydration effects (Clunie *et al.*, 1968).

As well as the monotonically decreasing hydration force, an *oscillatory* force is observed at very close approach of the mica surfaces. This extends to a distance of about 1.5 nm (about six water molecules) and has a periodicity of about 0.25 nm (roughly one water molecule). This type of behaviour is found with other liquids and is most likely a direct result of the 'packing' of molecules in layers between the surfaces. It may not be important for most surfaces, which are rough on a molecular scale, so that the force oscillations would tend to cancel each other out over quite small lateral distances. The monotonic force has only been found for interaction in water, but may possibly occur with other liquids having a high degree of molecular association (Gregory, 1988).

The range of these hydration forces is quite appreciable compared to the range of double layer repulsion and they would be expected to have an effect on colloid stability, especially at high ionic strength. The anomalous stability of latex particles at high salt concentration (Healy *et al.*, 1978) and the inhibition of bubble coalescence under similar conditions (Lessard and Zieminski, 1971) may provide indirect evidence for hydration repulsion.

With adsorbed layers of hydrophilic macromolecules, hydration can also lead to significant repulsion, but this will be considered along with steric interaction in Section 3.4.4.

3.4.3 Hydrophobic interaction

When a surface has no polar or ionic groups or hydrogen-bonding sites, there is no affinity for water and the surface is said to be *hydrophobic*. The nature of water in contact with such a surface will be different from that of ordinary, bulk water. Ordinary water is significantly structured because of hydrogen bonding between the molecules. The cooperative nature of this bonding (Symons, 1989) means that quite large clusters of hydrogen-bonded water molecules can form, although these are of a very transient nature, continually forming and breaking down in response to thermal energy fluctuations. The presence of a hydrophobic surface could restrict the natural structuring tendency of water, simply by imposing a barrier which prevents the growth of clusters in a given direction. Water confined in a gap between two such surfaces would thus be unable to form clusters larger than a certain size. For a narrow gap, this could be a serious limitation and result in an increased free energy of the water in relation to bulk water. In other words there would be an attraction between hydrophobic surfaces, as a consequence of water molecules migrating from the gap to the bulk water, where there are unrestricted hydrogen-bonding opportunities and a lower free energy.

Attraction between hydrophobic surfaces has been measured directly (Israelachvili and Pashley, 1984) and can be of surprisingly long range – up to about 80 nm (Claesson and Christenson, 1988). The systems used were mica sheets, with adsorbed hydrocarbon and fluorocarbon surfactants to give hydrophobic surfaces. The attraction was much stronger than the van der Waals force and of much greater range. The interaction of filaments of hydrophobized silica was measured by Rabinovich and Derjaguin (1988). They found an attractive force at large separation distances, one to two orders of magnitude greater than van der Waals attraction. Over the separation range 20–60 nm, the force decayed exponentially, with a decay length of about 12 nm.

For particles dispersed in water, it is possible for their surfaces to have some degree of hydrophobic character. For instance, negatively charged particles, with just enough adsorbed cationic surfactant to neutralize their charge, would present hydrophobic 'tails' of the surfactant to the water. As well as eliminating electrical repulsion between the particles, an extra attraction could be introduced as a result of the hydrophobic layer. By analogy with the measurements for mica, such an attraction could be of quite long range and play a large part in promoting flocculation. There has been a limited amount of work on the role of hydrophobic effects in flocculation (e.g. Lu and Song, 1991), although the position is far from clear. The process known as 'oil agglomeration' (Capes and Darcovich, 1984), where quite large particles, such as coal, are bound together by adhering oil films, is very dependent on the hydrophobicity of the particles. The attachment of bubbles to particles is of crucial importance in the flotation of minerals and is governed by hydrophobic interactions (Lu, 1991).

3.4.4 Steric interaction

Adsorbed layers, especially of polymers, can play a very important role in aggregation and deposition phenomena. In some cases, small amounts of adsorbed polymer can promote flocculation by a 'bridging' mechanism (see Section 3.4.5). In the case of colloidal particulate dispersions with larger adsorbed amounts, polymers can give greatly enhanced stability, by an effect which is usually known as steric *stabilization*. The most effective stabilizers are polymers which have some affinity for the surface, but adsorb in such a way that segments of polymer chains extend some distance into the dispersion medium. The simplest case is that of *terminally adsorbed block* copolymers, which have some segments which adsorb strongly on the particles and other, lyophilic segments which project into the continuous phase. These polymers form adsorbed layers like those shown schematically in Figure 3.2 and can give greatly enhanced stability against aggregation.

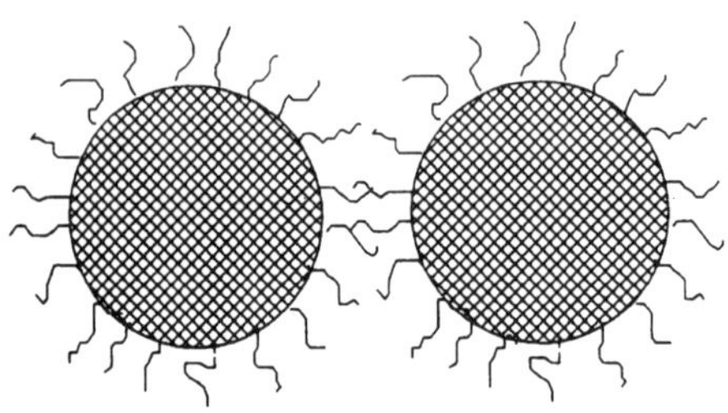

Figure 3.2 *Prevention of particle contact by terminally adsorbed polymer chains – steric stabilization*

The stabilizing action of such materials can be interpreted in fairly simple terms. As particles approach sufficiently close, the adsorbed layers come into contact and any closer approach would involve some interpenetration of the hydrophilic chains. Since these chains are hydrated, overlap of the layers would cause some dehydration and hence an increase in free energy and a repulsion between particles. To a first approximation, the repulsion can be assumed to become infinite as soon as the adsorbed layers begin to overlap, but zero at greater separations. This 'hard sphere' assumption is not strictly correct (Napper, 1983), but is good enough for most practical purposes and leads to some important conclusions.

It is reasonable to assume that the effective Hamaker constant for the adsorbed layer is rather low, so that there is little van der Waals attraction between the layers. In that case the most important effect of the adsorbed polymer is to limit the attraction between particles by keeping them a finite distance apart, where van der Waals attraction is lower. Contact between particles may then lead to fairly weak aggregates, which can easily be broken by shear. The most important factor determining the degree of steric stabilization is the thickness of the adsorbed layer relative to the particle size. Since van der Waals attraction energy is proportional to particle size, eqn (3.34), larger particles will need thicker stabilizing layers to confer the same degree of stability.

Sterically stabilized dispersions can be destabilized by changing the solvency of the medium for the stabilizing chains. With polyethylene oxide, adding certain salts (especially sulphates) and increasing temperature can cause dehydration of the chains, hence decreasing the stabilizing effect and causing flocculation (Napper, 1970). The case of block copolymers, illustrated in Figure 3.2, is not typical of most adsorbing polymers, which are attached at many points along the chain, giving features known as *trains*, *loops* and *tails* (Cosgrove, 1990). The effective thickness of the adsorbed layer is then more difficult to determine and can be dominated by the tails, although these represent a very small proportion of the adsorbed amount. Nevertheless, the principles outlined above are still generally valid; effective stabilization is achieved by fairly thick adsorbed layers, which are well solvated, so that overlap or interpenetration of the layers cannot occur.

Steric stabilization is a widespread phenomenon and, in the older colloid literature, materials acting in this way were known as 'protective colloids'. The classic example is the stabilization of gold sols by gelatin, discovered by Faraday, but the effect has been exploited, unknowingly, since ancient times, as in the preparation of stable dispersions of carbon black (ink) and of other pigments.

In the aquatic environment, most particles have adsorbed layers of natural organic material, such as humic substances (Tipping, 1988), which can have a major effect on their colloidal behaviour. Microorganisms produce extracellular polymers which may adsorb and have a great influence on particle interaction in biological systems. These natural polymers and organic materials are often weak acids and are anionic at neutral pH values. It is thought that the rather low negative zeta potentials found for most natural aquatic colloids are due to the adsorption of this anionic material. The stability of such colloids is often higher than expected on the basis of zeta potential and ionic strength, and it is likely that steric stabilization plays some role. Humic substances are known to enhance the stability of inorganic colloids (Jekel, 1986) and can cause greatly increased dosages of flocculants in water treatment.

3.4.5 Polymer bridging

Long-chain polymers generally adsorb on particles in the manner indicated in Figure 3.3 and, with large adsorbed amounts, can cause steric stabilization as discussed above. With much less adsorbed polymer, an individual chain can become attached to two or more particles, thus 'bridging' them together. In this way particles can form aggregates or attach themselves to macroscopic surfaces even though they may be charged and repel each other. This effect, previously known as 'sensitization', is now widely exploited in many industrial applications. In some texts the term 'flocculation' is restricted to cases where polymer bridging is the operative mechanism, and 'coagulation' refers to aggregation brought about by reducing double layer repulsion. This terminology is not employed here.

The essential requirements for polymer bridging are that there should be sufficient unoccupied particle surface for attachment of polymer segments from chains attached to other particles and that the polymer bridges should be of such an extent that they span the distance over which interparticle repulsion operates. Generally, the most effective bridging flocculation is found with linear polymers of very high molecular weight (several million), so that extended loops and tails can form, increasing the probability of attachment to other particles. Also, it is found that there is an *optimum dosage* range for effective flocculation. At lower dosages, there is insufficient polymer to form adequate bridging links between particles. With excess polymer, there is no longer enough bare particle surface for attachment of segments and the particles become restabilized, which may involve some steric repulsion. Figure 3.3 shows schematically the flocculation and restabilization of particles by adsorbed polymer. Bridging flocculation can give aggregates (flocs) which are much stronger than those produced just by the addition of salts (i.e. by reducing electrical repulsion).

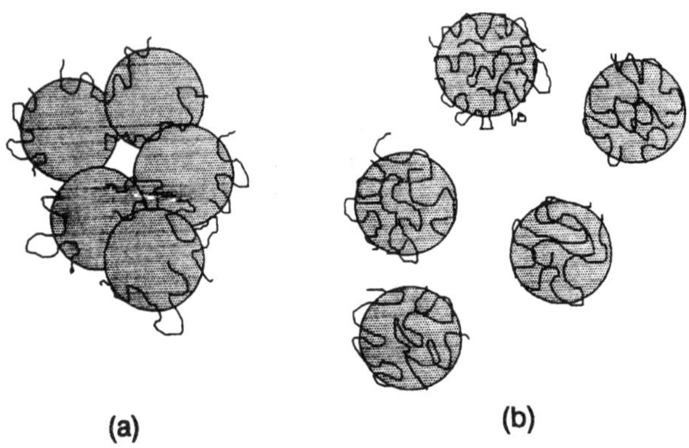

Figure 3.3 *Schematic diagram showing* (a) *bridging flocculation and* (b) *restabilization by adsorbed polymer*

Healy and La Mer (1964) proposed that optimum bridging flocculation would occur when the particle surfaces were half covered with adsorbed polymer, providing maximum opportunity for favourable particle–particle contacts. However, there is some ambiguity in the concept of 'surface coverage' in the case of adsorbing polymers and the condition of half surface coverage is not easy to establish. Some refinements to the La Mer model have been suggested (Hogg, 1984; Moudgil *et al.*, 1987). This will be discussed further in Chapter 6.

Many polymeric flocculants are ionic (polyelectrolytes) and the polymer charge introduces some complications. In some cases, anionic polymers are found effective for the flocculation of negatively charged particles, despite the like sign of charge. The most common flocculants used in practice are based on polyacrylamide, which may be hydrolysed so that amide groups are converted to carboxylic acid groups. At neutral pH values these are ionized to anionic carboxylate groups. The degree of hydrolysis, and hence the degree of anionic character, can be controlled during manufacture and has been found to have important effects on the flocculation efficiency.

Michaels (1954) showed that a certain degree of hydrolysis (about 30%) gave optimum flocculation of silts and clays. The explanation depends on the fact that polyelectrolyte chains are expanded to some extent, as a result of repulsion between charged segments. Below the optimum hydrolysis level, the polymer chain is less extended and hence less able to form extended bridges between particles. At higher charge densities adsorption of polyelectrolytes on similarly charged particles becomes restricted by electrical repulsion and hence flocculation is less effective. There are ionic strength effects also, although these have not been studied in detail. Higher ionic strengths cause polyelectrolyte chains to be less extended, but should also make adsorption easier, and allow closer approach of charged particles. There is also evidence of specific effects of certain ions. For instance, the presence of calcium ions can promote the adsorption of hydrolysed polyacrylamide on negatively charged particles, presumably by linking carboxylate groups with anionic sites on the particle surface.

When the particles and polyelectrolyte are of opposite charge (as in the important practical case of negative particles and cationic polymers), the possibility of destabilization simply by charge reduction arises. In fact, in many cases, the action of cationic polymers can be explained in terms of their strong adsorption on negative particles and the consequent reduction of double layer repulsion, allowing aggregation to occur. The rather flat configuration adopted by polyelectrolytes adsorbing on oppositely charged surfaces (Lyklema and Fleer, 1987) makes bridging contacts unlikely. For these reasons, the most effective cationic flocculants are often those of high charge density, with molecular weight playing only a minor role. There are important exceptions though, especially with concentrated dispersions, where kinetic factors increase the importance of bridging interactions.

An important refinement to the simple charge-neutralization picture is the 'electrostatic patch' model (Kasper, 1971, Gregory, 1973), in which particle

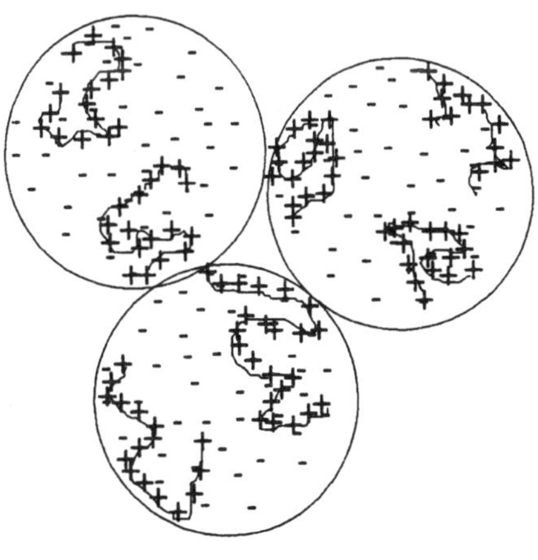

Figure 3.4 *'Electrostatic patch' model, in the case of negatively charged particles and cationic polymer*

charge is not neutralized uniformly by adsorbed polyelectrolyte. This condition can arise when the charge density of the polyelectrolyte is high and that of the particle surface is low. In such cases each charged site on the particle surface cannot be neutralized individually by an oppositely-charged polymer segment, simply for geometrical reasons (Figure 3.4). Polymer chains then must adsorb to give 'islands' or 'patches' of charge, surrounded by areas of opposite charge. Particles with polyelectrolyte adsorbed in this 'patchwise' manner can interact in such a way that oppositely charged areas of different particles come into contact, giving strong attraction. At low ionic strengths, this extra attraction can give measurable increases in flocculation rate due to the long range of the attractive double layer forces (Gregory, 1976).

3.5 DLVO description of colloidal stability

3.5.1 Basis of the DLVO theory

The ability of electrolytes to flocculate various aqueous dispersions has long been recognized (Schulze, 1882; Hardy, 1900; Freundlich, 1926) and expressions for rate of particle doublet formation in 'unstable' dispersions due to Brownian and shear-induced collisions were derived by Smoluchowski (1917) at the beginning of this century. However, at that time neither the origin of the attraction nor the functional form of the repulsion were known. Only after the establishment of the electrical double layer (EDL) theory (described in Chapter 2), could Derjaguin and Landau (1941) and Verwey and Overbeek (1948) synthesize their theory of colloid stability to rationalize the early observations, now widely known as the DLVO theory.

Essentially, van der Waals attraction and EDL repulsion are assumed to be additive and combined to give the total energy of interaction between particles as a function of separation distance. All other types of non-DLVO interaction, such as those discussed in Section 3.4, are ignored. The total interaction, usually expressed as potential energy V_T, is calculated as the sum of van der Waals attraction V_A and the EDL interaction V_R:

$$V_T = V_A + V_R \tag{3.48}$$

Depending on the relative strength of each of the two contributions, the energy versus distance profile typically displays one of the following three patterns or shapes shown in Figure 3.5(b), exhibiting:

(1) an infinitely deep minimum, called the primary minimum, at close separation, followed by a maximum, called the energy barrier, and another minimum, called the secondary minimum, occurring at a larger distance
(2) an infinitely deep (primary) minimum only
(3) an infinitely deep (primary) minimum plus a large maximum (the energy barrier) but no significant secondary minimum.

It will be shown in subsequent chapters that the shape of the interaction profile has a direct influence on the kinetics of colloidal aggregation and deposition.

Since the nature of the interaction profile is governed by various physicochemical parameters such as particle size, zeta potential, electrolyte composition and Hamaker constant, it is important to understand the role played by each of the parameters in controlling the nature of the interaction profile.

It should be noted that the superposition of van der Waals attraction and the EDL repulsion gives rise to an infinitely deep primary minimum which is physically unrealistic. A finite potential at the primary minimum can be obtained either by arbitrarily imposing a closest approach distance, e.g. $h_{min} = 0.1–0.2$ nm, to account for the molecular dimensions not recognized by the continuum theory, or by explicitly incorporating the short-ranged Born repulsion V_B in the calculation:

$$V_T = V_A + V_B + V_R \tag{3.49}$$

As mentioned previously, in practice infinitely strong attraction is unlikely, since short-range effects, such as those caused by ion hydration, may keep particles from coming into physical contact, making the primary minimum of finite depth. For two surfaces coming into close contact the roughness will also play a major role in determining the effective contact area and may restrict the minimum separation distance attainable.

One modification to the original DLVO approach is to recognize the existence of Stern layers and this has an effect on the choice of the appropriate separation distances. In the case of van der Waals attraction it seems reasonable to take the separation distance, h, as that between the true particle surfaces. For electrical repulsion, where zeta potentials are taken as the effective surface potentials, h in eqns (3.20) and (3.21) is strictly the separation between the shear planes, which may be up to 0.5 nm away from the true particle surface. Thus for a given particle separation, the repulsion will act over a distance $(h - 2\delta)$, which effectively increases the repulsion, especially at high ionic strength. This point has been discussed by Duckworth and Lips (1978), among others, and the modification is now quite commonly used (Penners and Koopal, 1987). For the purposes of illustrating the essential theory of colloid stability (below) we will not include this refinement, although serious workers may wish to include the effect of using a finite value of δ.

3.5.2 Stability criteria

When the zeta potential of the particles and the ionic strength are such that repulsion outweighs the attraction, then there is a potential energy barrier, which tends to prevent contact of particles and so prevent aggregation (Figure 3.5). As the ionic strength is increased and/or the zeta potential reduced (both effects usually result from an increased salt concentration), the energy barrier becomes lower and contact of the particles can occur more readily. Eventually, the barrier disappears altogether and, in principle, particles can then adhere each time they collide. Such particles are said to be fully destabilized and to undergo rapid aggregation. The latter term implies nothing about the *absolute rate* of aggregation, but that the rate cannot be further increased by the addition of more salt.

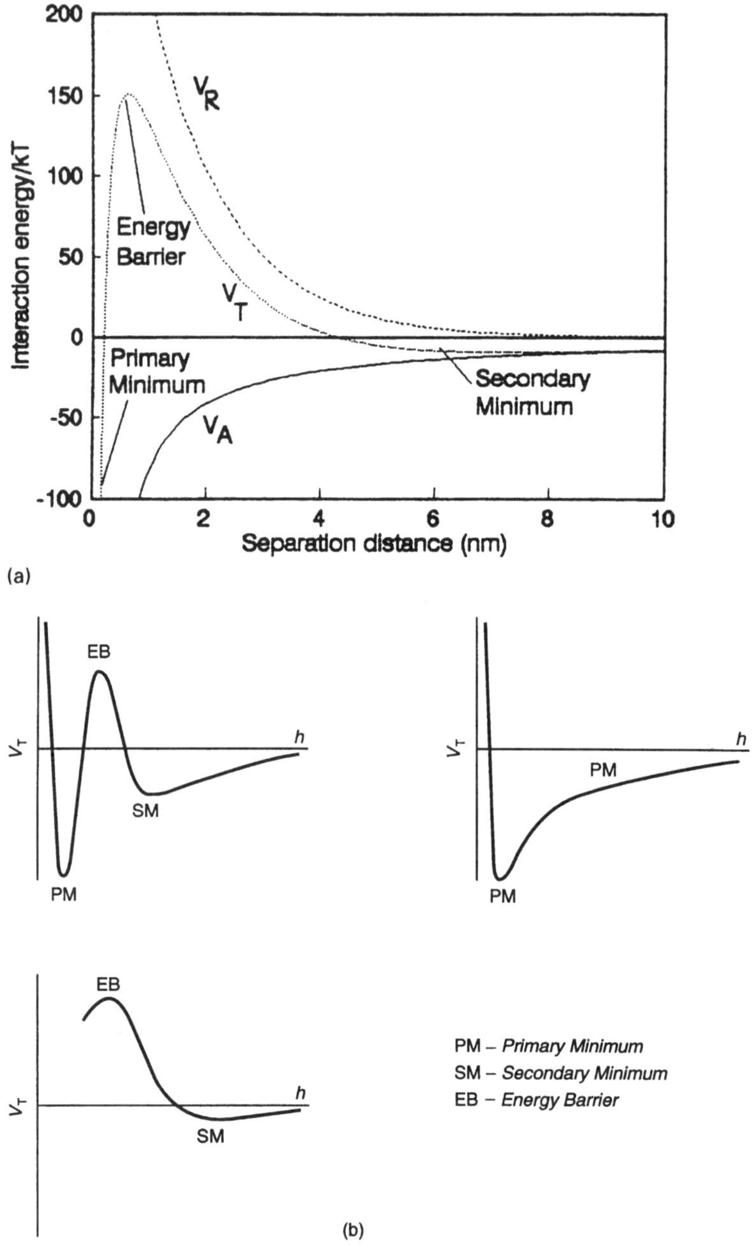

Figure 3.5 *Schematic representations of total energy of interaction versus surface-to-surface separation distance profiles:* (a) *quantitative example for equal spherical particles of 1 μm diameter in 1–1 electrolyte,* $\zeta = 30\,mV$, $C = 50\,mM$, $A = 8.3 \times 10^{-21}\,J$ (= 2kT), *and* (b) *the three characteristic types of profile*

There are several features of the potential energy diagram, Figure 3.5(a), which are important in a discussion of colloid stability. If there is a barrier, then colliding particles must overcome this in order to come into true contact. Potential energies are often expressed in kT units and, in Brownian motion, particles will rarely have an energy of more than a few kT. Thus, a barrier height of $20kT$ or more implies a very stable suspension, with only a minute fraction of colliding particles having sufficient energy to surmount the barrier. Once the barrier is overcome, then the particles are held by van der Waals attraction in a deep primary minimum, from which escape would be very unlikely.

Because of the different distance dependence of the van der Waals (power law) and electrical (exponential) energies, the former is always greater than the latter at sufficiently large separation distances. This gives a secondary minimum in the potential energy curve, which can be responsible for the formation of fairly weak aggregates. Since the interaction energies are directly proportional to particle size, secondary minimum effects are more significant with larger particles (greater than about 1 μm in diameter). They are also more important at moderately high salt concentrations, so that the range of electrical repulsion is reduced, allowing the particles to approach closer, where van der Waals attraction is greater. The presence of an adsorbed polymer layer giving steric repulsion (Section 3.4.4) leads to a similar type of long-range attraction, but with more control over the distance, and hence the depth, of the 'secondary minimum'.

In the presence of an energy barrier, only a certain fraction of collisions are effective and this fraction is known as the collision efficiency, α. The reciprocal of this value is known as the stability ratio, W. By measuring the rate of aggregation (as discussed in Chapters 6 and 9) as a function of salt concentration, it is possible to find the critical concentration at which the onset of rapid flocculation occurs. Simple theory (Reerink and Overbeek, 1954) suggests that a plot of log W against log (salt concentration), should show a linear decrease as far as the critical salt concentration, with $\log W$ being constant at higher concentrations. Although such plots do not always show the expected linearity, they are useful in determining the critical salt concentration at which rapid aggregation begins. Similar behaviour is observed in stability curves of particle deposition (Chapters 5 and 11).

The transition from kinetic stability to rapid aggregation occurs over a very narrow range of electrolyte concentrations, around the *critical coagulation concentration* (CCC) which may be estimated by assuming that no energy barrier exists. Since different conditions require different expressions to represent the van der Waals and the EDL forces, and different approximations applicable to the same range of conditions often give different results, the precise value for the CCC depends on the form of the expressions used for the colloidal interactions. In most practical cases, the conditions $\kappa a \gg 1$ and $\kappa h_{max} \geq 1$ are satisfied, so the linear superposition Gouy–Chapman expression, eqn (3.13), and the unretarded Hamaker expression, eqn (3.24), may be used. Bearing in mind that both the total energy of interaction V_T and the gradient dV_T/dh are equal to zero at the CCC, it follows that $\kappa h_{max} = 1$ and the CCC is given by:

$$\text{CCC} \propto z^{-6} A^{-2} \tanh^4 (ze\zeta/4kT) \qquad (3.50)$$

It is evident from the above equation that at large zeta potentials (i.e. $ze\zeta/4kT \gg 1$), CCC is proportional to z^{-6}. This dependency has become known as the *Schulze–Hardy rule*, which states that it is the valence of the counterions that has the principal effect on the stability of the colloid. In fact there are only a few cases where the rule is followed. One such case is in the flocculation of a synthetic latex by various salts (Greene and Saunders, 1970), where the critical values for Na^+, Ba^{2+} and La^{3+} were found to be in the ratios 1 : 0.014 : 0.0014, compared with the 'theoretical' ($1/z^6$) ratios of 1:0.016 : 0.0014. Tabulations of typical CCC values often refer to the early work of Freundlich (1926).

However, at small zeta potentials (i.e. $ze\zeta/4kT < 1$) as commonly encountered, the CCC has a much weaker dependence on the counterion charge. It becomes proportional to z^{-2}.

As has been noted in Chapter 2, the Gouy–Chapman theory is only applicable to symmetrical electrolytes. Multivalent ionic species can form complexes in solution or adsorb into the Stern layer to change the magnitude and even the sign of the surface charge. Therefore, the z^{-6} dependence is far from universal. Even with univalent electrolytes, eqn (3.50) only gives a semiquantitative estimation of the concentration at which coagulation should be observed. Experimentally observed values of CCC can differ from theoretical predictions by a factor of 2 (Russel *et al.*, 1989).

In many cases there are slight differences between different counterions of the same valence and these can be explained (Pieper *et al.*, 1975) on the basis of differences in size of the hydrated counterions and the consequent effect on the potential at the Stern plane (and hence on the zeta potential). Thus, for the alkali metal cations, CCC values decrease in the order $Li^+ > Na^+ > K^+ > Rb^+ > Cs^+$, which follows the well-known lyotropic series and shows that the most hydrated ion (Li^+) is the least effective in reducing double layer repulsion. This kind of effect has been explained (van de Ven, 1988) in terms of the friction coefficient of the ions and the consequent effect on the rate of approach of charged particles. However, these differences between ions of the same valence are very much less than those between ions of different charge and are of little concern from a practical standpoint.

In principle, neither particle size nor concentration should have any effect on CCC if the added salt is acting as an indifferent electrolyte. Particle size effects are not easy to study since it is difficult to obtain particles which differ only in size and not in some other important characteristic, such as surface charge density. The coagulation of hematite (α-Fe_2O_3) sols has been studied by Penners and Koopal (1987), who found a minimum CCC value for a particle diameter of about 400 nm. This was explained in terms of the effects of particle size on the depths of the primary and secondary minima.

If particle concentration is found to affect the CCC value, this is usually an indication of specific adsorption of counterions. With strong specific adsorption, there is a nearly direct proportionality between particle concentration and the critical salt concentration. Furthermore, addition of excess salt can cause restabilization of the particles because counterions are adsorbed to such an extent that the particle charge is reversed, causing a repulsion between the particles. Such effects are found with hydrolysing metal salts and other commonly used coagulants.

Although the above analysis can provide a qualitative understanding of colloid stability, it neither explains the fact that smaller particles are generally more difficult to stabilize, nor determines the degree of stability for a given set of conditions. These questions require the theory for the kinetics of aggregation which is described in later chapters.

Bibliography

Israelachvili, J. N. (1992) *Intermolecular and Surface Forces*, 2nd edn, Academic Press, London

Napper, D. H. (1983) *Polymeric Stabilization of Colloidal Dispersions*, Academic Press, New York

Russel, W. B., Saville, D. A. and Schowalter, W. R. (1989) *Colloidal Dispersions*, Cambridge University Press, Cambridge

Tanford, C. (1980) *The Hydrophobic Effect. Formation of Micelles and Biological Membranes*, 2nd edn, Wiley, New York

References

Bell, G. M. and Peterson, G. C. (1972) Calculation of the electrical double layer force between unlike spheres, *J. Colloid Interface Sci.*, **41**, 542–566

Bell, G. M., Levine, S. and McCartney, L. N. (1970) Approximate methods of determining the double-layer free energy of interaction between two charged colloidal spheres. *J. Colloid Interface Sci.*, **33**, 335–359

Capes, C. E. and Darcovich, K. (1984) A survey of oil agglomeration in wet fine coal processing. *Powder Technology*, **40**, 43–52

Churaev, N. V. and Derjaguin, B. V. (1985) Inclusion of structural forces in the theory of stability of colloids and films. *J. Colloid Interface Sci.*, **103**, 542–553

Claesson, P. M. and Christenson, H. K. (1988) Very long range attraction between uncharged hydrocarbon and fluorocarbon surfaces in water. *J. Phys. Chem.*, **92**, 1650–1655

Clayfield, E. J., Lumb, E. C. and Mackey, P. H. (1971) Retarded dispersion forces in colloidal particles — exact integration of the Casimir and Polder equation. *J. Colloid Interface Sci.*, **37**, 382–389

Clunie, J. S., Goodman, J. F. and Tate, J. R. (1968) Adsorption of inorganic ions in black foam films. *Trans Faraday Soc.*, **64**, 1965–1970

Cooper, W. D. and Marsden, R. S. (1978) Calculation of dispersion force interactions between colloidal particles in butan-1-ol. *J. Chem. Soc. Faraday I*, **74**, 2008–2016

Cosgrove, T. (1990) Volume fraction profiles of adsorbed polymers. *J. Chem. Soc. Faraday Trans*, **86**, 1323–1332

Czarnecki, J. (1979) Van der Waals attraction energy between sphere and half-space. *J. Colloid Interface Sci.*, **72**, 361–362

Czarnecki, J. and Dabros, T. (1980) Attenuation of the van der Waals attraction energy in the particle/semi-infinite medium system due to the roughness of the particle surface, *J. Colloid Interface Sci.*, **78**, 25–30

Czarnecki, J. and Itschenskij, V. (1984) Van der Waals attraction energy between unequal rough spherical particles. *J. Colloid Interface Sci.*, **98**, 590–591

Davies, B. and Ninham, B. W., (1972) Van der Waals forces in electrolytes. *J. Chem. Phys.*, **56**, 5797–5801

deGennes, P. G. (1987) Polymer at an interface: a simplified review. *Adv. Colloid Interface Sci.*, **27**, 189–209

Derjaguin, B. V. (1934) Friction and adhesion IV. The theory of adhesion of small particles. *Kolloid Z.*, **69**, 155–164
Derjaguin, B. V. and Landau, L. D. (1941) Theory of the stability of strongly charged lyophobic sols and of the adhesion of strongly charged particles in solutions of electrolytes. *Acta Physicochim. URSS*, **14**, 733–762
Devereux, O. F. and de Bruyn, P. L. (1963) *Interaction of Plane-Parallel Double Layers*, MIT Press, Cambridge, MA
Duckworth, R. M. and Lips, A. (1978) The influence of Stern layers on the calculation of Hamaker functions from coagulation data. *J. Colloid Interface Sci.*, **64**, 311–319
Dukhin, S. S. and Lyklema, J. (1987) Dynamics of colloid particle interaction. *Langmuir*, **3**, 94–98
Eriksson, J.C., Ljunggren, S. and Claesson, P. M. (1989) A phenomenological theory of long-range hydrophobic attraction forces based on a square-gradient variational approach. *J. Chem. Soc. Faraday Trans II*, **85**, 163–176
Feke, D. L. and Schowalter, W. R. (1985) The influence of Brownian diffusion on binary collision rates in colloidal dispersions. *J. Colloid Interface Sci.*, **106**, 203–214
Feke, D. L., Prabhu, N. D., Mann Jr, J. A. and Mann III, J. A. (1984) A formulation of the short-range repulsion between spherical colloidal particles. *J. Phys. Chem.*, **88**, 5735–5739
Fitts, D. D. (1966) Statistical mechanics: a study of intermolecular forces. *Annu. Rev. Phys. Chem.*, **17**, 59–82
Frens, G. and Overbeek, J. Th. G. (1972) Repetition and the theory of electrocratic colloids, *J. Colloid Interface Sci.*, **38**, 376–387.
Freundlich, H. (1926) *Colloid and Capillary Chemistry*, Methuen, London
Greene, B. W. and Saunders, F. L. (1970) In situ polymerization of surface active agents on latex particles. III The electrolyte stability of styrene–butadiene latexes. *J. Colloid Interface Sci.*, **33**, 393–404
Gregory, J. (1969) The calculation of Hamaker constants. *Adv. Colloid Interface Sci.*, **2**, 396–417
Gregory, J. (1973) Rates of flocculation of latex particles by cationic polymers. *J. Colloid Interface Sci.*, **42**, 448–456
Gregory, J. (1975) Interaction of unequal double layers at constant charge. *J. Colloid Interface Sci.*, **51**, 44–51
Gregory, J. (1976) The effect of cationic polymers on the colloidal stability of latex particles. *J. Colloid Interface Sci.*, **55**, 35–44
Gregory, J. (1977) Van der Waals forces between mica surfaces: comparison of theory and experiment. *J. Chem. Soc. Faraday I*, **73**, 1983–1987
Gregory, J. (1981a) Approximate expressions for retarded van der Waals interaction. *J. Colloid Interface Sci.*, **83**, 138–145
Gregory, J. (1981b) Flocculation in laminar tube flow. *Chem. Eng. Sci.*, **36**, 1789–1794
Gregory, J. (1988) Colloid interactions. *Mat. Sci. Forum*, **25–26**, 125–144
Gregory, J. (1989) Fundamentals of flocculation. *CRC Crit. Rev. Environ. Control,* **19**, 185–230
Gregory, J. and Wishart, A. J. (1980) Deposition of latex particles on alumina fibers. *Colloids Surfaces*, **1**, 313–334
Hamaker, H. C. (1937) The London–van der Waals attraction between spherical particles. *Physica*, **4**, 1058–1072
Hardy, W. B. (1900) A preliminary investigation of the conditions which determine the stability of irreversible hydrosols. *Proc. R. Soc. Lond.*, **66**, 110–125
Healy, T. W. and La Mer, V. K. (1964) Energetics of flocculation and redispersion by polymers. *J. Colloid Sci.*, **19**, 323–332
Healy, T. W., Homola, A. and James, R. O. (1978) Coagulation of amphoteric latex colloids: reversibility and specific ion effects. *Faraday Disc. Chem. Soc.*, **65**, 156–163
Hesselink, F. Th., Vrij, A. and Overbeek, J. Th. G. (1971) On the theory of stabilisation of dispersions

by adsorbed macromolecules. II. Interaction between two flat particles. *J. Phys. Chem.*, **75**, 2094–2103
Higashitani, K., Tanaka, T. and Matsuno, Y. (1978) A kinematic interpretation of coagulation mechanism of hydrophobic colloids. *J. Colloid Interface Sci.*, **63**, 551–560
Ho, N. F. H. and Higuchi, W. I. (1968) Preferential aggregation and coalescence in heterodispersed systems. *J. Pharm. Sci.*, **57**, 436–442
Hogg, R. (1984) Collision efficiency factors for polymer flocculation. *J. Colloid Interface Sci.*, **102**, 232–236
Hogg, R. I., Healey, T. W. and Fuerstenau, D. W. (1966) Mutual coagulation of colloidal dispersions. *Trans Faraday Soc.*, **62**, 1638–1651
Hough, D. B. and White, L. R. (1980) The calculation of Hamaker constants from Lifshitz theory with applications to wetting phenomena. *Adv. Colloid Interface Sci.*, **14**, 3–41
Israelachvili, J. N. (1992) *Intermolecular and Surface Forces*, 2nd edn, Academic Press, London
Israelachvili, J. N. and Adams, G. E. (1978) Measurement of forces between two mica surfaces in aqueous electrolyte solutions in the range 0–100 nm. *J. Chem. Soc. Faraday I*, **74**, 975–1001
Israelachvili, J. N. and Pashley, R. M. (1984) Measurement of the hydrophobic interaction between two hydrophobic surfaces in aqueous electrolyte solutions. *J. Colloid Interface Sci.*, **98**, 500–514
Israelachvili, J. N. and Tabor, D. (1972) The measurement of van der Waals dispersion forces in the range 1.5 to 130 nm. *Proc. R. Soc.*, **A331**, 19–38
James, R. O., Homola, A. and Healy, T. W. (1977) Heterocoagulation of amphoteric latex colloids. *J. Chem. Soc. Faraday I*, **73**, 1436–1445
Jekel, M. R. (1986) The stabilization of dispersed mineral particles by adsorption of humic substances. *Water Res.*, **20**, 1543–1554
Kar, G., Chander, S. and Mika, T. S. (1973) The potential energy of interaction between dissimilar electrical double layers. *J. Colloid Interface Sci.*, **44**, 347–355
Kasper, D. R. (1971) Theoretical and experimental investigations of the flocculation of charged particles in aqueous solutions by polyelectrolytes of opposite charge. PhD Thesis, California Institute of Technology
Kjellander, R. and Marcelja, S. (1988) Surface interactions in simple electrolytes. *J. Phys. France*, **49**, 1009–1015
Langbein, D. (1969) Van der Waals attraction between macroscopic bonds. *J Adhesion*, **1**, 237–245
Langmuir, I. (1938) The role of attractive and repulsive forces in the formation of tactoids, thixotropic gels, protein crystals and coacervates. *J. Chem. Phys.*, **6**, 873–896
Lessard, R. R. and Zieminski, S. A. (1971) Bubble coalescence and gas transfer in aqueous electrolyte solutions. *Ind. Eng. Chem. Fund.*, **10**, 260–269
Lifshitz, E. M. (1956) Theory of molecular attractive forces. *Soviet Phys. JETP*, **2**, 73–83
Lu, S. (1991) Hydrophobic interaction in flocculation and flotation. 3. Role of hydrophobic interaction in particle–bubble interaction. *Colloids Surfaces*, **57**, 73–81
Lu, S. and Song, S. (1991) Hydrophobic interaction in flocculation and flotation. I. Hydrophobic flocculation of fine mineral particles in aqueous solution. *Colloids Surfaces*, **57**, 49–60
Lyklema, J. (1980) Colloid stability as a dynamic phenomenon. *Pure Appl. Chem.*, **52**, 1221–1227
Lyklema, J. and Fleer, G. J. (1987) Electrical contribution to the effect of macromolecules on colloid stability. *Colloids Surfaces*, **25**, 357–368
Marra, J. (1986) Direct measurements of attractive van der Waals and adhesion forces between uncharged lipid bilayers in aqueous solutions. *J. Colloid Interface Sci.*, **109**, 11–20
McCartney, L. N. and Levine, S. (1969) An improvement on Derjaguin's expression at small potentials for the double-layer interaction energy of two spherical colloidal particles. *J. Colloid Interface Sci.*, **30**, 345–354
Michaels, A. S. (1954) Aggregation of suspensions by polyelectrolytes. *Ind. Eng. Chem.*, **46**,

1485–1490

Moudgil, B. M., Shah, B. D. and Soto, H. S. (1987) Collision efficiency factors in polymer flocculation of fine particles. *J. Colloid Interface Sci.*, **119**, 466–473

Napper, D. H. (1970) Flocculation of sterically stabilized dispersions. *J. Colloid Interface Sci.*, **32**, 106–114

Napper, D. H. (1983) *Polymeric Stabilization of Colloidal Dispersions*. New York, Academic Press

Nieuenhuis, E. A., Pathmamanoharan, C. and Vrij, A. (1981) Liquid-like structures in concentrated latex dispersions. A light-scattering study of PMMA particles in benzene. *J. Colloid Interface Sci.*, **81**, 196–213

Ohshima, H. and Kondo, T. (1988a) Comparison of three models on double layer interaction. *J. Colloid Interface Sci.*, **126**, 382–383

Ohshima, H. and Kondo, T. (1988b) Approximate analytic expression for double layer interaction at moderate potentials. *J. Colloid Interface Sci.*, **122**, 591–592

Overbeek, J. Th. G. (1952) *Colloid Science*, Vol. 1 (ed. Kruyt, H. R.) Elsevier, Amsterdam

Overbeek, J. Th. G. (1980) The rule of Schulze and Hardy. *Pure Appl. Chem.*, **52**, 1151–1161

Pashley, R. M. and Israelachvili, J. N. (1984) DLVO and hydration forces between mica surfaces in Mg^{2+}, Ca^{2+}, Sr^{2+} and Ba^{2+}chloride solutions. *J. Colloid Interface Sci.*, **97**, 446–455

Penners, N. H. G. and Koopal, L. K. (1987) The effect of particle size on the stability of haematite (α-Fe_2O_3) hydrosols. *Colloids Surfaces*, **28**, 67–83

Pieper, J. H. A., de Vooys, D. A. and Overbeek, J. Th. G. (1975) Lyotropic effects at the silver iodide/electrolyte solution interface. *J. Electroanal. Chem.*, **65**, 429–441

Prieve, D. C. and Russel, W. B. (1988) Simplified predictions of Hamaker constants for Lifshitz theory. *J. Colloid Interface Sci.*, **125**, 1–13

Rabinovich, Ya. I. and Derjaguin, B. V. (1988) Interaction of hydrophobized filaments in aqueous electrolyte solutions. *Colloids Surfaces*, **30**, 243–251

Rajagopalan, R. and Kim, J. S. (1981) Adsorption of Brownian particles in the presence of potential barriers: effect of different modes of double layer interaction. *J. Colloid Interface Sci.*, **83**, 428–448

Reerink, H. and Overbeek, J. Th. G. (1954) The rate of coagulation as a measure of the stability of silver iodide sols. *Disc. Faraday Soc.*, **18**, 74–84

Ruckenstein, E. and Prieve, D. C. (1976a) Adsorption and desorption of particles and their chromatographic separation. *AIChE J.*, **22**, 276–283

Ruckenstein and Prieve (1976b) On reversible adsorption of hydrosols and repeptisation. *AIChE J.*, **22**, 1145–1147

Russel, W. B., Saville, D. A. and Schowalter, W. R. (1989) *Colloidal Dispersions*, Cambridge University Press, Cambridge

Schenkel, J. H. and Kitchener, J. A. (1960) A test of the Derjaguin–Verwey–Overbeek theory with a colloidal suspension. *Trans Faraday Soc.*, **56**, 161–173

Schulze, H. (1882) Schwefelarsen im wasseriger Losung. *J. Prakt. Chem.*, **25**, 431–452

Smoluchowski, M. (1917) Versuch einer mathematischen Theorie der Koagulationskinetic kolloider Lsungen. *Z. Physik. Chem.*, **92**, 129–168

Symons, M. C. R. (1989) Liquid water — the story unfolds. *Chem. Britain*, **25**, 491–494

Tabor, D. and Winterton, R. H. S. (1969) The direct measurement of normal and retarded van der Waals forces. *Proc. R. Soc.*, **A312**, 435–450

Tanford, C. (1980) *The Hydrophobic Effect. Formation of Micelles and Biological Membranes* 2nd edn, Wiley, London

Tipping, E. (1988) Colloids in the aquatic environment. *Chem. Ind.*, **15**, 485–490

Usui, S. (1973) Interaction of electrical double layers at constant surface charge. *J. Colloid Interface Sci.*, **44**, 107–113

van de Ven, T. G. M. (1988) On the role of ion size in coagulation. *J. Colloid Interface Sci.*, **124**,

138–145

van Oss, C. J., Omenyi, S. N. and Neumann, A. W. (1979) Negative Hamaker coefficients. II. Phase separation of polymer solutions. *Colloid Polymer Sci.*, **257**, 737–744

Verwey, E. J. W. and Overbeek, J. Th. G. (1948) *Theory of the Stability of Lyophobic Colloids*, Elsevier, Amsterdam

Visser, J. (1972) On Hamaker constants: a comparison between Hamaker constants and Lifshitz–van der Waals constants. *Adv. Colloid Interface Sci.*, **3**, 331–363

Visser, J. (1981) The concept of negative Hamaker coefficients. Part 1. History and present status. *Adv. Colloid Interface Sci.*, **15**, 157–169

Wennerstrom, H and Jonsson, B. (1988) Amphiphile–water systems and electrostatic interactions. *J. Phys. France*, **49**, 1033–1041

Wiese, G. R. and Healy, T. W. (1970) Effect of particle size on colloid stability. *Trans Faraday Soc.*, **66**, 490–500

Wilemski, G. (1982) Weak repulsive interactions between dissimilar electrical double layers. *J. Colloid Interface Sci.*, **88**, 111–116

Xu, Z. and Yoon, R. H. (1989) The role of hydrophobic interaction in coagulation. *J. Colloid Interface Sci.*, **132**, 532–541

Chapter 4

Colloidal hydrodynamics and transport

Nomenclature

a	particle radius
a_1	radius of particle 1 (reference sphere)
a_2	radius of particle 2 (moving sphere)
a_c	radius of (spherical) collector
A, B, C	dimensionless functions used in trajectory equations (4.70)
C	numerical constants given in Table 4.6
b	half depth of a parallel-plate channel
$\bar{b}$	reduced mean size of ions
c	constant of order unity
C	numerical coefficient
D_0	diffusion coefficient for single particles in an unbounded fluid
D_0^r	rotary diffusion coefficient of single particles
D_0^s	short-time self-diffusion coefficient
D_∞^s	long-time self-diffusion coefficient
D_e	effective diffusion coefficient
D_m	mutual-diffusion coefficient
D_r	diffusion coefficient for rotation
D_s	diffusion coefficient for spin
$\mathfrak{D}$	diffusion tensor
$\mathfrak{D}_{ij}$	diffusion tensor for a pair of particles *i* and *j*
$\mathfrak{D}_m$	mutual-diffusion tensor
$\mathfrak{D}_s$	diffusion tensor for a single particle near solid wall
e	charge of electron
$\mathbf{e}$	unit vector
$\mathbf{f}$	external force per unit mass of fluid
F	value of net, non-hydrodynamic forces
f_1, f_2, f_3, f_4	universal hydrodynamic correction factors
$\mathbf{F}_1$	force on particle 1
F_D	value of frictional drag force
$\mathbf{F}_e$	external force
$\mathbf{F}_m$	effective external force field
F_S	net surface force
$\mathcal{g}(r)$	dimensionless function of r
G	shear rate

h	surface-to-surface separation distance
h_{min}	closest approach between two particles
h_{pm}	primary minimum separation distance
$\mathcal{h}$	dimensionless surface-to-surface separation ($=\mathcal{r} - 2$)
$\mathcal{h}_{pm}$	dimensionless primary minimum separation
$\mathfrak{I}$	unit tensor
$\mathbf{J}$	flux of particles
$j_{\perp}$	flux component normal to the collector surface
k	Boltzmann constant
$\mathcal{k}$	virial coefficient
$\mathcal{k}_2$	second virial coefficient
k_s	immobilization rate constant
$\mathcal{k}_m$	virial coefficient for mutual diffusion
L_0	reference length
ℓ_B	length scale of a random walk
m	mass
$\mathfrak{M}_{ij}$	mobility tensor for a pair of particles i and j
n	particle number density (or distribution function)
$\mathcal{n}$	dimensionless particle number density in three-dimensional space
$\mathcal{n}^*$	perturbation term
$\mathcal{n}_B$	dimensionless particle number density at energy well according to Boltzmann distribution
$\mathbf{n}$	unit vector normal to surface
$\mathcal{N}$	surface concentration of the adsorbed particles
$\mathcal{N}_0$	initial surface concentration of adsorbed particles
$\mathcal{N}_{\infty}$	steady-state surface concentration of adsorbed particles
$\mathcal{N}_s$	surface number density of immobilized particles
$\mathcal{N}_c$	surface number density corresponding to monolayer coverage
n_{∞}	particle number density in bulk suspension
n_s	particle number density in the immobilized phase
p	hydrostatic pressure
Pe	Peclet number
Q	number of charges
$\mathbb{Q}$	source term used in convection-diffusion equation
$\mathbb{Q}_s$	source term corresponding to particle immobilization
$\mathbb{Q}_r$	source term corresponding to particle detachment
r	centre-to-centre or centre-to-surface separation distance
$\mathfrak{r}$	scaled distance ($=r/a_c$)
$\mathcal{r}$	dimensionless centre-to-centre distance ($=2r/(a_1+a_2)$)
$(\mathfrak{r},\vartheta,\phi)$	polar coordinates
$\mathfrak{r}_e$	dimensionless distance between two spheres in a doublet
$\mathfrak{r}_{pm}$	dimensionless primary minimum distance
$\mathbf{r}_i$	position vector of sphere i
$\mathbf{r}_j$	position vector of sphere j
$\mathbf{r}_{ij}$	distance vector from sphere i to j
$\mathfrak{R}$	hydrodynamic resistance tensor
Re	Reynolds number

S_c	surface area of collector
Sh	Sherwood number
$\overline{Sh}$	global or average Sherwood number
Sh_0	Sherwood number obtained using Smoluchowski–Levich approximation
Sh_f	fast-coagulation Sherwood number
Sh_s	slow-coagulation Sherwood number
t	time
T	absolute temperature
T_R	period of rotation of doublets
$\mathbf{u}$	velocity of particle
u_0	initial value of particle velocity
U	relative speed of particle
U_0	reference speed
U_1	value of velocity of particle 1
U_2	value of velocity of particle 2
v	value of fluid velocity
$\mathbf{v}$	fluid velocity
v_m	value of mean velocity
$\mathbf{v}_m$	mean velocity vector
$\mathbf{v}_{max}$	maximum velocity
V_T	total colloidal interaction potential energy
W_d	stability ratio
x, y, z	coordinates
$(\mathscr{x}, \mathscr{y})$	dimensionless coordinates
z	valence
α	numerical function used in some expressions for λ
δ_{ij}	Kronecker delta
δ_F	thickness of surface-force boundary layer
δ_D	thickness of diffusion boundary layer
ε	dielectric constant
ϕ	volume fraction
Φ_s	reduced surface or zeta potential
γ	flow-type parameter
κ	Debye (reciprocal) length
λ	hydrodynamic correction function
Λ	particle-collector size ratio
ν	kinematic viscosity
μ	fluid viscosity
$[\mu]$	intrinsic viscosity
μ_e	effective viscosity of particle suspension
ρ	fluid mass density
τ_B	relaxation timescale of Brownian motion
υ	specific conductivity of electrolyte
ω	angular velocity
ξ	friction coefficient
ζ	zeta potential

Subscript

1,2	corresponding to particle 1 and 2, respectively
0	initial value or value at the collector surface
∞	steady-state value or value in the bulk suspension
pm	corresponding to the primary minimum
$\parallel$	parallel component
$\perp$	perpendicular component
+	positively charged
–	negatively charged
r	rotary diffusion

Superscript

~	dimensionless
TF	torque-free
RD	rigid dumbbell
r	rotary
s	self-diffusion

The colloidal interactions described in the preceding chapter generally have a rather small effective range, usually only a fraction of the particle radius, and mainly operate during particle collisions. This chapter deals with other types of forces, which act on or between colloidal particles and are responsible for bringing the particles together (and hence resulting in collisions). These are:

(1) Brownian motion of colloidal particles arising from the erratic bombardment of molecules of the surrounding fluid
(2) frictional drag force exerted on colloidal particles by the surrounding fluid
(3) hydrodynamic interactions mediated by the surrounding fluid and modifying the frictional drags on individual colloidal particles, and, if present,
(4) external forces such as gravity.

The interplay between all these different types of forces determines the properties or behaviour of colloidal dispersions, including their tendency to aggregate or become deposited onto a surface (or 'collector').

4.1 Basic concepts in fluid and particle dynamics

4.1.1 Navier–Stokes equation

A fluid whose mass density ρ remains constant in time *and* space is called an *incompressible* fluid. If the viscous properties of a fluid can be characterized by a single constant viscosity μ, the fluid is referred to as a *normal* or *Newtonian* fluid. Examples of Newtonian fluids include water and aqueous solutions of inorganic and some organic substances. Fundamental fluid mechanics shows that, under the continuum hypothesis, an incompressible Newtonian fluid can be fully described by the following equations:

$$\frac{\partial \mathbf{v}}{\partial t} + \mathbf{v} \cdot \nabla \mathbf{v} = -\frac{\nabla p}{\rho} + \frac{\mu}{\rho} \nabla^2 \mathbf{v} + \mathbf{f} \tag{4.1}$$

$$\nabla \cdot \mathbf{v} = 0 \tag{4.2}$$

where $\mathbf{v}$ is the fluid velocity vector, p the hydrostatic pressure, ρ the mass density of the fluid, and $\mathbf{f}$ the external body force exerted on a unit mass of the fluid. Eqn (4.1) is known as the *Navier–Stokes* equation which is, in essence, a reformulation of Newton's second law for a unit volume of the fluid. Eqn (4.2) is called the *continuity* equation and is actually the conservation law of matter, expressed specifically for the incompressible fluid.

Under steady state, i.e. $\partial \mathbf{v}/\partial t = 0$, and assuming that the external forces are absent, the Navier–Stokes equation, eqn (4.1), can be greatly simplified. For the sake of numerical analysis, the resulting equation may be written in a dimensionless form, by scaling the velocity and pressure with a reference length L_0 and velocity U_0, as follows:

$$Re\,\tilde{\mathbf{v}} \cdot \nabla\, \tilde{\mathbf{v}} = -\frac{\nabla \tilde{p}}{\rho} + \frac{\mu}{\rho} \nabla^2 \tilde{\mathbf{v}} \tag{4.3}$$

$$\tilde{\mathbf{v}} = \frac{\mathbf{v}}{U_0} \qquad \tilde{p} = \frac{p L_0}{\mu U_0} \tag{4.4}$$

$$Re = \frac{\rho L_0 U_0}{\mu} \tag{4.5}$$

Here, Re is called the *Reynolds number*, and is one of the most important parameters in fluid mechanics. Physically, it is interpreted as the ratio of inertial force to the viscous force, and its value often signifies the prevailing fluid flow regime. For instance, at small Reynolds numbers, the fluid flow is steady and well-behaved (the *laminar*-flow regime); at large Reynolds numbers beyond a critical value, the fluid flow becomes unsteady and chaotic (the *turbulent*-flow regime). Different flow regimes entail different treatments to solve the governing equations. It should be noted that the value of the Reynolds number, and hence the flow regime, is determined by the characteristic length scale and velocity used for reference. Thus, although a suspension flowing through a pipe at high speed may be turbulent, the local flow field around each suspended particle may still be laminar.

Mathematically, the Reynolds number can be used to justify simplifications of the Navier–Stokes equation. When $Re \ll 1$ and under steady state, the left-hand side terms in eqn (4.1) can all be neglected. By further assuming that external forces are absent or negligible, eqn (4.1) becomes:

$$\nabla^2 \mathbf{v} = \nabla p/\mu \tag{4.6}$$

This equation is known as the *Stokes* or *creeping flow* equation and is usually the starting point for theoretical analyses of colloidal hydrodynamics. It will be referred to frequently later in this chapter, and the important features should be noted. First, it should be recognized that the Stokes equation is for steady-state laminar flows with inertial effects being neglected (i.e. $Re \ll 1$). For example, for a spherical particle of 1 μm diameter in water, Re approaches unity only when the relative speed of the particle is as high as $1\ \mathrm{m\,s^{-1}}$. Thus, for colloidal systems, the condition that $Re \ll 1$ can be satisfied in almost all cases. Secondly, it is a linear differential equation. This is a very useful property, allowing the superposition technique to be invoked to decompose a complex flow field into simpler ones. This can greatly simplify the effort in deriving a solution to the equation for the complex flow field.

4.1.2 Flow field in simple geometries

In theoretical analyses of colloidal deposition systems, it is usually assumed that the presence of colloidal particles does not disturb the flow field of the carrying fluid. For these theoretical analyses, a quantitative description of the undisturbed flow field is essential. A selection of simple flow geometries is shown in Figure 4.1, and the corresponding flow fields near the surface of the collectors are given in Table 4.1.

In most cases, the flow field near the collector surface can be decomposed locally into a *stagnation-point flow* and a *simple shear flow* (see Figure 4.1). Thus, although they are rarely encountered in reality, the two flow fields have been extensively studied because of their theoretical importance.

The velocity of particles, **u**, can be calculated from the undisturbed flow field near the collector (Table 4.1) as follows:

$$u_{\parallel} = f_3 v_{\parallel} \tag{4.7}$$

$$u_{\perp} = f_1 \text{ coefficients } f_2 v_{\perp} \tag{4.8}$$

where $v_{\parallel}$ and $v_{\perp}$ are parallel and normal components of the undisturbed fluid flow field. The corresponding diffusion coefficients, $D_{\parallel}$ and $D_{\perp}$, can be found from:

$$D_{\parallel} = f_4 D_0 \tag{4.9}$$

$$D_{\perp} = f_1 D_0 \tag{4.10}$$

where D_0 is the diffusion coefficient for a single spherical particle in an unbounded fluid medium. In the above expressions, f_1, f_2, f_3 and f_4 are the universal correction factors for particle motion and diffusion near a planar solid wall (Brenner, 1961; Goldman *et al.*, 1967; Bart, 1968; Goren and O'Neill, 1971); their values are shown graphically in Figure 4.2.

It should be noted that the expressions given in Table 4.1 are only for the flow in the immediate vicinity of the collector surface, i.e. the region where deposition takes place. The full flow fields may be quite different from those indicated by the given expressions. For example, the full velocity profile of the flow through

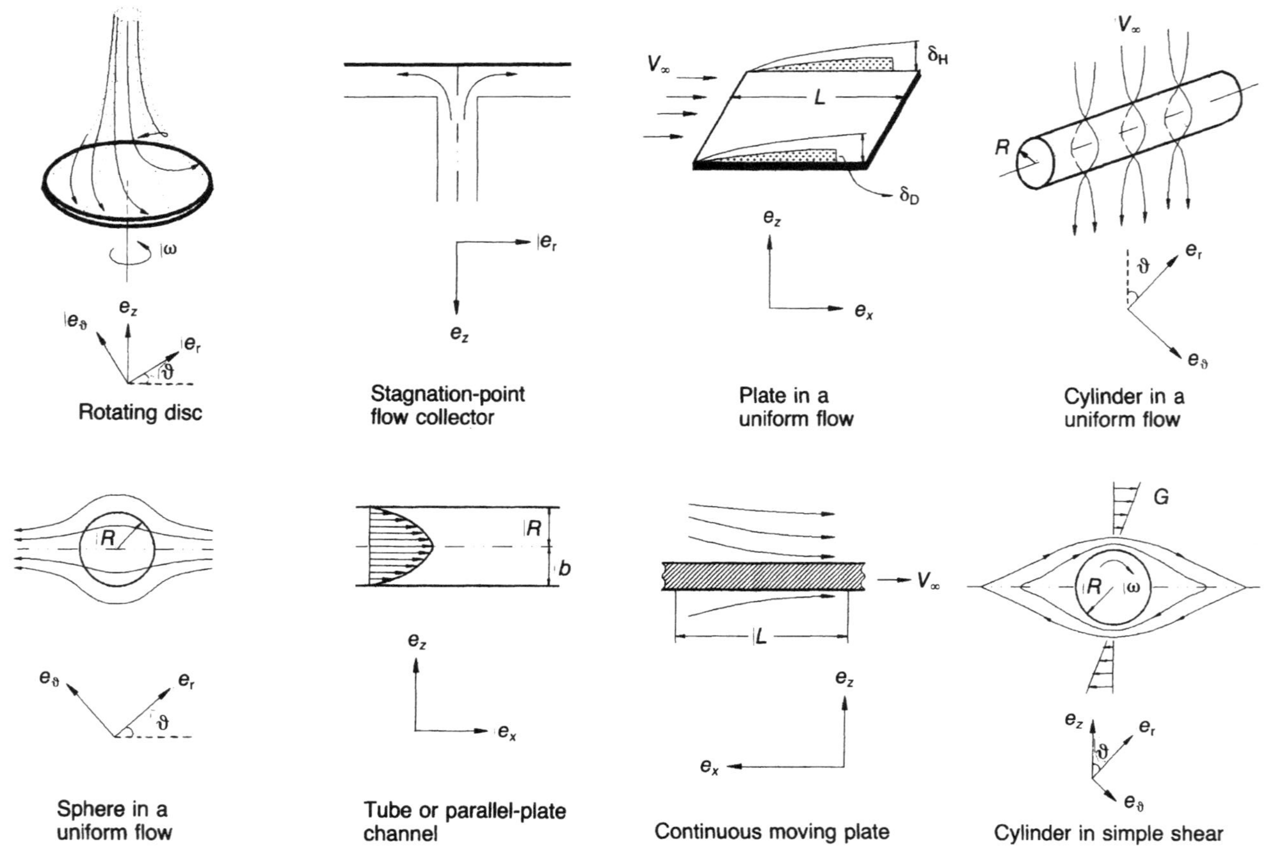

Figure 4.1 *Particle deposition-flow configurations for which solutions of the convective-diffusion equation are available (after Adamczyk, 1989)*

Table 4.1 *Undisturbed flow fields near various collector surfaces*

Collector	*Velocity*	Pe	*Reference*
Rotating disc	$v_r = 0.5102\omega^{3/2}\nu^{-1/2}rz$ $v_z = -0.5102\omega^{3/2}\nu^{-1/2}z^2$ $v_\theta = \omega r(1 - 0.616\omega^{1/2}\nu^{-1/2}z)$	$\frac{1.02\omega^{3/2}a^3}{\nu^{1/2}D_0}$	Levich (1962)
Stagnation-point flow collector	$v_r = \alpha_s rz$ $v_z = \alpha_s z^2$	$2\alpha_s a^3/D_0$	Dabros and van de Ven (1983)
Plate in uniform flow	$v_x = 0.332V_\infty^{3/2}\nu^{-1/2}zx^{-1/2}$ $v_z = 0.083V_\infty^{3/2}\nu^{-1/2}z^2x^{-3/2}$	$\frac{0.64V_\infty a^3}{\nu^{1/2}L^{3/2}D_0}$	Adamczyk and van de Ven (1982)
Cylinder in uniform flow	$v_r = -2V_\infty A_f(z/R)^2 \cos\vartheta$ $v_\theta = 4V_\infty A_f(z/R)\sin\vartheta$	$\frac{4A_f V_\infty a^3}{R^2 D_0}$	Adamczyk and van de Ven (1981a)
Sphere in uniform flow	$v_r = -1.5V_\infty A_f(z/R)^2\cos\vartheta$ $v_\theta = 1.5V_\infty A_f(z/R)\sin\vartheta$	$\frac{3A_f V_\infty a^3}{R^2 D_0}$	Levich (1962) Prieve and Ruckenstein (1974)
Parallel-plate channel	$v_x = 3V_m zb^{-1}$	$\frac{3V_m a^3}{2b^2 D_0}$	Adamczyk and van de Ven (1981b)
Cylindrical channel	$v_x = 3V_m zR^{-1}$	$\frac{3V_m a^3}{R^2 D_0}$	Adamczyk and van de Ven (1981b)
Continuously moving plate	$v_x = V_\infty - 0.444V_\infty^{-2/3}\nu^{-1/2}zx^{-1/2}$ $v_z = 0.111V_\infty^{3/2}\nu^{-1/2}z^2x^{-3/2}$	$\frac{0.222V_\infty^{3/2}a^3}{\nu^{1/2}L^{3/2}D_0}$	Adamczyk *et al.* (1982)
Cylinder in simple shear	$v_r = 2(G/R)z^2\sin(2\vartheta)$ $v_\theta = \frac{1}{2}G[1 + (4z/R)\cos(2\vartheta)]$	GR^2/D_0	Cox *et al.* (1968)

$\alpha_s = 0.164B_f^{3/2}\nu^{-1/2}$
A_f, B_f: flow parameters
V_∞: uniform flow velocity or speed of moving object
V_m: mean velocity
G: shear rate
R: curvature of collector surface
z: distance from collector surface
ν: kinematic viscosity

a parallel-plate channel is not linear, as implied in Table 4.1, but actually parabolic:

$$\mathbf{v} = \mathbf{v}_{max}\left[1 - \left(\frac{z}{b}\right)^2\right] \tag{4.11}$$

$$\mathbf{v}_m = \frac{2}{3}\mathbf{v}_{max} \tag{4.12}$$

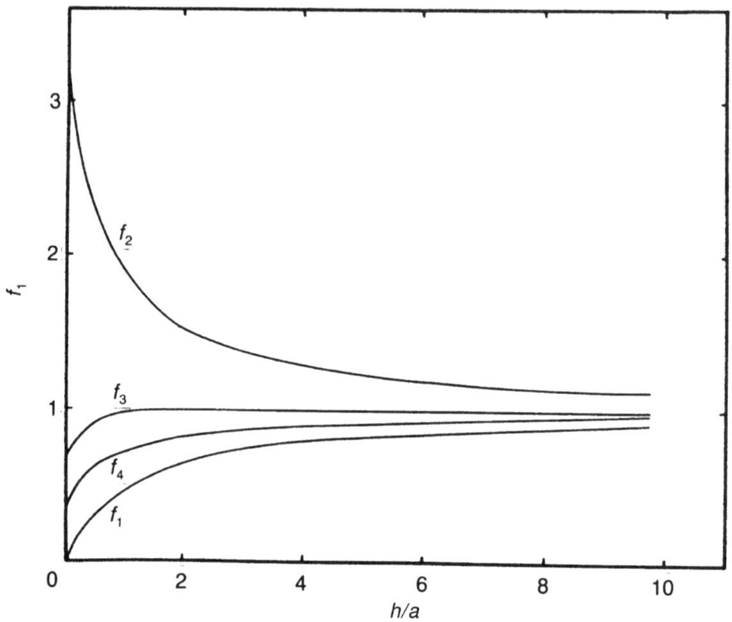

Figure 4.2 *Universal correction factors for particle motion and diffusion near a planar solid wall*

where z is the distance from the midplane of the channel, $\mathbf{v}_{max}$ is the maximum velocity and $\mathbf{v}_m$ the mean velocity.

Along with the expressions for the near-wall flow fields, Table 4.1 also gives definitions of another parameter, Pe, specific to the geometries listed. This parameter is known as the *Peclet number*, and it measures the relative strength of the fluid flow (convection) and Brownian motion (diffusion):

$$Pe = U_0 a / D_0 \tag{4.13}$$

where D_0 is the diffusion coefficient of a spherical particle of radius a subjected to a flow characterized by U_0. Physically, the magnitude of the Peclet number indicates which of the two mechanisms (convection and diffusion) is dominant in the process of particle transport. Mathematically, it can be used to justify simplifications to the equations governing the transport processes when carrying out theoretical calculations. A quantitative description of particle transport will be given later in this chapter.

Another reason why the simple shear flow is theoretically important is that the simple shear is often sufficient to model the actual local flow fields when studying the shear-induced aggregation or break-up of aggregates. For example, for Poiseulle flow in a parallel-plate channel, the shear rate (i.e. the velocity gradient) is given by:

$$G = \left| \frac{\mathrm{d}v}{\mathrm{d}z} \right| = \frac{3v_{\mathrm{m}}z}{b^2} \tag{4.14}$$

Since collision between two particles occurs in a very small space range (of the order of particle diameter) compared with the dimension of the collector, the local shear rate changes very little. As a result, it is often sufficient to consider the situation of constant shear rate (i.e. simple shear).

4.2. Brownian motion and diffusion

4.2.1 Brownian Motion

The chaotic, or Brownian, motion of colloidal particles is a direct result of collisions between the colloids and the fluid molecules surrounding them. The trajectory of a colloidal particle undergoing Brownian motion, obtained by tracking its movements at the usual experimental timescale intervals (e.g. seconds), is of the self-similar nature. That is, if any portion of a given Brownian trajectory is magnified (i.e. the sampling time interval is decreased), the magnified trajectory will look qualitatively similar to the original one. Thus, a Brownian trajectory is not a mathematically smooth curve, and the apparent velocity of a Brownian particle derived from it does not represent the true, physically well-defined, velocity of the particle. For this reason, the mean-square displacement is generally used to describe the motion of Brownian particles.

To estimate the order of magnitude of Brownian motion, the Brownian motion may be modelled by a series of independent *random walks*, with each walk being described by classical mechanics (such as Newton's second law and Stokes' law). If, at the start of a random walk, the particle acquires an initial velocity u_0 due to collisions with fluid molecules, then its subsequent motion may be described by:

$$m \frac{\mathrm{d}^2 x}{\mathrm{d}t^2} = -\xi \frac{\mathrm{d}x}{\mathrm{d}t} \tag{4.15}$$

where m is the mass of the particle, x the distance of the particle from the origin at time t, and ξ is called the friction coefficient, whose expression will be given later in Section 4.3. From eqn (4.15) it is readily seen that:

$$x(t) = \tau_{\mathrm{B}} u_0 \left[1 - \exp(-t/\tau_{\mathrm{B}})\right] \tag{4.16}$$

$$u(t) = u_0 \exp(-t/\tau_{\mathrm{B}}) \tag{4.17}$$

where $\tau_{\mathrm{B}} = m/\xi$ and has the dimension of time. τ_{B} can be regarded as the typical timescale for a single random walk, also known as the *relaxation time* for the particle's momentum. The length scale of a typical random walk, ℓ_{B}, can be considered to be equal to $\tau_{\mathrm{B}} u_0$. The initial velocity, u_0, may be estimated from the equipartition theorem of kinetic energy:

$$\tfrac{1}{2} m \langle u_0^2 \rangle = \tfrac{1}{2} kT \tag{4.18}$$

so that

$$u_0 \approx (kT/m)^{1/2} \tag{4.19}$$

$$\ell_B \approx \frac{(mkT)^{1/2}}{\xi} \tag{4.20}$$

where <> denotes statistical mean. Given a hydrosol of 1 μm radius and 1 kg dm^{-3} mass density in water at room temperature, it follows that the relaxation time is about 0.2 μs and the typical length of a single walk is about 0.3 nm.

By treating Brownian motion as a series of random walks, Einstein (1906) showed that the mean-square displacement of a Brownian particle in one-dimensional projection, averaged over a long period of time ($t \gg \tau_B$), can be expressed as:

$$<x^2> = 2D_0 t \tag{4.21}$$

Due to its isotropic nature, the mean-square displacement of Brownian motion in two dimensions is thus:

$$<r^2> = <x^2 + y^2> = 2<x^2> = 4D_0 t \tag{4.22}$$

Similarly, in three dimensions, we have:

$$<r^2> = <x^2 + y^2 + z^2> = 3<x^2> = 6D_0 t \tag{4.23}$$

The diffusion coefficient D_0 is a constant for a single particle in an unbounded fluid, and is related to the friction coefficient ξ by the Stokes–Einstein relation:

$$D_0 = kT/\xi \tag{4.24}$$

4.2.2 Diffusion

If a particle undergoing Brownian motion is given a sufficiently long time, it will zig-zag back and forth, and, eventually, visit every point within the space confining the particle. This ergodicity of Brownian particles is the underlying reason for the spontaneous equalization of particle concentration, a phenomenon generally known as *diffusion*. It should be pointed out that the equal probability for a Brownian particle to move back and forth at any one time does not conflict with the tendency of a Brownian particle to move away from its initial position over a period of time. The reference point, implied by the statement that a Brownian particle has an equal probability of moving back and forth, changes its position with time, but the initial position, from which the particle has a tendency to move away, is fixed over the period of time.

The diffusion of a collection of hydrodynamically non-interacting and equal-sized spherical particles is governed by the following equations:

$$\mathbf{J} = -D_0 \nabla n \tag{4.25}$$

$$\frac{\partial n}{\partial t} = D_0 \nabla^2 n \tag{4.26}$$

where **J** denotes the flux of the particles, and n the number density of particles. The two equations are sometimes referred to as *Fick's first and second law of diffusion.* It is from these equations that the diffusion coefficient derives its name.

For arbitrarily shaped and/or hydrodynamically interacting particles, the diffusion process can no longer be characterized by a constant D_0, because of the coupling between translation and rotation of each individual Brownian particle and between particles themselves. In such cases, a *diffusion tensor* $\mathfrak{D}$ must be used, and Fick's laws should be rewritten as:

$$\mathbf{J} = -\mathfrak{D} \bullet \nabla n \tag{4.27}$$

$$\frac{\partial n}{\partial t} = \nabla \bullet (\mathfrak{D} \bullet \nabla n) \tag{4.28}$$

The Stokes–Einstein relation may still be retained in the form analogous to eqn (4.24):

$$\mathfrak{D} = \mathfrak{R}^{-1} kT \tag{4.29}$$

with the friction coefficient now being replaced by a resistance tensor $\mathfrak{R}$. However, the validity of such an expression has yet to be established (van de Ven, 1989).

4.3 Motion of a single sphere

In many cases, colloidal particles can be treated as spheres. Therefore the motion of a sphere is of fundamental interest for colloidal systems. However, it should be stressed that, although a colloidal particle may be regarded as a sphere, the motion of a spherical colloidal particle can be quite different from that of an 'ordinary' sphere, because of the influence of the electrical double layer surrounding the colloidal particle. Consequently, it is necessary to distinguish between *hard* (i.e. electrically neutral) spheres and *soft* (i.e. electrically charged) spheres. Furthermore, the presence of a large solid wall (or other types of interfaces) will also have a considerable retardation effect on particles moving nearby. This section presents a brief summary of theoretical descriptions of the behaviour of a single sphere under these different conditions.

4.3.1 Motion of a single hard sphere

The frictional drag force F_D acting on a rigid sphere of radius a is linearly proportional to the particle's speed U relative to the surrounding fluid:

$$F_D = -\xi U \tag{4.30}$$

where ξ is the friction coefficient, given by:

$$\xi = 6\pi\mu a \tag{4.31}$$

Table 4.2 *Hydrodynamic correction functions for Stokes' law*

Correction function		*Validity*	*Source*
$1 + \frac{9}{8}\left(\frac{a}{r}\right)$	(4.34)	$a/r \ll 1$ Normal rigid plane	Lorentz (1907)
$\frac{4}{3}\sinh\alpha \sum_{n=1}^{\infty} \frac{n(n+1)}{(2n-1)(2n+3)} \times \left[\frac{2\sinh(2n+1)\alpha + (2n+1)\sinh 2\alpha}{4\sinh^2(n+\frac{1}{2})\alpha - (2n+1)^2\sinh^2\alpha} - 1\right]$	(4.35)	Normal rigid plane	Brenner (1961)
$1 + \frac{9}{16}\left(\frac{a}{r}\right)$	(4.36)	$a/r \ll 1$ parallel rigid plane	Lorentz (1907)
$\frac{a}{h} - 0.21\ln\left(\frac{h}{a}\right) + 0.97$	(4.37)	$h/a < 0.1$ Normal rigid plane	Cox and Brenner (1967)
$1 + 2.105\left(\frac{a}{R}\right)$	(4.38)	$a/R \ll 1$ Parallel cylindrical rigid wall	Hamberman and Sayre (1958)
$\frac{4}{3}\sinh\alpha \sum_{n=1}^{\infty} \frac{n(n+1)}{(2n-1)(2n+3)} \times \left[\frac{4\cosh^2(n+\frac{1}{2})\alpha + (2n+1)^2\sinh^2\alpha}{2\sinh(2n+1)\alpha - (2n+1)\sinh 2\alpha} - 1\right]$	(4.39)	Normal free surface	Brenner (1961)
$1 + \frac{3}{4}\gamma + \frac{9}{16}\gamma^2 + \frac{19}{64}\gamma^3 + \frac{93}{256}\gamma^4 + \frac{387}{1024}\gamma^5 + \frac{1197}{4096}\gamma^6 + \frac{5331}{16384}\gamma^7 + \frac{19821}{65536}\gamma^8 + \frac{76115}{262144}\gamma^9 + \ldots$	(4.40)	Parallel free surface	Faxen and Dahl (1925)
$\frac{4}{3}\sinh\alpha \sum_{n=1}^{\infty} \frac{n(n+1)}{(2n-1)(2n+3)} \times \left\{\frac{\mu_1\mu_2 X_n + [\mu_1\mu_3 + \mu_2^2]Y_n + 2\mu_2\mu_3 Z_n}{\mu_2(\mu_1+\mu_3)V_n + \mu_1\mu_3 T_n + 2\mu_2^2 S_n} - 1\right\}$	(4.41)[a]	Normal fluid–fluid interface; Fluid sphere	Bart (1968)
$\left[1 - \frac{9}{16}\gamma + \frac{1}{8}\gamma^3 - \frac{45}{256}\gamma^4 - \frac{1}{16}\gamma^5 \ldots\right]^{-1}$	(4.42)	$r/a > 1.04$ Parallel rigid plane	Faxen (1922)
$-\frac{8}{15}\ln(h/a) + 0.9588$	(4.43)	$h/a < 0.1$ Parallel rigid plane	Goldman *et al.* (1967)

Since the theoretical treatment of this subject has grown out of the early work of Stokes, the above relation is often referred to as Stokes' law. The relation provides good agreement with experimental results obtained at small Reynolds numbers, i.e. $Re < 1$. This expression can be derived from the Stokes equation, eqn (4.6), in which the inertia of the fluid is completely neglected. An extension to account for the inertial effect at large distances has been made by Oseen (1910):

$$F_D = 6\pi a\mu U\left(1 + \frac{3}{8}\frac{aU\rho}{\mu}\right) \tag{4.32}$$

which has been shown to be valid for $Re = 2aU\rho/\mu \leq 5$ (Schlichting, 1955).

Both expressions given above apply only to particles moving in an unbounded fluid medium. In most practical cases, however, the fluid will be bounded by solid walls and fluid–fluid interfaces. The presence of these boundaries at finite distances necessitates a modification to Stokes' law. By convention, such modifications are given in the form of a dimensionless correction factor, λ, so that:

$$F_D = 6\pi\mu a\lambda U \tag{4.33}$$

The correction factor is usually a function of the separation distance. Some analytical expressions corresponding to the translational motion of single particles are given in Table 4.2. Since rotational motion of particles is often considered to have negligible effects on particle transport leading to deposition and/or aggregation, expressions corresponding to the rotational motion are not included in the table.

4.3.2 Motion of a single soft sphere

From Chapters 2 and 3, it is known that particles will often become charged in a polar medium. The presence of an electrical double layer can hinder the motion of an otherwise hard sphere. This is because, first, the shear layer, which moves together with the sphere, effectively increases the particle radius, and secondly,

---

Footnote to Table 4.2

$r = h + a$, h is surface-to-surface distance
$\gamma = a/r$
$\alpha = \cosh^{-1}(r/a) = \ln\left\{\frac{r}{a} + \left[\left(\frac{r}{a}\right)^2 - 1\right]^{1/2}\right\}$
μ_1 = viscosity of sphere fluid
μ_2 = viscosity of fluid in which the sphere is immersed
μ_3 = viscosity of fluid forming the plane interface with μ_2
$X_n = (2n+1)^2 \sinh^2\alpha + 4\cosh^2(n+1/2)\alpha$
$Y_n = 2\sinh(2n+1)\alpha + (2n+1)\sinh 2\alpha$
$Z_n = \cosh(2n+1)\alpha + \cosh 2\alpha$
$V_n = 2\sinh(2n+1)\alpha - (2n+1)\sinh 2\alpha$
$T_n = 4\sinh^2(n+1/2)\alpha - (2n+1)^2\sinh^2\alpha$
$S_n = \cosh(2n+1)\alpha - \cosh 2\alpha$

Bart (1968) noted that eqn (4.41) applies to solid spheres only; for a fluid droplet settling towards a planar interface the correction function should be multiplied by another factor, $3(\mu_1 + \mu_2)/(3\mu_1 + 2\mu_2)$

the double layer tends to deform, thereby creating a dipole as the particle moves.

The correction factor for the settling velocity of a charged sphere in an unbounded electrolyte solution was first given by Smoluchowski (1921) for the case of $\kappa a \gg 1$:

$$\lambda = \left(1 + \frac{\varepsilon^2 \zeta^2}{\mu \upsilon a^2}\right)^{-1} \tag{4.44}$$

where υ denotes the specific conductivity of the electrolyte. Later the problem was looked at in some more detail by Booth (1954), who proposed that the correction factor may be expressed in the form of a power series of the total number of charges Q, or the zeta potential, carried by the sphere:

$$\lambda = 1 + \sum_{n=1}^{\infty} A_n Q^n = 1 + \sum_{n=1}^{\infty} B_n (e\zeta/kT)^n \tag{4.45}$$

Since in symmetrical electrolytes a positively charged particle behaves in the same way as an equally but negatively charged particle, the constants A_1 and B_1 must be zero. Therefore, the effective first order correction is actually in Q^2 or ζ^2. The expressions for the effective first order correction factors, given as rather complex integrals, will not be cited here, but a general conclusion is that the reduction due to dipole field usually amounts, at most, to a few per cent.

Proudman and Pearson (1957) considered the problem from a different perspective and showed that the correction factor may be approximated by:

$$\lambda = 1 + c_1 \kappa a \qquad (\kappa a \ll 1) \tag{4.46}$$

$$\lambda = 1 + c_2 (\kappa a)^{-2} \qquad (\kappa a \gg 1) \tag{4.47}$$

where κ is the Debye reciprocal length defined in eqn (2.13), and c_1 and c_2 are functions of the Peclet number, zeta potential and ion size.

More recently, Ohshima *et al.* (1984) obtained a general formula for the effective first-order correction factor, valid in the case of small Peclet numbers and zeta potentials in symmetrical electrolytes:

$$\lambda = 1 + \bar{b} \Phi_s^2 f(\kappa a) \tag{4.48}$$

where $\bar{b}$ is a measure of the average size of ions, Φ_s the reduced zeta potential ($= ze\zeta/kT$), and f a function of κa. $f(\kappa a) = 0$ for both $\kappa a = 0$ and $\kappa a = \infty$, and reaches the maximum value when $\kappa a \approx 1$. A different but simpler formula has also been suggested by Ohshima *et al.* (1984):

$$\lambda = 1 + \frac{12}{(\kappa a)^2} (b_+ A_+^2 + b_- A_-^2) \tag{4.49}$$

$$A_{\pm} = \ln \left[\frac{1 + \exp(\mp |\Phi_s| /2}{2} \right]$$

which can be shown to be identical to eqn (4.48) as $\kappa a \to \infty$ and $\zeta \to 0$.

The effects of such parameters as electrolyte concentration (via κa), surface potential (via ζ), fluid velocity (via Pe) and ion size (via b_i), on the correction factor is shown schematically in Figure 4.3 (van de Ven, 1989).

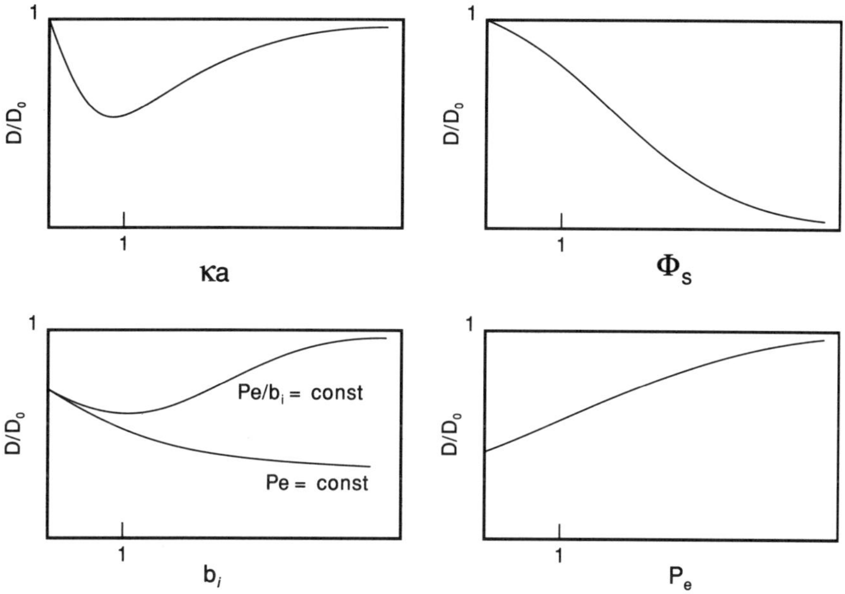

Figure 4.3 *Influence of* κa, Φ_s, b_i *and* Pe *on the diffusion coefficient ratio* D/D_0 (after van de Ven, 1989)

4.4 Relative motion of two spheres

4.4.1 In quiescent fluid

Although the general framework for many-body hydrodynamics has long been established (Brenner and O'Neill, 1972), solutions are only available for a few well-defined flow geometries. In this section, examples of two-sphere cases will be examined.

Relative motion in two-dimensional space

The first example is for the case of two equal spheres translating along the line of centres with a constant velocity in the same direction. In this case, Stimson and Jeffery (1926) showed that the hydrodynamic correction function λ can be expressed as:

$$\lambda = \frac{4}{3} \sinh \alpha \sum_{n=1}^{\infty} \frac{n(n+1)}{(2n-1)(2n+3)} \times \left[1 - \frac{4 \sinh^2 (n+1/2)\alpha - (2n+1)^2 \sinh^2 \alpha}{2 \sinh (2n+1)\alpha + (2n+1) \sinh 2\alpha}\right] \quad (4.50)$$

where $\alpha = \cosh^{-1} (r/a)$, and r is the distance between centres of the two spheres.

In the second case, the two equal spheres approach (or depart from) each other with equal speed. This is equivalent to the case of a sphere moving perpendicular to a free plane surface, and the correction function has the same form as eqn (4.39) but with α now being equal to $\cosh^{-1}(r/2a)$.

The third case of interest is the relative motion of two unequal spheres in a plane. In this case, it is possible to decompose the motion into parallel and perpendicular motions relative to the line of centres. As such, the force acting on the first particle has the form (Happel and Brenner, 1975):

$$\mathbf{F}_1 = 6\pi\mu a_1 \left[(U_{1_{//}}\lambda_{1_{//}} + U_{2_{//}}\lambda_{2_{//}})\,\mathbf{e}_{//} + (U_{1_\perp}\lambda_{1_\perp} + U_{2_\perp}\lambda_{2_\perp})\,\mathbf{e}_\perp \right] \tag{4.51}$$

where $\mathbf{e}$ is the unit vector, the subscript $''$ denotes the parallel component, and the subscript $\perp$ denotes the perpendicular component. If the two particles do not rotate as they move, the approximate expressions for the correction functions are given by:

$$\lambda_{1_{//}} = 1 + \frac{9}{4}\frac{a_1 a_2}{r^2} + \frac{3}{4}\left(-2\frac{a_1^3 a_2}{r^4} + \frac{27}{4}\frac{a_1^2 a_2^2}{r^4} + 3\frac{a_1 a_2^3}{r^4}\right) \tag{4.52}$$

$$\lambda_{2_{//}} = -\frac{3}{2}\frac{a_2}{r} + \frac{1}{2}\left(\frac{a_1^2 a_2}{r^3} - \frac{27}{4}\frac{a_1 a_2^2}{r^3} + \frac{a_2^3}{r^3}\right) - \frac{9}{4}\left(\frac{a_1^3 a_2^2}{r^5} + \frac{27}{8}\frac{a_1^2 a_2^3}{r^5} + \frac{a_1 a_2^4}{r^5}\right) \tag{4.53}$$

$$\lambda_{1_\perp} = 1 + \frac{9}{16}\frac{a_1 a_2}{r^2} + \frac{3}{8}\left(\frac{a_1^3 a_2}{r^4} + \frac{27}{32}\frac{a_1^2 a_2^2}{r^4} + 3\frac{a_1 a_2^3}{r^4}\right) \tag{4.54}$$

$$\lambda_{2_\perp} = \frac{3}{4}\frac{a_2}{r} + \frac{1}{4}\left(\frac{a_1 a_2^2}{r^3} + \frac{27}{16}\frac{a_1 a_2^2}{r^3} + \frac{a_2^3}{r^3}\right) + \frac{63}{64}\left(\frac{a_1^3 a_2^2}{r^5} + \frac{27}{112}\frac{a_1^2 a_2^3}{r^5} + \frac{a_1 a_2^4}{r^5}\right) \tag{4.55}$$

If the two spheres are free to rotate, the parallel correction functions will remain unchanged, but the perpendicular correction functions will have a slightly different form:

$$\lambda_{1_\perp} = 1 + \frac{9}{16}\frac{a_1 a_2}{r^2} + \frac{3}{8}\left(\frac{a_1^3 a_2}{r^4} + \frac{27}{32}\frac{a_1^2 a_2^2}{r^4} + \frac{a_1 a_2^3}{r^4}\right) \tag{4.56}$$

$$\lambda_{2_\perp} = \frac{3}{4}\frac{a_2}{r} + \frac{1}{4}\left(\frac{a_1 a_2^2}{r^3} + \frac{27}{16}\frac{a_1 a_2^2}{r^3} + \frac{a_2^3}{r^3}\right) + \frac{27}{64}\left(\frac{a_1^3 a_2^2}{r^5} + \frac{9}{16}\frac{a_1^2 a_2^3}{r^5} + \frac{a_1 a_2^4}{r^5}\right) \tag{4.57}$$

It can be seen, by comparing eqns (4.54) and (4.55) with eqns (4.56) and (4.57), that the freedom to rotate results in a slight reduction of the drag force.

Relative motion in three-dimensional space

The problem of relative motion of two spheres in three-dimensional space has been studied by several authors (Batchelor, 1976; Adler, 1981; Jeffrey and Onishi, 1984). By decomposing the relative motion of two spheres into two components – one along the line of centres and the other normal to it – the diffusion tensor $\mathfrak{D}$ can be expressed as:

$$\mathfrak{D} = (D_1 + D_2)\begin{pmatrix} \lambda_{/\!/} & 0 \\ 0 & \lambda_{\perp} \end{pmatrix} \tag{4.58}$$

$$D_1 = \frac{kT}{6\pi\mu a_1} \qquad D_2 = \frac{kT}{6\pi\mu a_2}$$

$\lambda_{/\!/}$ and $\lambda_{\perp}$ are both functions of the size ratio, a_2/a_1, and the dimensionless surface-to-surface separation distance defined as $h = r - 2$, where $r = 2r/(a_1 + a_2)$ being the dimensionless centre-to-centre separation distance between particles 1 and 2. Expressions for $\lambda_{/\!/}$ and $\lambda_{\perp}$ have been given by Batchelor (1976), and will not be included here. For equal-sized spheres, variations of $\lambda_{/\!/}$ and $\lambda_{\perp}$ with r are shown in Figure 4.4.

As far as particle aggregation is concerned, $\lambda_{/\!/}$ is more important than $\lambda_{\perp}$, since it is a direct measure of the increase in the hydrodynamic resistance as the

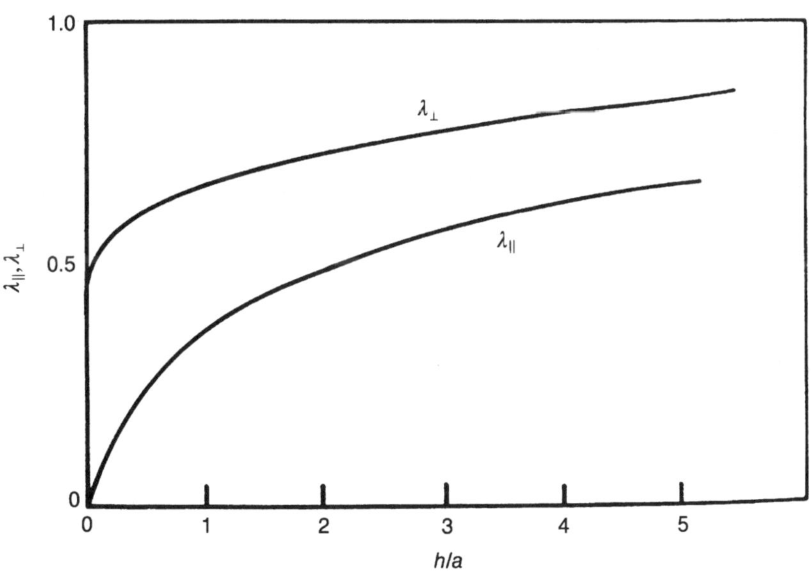

Figure 4.4 *Variation of $\lambda_{/\!/}$ and $\lambda_{\perp}$ with the dimensionless gap width between two equal-sized spheres (after van de Ven, 1989)*

two spheres approach each other along the line centres. For equal-sized spheres, $\lambda_{\parallel}$ can be expressed as an infinite series (van de Ven, 1989):

$$\lambda_{\parallel}^{-1} = \frac{4}{3}\sinh\alpha \sum_{n=1}^{\infty} \frac{n(n+1)}{2\Lambda}\left[\frac{2n+1}{2(2n-1)}\exp(2\alpha) + \frac{2n+1}{2(2n+3)}\exp(-2\alpha) + \frac{4\exp[-(2n+1)\alpha]}{(2n-1)(2n+3)} - 1\right] \tag{4.59}$$

$$\alpha = \cosh^{-1}(r/2a) \qquad \Lambda = 2\sinh(2n+1)\alpha - (2n+1)\sinh 2\alpha$$

Alternatively, it may be estimated using a simple formula (with an error less than 1%) given by Honig *et al.* (1971):

$$\lambda_{\parallel} \approx \frac{6\hbar^2 + 4\hbar}{6\hbar^2 + 13\hbar + 2} \tag{4.60}$$

where $\hbar = \imath - 2$, is the dimensionless gap width between the two spheres.

When studying deposition or aggregation in a *dilute* dispersion using the simulation methods to be described in Chapter 7, it is convenient to combine the fractional and diffusive effects into a single diffusion tensor, $\mathfrak{D}_{ij}$, or the so-called *mobility tensor* $\mathfrak{M}_{ij} = \mathfrak{D}_{ij}/kT$. Mazur and Saarloos (1982) have shown that the dominant contributions to the diffusion tensor for a N-sphere system are of the order of $r^{-(3N-5)}$. Therefore, three-body interactions first appear at order r^{-4}, and four-body interactions at order r^{-7}. Explicit expressions for the diffusion tensor are available up to order r^{-12} (Schmitz and Felderhof, 1983). However, for computational reasons, it is often only feasible, but none the less acceptable, for computer simulations to use diffusion tensors of up to order r^{-3}, which contain contributions from single (frictional) and pair (diffusive) hydrodynamic interactions. Examples of such diffusion tensors include the first-order expression for equal-sized spheres given by Oseen (1910):

$$\mathfrak{D}_{ij} = \delta_{ij}\frac{kT}{6\pi\mu a}\mathfrak{I} + (1-\delta_{ij})\frac{kT}{8\pi\mu r}\left(\mathfrak{I} + \frac{\mathbf{r}_{ij}\mathbf{r}_{ij}}{r^2}\right) \tag{4.61}$$

$$\delta_{ij} = \begin{cases} 1 & (i = j) \\ 0 & (i \neq j) \end{cases}$$

and the third-order expression given by Rotne and Prager (1969):

$$\mathfrak{D}_{ij} = \frac{kT}{6\pi\mu a}\left[\delta_{ij}\mathfrak{I} + (1-\delta_{ij})\frac{3}{4}\left(\frac{a}{r}\right)\left(\mathfrak{I} + \frac{\mathbf{r}_{ij}\mathbf{r}_{ij}}{r^2}\right) - (1-\delta_{ij})\frac{1}{2}\left(\frac{a}{r}\right)^3\left(3\frac{\mathbf{r}_{ij}\mathbf{r}_{ij}}{r^2} - \mathfrak{I}\right)\right] \tag{4.62}$$

where $\mathfrak{I}$ is the unit tensor, $\mathbf{r}_{ij} = \mathbf{r}_j - \mathbf{r}_i$, $r = |\mathbf{r}_j - \mathbf{r}_i|$, $\mathbf{r}_i$ and $\mathbf{r}_j$ are position vectors of sphere i and j, respectively, and δ_{ij} is the Kronecker delta.

Motion of doublets

Having considered the motion of two spheres in general, it is now necessary to examine a special case of interest, i.e. the motion of a doublet. This corresponds to the situation where the two spheres travel at the same velocity. The translational diffusion coefficients corresponding to motion along and normal to the line of centres, respectively, may be obtained by assuming $\mathbf{U}_1 = \mathbf{U}_2$ in eqns (4.52) and (4.54) for rigid dumbbells, or eqns (4.52) and (4.56) for torque-free doublets. Alternative expressions for the translational motion of torque-free doublets are also available (Jeffrey and Onishi, 1984). As has been noted above, freedom to rotate for each of the spheres translating normal to the line of centres leads to a reduction in drag force. Hence the normal component of the translational diffusion tensor for a rigid dumbbell is smaller than that for a torque-free doublet.

The rotation of a doublet can be decomposed into the rotation D_r about an axis perpendicular to the line of centres D_r and the spin D_s about the line of centres. The rotary diffusion coefficient for a torque-free doublet is larger than that for a rigid dumbbell, especially at short distances. However, the spin diffusion coefficients for both types of doublets are the same, since the spin about the line of centres does not induce translation of the two spheres.

For torque-free (TF) doublets of equal-sized spheres, the rotary diffusion coefficient D_r^{TF} can be approximated by:

$$D_r^{TF} = D_0^r \frac{1}{(1 + 0.5h)^2} \left(\frac{0.2673 \ln h - 0.724}{\ln h - 1.350} \right) \tag{4.63}$$

at small h values, and by:

$$D_r^{TF} = D_0^r \left[\frac{8}{3} r^{-2} - 2r^{-3} - \frac{5}{2} r^{-5} \right] \tag{4.64}$$

at large h (or r) values (van de Ven, 1989). Here, $D_0^r = kT/8\pi\mu a^3$ denotes the rotary diffusion coefficient for a single isolated sphere.

The corresponding diffusion coefficient for rigid dumbbells (RD) may be approximated as:

$$D_r^{RD} = D_0^r (0.2673 - 0.153h - 0.0341h^2 + 0.0089h^2 \ln h + \ldots) \tag{4.65}$$

at small h values, and by

$$D_r^{RD} = \frac{8}{3} D_0^r \left(r^{-2} - \frac{3}{4} r^{-3} - 4r^{-4} + 4r^{-5} \right) \tag{4.66}$$

at large h (or r) values (van de Ven, 1989).

Motion of spheres in the vicinity of a wall

So far, the motion of a sphere in the presence of a second sphere in an unbounded fluid medium has been discussed. For particle deposition processes it is obviously necessary to take wall effects into account as well. Explicit expressions for mobility tensors in the presence of a solid plane wall have been derived by Beenakker *et al.* (1984) up to order $(a/r)^n(a/r_s)^m$ with $n + m \leq 3$, which account for hydrodynamic interactions of one and two spheres and the wall:

$$6\pi\mu a_i \mathfrak{M}_{ij}^{TT} = \mathfrak{I}\delta_{ij} + \left[\frac{3}{4} a_i r^{-1} \left(\mathfrak{I} + \mathbf{e}_{ij}\mathbf{e}_{ij}\right) - \frac{3}{4} a_i (a_i^2 + a_j^2) r^{-3} \left(\mathbf{e}_{ij}\mathbf{e}_{ij} - \frac{1}{3}\mathfrak{I}\right)\right]\left(1 - \delta_{ij}\right)$$
$$- \frac{3}{4} a_i r_s^{-1} \left[\mathfrak{I} + \mathbf{e}_{ij_s}\mathbf{e}_{i_sj} - 2l_i r_s^{-1} \mathbf{e}_{ij_s}\mathbf{n} + 2l_j r_s^{-1} \mathbf{n}\mathbf{e}_{i_sj} + 2l_i l_j r_s^{-2} \left(\mathfrak{I} - 2\mathbf{nn} - 3\mathbf{e}_{ij_s}\mathbf{e}_{i_sj}\right)\right]$$
$$+ \frac{3}{4} a_i (a_i^2 + a_j^2) r_s^{-3} \left(\mathbf{e}_{ij_s}\mathbf{e}_{i_sj} - \frac{1}{3}\mathfrak{I}\right)$$
$$- \frac{3}{2} a_i (a_i^2 - a_j^2) r_s^{-4} \left(l_i \mathbf{e}_{ij_s}\mathbf{n} + l_j \mathbf{n}\mathbf{e}_{i_sj}\right)$$
$$+ \frac{3}{2} a_i r_s^{-5} \left(a_i^2 l_j + a_j^2 l_i\right)\left(l_i + l_j\right)\left(\mathfrak{I} - 2\mathbf{nn} - 5\mathbf{e}_{ij_s}\mathbf{e}_{i_sj}\right) \quad (4.67)$$

$$8\pi\mu a_i \mathfrak{M}_{ij}^{RR} = \mathfrak{I}\delta_{ij} + \frac{3}{2} a_i^2 a_j r^{-3} \left(\mathbf{e}_{ij}\mathbf{e}_{ij} - \frac{1}{3}\mathfrak{I}\right)\left(1 - \delta_{ij}\right)$$
$$- \frac{3}{2} a_i^2 a_j r_s^{-3} \left[\mathbf{e}_{i_sj}\mathbf{e}_{ij_s} - \frac{1}{3}\mathfrak{I} + 2\left(l_i + l_j\right)^2 r_s^{-2}\mathfrak{I} - 2\left(\mathbf{e}_{ij_s} \times \mathbf{n}\right)\left(\mathbf{e}_{ij_s} \times \mathbf{n}\right)\right] \quad (4.68)$$

$$12\pi\mu a_i^2 \mathfrak{M}_{ij}^{RT} = -\frac{3}{2} a_i^2 r^{-2}\, \mathfrak{E}\bullet\mathbf{e}_{ij}\left(1 - \delta_{ij}\right)$$
$$+ \frac{3}{2} a_i^2 r_s^{-2} \left[\mathfrak{E}\bullet\mathbf{e}_{ij_s} + 2l_j r_s^{-1} \left(\mathfrak{E}\bullet\mathbf{n} - 3(\mathbf{e}_{ij} \times \mathbf{n})\mathbf{e}_{i_sj}\right)\right] \quad (4.69)$$

$$r = |\mathbf{r}_j - \mathbf{r}_i|, \qquad l_i = \mathbf{n}\bullet\mathbf{r}_i, \qquad l_j = \mathbf{n}\bullet\mathbf{r}_j,$$
$$\mathbf{e}_{ij} = (\mathbf{r}_j - \mathbf{r}_i)/r, \qquad r_s = |\mathbf{S}\bullet\mathbf{r}_j - \mathbf{r}_i| = (r^2 + 4l_i l_j)^{1/2},$$
$$\mathbf{e}_{ij_s} = (\mathbf{S}\bullet\mathbf{r}_j - \mathbf{r}_i)/r_s, \qquad \mathbf{e}_{i_sj} = (\mathbf{r}_j - \mathbf{S}\bullet\mathbf{r}_i)/r_s = \mathbf{e}_{ij_s} + 2\mathbf{n}(l_i + l_j)/r_s,$$
$$\mathbf{S} = \mathfrak{I} - 2\mathbf{nn} \qquad \mathfrak{E}:\mathfrak{E} = -2\mathfrak{I}$$

where the superscripts TT, RR and RT denote the translation, rotational and coupling tensors, respectively. The expressions given above reduce, for $i=j$, to the diffusion tensor for a single particle near a planar wall, and, for l_i and $l_j \to \infty$, to the Rotne–Prager tensor for two spheres in an unbounded medium.

4.4.2 In shear flow

This problem has been solved by Kao *et al.* (1977), Arp and Mason (1977) and Adler (1981) under various conditions. For a linear two-dimensional shear flow field, the trajectory of one sphere with respect to the other can be obtained from the following equations (van de Ven, 1989):

$$\begin{aligned}\frac{\mathrm{d}r}{\mathrm{d}t} &= (1+\gamma)AG\sin^2\vartheta\sin 2\phi + CF_e \\ \frac{\mathrm{d}\vartheta}{\mathrm{d}t} &= \frac{1}{4}(1+\gamma)BG\sin 2\vartheta\sin 2\phi \\ \frac{\mathrm{d}\phi}{\mathrm{d}t} &= \frac{1}{2}[(1-\gamma)+(1+\gamma)B\cos 2\phi]G\end{aligned} \tag{4.70}$$

The relation between the polar coordinates (r,ϑ,ϕ) of one sphere with respect to the other are shown in Figure 4.5. γ is a parameter used to denote the type of the flow. $\gamma=0$ symbolizes a simple shear flow; $\gamma=-1$ a pure rotational flow; and $\gamma=1$ a pure shear flow. G is the shear rate. F_e is the net colloidal force between the two spheres (e.g. the sum of the electrical double layer and van der Waals forces) scaled by $3\pi\mu a_m^2$, where $a_m=(a_1+a_2)/2$ is the mean size of the two spheres. A, B and C are known dimensionless functions of the dimensionless distance $\imath$ and the radius ratio a_2/a_1. Values or asymptotic expressions for A, B and C can be found in Kao *et al.* (1977) and Arp and Mason (1977) for equal spheres, and in Adler (1981) for unequal spheres. The full details will not be laboured here, but some of the more significant general trends will be discussed and illustrated below.

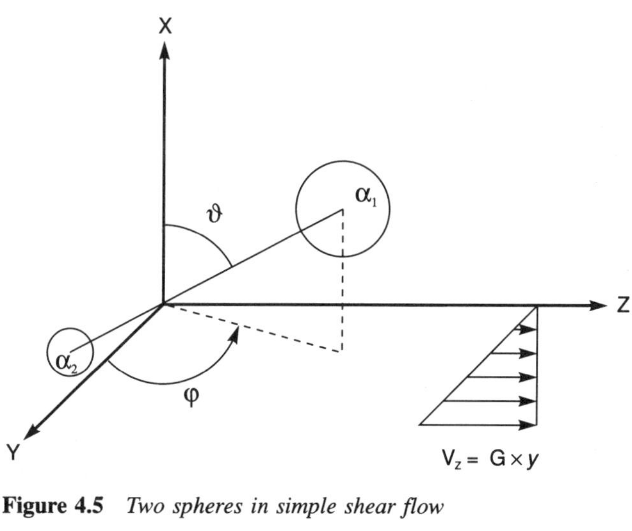

Figure 4.5 *Two spheres in simple shear flow*

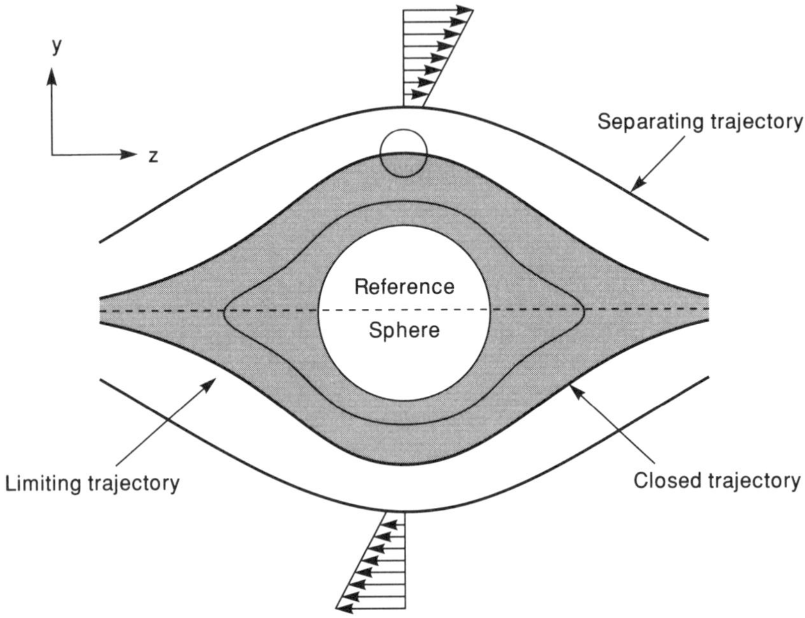

Figure 4.6 *Schematic representation of equatorial trajectories of two spheres in simple shear*

Relative motion of hard spheres

By integrating eqns (4.70), the trajectory of one sphere relative to the other can be obtained. Figure 4.6 shows schematically the equatorial trajectories of a sphere (of radius a_2) relative to another (the reference sphere of radius a_1) in a simple shear flow under different initial conditions.

Two types of trajectories can be distinguished. One consists of open trajectories, and the other of closed trajectories. The two types of trajectories are separated by a so-called *limiting trajectory*. If Brownian motion, external forces and colloidal interactions were all absent, the two spheres approaching each other from infinity would not be able to penetrate the region (hatched in Figure 4.6) enclosed by the limiting trajectory. The closest approach h_{min} depends strongly on the radius ratio. For instance:

when $a_2/a_1 = 1$, $h_{min} = 4.2 \times 10^{-5} a_1$ (Adler, 1981)

and

when $a_2/a_1 = 0$, $h_{min} = 0.16a_1$ (Cox *et al.* 1968)

These examples clearly indicate that under simple shear, equal spheres are more likely to aggregate than unequal spheres, because they can approach each other more closely.

Rotation of a doublet

If the two spheres are bound (e.g. by colloidal forces) to form a doublet, then the doublet will rotate indefinitely in simple shear about the vorticity axis (the X-axis in Figure 4.5) with a well-defined period of rotation. The period of rotation, T_R, depends on the particle separation in the following manner:

$$\frac{GT_R}{2\pi} = r_e + r_e^{-1} \tag{4.71}$$

where r_e is a characteristic dimensionless separation distance, related to the function B in eqn (4.70) by:

$$r_e = \left(\frac{2}{B} - 1\right)^{1/2} \tag{4.72}$$

Equation (4.71) has been shown to be very accurate for equal or similarly sized spheres, with an error of less than 1% under most conditions (van de Ven and Mason, 1976). However, for spheres of widely different size, it does not always hold true.

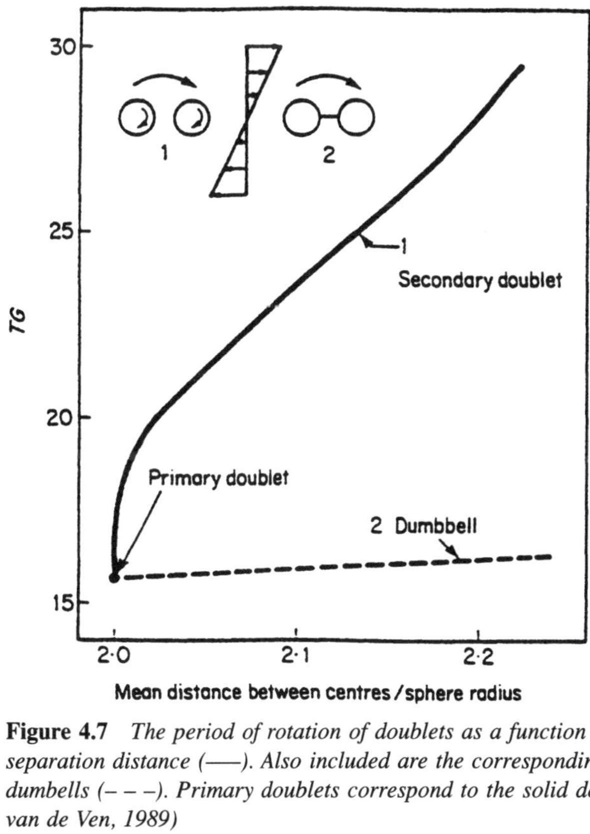

Figure 4.7 *The period of rotation of doublets as a function of the equilibrium separation distance (——). Also included are the corresponding values for rigid dumbells (– – –). Primary doublets correspond to the solid dot at T_RG = 15.62. (From van de Ven, 1989)*

For loosely bound doublets whose primary particles are free to spin, $T_R G$ is a steep function of r_e; in contrast, for rigid doublets, the variation of $T_R G$ with r_e is much more gradual. This can be seen in Figure 4.7. In reality, an intermediate situation is more likely to be true because the primary particles are unlikely to be entirely free to spin due to surface heterogeneity effects.

Relative motion of soft spheres

Up to now, the discussion has been presented under the assumption that particles are hydrodynamically hard spheres, although they may be subjected to colloidal forces (in the case of doublets). In fact, the presence of electrical double layers not only gives rise to an interparticle interaction but may also alter the local flow field around each particle. As a result, A, B and C in eqn (4.70) are no longer just functions of r and a_2/a_1; they will also depend on parameters that define the total colloidal interaction energy (e.g. zeta-potential, Debye length, Hamaker constant, ion size). This dependency makes it very difficult to predict the trajectories of two soft spheres. Fortunately, as has been demonstrated by van de Ven (1989), for both $\kappa a \gg 1$ and $\kappa a \ll 1$, neglecting the explicit dependency of these extra parameters does not incur too much error.

Even without explicit consideration of the electroviscous effects due to the electrical double layer (discussed later in Section 4.5), the presence of colloidal forces drastically changes the nature of the trajectories. The most obvious change is that the trajectories are no longer symmetrical about the Y-axis (Figure 4.8). If the net colloidal force is attractive, an incoming sphere is either captured by the other sphere or carried away by the flow. It is possible for the incoming sphere

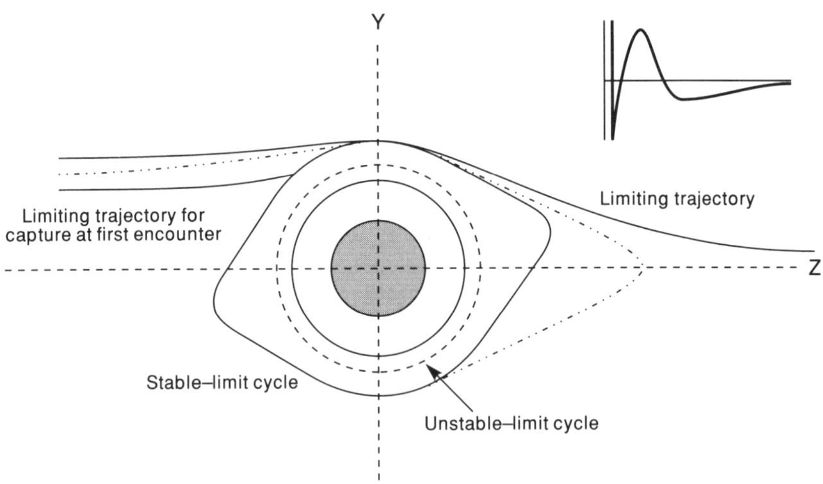

Figure 4.8 *Relative trajectories of two spheres with an interaction profile (shown in inset). The dotted line is an example of a trajectory leading to capture in the secondary minimum during the second encounter (after van de Ven, 1989)*

to orbit the second (reference) sphere several times before being captured. If there is a net repulsive interaction, no capture takes place anyway. When the total interaction has an energy barrier, two limiting cycles can exist around the reference sphere: one is stable and corresponds to an orbit around the secondary-minimum distance; and the other is unstable and corresponds to an orbit around the energy-barrier distance. In the latter case, the sphere will either fall into the primary minimum or move towards the secondary minimum. These three situations are depicted in Figure 4.8.

Adler (1981) has shown that, under fixed physicochemical conditions, the influence of the colloidal interactions will decrease as the radius ratio becomes smaller. Moreover, the influence is only in effect after the collision. It may be noted that the range of the colloidal interactions is usually much smaller than that of the hydrodynamic interactions, but when the colloidal interactions are active, a very slight perturbation can dramatically change the fate of the incoming sphere after collision.

4.5 Concentration dependence of diffusion coefficients

In a dispersion of hydrodynamically interacting particles, the hydrodynamic interactions not only affect the behaviour of each individual particle but also change the properties of the suspension as a whole. The former effect operates on a microscopic level, whereas the latter manifests itself at a macroscopic level. Both can be, and often are, described using diffusion coefficients. Thus, it is important to distinguish them conceptually. There are three types of diffusion coefficient:

- a short-time self-diffusion coefficient
- a long-time self-diffusion coefficient
- a mutual-diffusion coefficient

The self-diffusion coefficients characterize the motion of a Brownian particle in a swarm of other particles. When observed for a short time, the particle is seen to move *among* other particles; when observed for a long time, the particle is seen to move *through* the swarm of particles. The self-diffusion coefficients associated with these two timescales are denoted by D_0^s and D_∞^s, respectively. As each particle moves randomly, local fluctuations in particle concentration take place in the dispersion, giving rise to local concentration gradients. This leads to particle diffusion from high-concentration regions to low-concentration regions, and the diffusion coefficient corresponding to such a *collective* process is symbolized by D_m. The self-diffusion coefficients are useful to evaluate the effect of hydrodynamic interactions on the motion of individual particles (e.g in Brownian dynamics simulations). The mutual diffusion constant is needed to describe transport properties of the dispersions (e.g. in convective-diffusion theory).

Since the strengths of hydrodynamic interactions depend on the interparticle separations which, on average, are related to the volume fraction ϕ of particles in the suspension, in practice the diffusion coefficients are usually expressed in

Table 4.3 *The virial coefficient k for hard sphere dispersions*

Type	k	*Source*
Mutual-diffusion	$k_m = 1.45$	Batchelor (1976)
	$k_m = 1.56$	Felderhof (1978)
	$k_m = -2.0$	Allison *et al.* (1979)
	$k_m = -0.898$	Carter and Phillies (1985)
Short-time self-diffusion (translational)	$k_0^s = -1.83$	Batchelor (1976)
	$k_0^s = -1.73$	Felderhof (1978)
	$k_0^s = -2$	Yoshida (1983)
Long-time self-diffusion (translational)	$k_\infty^s = -2.68$	Batchelor (1976)
	$k_\infty^s = -1.89$	Felderhof (1977)
Short-time self-diffusion (rotary)	$k_r^s = -5/16$	Muthukumar and Freed (1983)
Long-time self-diffusion (rotary)	$k_r^s = -1$	Masliyah and van de Ven (1987)

Judging from the theoretical procedure used in estimating k_m and from available experimental data, Batchelor's results seem to be the most reliable.

terms of ϕ. For dilute monodisperse systems, the effective diffusion coefficient, D_e, can be expressed in the form:

$$D_e = D_0(1 + k\phi) \tag{4.73}$$

where k is a type-specific constant, called a virial coefficient. D_e may either represent the self-diffusion, mutual diffusion, or the coefficients for short-time or long-time, translational or rotary, motions. Different values have been suggested for each type of virial coefficient and these values differ not only in magnitude but also in sign. Some of the values are given in Table 4.3.

When colloidal forces are taken into account, the virial coefficient for mutual diffusion may be calculated from the following expression (van de Broeck *et al.*, 1981):

$$k_m = 1.45 + \frac{1}{a^3}\int_{2a}^{\infty} \left[g(r) - 6\right] \exp\left[(-V_T/kT) - 1\right] r^2 \, dr \tag{4.74}$$

where V_T is the total colloidal interaction, $g(r)$ is a known function of particle separation r and $[g(r)-6]$ is negative for all r. It can be seen that for repulsive interactions k_m is greater than 1.45 whilst for attractive interactions k_m is smaller than 1.45. If the electroviscous effects due to electrical double layers are considered, the values of both the constant 1.45 and the function $g(r)$ need to be modified accordingly (see van de Ven, 1989).

When particle suspensions are subjected to an external flow field, the influence of the hydrodynamic interactions is reflected by the change in the apparent viscosity of the fluid. In very dilute dispersions, interparticle hydrodynamics may be neglected and the viscosity is determined solely by the disturbance flow around a single isolated particle. In this case, the effective viscosity μ_e is given by (Einstein, 1906):

$$\mu_e = \mu(1 + 2.5\phi) \tag{4.75}$$

This expression is applicable for hard sphere dispersions. In the presence of electrical double layers, the constant 2.5 is replaced by the intrinsic viscosity $[\mu]$ which, for $\kappa a \gg 1$ and small zeta potentials, is given by (Krasney-Ergen, 1936):

$$[\mu] = 2.5 + \frac{6\varepsilon^2\xi^2}{\upsilon\mu a^2} \tag{4.76}$$

The increase in the viscosity due to the presence of electrical double layers around the colloidal particles is called the *primary electroviscous effect.*

For dilute dispersions in which pair interactions are important but interactions involving three or more bodies can be neglected, a second-order term appears in the expression for the effective viscosity:

$$\mu_e = \mu(1 + [\mu]\phi + k_2\,\phi^2) \tag{4.77}$$

Values for the second virial coefficient k_2 for dilute monodispersions of hard spheres are listed in Table 4.4.

For small zeta-potentials and $Pe \ll 1$, the second virial coefficient k_2 of soft sphere dispersions is of the form (van de Ven, 1989):

$$k_2 = 6.2 + \frac{3}{40(\kappa a)^5}\left[\ln\left(\frac{\theta}{\ln(\theta/\ln\theta)}\right)\right]^4 \ln\left(\frac{\theta}{\ln\theta}\right) + \bar{b}\Phi_s^2 g(\kappa a)$$

$$\theta = 4\pi\varepsilon\zeta^2 a\kappa \exp(2\kappa a)/kT \tag{4.78}$$

where θ is a parameter measuring the relative importance of electrical double layer with respect to Brownian motion, and $\bar{b}$ the reduced mean ion size. The

Table 4.4 *Second virial coefficient for hard sphere dilute dispersions*

Type of flow	*Peclect number*	k_2	*Source*
$-1 \le \gamma \le 1$	0	6.2	Batchelor (1977)
$\gamma = 0$	∞	~5.2	Batchelor and Green (1972)
$\gamma = 1$	∞	7.6	Batchelor and Green (1972)

exact form of the function $g(\kappa a)$ is unknown but it is considered to be of minor significance compared with the second term in the above expression (Russel, 1980). The increase in the second virial coefficient due to the double layer interactions is usually referred to as the *secondary electroviscous effect*. From eqn (4.78), it can be seen that the secondary electroviscous effect consists of two contributions, one arising from the electrostatic repulsion between two charged spheres (electrostatic forces) and the other due to the modification of the flow field around the two spheres (electroviscous forces).

4.6 Quantitative description of deposition phenomena

4.6.1 Overview

In previous sections we have described the effect of Brownian motion (diffusion) and the influence of fluid flow (convection) on the behaviour of colloidal particles in dilute suspensions. The real purposes of studying colloidal hydrodynamics are to gain a better understanding of the mechanisms underlying such colloidal processes as deposition and aggregation, and, more importantly, to use the knowledge to tackle practical design problems involving these processes. Theoretical prediction of the kinetics of colloidal processes such as particle deposition and aggregation requires intimate knowledge of colloidal forces, flow field and associated hydrodynamic influences within the colloidal systems. Since many of the fundamental aspects have been described in the preceding chapters, much of the remaining part of this book will be devoted to the applications of the knowledge to specific processes that involve colloidal deposition and/or aggregation. This section will concentrate on some conventional theories of colloidal deposition processes.

Two principal approaches are available for modelling purposes. These will be described in detail in subsequent sections, but before going into details, a brief overview of these approaches will be given here. The *Lagrangian* approach considers the behaviour of individual particles, or their trajectories, at a microscopic level. The other method, known as the *Eulerian* approach, describes the particles collectively in terms of their distribution, or probability density, in time and space. Thus, it is a typical macroscopic approach. These two basic methodologies are summarized in Figure 4.9.

Until recently, the Lagrangian approach has been extensively used to describe the capture of non-Brownian particles where particle trajectories are deterministic and easy to express analytically, although in principle it is capable of dealing with Brownian particles. The reason may be that incorporation of Brownian motion gives the method a stochastic feature, which entails laborious and time-consuming step-by-step integration of the stochastic equation of motion. This also explains why the Lagrangian approach was less attractive than the Eulerian approach, before the widespread availability of high-performance computers. Indeed, Eulerian methods have been the most widely employed for describing transport and deposition phenomena of Brownian particles. The methodology is not restricted to solid–liquid mixtures but can be applied to

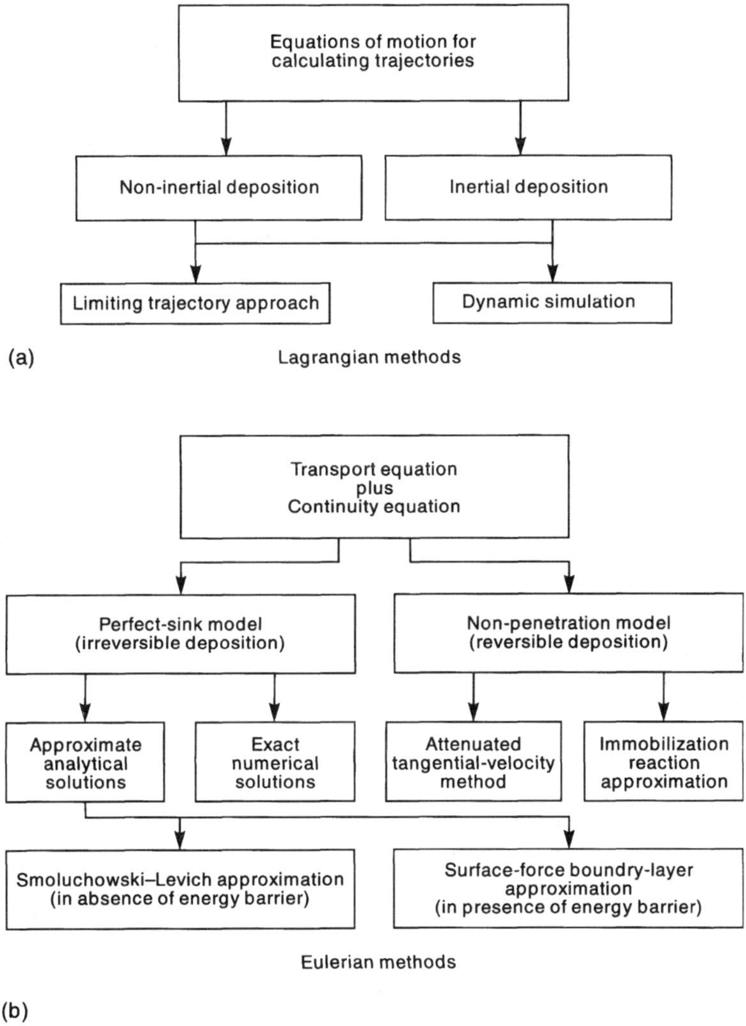

Figure 4.9 *Classification of theoretical methods for calculating the rates of particle deposition:* (a) *Lagrangian methods*, (b) *Eulerian methods*

different colloidal dispersions (aerosols, hydrosols etc.) and to some types of non-colloidal particulate systems.

There are quite a number of comprehensive reviews on both approaches, including, notably, those by Hidy and Brock (1970), Davies (1973) and Friedlander (1977) on aerosols; and Spielman (1977), Tien and Payatakes (1979) and Adamczyk *et al.* (1983) on hydrosols. In keeping with the scope set out in Chapter 1, colloidal deposition at a solid–liquid interface will be emphasized here.

Since the Lagrangian approach is considered in detail in Chapters 5 and 7, this section will be confined to the various approximations based on the Eulerian approach. Further application examples using both approaches are described in Section IV of this book.

4.6.2 Basic equations and boundary conditions

When dealing with colloidal deposition, it is often convenient to consider the process to be accomplished in two stages, namely the transport stage and the actual deposition stage. The mechanisms that may be involved in the transport stage include diffusion, convection, and perhaps the presence of a temperature and/or density gradient in the fluid. Once particles arrive in the vicinity of the collector surface, the colloidal forces, surface conditions and local flow fields take over to control the actual deposition stage. The timescales involved in the two stages usually differ by orders of magnitude, so the speed of the whole process, or in this case the rate of deposition, is effectively determined by the slower stage. Therefore, if the conditions are favourable to deposition (e.g. the net colloidal force is attractive), the transport stage dictates the rate of deposition. If the conditions are unfavourable (e.g. there is an energy barrier for particles to overcome before being deposited), the deposition stage decides the rate of deposition.

Due to the lack of detailed information on surface conditions and uncertainty about the validity of colloidal force expressions at very close separations, the actual attachment stage is often treated with highly idealized models which, in Eulerian methodology, are employed as boundary conditions for the equations that govern the transport stage. Therefore, the equations used to describe deposition processes are in fact transport equations, often referred to as the *convective-diffusion* (C-D) equations. These equations are much like Fick's laws (Section 4.2) except that the deposition equations incorporate convection as well as diffusion, since both diffusion and convection are likely to take part in deposition processes.

For interacting Brownian particles, the C-D equation may be written as (Peters, 1990):

$$\frac{\partial n}{\partial t} + \nabla \cdot (n\mathbf{v}) = \nabla \cdot (\mathfrak{D}_{\mathrm{m}} \cdot \nabla n - n\mathbf{F}_{\mathrm{m}}) \tag{4.79}$$

where n is a particle distribution function, $\mathbf{v}$ is a velocity vector representing an external flow field, $\mathfrak{D}_{\mathrm{m}}$ is a mutual diffusion tensor which incorporates the hydrodynamic effects arising from friction, pairwise diffusion and the presence of a solid collector surface, and $\mathbf{F}_{\mathrm{m}}$ is an effective force used to describe the effect of an externally applied force field on mutual diffusion. In deriving the above equation, it is assumed that (a) superposition principles can be applied to both hydrodynamic and colloidal forces, (b) the only non-uniformity in the host fluid arises from convection (other non-uniformities such as temperature and density gradients are absent), and (c) particles are spherically shaped, or spherically isotropic, and non-rotating.

For non-interacting Brownian particles in very dilute suspensions, the mutual diffusion tensor is replaced by a diffusion tensor $\mathfrak{D}_s$ for single particles near the collector surface, and the C-D equation may be written as (van de Ven, 1989):

$$\frac{\partial n}{\partial t} + \nabla \bullet \mathbf{j} = \mathbb{Q} \tag{4.80}$$

with the particle flux $\mathbf{j}$ given by:

$$\mathbf{j} = -\mathfrak{D}_s \bullet \nabla n + n\mathbf{v} + n(\mathfrak{D}_s \bullet \mathbf{F}_e / kT) \tag{4.81}$$

Here $\mathbb{Q}$ is added to represent a possible sink or source not included in the flux expression, and $\mathbf{F}_e$ denotes the external force field. Equation (4.80), together with eqn (4.81), is the starting point for most of the theoretical predictions given in the published literature.

In order to solve the C-D equation, appropriate boundary conditions must be formulated. Since our knowledge about the conditions near the collector surface is far from complete, various simplifying assumptions have to be made. Early work in this subject area employed the so-called *perfect-sink* model to calculate the deposition rates. This model assumes that all particles arriving at the collector surface, or at a separation distance corresponding to the primary energy minimum h_{pm}, will be irreversibly adsorbed immediately and subsequently disappear from the system, i.e.

$$n(h = 0) = 0 \qquad \text{or} \qquad n(h = h_{pm}) = 0 \tag{4.82}$$

This model ignores the formation of a separate phase (deposit) formed by the adsorbed particles at the interface, and precludes the possibility of immobilization reaction at a finite rate. Therefore, predictions based on this model are only applicable to the *initial* stage of the deposition process.

To overcome these limitations, a so-called *non-penetration* model has been proposed by Adamczyk and van de Ven (1984), which specifies a vanishing normal flux component at the interface:

$$j_\perp (h = h_{pm}) = 0 \tag{4.83}$$

and uses the source term $\mathbb{Q}$ in the C-D equation to represent the effects that cannot be described (or known) explicitly.

Both models and representative results obtained from these models are considered below.

4.6.3 Perfect-sink model

Before considering the various approximations for particle deposition rates, it is helpful to introduce some commonly used dimensionless groups (Table 4.5). These dimensionless groups convey the relative importance of various terms in the governing equations and their solutions.

Even under the simple assumption of the perfect-sink model, analytical solutions to the C-D equation, which are of the most practical interest, cannot be obtained when both hydrodynamic and some other specific interactions are taken

Table 4.5 *Some important dimensionless parameters*

Parameter	*Symbol*	*Definition*
Double layer thickness	τ	κa
Adhesion number	Ad	$A/6kT$
Double layer number	Dl	$4\pi\varepsilon\zeta_1\zeta_2 a/kT$
Gravity number	Gr	$4\pi(\rho_p - \rho_f)ga^4/3kT$
Electric field number	El	$E_0 Q_e a/kT$
External forcefield number	Ex	Gr or El

into account. To accommodate these interactions, numerical solutions have to be employed.

Smoluchowski–Levich approximation

In this approximation, it is assumed that (a) the particle-wall hydrodynamic interaction is counterbalanced by van der Waals attractions between particles and the wall, and all other colloidal and external forces are absent or simply neglected, (b) particles move with the undisturbed fluid velocity, and (c) interception plays no role in deposition. With these assumptions and under steady state, the C-D equation reduces to:

$$D_0 \nabla^2 n = \mathbf{v} \bullet \nabla n \tag{4.84}$$

This equation has been solved analytically for some well-defined geometries as listed in Table 4.6. The results are usually given in terms of the *Sherwood number*, *Sh*, defined as:

$$Sh = j_\perp a / D_0 n_\infty \tag{4.85}$$

where $j_\perp$ denotes the normal component of the particle flux, and n_∞ is particle number density in the bulk dispersion. In general, $j_\perp$ depends on the position on the collector surface, so *Sh* only reflects the local particle flux. An overall deposition rate may be obtained by averaging *Sh* over the entire collector surface (if it has a finite dimension) or over a representative area (if it is infinitely large):

$$\overline{Sh} = \frac{1}{S_c} \int_{S_c} Sh \, \mathrm{d}S \tag{4.86}$$

where S_c is the surface area of the collector.

It is notable that for all the geometries listed in Table 4.6 the Sherwood number is proportional to $Pe^{1/3}$, and that such dependency is valid for barrier-less depositions at $Pe < 10^{-2}$ (Dabros and Adamczyk, 1979).

From Table 4.6, it can be seen that the coefficient in the Sherwood number expressions for rotating discs or stagnation-point flow collectors is a constant. This means that these collectors are uniformly accessible to the particles. This property makes the rotating disc or the stagnation-point flow collector an ideal

Table 4.6 *Deposition rates from Smoluchowski–Levich approximation*

Collector	Sh	$\overline{\text{Sh}}$		*Source*
Rotating disc or stagnation-point flow collector	$0.616Pe^{1/3}$	$0.616Pe^{1/3}$	(4.87)	Levich (1962)
Cylinder in a uniform flow	$0.583f_1(\varkappa)Pe^{1/3}$	$0.460Pe^{1/3}$	(4.88)	Levich (1962)
Sphere in a uniform flow	$0.682f_2(\varkappa)Pe^{1/3}$	$0.435Pe^{1/3}$	(4.89)	Natanson (1957)
Plate in a uniform flow	$0.616f_3(\varkappa)Pe^{1/3}$	$1.232Pe^{1/3}$	(4.90)	Levich (1962)
Parallel-plate channel or cylinder	$0.678f_4(\varkappa)Pe^{1/3}$	$\dfrac{1.017Pe^{1/3}}{\overline{L}^{1/3}}$	(4.91)	Adamczyk and van de Ven (1981a, b)
Moving wire or plate	$0.932f_5(\varkappa)Pe^{1/3}$	$1.964Sc^{1/6}Pe^{1/3}$	(4.92)	Adamczyk *et al.* (1982)

$f_1 = \sin^{1/2} z/\left(\int_0^z \sin^{1/2} z\, dz\right)^{1/3}$; $f_2 = \sin z/(z - \sin 2z/2)^{1/3}$;
$f_3 = (x/L)^{-1/2}$; $f_5 = (x/L)^{-1/6} Sc^{1/2}$;
$f_4 = (x/b)^{-1/3}$ for channel; $f_4 = (x/R)^{-1/3}$ for tube

arrangement for experiments, and consequently these two geometries are more extensively studied than others (see Chapters 5 and 10). The simple $Pe^{1/3}$ relation is only valid for small values of Pe. At larger Pe values, the Sherwood number for a rotating disc is given by (Dabros and Czarnecki, 1975):

$$Sh = \overline{Sh} = \frac{0.616Pe^{1/3}\exp(-\beta^3)}{1 - 1.12[\beta - (\beta^4/4) + (\beta^7/14) - (\beta^{10}/60) + \ldots]} \tag{4.93}$$

where $\beta = 0.550Pe^{1/3}$. In the presence of an external force field and/or particle interception, the Sherwood number becomes:

$$Sh = \overline{Sh} = \frac{\exp\left[Ex\hbar_{pm} - \dfrac{Pe}{6}(\hbar_{pm} + 1)^3\right]}{\displaystyle\int_{\hbar_{pm}}^{\infty} \exp\left[Ex\hbar - \dfrac{Pe}{6}(\hbar + 1)^3\right] d\hbar} \tag{4.94}$$

where $\hbar_{pm} = h_{pm}/a$ is the dimensionless primary-minimum distance, and Ex represents the external uniform force field such as Gr and El. This expression has been shown to be valid for values of Pe and Ex up to 10 (Dabros and Adamczyk, 1979).

Limiting solutions for zero Peclet number

In the limiting case where convection effects can be neglected (i.e. $Pe \ll 1$) and particles are subjected only to potential forces (i.e. derivable from the total potential energy V_T), analytical solutions to the steady-state C-D equations can be obtained for some well-defined collectors of finite dimensions.

For example, given a spherical collector of radius a_c immersed in a quiescent suspension, the steady-state C-D equation may be written as:

$$\frac{d}{dr}\left[r^2 \lambda \left(\frac{dn}{dr} + n\frac{dV_T}{dr}\right)\right] = 0 \qquad (4.95)$$

where r is the dimensionless distance (scaled by a_c), and λ is the hydrodynamic correction function. A solution to this equation has been given by van de Ven and Mason (1977), which may be expressed as:

$$Sh = \frac{\frac{1}{2}\left(\frac{a}{a_c}\right)}{\displaystyle\int_{r_{pm}}^{\infty} \frac{\exp(V_T/kT)}{\lambda r^2}\, dr} \qquad (4.96)$$

For other finite-sized collectors, similar expressions for Sh are expected, but Sh will obviously depend on the precise geometries of these collectors. No steady deposition rate exists for infinite collectors, since it would take an infinitely long time to develop a steady state.

If deposition is caused by the external force alone (i.e. both Dl and Pe are much smaller than Ex), the Sherwood number can be expressed as:

$$\overline{Sh} = Ex \qquad (4.97)$$

for rotating disc, plate in uniform flow, parallel-plate channel and moving plate, or as:

$$\overline{Sh} = Ex/\pi \qquad (4.98)$$

for cylinder in uniform flow, circular tube and moving wire (Adamczyk and van de Ven, 1981a,b, 1982; Prieve and Ruckenstein, 1974; Adamczyk *et al.*, 1982).

Surface-force boundary-layer approximation

The Smoluchowski–Levich approximation does not consider explicitly any of the interactions that may be involved in a deposition process, except the effect of diffusion. Thus, it is only applicable to barrier less depositions. When there is a significant energy barrier at the fluid–collector interface, the problem may be tackled using a method called the *surface–force boundary-layer* (SFBL) approximation, also known as the *interaction–force boundary-layer* (IFBL) approximation. The principal assumption underlying this method is that the surface-force boundary layer (where the energy barrier is present and the actual deposition takes place) is much thinner than the diffusion boundary layer (within which particles are transported from the bulk suspension to the vicinity of the collector surface). With this assumption, effects of the energy barrier can be localized and treated separately from the transport process in the diffusion boundary layer.

Within the surface-force boundary layer, convection can be neglected and the only transport mechanism is diffusion in a potential force field. This problem corresponds to the limiting case of zero Peclet number, where eqn (4.95) applies. For the diffusion boundary layer, the usual C-D equation such as eqn (4.84) can still be used. The boundary conditions, however, should now be specified so as to match the solution for the surface-force boundary layer.

Applying this scheme to the rotating disc, Ruckenstein and Prieve (1973) obtained:

$$Sh = \overline{Sh} = Sh_0 \left(\frac{\mathbb{K}}{\mathbb{K} + 1} \right) \tag{4.99}$$

$$\mathbb{K}^{-1} = \int_0^{\delta_F} \frac{\exp(V_{max}/kT)}{\lambda} \, dh \tag{4.100}$$

where Sh_0 is the corresponding Smoluchowski–Levich Sherwood number given in Table 4.6, and δ_F denotes the thickness of the surface force boundary layer.

Since its inception by Ruckenstein and Prieve (1973) and Spielman and Friedlander (1974), the scheme has been applied to other collectors (Kim and Rajagopalan, 1982; Chari and Rajagopalan, 1985). Improvements have also been attempted to account for the presence of the secondary-minimum, external forces and radial loss of particles (Prieve and Lin, 1980), and for omission of interactions in the outer region (Sjollema and Busscher, 1989). The resulting expressions for other collectors are more complicated, and will not be quoted here. However, the application of the SFBL (or IFBL) approximation is considered in more detail in Chapter 5 for deposition onto spherical collectors.

Additivity of deposition rates

In the presence of an energy barrier or when Dl and Ex are comparable in magnitude with Pe, no exact or approximate analytical solutions to the C-D equations are available. In these cases, an analytical approach is sometimes attempted by assuming that the deposition rates arising from different factors are additive (van de Ven, 1989):

$$Sh = Sh_0 + CPe^{1/3} \tag{4.101}$$

where Sh_0 represents the limiting solutions at zero Peclet number (i.e. eqns (4.96) to (4.98)), and C corresponds to the numerical coefficients given in Table 4.6. It has been suggested that the additivity rule yields results which agree reasonably well with those obtained from more rigorous (and arduous) considerations of coupling effects. However, since there is no real theoretical foundation for eqn (4.101), it is hardly surprising that in some cases it can be wrong by more than an order of magnitude (Adamczyk and van de Ven, 1982). Furthermore, the $Pe^{1/3}$ dependence does not apply to free-moving collectors (van de Ven, 1989).

4.6.4 Non-penetration model

The major difference between the perfect-sink and non-penetration models is that, whilst the former assumes a zero number density, the latter assumes a zero flux, of the particles at the collector surface(s). The immobilization-reaction approximation, developed by Adamczyk and van de Ven (1984), will be used in this section to illustrate the capability of the non-penetration model.

Deposition kinetics

In the immobilization-reaction approximation, the source term in the C-D equation is used to account for all the specific interactions, which are at present not explicitly known, and the presence of the immobilized phase formed by adsorbed particles. The specific interactions, generally attributed to surface heterogeneities, are responsible both for holding the adsorbed particles on the collectors' surface and for possible detachment of particles after they have been adsorbed. Accordingly, the deposition rate is determined by the kinetics of particle accumulation, immobilization and removal, i.e.

$$\frac{\partial n}{\partial t} + \nabla \cdot \mathbf{j} = -\mathbb{Q}(n, n_s, \mathbf{r}, t) \tag{4.102}$$

$$\frac{\partial n_s}{\partial t} = \mathbb{Q}(n, n_s, \mathbf{r}, t) \tag{4.103}$$

$$\mathbb{Q} = \mathbb{Q}_s(n, n_s, \mathbf{r}, t) - \mathbb{Q}_r(n, n_s, \mathbf{r}, t) \tag{4.104}$$

where n is the number density of the immobilized phase, $\mathbb{Q}_s$ the immobilization term, and $\mathbb{Q}_r$ the particle detachment term.

Since particle accumulation occurs around energy minima, which usually only extend over a fraction of particle radius, and since collectors considered in theoretical models are geometrically well-defined and symmetrical at least about one axis, the analysis can be reduced into a two-dimensional problem and the surface concentration can be used instead of the usual three-dimensional number density. With these considerations, the above equations can be greatly simplified by an integration step over the energy well (i.e. the accumulation region). The resulting equations represent a general description of the deposition process, allowing for both adsorption and desorption to occur, not only in the primary minimum but also in the secondary minimum. Under certain conditions and with some additional assumptions about the form of the particle immobilization and detachment terms, it is possible to obtain an analytical solution to the simplified equations. This can be illustrated by the following example.

Applying the model to rotating disc geometry for the case of an isolated deep energy minimum, the surface concentration of the immobilized phase, N, is given by (Adamczyk and van de Ven, 1984):

$$\mathcal{N} = \mathcal{N}_\infty \left[1 - \exp\left(-t/\tau_D\right)\right] + \mathcal{N}_0 \exp\left(-t/\tau_D\right) \quad (4.105)$$

$$\mathcal{N}_s = k_s \mathcal{N}_\infty \left\{t + \tau_D \left[\exp\left(-t/\tau_D\right) - 1\right] + k_s \mathcal{N}_0 \tau_D \left[1 - \exp\left(-t/\tau_D\right)\right]\right\} \quad (4.106)$$

where $\mathcal{N}_0$ is the initial surface concentration, $\mathcal{N}_\infty = Sh_0/(cPe + k_s)$ is the steady-state surface concentration of adsorbed particles, c a constant of order unity, k_s the immobilization rate constant, $\tau_D = 1/(cPe + k_s)$, and Sh_0 the Sherwood number obtained from the perfect-sink boundary condition.

If $Pe < 10^{-2}$ and no external forces exist, Sh_0 is then given by the Smoluchowski–Levich solution in Table 4.6. It follows from eqn (4.105) that the maximum possible monolayer coverage (or the surface concentration at dynamic equilibrium) on the rotating disc is given by (van de Ven, 1989):

$$\mathcal{N}_c = C_d n_\infty a^{1/3} \omega^{-1} \quad (4.107)$$

$$C_d = \frac{2.43}{c} \left(\frac{\nu^{1/2} kT}{6\pi\mu}\right)^{2/3}$$

which, for a fixed value of n_∞, is directly proportional to $a^{1/3}$ but inversely proportional to ω (the angular velocity of the rotating disc).

If there is a relatively strong external force so that $Sh = Ex$, then the monolayer coverage takes the form (van de Ven, 1989):

$$\mathcal{N}_c = n_\infty a^3 Ex/cPe \quad (4.108)$$

From the above results, it can be seen that particle accumulation in the energy well increases linearly with time during the initial stages of deposition and eventually attains a stationary value independent of the initial coverage. The number of immobilized particles increases essentially linearly with time after a short transition period T_d (usually of the order of a few seconds). An increase in the rotating speed leads to a decrease in the particle accumulation in the energy well. Both the number of accumulation and that of immobilization ($\mathcal{N}$ and $\mathcal{N}_s$) are virtually independent of the depth and extension of the energy well, except, that of course, it must be rather deep. For all Sherwood numbers, the desorption kinetics (derived by letting $\mathcal{N}_0 \gg \mathcal{N}_\infty$ and $k_s = 0$) is characterized by an exponential decay with a transition time given $T_d = cPea^2/D_0$ which is, again, independent of the depth of the energy well.

Particle accumulation in energy wells

The introduction of the surface concentration not only simplifies the procedure for calculating the deposition rate but also facilitates the explicit formulation of particle distribution in the energy minima. For example, assuming only the primary minimum is present, it has been shown (van de Ven, 1989) that the distribution of the mobile particles in the energy well only slightly deviates from the Boltzmann distribution:

$$\mathscr{n}_{pm} = \left(\mathscr{N}_m \mathbb{I}_{pm}^{-1} + \mathbb{J}_{pm} Sh\right) \exp(-V_{pm}/kT) \tag{4.109}$$

$$\mathbb{I}_{pm} = \int_{\ell} \exp(-V_T/kT)\, d\mathscr{y}$$

$$\mathbb{J}_{pm} = \int_{\mathscr{y}_1}^{\mathscr{y}_2} \Gamma(\mathscr{y}) \exp(V_T/kT)/F(\mathscr{y})\, d\mathscr{y}$$

$$\Gamma(\mathscr{y}) = \mathbb{I}_{pm}^{-1} \int_{\mathscr{y}_1}^{\mathscr{y}_2} \exp(-V_T)\, d\mathscr{y}$$

The first term represents the quasi-Boltzmann contribution $\mathscr{n}_B$, the second is small a perturbation term $\mathscr{n}^* \bullet \mathscr{n}^*/\mathscr{n}_B$ is usually about $10^{-4} \sim 10^{-6}$ for $\mathscr{N} \approx 1$.

The model has also been applied to the case where both energy minima are present. By comparing with numerical solutions, the above equations have been found to be valid if $(Pe + Ex) < 1$ and the energy wells are deeper than $5kT$ (Adamczyk *et al.*, 1984). Similar kinetic equations have also been formulated for other collectors but are much more complicated. However, all collectors have the following features in common:

(1) the number of particles accumulated dynamically in the primary minimum region $\mathscr{N}$ is of the order of $Sh(\mu_T + k_s)$, μ_T being the dimensionless tangential component of the velocity at the primary minimum, and which can be assumed to be constant since the range of the primary-minimum region is very small
(2) the steady-state value of $\mathscr{N}$ attains a stationary value after a transition time of the order of $1/(\mu_T + k_s)$
(3) if the immobilization process is very fast, i.e. $k_s \gg \mu_T$, then $\mathscr{N}_s$ is a linear function of time and thus the perfect-sink approximation is recovered.

4.7 Bibliography

Edwards, D. A., Brenner, H. and Wasan, D. T. (1991) *Interfacial Transport Processes and Rheology*, Butterworth-Heinemann, Newton, MA, USA

Happel, J. and Brenner, H. (1986) *Low Reynolds Number Hydrodynamics*, Martinus Nijhoff, Dordrecht

Levich, V. G. (1962) *Physicochemical Hydrodynamics*, Prentice-Hall, Englewood Cliffs, New Jersey

Probstein, R. F. (1989) *Physicochemical Hydrodynamics: An Introduction*, Butterworth-Heinemann, Newton, MA, USA

Rosner, D. E. (1990) *Transport Processes in Chemically Reacting Flow Systems*, 3rd edn, Butterworth-Heinemann, Newton, MA, USA

Russel, W. B., Saville, D. A. and Schowalter, W. R. (1989) *Colloidal Dispersions*, Cambridge University Press Cambridge

Tien, C. (1989) *Granular Filtration of Aerosols and Hydrosols*, Butterworth-Heinemann, Oxford

van de Ven, T. G. M. (1989) *Colloidal Hydrodynamics*, Academic Press, London

4.8 References

Adamczyk, Z. (1989) Particle deposition from flowing suspensions. *Colloid Surface,* **39**, 1–37

Adamczyk, Z. and van de Ven, T. G. M. (1981a) Deposition of Brownian particles onto cylindrical collectors. *J. Colloid Interface Sci.*, **84**, 497–518

Adamczyk, Z. and van de Ven, T. G. M. (1981b) Deposition of particles under external forces in laminar flow through parallel-plate and cylindrical channels. *J. Colloid Interface Sci.*, **80**, 340–356

Adamczyk, Z. and van de Ven, T. G. M. (1982) Particle transfer to a plate in uniform flow. *Chem. Eng. Sci.*, **37**, 869–880

Adamczyk, Z. and van de Ven, T. G. M. (1984) Kinetics of particle accumulation at collector surfaces I. Approximate analytical solutions, II. *J. Colloid Interface Sci.*, **97**, 68–90

Adamczyk, Z., Dabros, T. and van de Ven, T. G. M. (1982) Transfer of Brownian particles to continuous moving surfaces. *Chem. Eng. Sci.*, **37**, 1513–1522

Adamczyk, Z., Dabros, T., Czarnecki, J. and van de Ven, T. G. M. (1983) Particle transfer to solid surfaces. *Adv. Colloid Interface Sci.*, **19**, 183–252

Adamczyk, A., Dabos, T., Czarnecki, J. and van de Ven, T. G. M. (1984) Kinetics of particle accumulation at collector surfaces, II Exact numerical solutions. *J. Colloid Interface Sci.*, **97**, 91–104

Adler, P. M. (1981) Interaction of unequal spheres I. Hydrodynamic interaction: colloidal forces II. Conducting spheres III: Experimental. *J. Colloid Interface Sci.*, **84**, 461–474, 475–488, 489–496

Allison, S. A., Chang, E. L. and Schurr, J. M. (1979) The effects of direct and hydrodynamic forces on macromolecular diffusion. *Chem. Phys.*, **38**, 29

Arp, P. A. and Mason, S. G (1977) The kinetic of flowing dispersions. VIII. Doublets of rigid spheres: Theoretical. XI: Experimental. *J. Colloid Interface Sci.*, **61**, 21–43, 44–61

Bart, E. (1968) The slow unsteady settling of a fluid sphere towards a flat fluid interface. *Chem. Eng. Sci.*, **23**, 193–210

Batchelor, G. K. (1976) Brownian diffusion of particles with hydrodynamic interaction. *J. Fluid Mech.*, **74**, 1–29

Batchelor, G. K. (1977) The effect of Brownian motion on the bulk stress in a suspension of spherical particles. *J. Fluid Mech.*, **83**, 97–117

Batchelor, G. K. and Green, J. T. (1972) The hydrodynamic interaction of two small freely-moving spheres in a linear flow field. *J. Fluid Mech.*, **56**, 375–400

Beenakker, C. W. J., van Saarloos,W. and Mazur, P. (1984) Many-sphere hydrodynamic interactions III. The influence of a plane wall. *Physica*, **127A**, 451–472

Booth, F. (1954) Sedimentation potential and velocity of solid spherical particles. *J. Chem. Phys.*, **22**, 1956–1968

Brenner, H. (1961) The slow motion of a sphere through a viscous fluid towards a plane surface. *Chem. Eng. Sci.*, **16**, 242–251

Brenner, H. and O'Neill, M. E. (1972) On the Stokes resistance of multi-particle systems in a linear shear field. *Chem. Eng. Sci.* **27**, 1421–1439

Carter, J. M. and Phillies, G. D. J. (1985) Second-order concentration correction to the mutual diffusion coefficient of a suspension of hard Brownian spheres. *J. Phys. Chem.*, **89**, 5118–5124

Chari, K. and Rajagopalan, R. (1985) Transport of colloidal particles over energy barriers *J. Colloid Interface Sci.*, **107**, 278–282

Cox, R. G. and Brenner, H. (1967) The slow motion of a sphere through viscous fluid towards a plane surface — II. Small gap widths, including inertial effects. *Chem. Eng. Sci.*, **22**, 1753–1777

Cox, R. G.; Zia, I. Y. Z. and Mason, S. G. (1968) Particle motion in sheared suspensions XXV. Streamlines around cylinders and spheres. *J. Colloid Interface Sci.*, **27**, 7

Dabros, T. and Adamczyk, Z. (1979) Non-inertial particle transfer to the rotating disc under an

external force field (laminar flow). *Chem. Eng. Sci.*, **34**, 1041–1049
Dabros, T. and Czarnecki, J. (1975) Transport of particles with finite dimensions to a rotating disc surface. *J. Colloid Interface Sci.*, **53**, 335–336
Dabros, T. and van de Ven, T. G. M. (1983) A direct method for studying particle deposition onto solid surfaces. *Colloid Polymer Sci.*, **261**, 694–707
Dabros, T. and van de Ven, T. G. M. (1987) Deposition of latex particles on glass surfaces in an impinging jet. *Physicochem. Hydrodynam.*, **8**, 161–172
Davies, C. N. (1973) *Air Filtration*, Academic Press, Orlando, FL, USA
Einstein, A. (1906) On the theory of Brownian movement. *Ann. d. Phys.*, **19**, 371–381
Faxen, H. (1922) The resistance to the motion of a solid sphere in a viscous liquid enclosed between parallel walls. *Ann. Physik*, **68**, 89
Faxen, H. and Dahl, H. (1925) Mutual action of two spheres falling in a viscous liquid. *Arkiv Mat. Astron. Fys.*, **19A**, No. 13
Felderhof, B. U. (1977) Hydrodynamic interaction between two spheres. *Physica*, **A89**, 373–384
Felderhof, B. U. (1978) Diffusion of interacting Brownian particles. *J. Phys.*, **A11**, 929
Friedlander, S. K. (1977) *Smoke, Dust and Haze*, Wiley, New York
Goldman, A. J., Cox, R. G. and Brenner, H. (1967) Slow viscous motion of a sphere parallel to a plane wall I. Motion through a quiescent fluid. II. Couette flow. *Chem. Eng. Sci.*, **22**, 637–651, 653–660
Goren, S. L. and O'Neil, M. E. (1971) On the hydrodynamic resistance to a particle of a dilute suspension when in the neighbourhood of a large obstacle. *Chem. Eng. Sci.*, **26**, 325–338
Hamberman, W. and Sayre, R. M. (1958) *Motion of Rigid and Fluid Spheres in Stationary and Moving Liquids inside Cylindrical Tubes*. Report 1143, David W. Taylor, Model Basin, US Navy Department, Washington DC
Happel, J. and Brenner, H. (1975) *Low Reynolds Number Hydrodynamics*, Noordhoff, Leyden, Netherlands
Hidy, G. M. and Brock, J. R. (1970) *The Dynamics of Aerocolloidal Systems, Vol. 1*, Pergamon Press, London
Honig, E. O., Roebersen, G. J. and Wiersema, P. H. (1971) Effect of hydrodynamic interaction on the coagulation rate of hydrophobic colloids. *J. Colloid Interface Sci.*, **36**, 97–109
Jeffrey, D. J. and Onishi, Y. (1984) Calculations of the resistance and mobility functions for two unequal rigid spheres in low-Reynolds number flow. *J. Fluid Mech.*, **139**, 261–290
Kao, S. V., Cox, R. G. and Mason, S. G. (1977) Streamlines around single spheres and trajectories of a pair of spheres in two-dimensional creeping flow. *Chem. Eng. Sci.*, **32**, 1505
Kim, J. S. and Rajagopalan, R. (1982) A comprehensive equation for the rate of adsorption of colloidal particles and for stability ratios. *Colloids Surfaces*, **4**, 17–31
Krasney-Ergen, W. (1936) Viscosity of suspensions. Part II. *Kolloid Z.*, **74**, 172–178
Levich, V. G. (1962) *Physicochemical Hydrodynamics*, Prentice-Hall, Englewood Cliffs, New Jersey
Lorentz, H. A. (1907) A general theorem concerning the motion of a viscous fluid and a few consequences derived from it. *Alhandl. Theoret. Phys.*, **1**, 23
Masliyah, J. H. and van de Ven, T. G. M. (1987) Concentration dependence of spin friction coefficient in suspensions, parallel cylinder and spheres. *J. Chem. Soc. Faraday Trans I*, **87**, 547–557
Mazur, P. and van Saarloos, W. (1982) Many-sphere hydrodynamic interactions and mobilities in a suspension. *Physica*, **115A**, 21–57
Muthukumar, M. and Freed, K. F. (1983) Cluster expansion for concentration dependence of self-friction coefficients for suspensions of interacting spheres. *J. Chem. Phys.*, **78**, 497–510
Natanson, G. L. (1957) Deposition of aerosol particles from a gas stream flowing around a cylinder. *Dokl. Akad. Nauk. USSR*, **112**, 696
Ohshima, H., Healy, T. W., White, L. R. and O'Brien, R. W. (1984) Sedimentation velocity and potential in a dilute suspension of charged spherical colloidal particles. *J. Chem. Soc. Faraday Trans II*, **80**, 1299–1317

Oseen, C. W. (1910) Über die Stokes'sche Formel und über die verwandte Aufgabe in der Hydrodynamik. *Arkiv. Mat. Astron. Fys.*, **6**, 75

Peters, M. H. (1990) Adsorption of interacting Brownian particles onto surfaces II. Results for hydrodynamic interactions. *J. Colloid Interface Sci.*, **138**, 451–464

Prieve, D. C. and Lin, M. M. J. (1980) Adsorption of Brownian hydrosols onto a rotating disc aided by a uniform applied force. *J. Colloid Interface Sci.*, **76**, 32–47

Prieve, D. C. and Ruckenstein, E. (1974) Effect of London forces upon the rate of deposition of Brownian particles. *AIChE J.*, **20**, 1178–1187

Proudman, I. and Pearson, J. R. A. (1957) Expansions at small Reynolds numbers for the flow past a sphere and a circular cylinder. *J. Fluid Mech.*, **2**, 237–262

Rotne, J, and Prager, S (1969) Variational treatment of hydrodynamic interaction in polymers. *J. Chem. Phys.*, **50**, 4831–4837

Ruckenstein, E. and Prieve, D. C. (1973) Rate of deposition of Brownian particles under the action of London and double-layer forces *J. Chem. Soc. Faraday Trans II*, **69**, 1522–1536

Russel, W. B. (1980) A review of the role of colloidal forces in the rheology of suspensions. *J. Rheol.*, **24**, 287

Russel, W. B., Saville, D. A. and Schowalter, W. R. (1989) *Colloidal Dispersions*, Cambridge University Press, Cambridge

Schlichting, H. (1955) *Boundary Layer Theory*, McGraw-Hill Book Co. Inc., New York

Schmitz, R. and Felderhof, B. U. (1983) Mobility matrix for two spherical particles with hydrodynamic interaction *Physica.*, **A116**, 163–177

Sjollema, J. and Busscher, H. J. (1989) Deposition of polystyrene latex particles towards polymethylmethacrylate in a parallel plate flow cell. *J. Colloid Interface Sci.*, **132**, 382–394

Smoluchowski, M. (1921) *Graetz Handbuch der Elektrizitat und das Magnetismus*, Vol. II, Leipzig

Spielman, L. A. (1977) Particle capture from low speed flows. *Annu. Rev. Fluid Mech.*, **9**, 297–319

Spielman, L. A. and Friedlander, S. K. (1974) Role of the electrical double layer in particle deposition by convective diffusion. *J. Colloid Interface Sci.*, **46**, 22–31

Stimson, M. and Jeffery, G. B. (1926) The motion of two spheres in a viscous fluid. *Proc. R. Soc. A.*, **111**, 110–116

Tien, C and Payatakes, J. P. (1979) Advances in deep bed filtration. *AIChE J.*, **25**, 737–759

van den Broeck, C.; Lostak, F. and Lekkerkerker, H. N. W. (1981) The effect of direct interactions on Brownian diffusion. *J. Chem. Phys.*, **74**, 2006–2010

van de Ven, T. G. M. (1989) *Colloidal Hydrodynamics*, Academic Press, London

van de Ven, T. G. M. and Mason, S. G. (1976) The microrheology of colloidal dispersions IV: Pairs of interacting spheres in shear flow. V: Primary and secondary doublets of spheres in shear flow. *J. Colloid Interface Sci.*, **57**, 505–516, 517–534

van de Ven, T. G. M. and Mason, S. G. (1977) The microrheology of colloidal dispersions VIII. The effect of shear on perikinetic doublet formation. *Colloid Polymer Sci.*, **255**, 794–804

van de Ven, T. G. M.; Dabros, T. and Czarnecki, J. (1983) Flexible bonds between latex particles and solid surfaces. *J. Colloid Interface Sci.*, **93**, 580–581

Yoshida, N (1983) Concentration dependence of the self-diffusion coefficient of hard spheres in solution. *Chem. Phys. Lett.*, **101**, 555–561

Part II

Modelling and simulation

Chapter 5

Modelling of particle deposition onto ideal collectors

Nomenclature

A	Hamaker constant
$\hat{A}$	constant used in eqn (5.58)
A^*	normalized particle size; $A^* = a_p / B$
A_1	constant used in eqn (5.118)
A_1^+	dimensionless coefficient used in eqn (5.127) (defined in Table 5.2)
A_s	porosity-dependent parameter of Happel's model
a_c	radius of a spherical collector
a_i	variable coefficients used in eqns (5.21) and (5.85); $i=1, 2, 3$
a_p	particle radius
B	midway depth of a parallel-plate channel
B_1	coefficient used in eqn (5.119)
B_2	coefficient used in eqn (5.119)
B_1^+	dimensionless coefficient used in eqn (5.128) (defined in Table 5.2)
B_2^+	dimensionless coefficient used in eqn (5.128) (defined in Table 5.2)
b	radius of the fluid envelope in Happel's model; $a_c/b = (1-f)^{1/3}$
b_i	variable coefficients used in eqn (5.43); $i=1, 2, 3$
C	particle concentration
C_0	bulk particle concentration
C^*	dimensionless particle concentration; $C^* = C/C_0$
$\mathbf{D}$	particle diffusion coefficient tensor
D_∞	diffusion coefficient in an infinite medium; $D_\infty = kT/(6\pi\mu a_p)$
D_i	component of the diffusion coefficient in direction i; $i = x, y, z, r$, or θ
d_c	collector diameter
d_p	particle diameter
$\mathbf{F}$	force vector
F^*	dimensionless normal force
F^D	fluid drag force
F^{EDL}	electrical double layer force
F^{VDW}	van der Waals' force
F_G	gravitational force
F_G^*	dimensionless gravitational force
F_i	force acting on a spherical particle
F_j	component of external force in direction j; $j = x, y, z, r$, or θ
f	porosity of a porous medium
$f_i(H)$	universal functions to account for hydrodynamic interaction; $i = 1, 2, 3$, or 4

g	gravitational acceleration, 9.81 m/s^2
$g_1(H)$	universal hydrodynamic function, $g_1(H) = 1/f_1(H)$
H	dimensionless separation distance; $H = h/a_p$ or y/a_p for a spherical collector
H_u	dimensionless distance of closest approach measured from the surface of the upper plate in a parallel-plate channel
h	surface to surface separation distance
I	overall rate of particle deposition
$\mathbf{J}$	particle flux vector
J	local flux of particles
J^*	dimensionless local flux of particles; $J^* = J/UC_0$
J_z	component of particle flux in direction z
K_i	porosity-dependent functions used in Eqn 5.58; $i = 1, 2, 3,$ or 4
K_F	pseudo first order rate constant given Eqn 5.102
k	Boltzmann constant, 1.3805×10^{-23} J K^{-1}
L	vertical distance of the impinging jet from the stagnation point (Figure 5.3)
m	mass of particle
N_{Pe}	particle Peclet number; $N_{Pe} = 2Ua_p/D_\infty$
N_R	dimensionless ratio used in eqn (5.84); $N_R = a_p/r$
Pe	particle Peclet number given by eqns (5.19), (5.30) or (5.40)
p	ratio given by a_c/b
Q	volumetric flow
R	aperture radius of the impinging jet (Figure 5.3)
Re	Reynolds number
r	radial coordinate
r^*	dimensionless radial coordinate; $r^* = r/a_c$
r^+	dimensionless radial coordinate; $r^+ = r/a_p$
$S(\beta)$	function used in eqn (5.108)
Sh	Sherwood number
Sh_L	Sh to the lower plate in a parallel-plate channel
Sh_u	Sh to the upper plate in a parallel-plate channel
s_i	drag correction terms defined by eqns (5.124)–(5.126), $i = 1, 2$ or 3
T	absolute temperature
T_i	torque acting on a spherical particle
t	time
U	approach (superficial) velocity
$\mathbf{u}$	particle velocity vector
u_θ	tangential velocity component of particles
u_r	radial velocity component of particles
u_θ^*	dimensionless tangential velocity described by eqn (5.128)
u_r^*	dimensionless radial velocity described by eqn (5.127)
u_x	velocity component of particles in the x direction
u_y	velocity component of particles in the y direction
u_z	axial velocity component of particles
V_θ	dimensionless tangential velocity; $V_\theta = v_\theta/U$
V_m	mean fluid velocity in a parallel-plate channel

V_r dimensionless radial velocity; $V_r = v_r/U$
V_∞ velocity through an impinging jet defined by eqn (5.33)
v_θ tangential velocity component of fluid
v_r radial velocity component of fluid
v_z axial velocity component of fluid
w porosity-dependent parameter defined by eqn (5.65)
x horizontal coordinate
x^* dimensionless horizontal coordinate; $x^* = x/B$
y coordinate perpendicular to the surface of a spherical collector; $y = r - a_c - a_p$
y' distance from the collector surface to the centre of the spherical particle; $y' = r - a_c$
y^+ dimensionless distance; $y^+ = y'/a_p$
z axial coordinate

α collision (attachment) efficiency
α_s constant used in eqn (5.27)
α_s^* parameter defined by eqn (5.31)
β dimensionless parameter defined by eqn (5.109)
δ_D thickness of diffusion boundary layer
δ_F thickness of interaction force boundary layer
η dimensionless colloid removal rate (single collector efficiency)
η_0 η calculated in the absence of double layer interaction
θ tangential coordinate
θ_s value of θ where the limiting trajectory intersects the outer edge of the Happel shell at $r = b$
κ inverse Debye length
λ characteristic wavelength of van der Waals interaction
μ absolute viscosity of fluid
ρ density
ϕ_T total interaction energy
Φ dimensionless interaction energy; $\Phi = \phi/kT$
Ψ stream function
Ψ_L the value of the stream function (based on Happel's model) where the limiting trajectory intersects the fluid shell at $r = b$ and $\theta = \theta_s$
ω angular velocity of a rotating disc
ω_p rotation of a particle

This chapter describes in detail the quantitative formulation of the particle transport equation in various well-defined deposition systems. Calculations of particle deposition rate from the particle transport equation are elaborated. The analysis in this chapter combines fundamental theories of colloidal interactions and hydrodynamics described previously in Chapters 3 and 4. Representative simulations for the effect of several chemical-colloidal variables on the rate of particle deposition are also presented.

5.1 Rotating disc system

The rotating disc is a popular tool for studying mass transfer and surface reactions in various engineering and applied science disciplines. The main reason for the broad use of this technique is the fact that the hydrodynamics of fluid motion near the disc surface are relatively simple and well known. A unique feature of this system is that the thicknesses of the hydrodynamic and diffusion boundary layers are constant over the entire surface of the rotating disc. In addition, the convective-diffusion equation near the surface of the disc is one-dimensional, so that analytical solutions for mass transfer are available. As a result, the rotating disc technique has been used extensively in numerous theoretical and experimental studies of particle deposition (e.g. Marshal and Kitchener, 1966; Dabros and Adamczyk, 1979; Prieve and Lin, 1980; Rajagopalan and Kim, 1981).

5.1.1 Hydrodynamics

The problem of laminar flow near the surface of a flat disc rotating about an axis perpendicular to its plane with a uniform angular velocity ω, in a fluid otherwise at rest, was solved by Von Kármán (1921) and Cochran (1934) more than half a century ago. The exact solution of the Navier–Stokes equation to obtain the flow field adjacent to the rotating disc surface is described in detail in the classic books of Schlichting (1960) and Levich (1962).

The fluid flow in the neighbourhood of the surface of a rotating disc is described in Figure 5.1. Far from the disc surface, the fluid moves in the axial

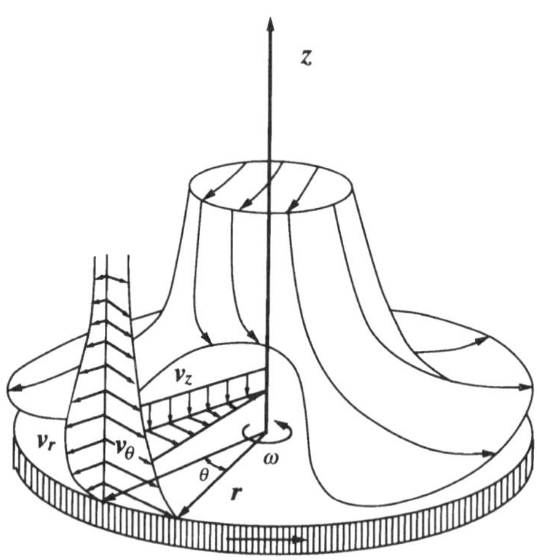

Figure 5.1 *A schematic description of the flow near a disc rotating in a fluid at rest (adapted from Schlichting, 1960). The fluid velocity components are: radial* = v_r*; tangential* = v_u*; and axial* = v_z

direction, z, towards the rotating disc. In a thin layer adjacent to the surface, the liquid acquires a rotating motion. The angular velocity of the fluid increases and eventually reaches that of the disc as the surface of the disc is approached. Furthermore, the fluid also acquires a radial velocity under the influence of the centrifugal force. Thus, within the momentum boundary layer, radial, axial and tangential components exist, whereas outside the momentum boundary layer, the radial and tangential velocity components vanish.

In order to obtain the velocity distribution of a viscous, incompressible fluid in the rotating disc system, the Navier–Stokes equation under steady conditions is solved using a cylindrical coordinate system (with r, z and θ as shown in Figure 5.1). The following assumptions are made: (1) the disc is assumed to be sufficiently large so that edge effects are negligible, and (2) changes in the tangent (circumferential) coordinate are negligible due to axial symmetry.

When dealing with problems of mass (or particle) transfer toward surfaces in liquids, the velocity distribution within the diffusion boundary layer is of interest. In water, the Schmidt number is always sufficiently large so that the momentum boundary layer is very thick compared to the diffusion boundary layer. Under these conditions, only the leading term in the Taylor series expansion about $z=0$ of each of the fluid components is required. With this procedure, the fluid velocity components in the radial (r), tangential (θ) and axial (z) coordinates can be described by (Cochran, 1934; Sparrow and Gregg, 1959; Prieve and Lin, 1980):

$$v_r(r, z) = 0.510(\omega^3\rho/\mu)^{1/2} rz \tag{5.1}$$

$$v_\theta(r) = r\omega \tag{5.2}$$

$$v_z(z) = -0.510(\omega^3\rho/\mu)^{1/2} z^2 \tag{5.3}$$

where ρ and μ are the density and the viscosity of the fluid, respectively.

In the vicinity of the disc surface, the velocity of particles induced by fluid motion is different from that of the fluid due to the effect of hydrodynamic interaction. Following Prieve and Lin (1980), the particle velocity components are related to the fluid velocity components by:

$$u_r(r, z) = f_3(H) v_r(r, z) \tag{5.4}$$

$$u_\theta(r, z) = f_3(H) v_\theta(r) \tag{5.5}$$

$$u_z(z) = f_1(H) f_2(H) v_z(z) \tag{5.6}$$

Here $f_1(H)$, $f_2(H)$ and $f_3(H)$ are hydrodynamic correction factors described previously in Chapter 4. These correction factors depend on the dimensionless particle–collector separation distance H.

5.1.2 Particle transport equation

The transport of particles in the vicinity of the rotating disc surface is described by the steady convective-diffusion equation (see Section 4.6.2):

$$\nabla\cdot(\mathbf{u}C) = \nabla\cdot(\mathbf{D}\cdot\nabla C) - \nabla\cdot\left(\frac{\mathbf{D}\cdot\mathbf{F}}{kT}C\right) \tag{5.7}$$

Here C is the concentration of the suspended particles, $\mathbf{u}$ is the particle velocity vector, $\mathbf{D}$ is the particle diffusion tensor, k is the Boltzmann constant, T is the absolute temperature, and $\mathbf{F}$ is the external force vector. The external forces that are usually considered in particle deposition in polar liquids are van der Waals, electrical double layer (usually referred to as colloidal interaction forces) and gravitational forces. In principle, other forces, (e.g. non-DLVO forces such as those described in Chapter 3) can be included if quantitative expressions for them are available.

For the rotating disc system, cylindrical coordinates (r, θ and z) are utilized as described in Figure 5.2. The various components of the convective-diffusion equation, in this coordinate system, can be described as:

$$\nabla \bullet (\mathbf{u}C) = u_z \frac{\partial C}{\partial z} + \left[f_3(H) \frac{\partial v_r}{\partial r} + \frac{f_3(H)}{r} v_r + f_1(H) f_2(H) \frac{\partial v_z}{\partial z} + f_1(H) v_z \frac{\partial f_2(H)}{\partial z} + f_2(H)\, v_z \frac{\partial f_1(H)}{\partial z} \right] C \tag{5.8}$$

$$\nabla \bullet (\mathbf{D} \bullet \nabla C) = \left[f_1(H) \frac{\partial^2 C}{\partial z^2} + \frac{\partial f_1(H)}{\partial z} \frac{\partial C}{\partial z} \right] D_\infty \tag{5.9}$$

$$\nabla \bullet (\mathbf{D} \bullet \mathbf{F}C) = \left[f_1(H)\, F_z \frac{\partial C}{\partial z} + f_1(H)\, C \frac{\partial F_z}{\partial z} + F_z C \frac{\partial f_1(H)}{\partial z} \right] D_\infty \tag{5.10}$$

Where F_z is the component of the external forces in the axial direction. Equations (5.8), (5.9) and (5.10) describe the spatial change of the convective, diffusive and drift (migration) fluxes of particles, respectively. In the derivation of these equations, the following assumptions were made: (1) the particle

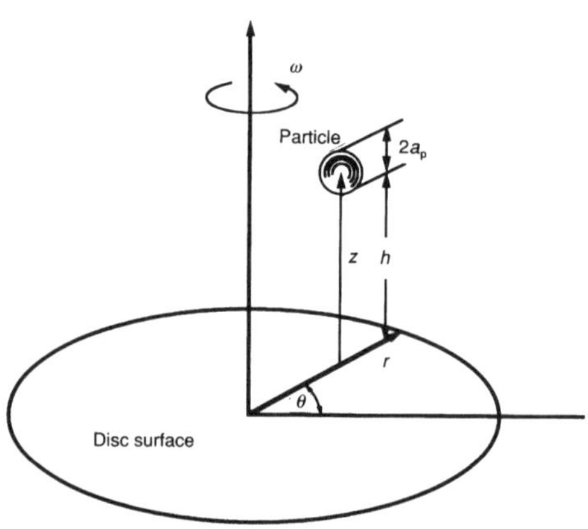

Figure 5.2 *A schematic description of a spherical particle of radius a_p approaching a rotating disc collector, illustrating the coordinate system used (adapted from Prieve and Lin, 1980)*

concentration is a function of the axial coordinate (z) alone, as suggested by Levich (1962) and Prieve and Lin (1980); (2) the diffusion coefficient in the axial coordinate is $D_z = f_1(H)D_\infty$, D_∞ being the bulk particle diffusion coefficient; and (3) the colloidal interaction forces act perpendicular to the disc surface.

In order to calculate particle deposition rates, the following boundary conditions are employed:

$$C = 0 \text{ at } z = a_p \tag{5.11}$$

and

$$C = C_0 \text{ at } z = \infty \tag{5.12}$$

The first boundary condition states that all particles arriving at the collector surface are no longer part of the dispersed phase (perfect-sink boundary condition). In the second boundary condition, the concentration of particles at large distance from the disc surface is assumed to be equal to the bulk concentration, C_0.

Introducing the following dimensionless variables:

$$C^* = C/C_0 \tag{5.13}$$

and

$$H = h/a_p \tag{5.14}$$

with

$$h = z - a_p \tag{5.15}$$

the convective-diffusion equation becomes:

$$f_1(H)\frac{d^2C^*}{dH^2} + \left[\frac{df_1(H)}{dH} + f_1(H)\,f_2(H)\,Pe(H+1)^2 - f_1(H)F^*\right]\frac{dC^*}{dH}$$
$$+ \Bigg[2f_1(H)\,f_2(H)\,Pe(H+1) + f_1(H)\,Pe(H+1)^2\frac{df_2(H)}{dH}$$
$$+ f_2(H)\,Pe(H+1)^2\frac{df_1(H)}{dH} - 2f_3(H)\,Pe(H+1)$$
$$- f_1(H)\frac{dF^*}{dH} - F^*\frac{df_1(H)}{dH}\Bigg]C^* = 0 \tag{5.16}$$

subject to:

$$C^* = 0 \text{ at } H = 0 \tag{5.17}$$

$$C^* = 1 \text{ at } H = \infty \tag{5.18}$$

Here Pe is a particle Peclet number defined as:

$$Pe = 0.510\left(\frac{\omega^3\rho}{\mu}\right)^{1/2}\frac{a_p^3}{D_\infty} \tag{5.19}$$

The net perpendicular force exerted on the particle (sum of colloidal and gravitational forces) in dimensionless form is denoted as F^*; it is defined as:

$$F^* = \frac{F_z a_p}{kT} \tag{5.20}$$

where k is the Boltzmann constant and T is the absolute temperature. In the definition of the Peclet number, the characteristic length is selected as the radius of the particles in suspension (a_p) and the characteristic velocity as the axial fluid velocity (v_z) evaluated at $z=a_p$.

For convenience in numerical calculations, eqn (5.16) can be rearranged as:

$$a_1(H)\frac{d^2C^*}{dH^2} + a_2(H)\frac{dC^*}{dH} + a_3(H)C^* = 0 \tag{5.21}$$

where $a_1(H)$, $a_2(H)$ and $a_3(H)$ are variable coefficients. This equation is a stiff two-point boundary value problem, and can be solved only by numerical means.

The dimensionless flux of particles at the disc surface is usually expressed in terms of the Sherwood number:

$$Sh = -\frac{a_p}{D_\infty C_0} J_z \tag{5.22}$$

The flux in the axial coordinate, J_z, is given by:

$$J_z = u_z C - D_z \frac{\partial C}{\partial z} + \frac{D_z C}{kT} F_z \tag{5.23}$$

Using the dimensionless variables defined previously, the dimensionless flux of particles at the disc surface can be expressed as:

$$Sh = \lim_{H\to 0}\left[f_1(H)\frac{dC^*}{dH} - f_1(H) f_2(H)(H+1)^2 PeC^* - f_1(H)F^*C^* \right] \tag{5.24}$$

C^* and dC^*/dH, in this equation, are obtained from a numerical solution of eqn (5.21) with the boundary conditions described by eqns (5.17) and (5.18).

It can be shown (Dabros and Adamczyk, 1979) that, for Brownian particles, Sh can be expressed as:

$$Sh = 0.776(Pe)^{1/3} \text{ as } Pe \to 0 \tag{5.25}$$

in the absence of colloidal and hydrodynamic interactions. This is identical to the analytical solution of Levich (1962) for mass transfer of molecular solutes towards the surface of a rotating disc. This analytical expression can be used to test the numerical solution of the particle transport equation.

5.2 Stagnation-point flow

The rotating disc technique has been used extensively in particle deposition studies due to the relatively simple and well-defined hydrodynamics and mass transfer. Mass transfer in stagnation-point flow has the same advantages; the thickness of the hydrodynamic and diffusion boundary layers remain constant in the vicinity of the stagnation point. In addition, in a stagnation-point flow system, it is possible to observe the deposition of particles directly by microscopic means when a transparent collector is used. This observation technique cannot be applied to the rotating disc system because of the motion of the disc. Furthermore, particle deposition in a stagnation-point flow can provide a first approximation for deposition on more complex systems such as spherical and cylindrical collectors. Theoretical and experimental studies of particle deposition in a stagnation-point flow are available in the literature (e.g. Dabros and van de Ven, 1983, 1987; Chari and Rajagopalan, 1985a,b; Adamczyk *et al.*, 1986).

5.2.1 Hydrodynamics

There are two types of stagnation-point flow, namely plane stagnation-point flow and axisymmetrical stagnation point flow. The laminar flow in a plane stagnation-point flow (also referred to as Hiemenz flow) was discussed by Schlichting (1960). Analysis of particle deposition in such a system was carried out by Chari and Rajagopalan (1985a). The axisymmetrical stagnation-point flow (also referred to as impinging jet or symmetrical radial stagnation flow) is similar in many aspects to the former system. The hydrodynamics of this system are discussed in detail by Schlichting (1960) and Dabros and van de Ven (1983, 1987). In this chapter, we will deal with the hydrodynamics and particle deposition in axisymmetrical stagnation-point flow.

A schematic representation of fluid motion in a stagnation-point flow is shown in Figure 5.3a. A description of the impinging jet system for creating the

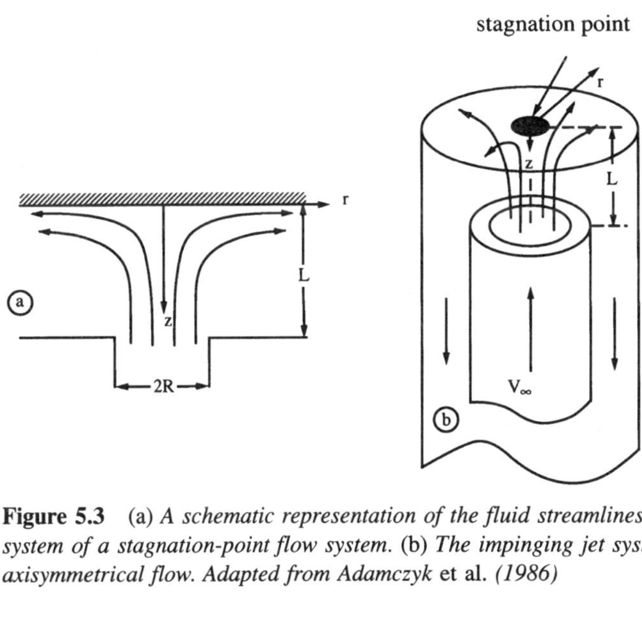

Figure 5.3 (a) *A schematic representation of the fluid streamlines and coordinate system of a stagnation-point flow system.* (b) *The impinging jet system for an axisymmetrical flow. Adapted from Adamczyk* et al. *(1986)*

axisymmetrical flow is presented in Figure 5.3b. In this system, a fluid stream impinges on a flat surface at a right angle to it and flows away radially in all directions. Due to the radial symmetry, the flow can be described in terms of the spatial coordinates r and z of the cylindrical coordinate system. Dabros and van de Ven (1983, 1987) solved the stationary Navier–Stokes equation for laminar flow in the neighbourhood of the stagnation point. Based on their analysis, the flow field can be approximated by:

$$v_r = \alpha_s r z \tag{5.26}$$

$$v_z = -\alpha_s z^2 \tag{5.27}$$

where α_s is a constant that can be calculated numerically. This constant depends on the Reynolds number and the geometry of the system. Numerical values of α_s as a function of the Reynolds number were presented by Dabros and van de Ven (1983, 1987) and Adamczyk *et al.* (1986). In those calculations, the geometry of the stagnation point was characterized by L/R; L and R are described in Figure 5.3. The particle velocity components are related to the fluid velocity by:

$$u_r = f_3(H) v_r \tag{5.28}$$

$$u_z = f_1(H) f_2(H) v_z \tag{5.29}$$

where $f_1(H)$, $f_2(H)$ and $f_3(H)$ are the hydrodynamic correction factors described earlier.

5.2.2 Particle transport equation

The formulation of the particle transport equation near the stagnation-point flow is similar to that described previously with the rotating disc system. The cylindrical coordinate system, with r and z as spatial coordinates, is shown in Figure 5.3a. The dimensionless variables C^*, H and F^* (defined previously in the rotating disc system) are used in the derivation of the particle transport equation. In addition, a particle Peclet number is defined

$$Pe = \alpha_s^* a_p^3 / D_\infty \tag{5.30}$$

where

$$\alpha_s^* = \alpha_s Re \mu / (\rho R^3) \tag{5.31}$$

$$Re = V_\infty R \rho / \mu \tag{5.32}$$

and

$$V_\infty = Q / (\pi R^2) \tag{5.33}$$

Here Re is the Reynolds number, Q is the volumetric flow rate of the fluid, and R is the aperture radius of the impinging jet apparatus. Following the procedure used in the derivation of the particle transport equation in the rotating disc system (described earlier), an *identical* dimensionless particle transport equation (eqn (5.16)) and boundary conditions (eqns (5.17) and (5.18)) are obtained for the stagnation-point flow. It is worthwhile mentioning that this dimensionless transport equation is also identical to that describing particle deposition onto a spherical collector near the forward stagnation point, when $Pe = 3Ua_p^3/2a_cD_\infty$ (U is the approach velocity of the fluid and a_c is the radius of the spherical collector). The dimensionless flux of particles towards the stagnation point is calculated similarly to that of the rotating disc system (described previously).

5.3 Parallel-plate channel

One of the means to investigate the kinetics of particle deposition onto stationary surfaces is by the use of a parallel-plate channel. Parallel-plate channels have been used in several theoretical and experimental studies of particle deposition (e.g. Bowen *et al.*, 1976; Bowen and Epstein, 1979; Adamczyk and van de Ven, 1981; Sjollema and Busscher, 1989, 1990).

5.3.1 Hydrodynamics

Consider a parallel-plate channel of a depth $2B$ (Figure 5.4) through which an incompressible fluid flows with a steady laminar motion. The fluid velocity field and the transport equation are expressed relative to a fixed cartesian coordinate system (x, y and z) having its origin at a point sufficiently far from the entrance region, at a distance where the fluid velocity profile is assumed to be fully developed. The fluid velocity can be obtained from a solution of the Navier–Stokes equation with no slip boundary conditions at both the lower and upper channel walls. The undisturbed fluid velocity is expressed as:

$$v = \frac{3}{2} V_m \frac{z}{B}\left(2 - \frac{z}{B}\right)\mathbf{i}_x \tag{5.34}$$

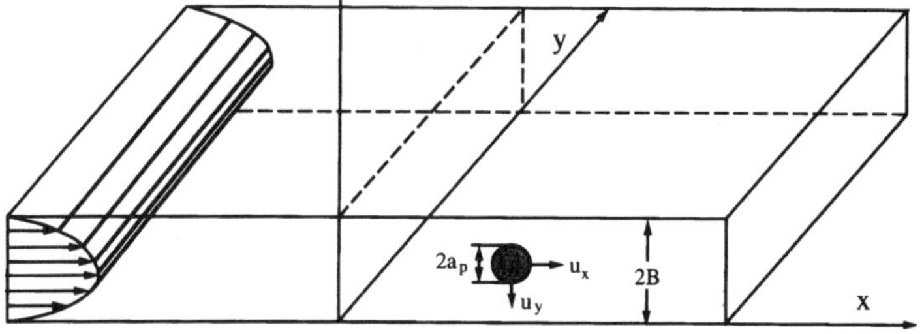

Figure 5.4 *A schematic description of a parallel-plate channel system (adapted from Adamczyk and van de Ven, 1981). The velocity field of the fluid is assumed to be fully developed at* x = 0*. No deposition of particles is assumed for* x < 0

where V_m is the mean fluid velocity (scalar quantity) and $\mathbf{i}_x$ is a unit vector in the x direction. Due to the effect of hydrodynamic interaction, the velocity of the particles in the vicinity of the upper and lower walls of the channel deviates from the fluid velocity field. The hydrodynamics in this case are much more complex than in the rotating disc and stagnation-point flow systems. A rigorous treatment of particle hydrodynamics in a parallel-plate channel was presented by Adamczyk and van de Ven (1981). Approximations to the particle flow field, as introduced by Adamczyk and van de Ven, will be employed in the derivation of the particle transport equation.

5.3.2 Particle transport equation

In the absence of sources and sinks and under steady conditions, the particle flux components in x, y and z can be expressed as:

$$J_x = -D_x \frac{\partial C}{\partial x} + u_x C + \frac{D_x F_x C}{kT} \tag{5.35}$$

$$J_y = -D_y \frac{\partial C}{\partial y} + u_y C + \frac{D_y F_y C}{kT} \tag{5.36}$$

$$J_z = -D_z \frac{\partial C}{\partial z} + u_z C + \frac{D_z F_z C}{kT} \tag{5.37}$$

Here D_z is the normal component of the diffusion tensor, while D_x and D_y are the tangential components; u_x, u_y and u_z are the particle velocity components induced by the fluid flow; and F_x, F_y and F_z are the components of the external force vector.

When J_x, J_y and J_z are incorporated in the continuity equation (i.e. $\nabla \cdot \mathbf{J} = 0$), the resultant particle transport equation is three-dimensional. This can be further simplified if we assume that the width of the channel (in the y direction) is much larger than the diffusion boundary layer (i.e. we can assume no change in the y coordinate), and that the colloidal interaction forces act only in the z direction. Following the procedure described by Adamczyk and van de Ven (1981), the particle transport equation can be simplified by introducing the following dimensionless variables:

$$H = \frac{z}{a_p} - 1 \tag{5.38}$$

$$A^* = a_p/B \tag{5.39}$$

$$Pe = \frac{3V_m a_p^3}{2B^2 D_\infty} \tag{5.40}$$

$$x^* = x/B \tag{5.41}$$

With these dimensionless variables, the particle transport equation can be presented in the general form as:

$$\left\{Pef_3(H)(H+1)\left[2-(H+1)A^*\right]\right\}\frac{\partial C^*}{\partial x^*} + \frac{\partial}{\partial H}\left[f_1(H)F^*C^* - f_1(H)\frac{\partial C^*}{\partial H}\right] = 0 \quad (5.42)$$

where F^* is the dimensionless external force in the z direction, defined by eqn (5.20). In the derivation of this equation, it was assumed that the direction of gravity coincides with the z axis.

For convenience in numerical calculations, eqn (5.42) can be written as:

$$\frac{\partial C^*}{\partial x^*} = b_1(H)\frac{\partial^2 C^*}{\partial H^2} + b_2(H)\frac{\partial C^*}{\partial H} + b_3(H)C^* \quad (5.43)$$

where $b_1(H)$, $b_2(H)$ and $b_3(H)$ are coefficients that can be determined from eqn (5.42). The usual perfect-sink boundary conditions are applied to the upper and lower surfaces of the channel:

$$C^* = 0 \text{ at } H = 0 \text{ for } x^* > 0 \quad (5.44)$$

$$C^* = 0 \text{ at } H = H_u = \frac{2B}{a_p} - 2 \text{ for } x^* > 0 \quad (5.45)$$

In addition, a third boundary condition is given for the channel midway:

$$\left.\frac{\partial C^*}{\partial \mathrm{H}}\right|_{H=(B/a_p)-1} = 0 \quad (5.46)$$

However, this symmetry boundary condition at the channel midway may not be accurate when the gravitational force is significant. In such a case, Adamczyk and van de Ven (1981) used:

$$C^* = C^*(H) \text{ at } x^* = 0 \quad (5.47)$$

This boundary condition describes the inlet particle concentration distribution in the suspension as a function of H.

From the boundary conditions described by eqns (5.44) and (5.45), it follows that $\partial C^*/\partial x^* = 0$ at the upper and lower surfaces of the channel. Thus, eqn (5.42) is reduced to:

$$\frac{\partial}{\partial H}\left\{f_1(H)\left[\frac{\partial C^*}{\partial H} - F^*C^*\right]\right\}_{H=0} = \frac{\partial}{\partial H}\left\{f_1(H)\left[\frac{\partial C^*}{\partial H} - F^*C^*\right]\right\}_{H=H_u} = 0 \quad (5.48)$$

Integration of this equation results in:

$$Sh_{\mathrm{L}} = f_1(H)\left(\frac{\partial C^*}{\partial H} - F^*C^*\right)\Bigg|_{H=0} \tag{5.49}$$

$$Sh_{\mathrm{u}} = f_1(H)\left(\frac{\partial C^*}{\partial H} - F^*C^*\right)\Bigg|_{H=H_{\mathrm{u}}} \tag{5.50}$$

Here Sh_{L} and Sh_{u} are the dimensionless deposition rates (expressed as a Sherwood number) for the lower and upper surfaces of the channel, respectively. The dimensionless concentration $C^*(H, x^*)$ is obtained from a numerical solution of eqn (5.43) with the specified boundary conditions. The dimensionless local fluxes of particles are obtained from eqns (5.49) and (5.50) using this numerically calculated concentration profile. It should be noted that Sh_{L} and Sh_{u} depend on the longitudinal coordinate x^*. In the absence of gravitational forces, one obtains $Sh_{\mathrm{L}} = Sh_{\mathrm{u}}$.

It can be shown (Adamczyk and van de Ven, 1981) that for Brownian particles, in the absence of colloidal and hydrodynamic interactions:

$$Sh_{\mathrm{L}} = Sh_{\mathrm{u}} = \frac{1}{\Gamma\left(\frac{4}{3}\right)}\left(\frac{2Pe}{9x^*}\right)^{1/3} \tag{5.51}$$

This equation is similar to that derived by Leveque (1928) for heat transfer in a channel. This analytical expression can be used to test the numerical solution of the particle transport equation.

5.4 Spherical collector

Transport and deposition of colloidal particles in granular media are of paramount importance in many technological and natural processes. In many cases, an individual grain in a granular medium can be approximated as a sphere; hence, the granular medium can be viewed as an assemblage of spherical collectors. Numerous theoretical and experimental studies of particle deposition onto spherical collectors are available in the literature (e.g. Yao *et al.*, 1971; Rajagopalan and Tien, 1976; Matijevic and Kallay, 1983; Tien, 1989; Elimelech, 1991).

5.4.1 Hydrodynamics

The fluid flow field around a spherical collector can be obtained from a solution of the Navier–Stokes equation in spherical coordinates. A schematic description of a spherical collector and the coordinate system is given in Figure 5.5. For an incompressible fluid, the stream function (Ψ) of a steady, axisymmetrical, low Reynolds number flow satisfies the following equation:

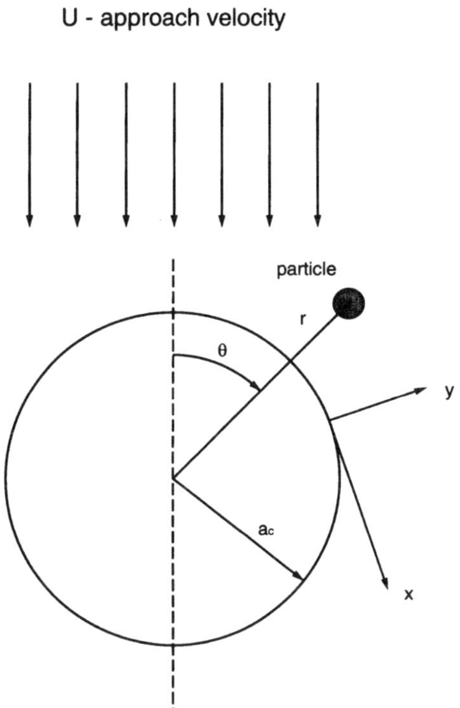

Figure 5.5 *A schematic description of a fluid approaching a spherical collector of radius* a_c *and the coordinate system used*

$$\left[\frac{\partial^2}{\partial r^2} + \frac{\sin\theta}{r^2}\frac{\partial}{\partial\theta}\left(\frac{1}{\sin\theta} - \frac{\partial}{\partial\theta}\right)\right]^2 \Psi = 0 \tag{5.52}$$

The fluid velocity components in the radial and tangential coordinates can be found from the stream function using:

$$v_r = -\frac{1}{r^2 \sin\theta}\frac{\partial\Psi}{\partial\theta} \tag{5.53}$$

$$v_\theta = \frac{1}{r \sin\theta}\frac{\partial\Psi}{\partial r} \tag{5.54}$$

Applying no slip-boundary conditions at the collector surface results in an analytical expression for the stream function around a spherical collector (Tien, 1989):

$$\Psi = \frac{1}{2} U a_c^2 \sin^2\theta \left[\frac{1}{2}\frac{a_c}{r} - \frac{3}{2}\frac{r}{a_c} + \left(\frac{r}{a_c}\right)^2\right] \tag{5.55}$$

where U is the approach velocity of the fluid and a_c is the radius of the spherical collector. Using eqns (5.53) and (5.54), analytical expressions for the radial and tangential fluid velocity components can be derived. These are:

$$v_r = -U \cos\theta \left[\frac{1}{2}\left(\frac{a_c}{r}\right)^3 - \frac{3}{2}\frac{a_c}{r} + 1\right] \tag{5.56}$$

$$v_\theta = U \sin\theta \left[-\frac{1}{4}\left(\frac{a_c}{r}\right)^3 - \frac{3}{4}\frac{a_c}{r} + 1\right] \tag{5.57}$$

In a granular medium, however, the flow field around a spherical collector is influenced by the neighbouring collectors. An appropriate flow model has to be used to account for the disturbance of the flow field around the individual collectors. The general solution for the stream function, satisfying eqn (5.52), may be written as (Tien, 1989):

$$\Psi = \hat{A}\left[K_1\left(\frac{a_c}{r}\right) + K_2\left(\frac{r}{a_c}\right) + K_3\left(\frac{r}{a_c}\right)^2 + K_4\left(\frac{r}{a_c}\right)^4\right]\sin^2\theta \tag{5.58}$$

where $\hat{A}$, K_1, K_2, K_3 and K_4 are arbitrary constants which can be determined by the boundary conditions imposed by the flow model.

Various models that describe the flow field in a packed bed of spherical collectors are available (Brinkman, 1947; Happel, 1958; Kuwabara, 1959). An excellent summary of these models is given by Tien (1989). Among these models, Happel's sphere-in-cell model (Happel, 1958) is the most commonly used. This model will be used here to describe the flow field in a granular medium.

In Happel's model, the porous medium is treated as an assemblage of identical spherical collectors, each of which is enveloped in a shell of fluid. The thickness of the shell, b, is determined so that the overall porosity of the porous medium is maintained for the single collector:

$$b = a_c(1-f)^{-1/3} \tag{5.59}$$

where f is the porosity of the porous medium. Happel's model has been used successfully in numerous particle deposition and filtration studies (e.g. Spielman and FitzPatrick, 1973; Rajagopalan and Tien, 1976; Elimelech and O'Melia, 1990a,b; Elimelech and Song, 1992a).

The flow field in Happel's model can be described by eqn (5.58) with the following coefficients (Payatakes *et al.*, 1974; Tien, 1989):

$$\hat{A} = Ua_c^2/2 \tag{5.60}$$

$$K_1 = 1/w \tag{5.61}$$

$$K_2 = -(3 + 2p^5)/w \tag{5.62}$$

$$K_3 = (2 + 3p^5)/w \tag{5.63}$$

$$K_4 = -p^5/w \tag{5.64}$$

$$w = 2 - 3p + 3p^5 - 2p^6 \tag{5.65}$$

$$p = a_c/b = (1-f)^{1/3} \tag{5.66}$$

Having this expression for Ψ, the velocity components of the fluid are easily obtained from eqns (5.53) and (5.54). Considering hydrodynamic interaction, the radial and tangential particle velocity components (u_r and u_θ, respectively) are related to the fluid velocity components by:

$$u_r = f_1(H)f_2(H)v_r \tag{5.67}$$

$$u_\theta = f_3(H)v_\theta \tag{5.68}$$

Here $f_1(H)$, $f_2(H)$ and $f_3(H)$ are the universal correction factors for hydrodynamic interaction described previously in this chapter. In addition to the retarding effect of hydrodynamic interaction on the particle velocity, the mobility of the particles in the vicinity of the collector surfaces is also reduced. The diffusion coefficients in r and θ, with correction for hydrodynamic interaction, are given by (Adamczyk *et al.*, 1983):

$$D_r = f_1(H)D_\infty \tag{5.69}$$

$$D_\theta = f_4(H)D_\infty \tag{5.70}$$

where $f_4(H)$ is a universal hydrodynamic function described previously in Chapter 4.

5.4.2 Particle transport equation

The components of the convective-diffusion equation can be written in spherical coordinates as:

$$\nabla\cdot(\mathbf{u}C) = f_1(H)f_2(H)v_r\frac{\partial C}{\partial r} + \frac{f_3(H)v_\theta}{r}\frac{\partial C}{\partial \theta} + \left[f_1(H)f_2(H)\frac{\partial v_r}{\partial r} + f_1(H)v_r\frac{\partial f_2(H)}{\partial r} + f_2(H)v_r\frac{\partial f_1(H)}{\partial r} + \frac{2}{r}f_1(H)f_2(H)v_r + \frac{2}{r}f_3(H)v_\theta \cot\theta\right]\mathbf{C} \tag{5.71}$$

$$\nabla\cdot(\mathbf{D}\cdot\nabla C) = \left[f_1(H)\frac{\partial^2 C}{\partial r^2} + \frac{2f_1(H)}{r}\frac{\partial C}{\partial r} + \frac{\partial f_1(H)}{\partial r}\frac{\partial C}{\partial r} + f_4(H)\frac{\cot\theta}{r^2}\frac{\partial C}{\partial \theta} + \frac{f_4(H)}{r^2}\frac{\partial^2 C}{\partial \theta^2}\right]D_\infty \tag{5.72}$$

$$\nabla\cdot\left(\frac{\mathbf{D}\cdot\mathbf{F}}{kT}C\right) = \left\{\left[\frac{2f_1(H)F_r}{rkT} + \frac{F_r}{kT}\frac{\partial f_1(H)}{\partial r} + \frac{f_1(H)}{kT}\frac{\partial F_r}{\partial r} + \frac{f_4(H)F_\theta\cot\theta}{rkT} + \frac{f_4(H)}{rkT}\frac{\partial F_\theta}{\partial \theta}\right]C + \frac{f_1(H)F_r}{kT}\frac{\partial C}{\partial r} + \frac{f_4(H)F_\theta}{rkT}\frac{\partial C}{\partial \theta}\right\}D_\infty \tag{5.73}$$

Here F_r and F_θ are the radial and tangential components of the external force vector, respectively. As opposed to the deposition systems described previously, the gravity force has radial and tangential components. The colloidal interaction forces, on the other hand, are assumed to have only a perpendicular (radial) component. Therefore, it is more convenient to express the external force vector as the sum of colloidal and gravitational forces as follows:

$$F = -\nabla\phi_T + \frac{4}{3}\pi a_p^3 \Delta\rho \mathbf{g} \tag{5.74}$$

Here ϕ_T is the total energy of interaction originating from the colloidal interaction forces, $\Delta\rho$ is the difference between the particle and fluid densities, and $\mathbf{g}$ is the gravitational acceleration vector. Hence, the radial and tangential components of the external force vector are:

$$F_r = -\frac{d\phi_T}{dr} - F_G \cos\theta \tag{5.75}$$

$$F_\theta = F_G \sin\theta \tag{5.76}$$

with F_G given by

$$F_G = \frac{4}{3}\pi a_p^3 \Delta\rho g \tag{5.77}$$

In order to write the particle transport equation in non-dimensional form, the following new dimensionless variables are introduced:

$$V_r = v_r/U \tag{5.78}$$

$$V_\theta = v_\theta/U \tag{5.79}$$

$$N_{Pe} = (2Ua_p)/D_\infty \tag{5.80}$$

$$\Phi = \phi_T/kT \tag{5.81}$$

$$N_R = a_p/r \tag{5.82}$$

$$F_G^* = \frac{F_G a_p}{kT} \tag{5.83}$$

Using these variables, the particle transport equation is written as:

$$\frac{N_{Pe}}{2}\Bigg\{f_1(H)f_2(H)V_r\frac{\partial C^*}{\partial H} + f_3(H)N_R V_\theta \frac{\partial C^*}{\partial\theta}$$

$$+\Bigg[f_1(H)f_2(H)\frac{\partial V_r}{\partial H} + f_1(H)V_r\frac{\partial f_2(H)}{\partial H} + f_2(H)V_r\frac{\partial f_1(H)}{\partial H}$$

$$+ 2f_1(H)f_2(H)N_R V_r + 2f_3(H)N_R V_\theta \cot\theta\Bigg]C^*\Bigg\}$$

$$= f_1(H)\frac{\partial^2 C}{\partial H^2} + 2f_1(H)N_R\frac{\partial C^*}{\partial H} + \frac{\partial f_1(H)}{\partial H}\frac{\partial C^*}{\partial H}$$

$$+\left\{\left[2f_1(H)N_{\rm R}+\frac{\partial f_1(H)}{\partial H}\right]\left[\frac{\partial\Phi}{\partial H}+F_{\rm G}^*\cos\theta\right]+f_1(H)\frac{\partial^2\Phi}{\partial H^2}\right.$$

$$\left.-\,2f_4(H)N_{\rm R}F_{\rm G}^*\cos\theta\right\}C^*+f_1(H)\left[\frac{\partial\Phi}{\partial H}+F_{\rm G}^*\cos\theta\right]\frac{\partial C^*}{\partial H}$$

$$-f_4(H)N_{\rm R}F_{\rm G}^*\sin\theta\frac{\partial C^*}{\partial\theta} \tag{5.84}$$

For convenience in numerical calculations, the particle transport equation can be reduced to:

$$\frac{\partial C^*}{\partial\theta}=a_1(H,\theta)\frac{\partial^2C^*}{\partial H^2}+a_2(H,\theta)\frac{\partial C^*}{\partial H}+a_3(H,\theta)C^* \tag{5.85}$$

where $a_1(H,\theta)$, $a_2(H,\theta)$ and $a_3(H,\theta)$ are variable coefficients which can be determined from eqn (5.84). The following boundary conditions are used (Prieve and Ruckenstein, 1974; Song and Elimelech, 1992):

$$C^*(H=0,\theta)=0 \tag{5.86}$$

$$C^*(H\to\infty,\theta)=1 \tag{5.87}$$

$$\left(\frac{\partial C^*}{\partial\theta}\right)_{\theta=0}=0 \tag{5.88}$$

The first and second boundary conditions are similar to those used previously with the rotating disc, stagnation-point flow, and parallel-plate channel systems. The third boundary condition arises from the symmetry around the forward stagnation path (Prieve and Ruckenstein, 1974). For particle deposition at very small Peclet numbers, the second boundary condition should be modified as suggested by Song and Elimelech (1992).

Once the dimensionless concentration distribution of particles around the collector, $C^*(H,\theta)$, is determined numerically, the perpendicular flux of particles at the collector surface can be evaluated. When the local flux of particles at the surface is integrated over the entire surface of the collector, the overall rate of particle deposition is obtained. The local particle flux perpendicular to the surface can be expressed as:

$$J(H,\theta)=UC_0J^*(H,\theta) \tag{5.89}$$

where $J^*(H,\theta)$ is a dimensionless flux given by:

$$J^*(H,\theta)=-\frac{2f_1(H)}{N_{Pe}}\frac{\partial C^*}{\partial H}+f_1(H)f_2(H)V_rC^*-\frac{2f_1(H)}{N_{Pe}}\left[\frac{\partial\Phi}{\partial H}+F_{\rm G}^*\cos\theta\right]C^* \tag{5.90}$$

Since the radial coordinate is away from the collector surface, the local flux towards the collector calculated from eqn (5.90) is negative. Hence, in determining the total particle deposition rate (as shown later by eqns (5.91)–(5.93)), a positive value of $J(H,\theta)$ is used.

In this equation, the first term describes the diffusive flux of particles, the second term describes the convective flux of particles, and the third term describes the drift flux of particles due to migration (drift) velocity originating from the colloidal interaction potentials (i.e. van der Waals and electrical double layer) and gravitational forces. The dimensionless local flux of particles at the collector surface is evaluated from eqn (5.90) at $H \to 0$.

The overall rate of particle deposition on the collector, I, can be obtained from integration of the local flux (at $H \to 0$) over the entire surface of the collector as follows:

$$I = 2\pi r^2 \int_0^\pi J(H \to 0, \theta) \sin\theta \, d\theta \tag{5.91}$$

In studies concerning the collection of colloidal particles in porous media, it is convenient to use a dimensionless particle removal rate, η, known as the single collector efficiency. When the overall particle deposition rate, I, is normalized by the convective flux of particles towards the projected area of the Happel fluid shell, the single collector efficiency is defined as:

$$\eta = \frac{I}{\pi b^2 U C_0} \tag{5.92}$$

where b is the radius of the fluid envelope in Happel's model. In this model $b = a_c(1-f)^{-1/3}$, with f being the porosity of the porous medium and a_c the radius of the collector. When the overall particle deposition rate is normalized by the convective flux of particles towards the projected area of the collector, the single collector efficiency is defined as:

$$\eta = \frac{I}{\pi a_c^2 U C_0} \tag{5.93}$$

5.5 Interaction-force boundary-layer approximation

Although particle deposition rates can be found by a numerical solution of the convective-diffusion equation, the procedure may be tedious, and, in addition, no general solution is obtained. An approximate analytical solution for calculating the deposition rate of particles when electrical double layer repulsive forces are significant may be useful. In this section, the interaction-force boundary-layer (IFBL) approximation will be presented in detail. The IFBL approximation was developed by Ruckenstein and Prieve (1973) and Spielman and Friedlander (1974). It is useful for predicting deposition rates of Brownian particles in the presence of repulsive double layer interactions. The derivation will be presented for deposition on a spherical collector (in the presence of neighbouring collectors). This approximation can also be applied, in a similar way, to other systems such as parallel-plate and cylindrical channels (Bowen *et al.*, 1976), and rotating discs (Prieve and Lin, 1980; Rajagopalan and Kim, 1981).

5.5.1 Assumptions and principles

Colloidal interaction forces operate over very short distances from the surface of the collector (usually several tens of nanometres). It is possible to identify two regions adjacent to the collector surface, characterized by the relative magnitude of the colloidal interaction forces. In the inner region, in the vicinity of the collector surface, the colloidal interaction forces predominate while the convective transport components are negligible. This region is referred to as the interaction-force boundary layer (or surface interaction boundary layer), and is taken to be on the order of the double layer thickness. In the outer region, the colloidal interaction forces vanish while the convective transport components become important. The thickness of this region is taken to be on the order of the diffusion boundary layer.

The analytical approximation which will be presented in this section is based on the assumption that the diffusion boundary layer is very thick compared to the IFBL. The thickness of the IFBL (δ_F) is on the order of κ^{-1} (Debye length), while the thickness of the diffusion boundary layer (δ_D) is on the order of $a_c(D/Ua_c)^{1/3}$ (Levich, 1962). Accordingly, the necessary condition for the IFBL approximation is:

$$\frac{\delta_D}{\delta_F} = \kappa a_c \left(\frac{D_\infty}{Ua_c}\right)^{1/3} \gg 1 \tag{5.94}$$

Additional assumptions in this approximation are that interception due to the finite size of the particles and gravitational forces play no role in the deposition of the Brownian particles.

5.5.2 Quantitative formulation

In the inner region, where the fluid velocity is negligible, the transport of particles is controlled by diffusion and colloidal forces. The steady convective-diffusion equation becomes:

$$\nabla \bullet \left(\mathbf{D} \bullet \nabla C + \frac{C\mathbf{D}}{kT} \nabla \phi_T\right) = 0 \tag{5.95}$$

With the assumptions that the total interaction energy ϕ_T is only a function of the boundary-layer coordinate y (perpendicular to the collector surface), and that the tangential diffusion is small compared to the diffusion perpendicular to the surface, eqn (5.95) can be written as:

$$\frac{\partial}{\partial y}\left(D_y \frac{\partial C}{\partial y} + \frac{CD_y}{kT}\frac{\partial \phi_T}{\partial y}\right) = 0 \tag{5.96}$$

Integrating this equation twice with respect to y, with the boundary condition of $C=0$ at $y=0$ (perfect sink) in the second integration, yields:

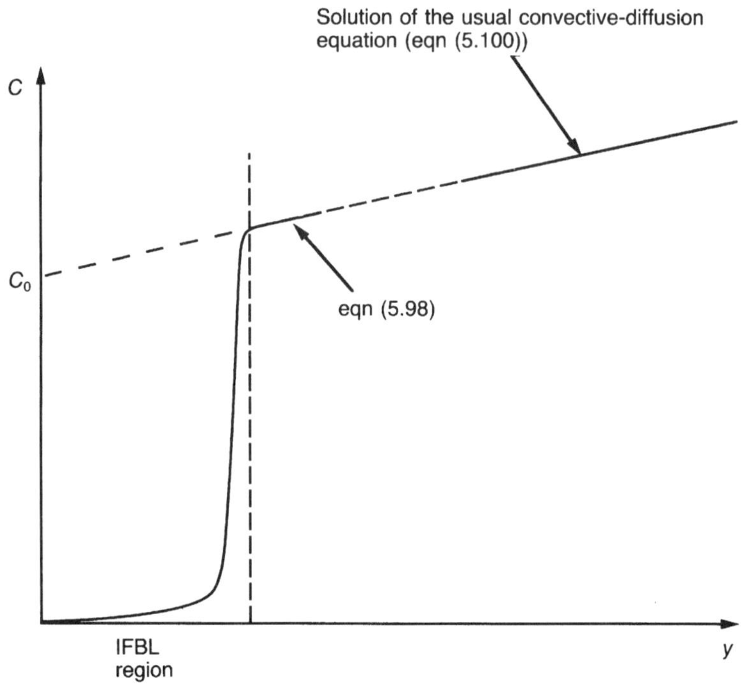

Figure 5.6 *A qualitative description of the concentration profiles of particles as a function of the separation distance from a collector surface, as expected from the interaction-force boundary-layer approximation (adapted from Spielman and Friedlander, 1974)*

$$C = -\frac{J(x)}{D_\infty} \exp(-\phi_T/kT) \left\{ \int_0^y [g_1(H) \exp(\phi_T/kT) - 1] \, dy + y \right\} \tag{5.97}$$

Here $g_1(H)$ is a hydrodynamic correction factor to account for hydrodynamic interaction, $g_1(H) = 1/f_1(H)$; and $J(x)$ is a local particle flux which is constant with respect to y. The last equation describes the concentration profile of particles as a function of the separation distance from the collector surface.

At distances much larger than the thickness of the double layer ($y \gg \kappa^{-1}$), the total interaction energy vanishes ($\phi_T = 0$), and eqn (5.97) is reduced to:

$$C = -\frac{J(x)}{D_\infty} \left\{ \int_0^{\delta_D} [g_1(H) \exp(\phi_T/kT) - 1] \, dy + y \right\} \tag{5.98}$$

The upper limit of the integration has been taken arbitrarily as δ_D, the thickness of the diffusion boundary layer, since this ensures that ϕ_T is equal to zero (note that $\delta_D \gg \kappa^{-1}$). Based on eqns (5.97) and (5.98), the concentration profile of particles is described *qualitatively* in Figure 5.6. By matching the linear portion of the concentration distribution outside the inner region (described by eqn (5.98)) with the solution of the convective-diffusion equation in the absence of colloidal forces, an approximate solution to the deposition in the presence of colloidal interactions can be obtained. The matching procedure is illustrated in Figure 5.6. Spielman and Friedlander (1974) suggested that the matching procedure is equivalent to taking the following surface boundary condition to the convective-diffusion equation:

$$C(y = 0,x) = -\frac{J(x)}{D_\infty}\left\{\int_0^{\delta_D}[g_1(H)\exp(\phi_T/kT) - 1]\,dy\right\} \tag{5.99}$$

Hence, to calculate the deposition rate of Brownian particles in the presence of repulsive double layer interactions, one needs to solve the steady convective-diffusion equation:

$$\mathbf{v}\bullet\nabla C = D_\infty\nabla^2 C \tag{5.100}$$

with the following pseudo-first-order surface reaction as a boundary condition:

$$J_0(x) = -K_F C(y = 0,x) = -D_\infty \left.\frac{\partial C}{\partial y}\right|_{y=0} \tag{5.101}$$

Here $\mathbf{v}$ is the fluid velocity vector, $J_0(x)$ is the local flux of particles at the collector surface ($y=0$), and K_F is a pseudo-first-order rate constant given by:

$$K_F = D_\infty\left\{\int_0^{\delta_D}[g_1(H)\exp(\phi_T/kT) - 1]\,dy\right\}^{-1} \tag{5.102}$$

This boundary condition at the surface implicitly accounts for the retarding effect of electrical double layer repulsive forces.

5.5.3 Available analytical solution

Under the assumptions of (1) axisymmetrical flow, (2) negligible tangential diffusion, and (3) $r \approx \alpha_c$, eqn (5.100) simplifies to:

$$v_y\frac{\partial C}{\partial y} + \frac{v_\theta}{a_c}\frac{\partial C}{\partial \theta} = D_\infty\left(\frac{2}{a_c}\frac{\partial C}{\partial y} + \frac{\partial^2 C}{\partial y^2}\right) \tag{5.103}$$

The particle velocity components (u_θ and u_y) are replaced by the expressions for the fluid velocity components. In doing so, it is assumed that the particles follow the fluid streamlines, and that hydrodynamic interaction has no effect on the particle trajectories adjacent to the collector surface. Following Levich (1962), the stream function within the diffusion boundary layer can be approximated by:

$$\Psi \approx \frac{3}{4}A_s U y^2 \sin^2\theta \tag{5.104}$$

where A_s is a dimensionless flow parameter of the Happel porous medium model. This porosity-dependent parameter is defined as:

$$A_s = \frac{2(1 - p^5)}{w} \tag{5.105}$$

The variables p and w in this equation were described earlier in this chapter. Hence, the fluid velocity components can be derived from the stream function to yield:

$$v_\theta = \frac{3}{2} A_s U \frac{y}{a_c} \sin\theta \tag{5.106}$$

$$v_y = -\frac{3}{2} A_s U \left(\frac{y}{a_c}\right)^2 \cos\theta \tag{5.107}$$

An analytical solution for eqn (5.103) with the surface boundary condition, eqn (5.101), was derived by Spielman and Friedlander (1974), and is described in detail in that reference. The analytical solution for η, the single collector efficiency, is given by:

$$\eta = 4.0 A_s^{1/3} \left(\frac{D_\infty}{2a_c U}\right)^{2/3} \left(\frac{\beta}{1+\beta}\right) S(\beta) \tag{5.108}$$

$$\beta = \frac{1}{3} (2)^{1/3} \Gamma\left(\frac{1}{3}\right) A_s^{-1/3} \left(\frac{D_\infty}{U a_c}\right)^{1/3} \left(\frac{K_F a_c}{D_\infty}\right) \tag{5.109}$$

where $S(\beta)$ is a slowly varying function of β with numerical values given in Table 5.1.

In the absence of colloidal and hydrodynamic interactions (i.e. $\phi_T = 0$ and $g_1(H) = 1$), β approaches infinity while $S(\beta)$ approaches 1. Under these conditions, the single collector efficiency is reduced to:

Table 5.1 *Numerical values of the function* S(β) *(after Spielman and Friedlander, 1974)*

β	S(β)
0.00	1.423
0.01	1.418
0.02	1.414
0.05	1.400
0.10	1.380
0.20	1.345
0.50	1.269
1.00	1.197
2.00	1.128
5.00	1.062
10.0	1.033
20.0	1.018
50.0	1.007
100.0	1.004
∞	1.000

$$\eta = 4.0A_s^{1/3}\left(\frac{D_\infty}{2a_c U}\right)^{2/3} \tag{5.110}$$

This expression is identical to Levich's solution for mass transfer of small particles (molecules) to a sphere with Happel's flow model (Levich, 1962). Equation (5.110) can be used to test the accuracy of the numerical solution of the particle transport equation.

5.5.4 Evaluation of K_F

In order to evaluate the pseudo rate constant K_F (eqn (5.102)), a proper form of the hydrodynamic function $g_1(H)$ should be available. An exact solution for $g_1(H)$ was given by Brenner (1961). A simple approximation for estimating $g_1(H)$ was suggested by Dahneke (1974) and can be used satisfactorily to determine K_F. This expression deviates from the exact solution of Brenner by no more than 6% and is given by:

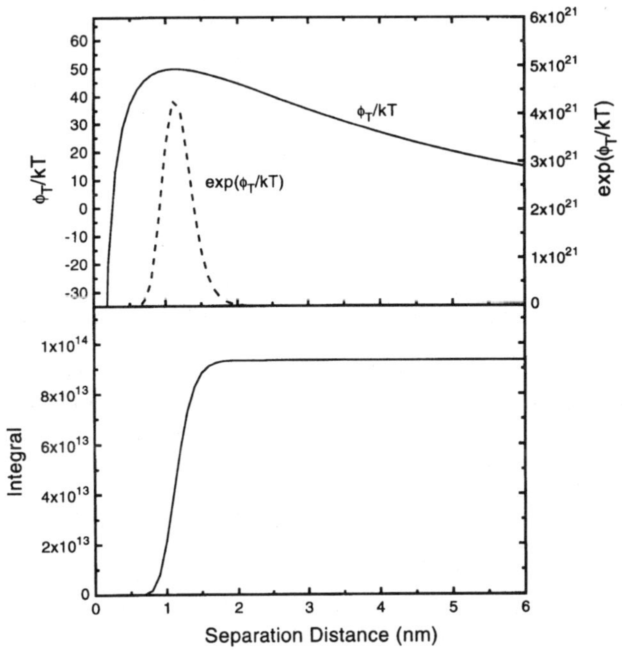

Figure 5.7 *The total interaction energy $\phi_T(y)$, the function exp $[\phi_T(y)/kT]$, and the value of the definite integral in eqn 5.102, as a function of the separation distance between a particle and a spherical collector. The following parameters were used in the calculations: surface potentials of particles and collectors = –40 and –30 mV, respectively; ionic strength = 0.01 M (1–1 electrolyte); Hamaker constant = 1×10^{-20} J; particle diameter = 200 nm; and temperature = 25°C*

$$g_1(H) = 1 + \frac{1}{H} = 1 + \frac{a_p}{y} \tag{5.111}$$

where y is the particle–collector (surface to surface) separation.

The definite integral in K_F can be evaluated numerically. In the following example, the upper limit of the integral has been taken arbitrarily as 100 nm, when an energy barrier in the total interaction energy profile was present. Larger values of the upper limit did not change the value of the integral. Representative results of ϕ_T, exp (ϕ_T/kT), and the value of the definite integral in K_F as a function of the upper limit of the integral, are shown in Figure 5.7. The expression of Hogg *et al.* (1966) and Gregory (1981) for the double layer and van der Waals interactions, respectively, were used to calculate ϕ_T. It is demonstrated that the contribution of the interaction energy to the integral at short separations is dramatic, and that the value of the integral is practically constant for $y > 3$ nm.

At high salt concentrations or low surface potentials, when an energy barrier in the total interaction profile is absent, the integral slowly diverges as the upper limit of the integration is increased. This effect is more pronounced for large particles since the hydrodynamic interaction function, $g_1(H)$, depends on particle size. This phenomenon has also been reported by others (e.g. Prieve and Ruckenstein, 1976). It is widely accepted that the approximate analytical solution for the rate of deposition, presented in this section, is valid only in the presence of repulsive forces, i.e. when an energy barrier exists (Spielman and Friedlander, 1974; Ruckenstein and Prieve, 1973; Prieve and Ruckenstein, 1976).

5.6 Trajectory analysis

Trajectory analysis refers to a methodology used to determine the rate of particle deposition from particle trajectories. The trajectory of a particle over a given collector is determined by the various forces acting on the particle. Thus, by knowing the magnitude and direction of these forces, one can calculate the trajectories of the particles in suspension. In order to determine the collection efficiency of particles, it is sufficient to calculate the so-called 'limiting trajectory'. The limiting trajectory separates trajectories that result in the interception of particles by the collector from those which do not.

The trajectory analysis is commonly applied to non-Brownian particles, i.e. particles which are not significantly affected by Brownian motion. The method of trajectory analysis was first introduced by Sell (1931) and Albrecht (1931) in their studies on aerosol filtration. O'Melia and Stumm (1967) suggested its use in water filtration. The first successful calculations for hydrosol deposition onto an isolated spherical collector were presented by Yao (1968). Since then, refinements have been introduced by other investigators to include all important mechanisms and various porous media models. The trajectory approach has been used most extensively by Tien and coworkers (Tien (1989) and references therein) for various porous media models. This analysis was also applied to a rotating disc system by Spielman and FitzPatrick (1973). In the following

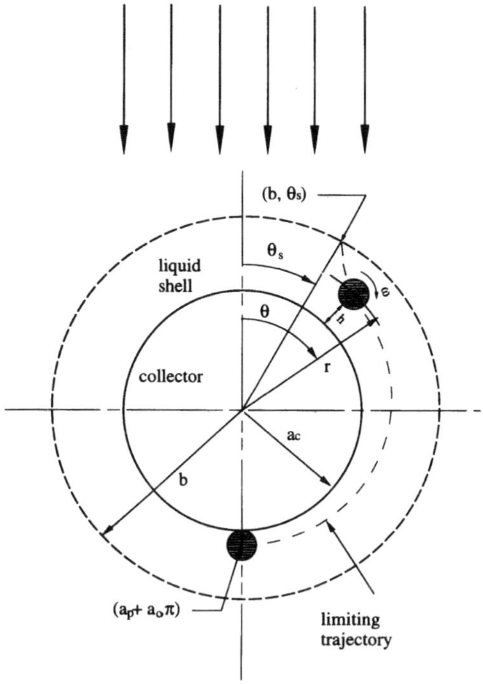

Figure 5.8 *A schematic description of Happel's sphere-in-cell model as used in the trajectory analysis (adapted from Rajagopalan and Tien, 1976). The coordinate system and the limiting trajectory are shown in this figure*

paragraphs, we will present the development of the trajectory equation of non-Brownian particles in porous media, composed of spherical collectors, following the approach of Rajagopalan and Tien (1976).

5.6.1 Collector and flow model

In the derivation of the trajectory equation in this work, Happel's sphere-in-cell model will be used (Happel, 1958). A general description of Happel's flow model, including the spherical coordinate system and relevant parameters, is given in Figure 5.8. Due to radial symmetry, the problem can be described in the radial (r) and polar (θ) coordinates alone. Following the notation used by Rajagopalan and Tien (1976) in their mathematical development, length variables that are made non-dimensional with respect to the collector radius (a_c) or the particle radius (a_p) are designated with '*' and '+' superscripts, respectively.

The creeping-flow field in the Happel model is described by the following stream function (Tien, 1989):

$$\Psi = \frac{1}{2} Ua_c^2 \sin^2 \theta \left(\frac{K_1}{r^*} + K_2 r^* + K_3 r^{*2} + K_4 r^{*4} \right) \tag{5.112}$$

Here $r^* = r/a_c$; and K_1, K_2, K_3 and K_4 are porosity-dependent parameters described earlier in this chapter (eqns (5.61)–(5.64)). As shown in Figure 5.8, the length r is measured from the centre of the spherical collector to the centre of the particle, and b is the radius of the fluid shell surrounding the collector. The fluid velocity components in the radial and tangential coordinates (v_r and v_θ respectively) can be obtained from the stream function using eqns (5.53) and (5.54). This results in:

$$v_r = - U \cos \theta \left(\frac{K_1}{r^{*3}} + \frac{K_2}{r^*} + K_3 + K_4 r^{*2} \right) \tag{5.113}$$

$$v_\theta = \frac{1}{2} U \sin \theta \left(\frac{-K_1}{r^{*3}} + \frac{K_2}{r^*} + 2K_3 + 4K_4 r^{*2} \right) \tag{5.114}$$

5.6.2 Force balance and basic formulations

The trajectory analysis is derived by applying Newton's second law to a suspended particle in the flowing fluid:

$$m \frac{d\mathbf{u}}{dt} = \sum_i F_i \tag{5.115}$$

$$m a_p \frac{d\omega_p}{dt} = \sum_i T_i \tag{5.116}$$

where m is the mass of the particle, $\mathbf{u}$ is the velocity vector of the particle, ω_p is the rotation of the particle, and F_i and T_i are the forces and torques acting on the particle, respectively. In the analysis presented here, the forces that will be considered are fluid drag (F^D), electrical double layer (F^{EDL}) and van der Waals (F^{VDW}). Expressions for the colloidal forces, F^{EDL} and F^{VDW}, can be derived from the corresponding expressions for the interaction energy (described in Chapter 3) by using:

$$F_i = -\nabla \phi_i \tag{5.117}$$

where F_i and ϕ_i are the force and interaction energy of type i, respectively. Since the colloidal forces act perpendicular to the surface, only radial forces are considered. Furthermore, the inertial, gravitational and colloidal interaction forces do not give rise to a net torque on the particle due to its sphericity. A summary of analytical expressions for the various forces and torques was presented by Rajagopalan and Tien (1976, 1979) and Tien (1989).

In order to include the effects of hydrodynamic interaction, the net fluid drag force is calculated by considering a linear combination of three separate flow systems (Goldman *et al.*, 1967a,b; Goren and O'Neill, 1971). These flows are: (1) the translation of the particle in an otherwise stationary fluid in the presence of the collector; (2) the rotation of the particle in an otherwise stationary fluid in the presence of the collector; and (3) the motion of the fluid in the presence of the stationary particle and collector. This analysis is described in more detail in Rajagopalan (1974) and Tien (1989).

The hydrodynamic correction factors were calculated for the interaction of a spherical particle with a flat plate with the assumption that the bulk fluid velocity field can be described by (Rajagopalan and Tien, 1976):

$$v_r = -A_1 y'^2 \tag{5.118}$$

$$v_\theta = B_1 y' + B_2 y'^2 \tag{5.119}$$

where A_1, B_1 and B_2 are coefficients and y' is the distance from the collector surface to the centre of the spherical particle ($y' = r - a_c = y + a_p$). These coefficients are evaluated by matching eqns (5.118) and (5.119) to those calculated from Happel's model (eqns (5.113) and (5.114)). The inertial forces were not included in the trajectory analysis. It was shown that these forces are negligible for deposition of particles from flowing suspension under low Reynolds number flow conditions (Yao, 1968; Rajagopalan and Tien, 1976).

The torque balance yields the following equation for the particle rotation:

$$\omega_p = u_\theta \left(\frac{g_\phi^t}{a_p g_\phi^r} \right) + \left(\frac{B_1 g_{1\phi}^m + B_2 y' g_{2\phi}^m}{g_\phi^r} \right) \tag{5.120}$$

Here u_θ is the particle velocity in the θ direction. Rotation of the particles is caused by variations in the θ component of the fluid velocity and by hydrodynamic interaction effects. Tabulated values for the rotational hydrodynamic correction factors (g terms above) were presented by Rajagopalan (1974) and Rajagopalan and Tien (1979).

A balance of forces acting in the radial direction yields:

$$6\pi\mu a_p u_r f_r^t = -6\pi\mu a_p A_1 y'^2 f_r^m + F_r^G + F_r^{VDW} + F_r^{EDL} \tag{5.121}$$

Hence, the equation for the radial velocity of the particle, u_r, is:

$$\frac{dr}{dt} = u_r = \frac{1}{f_r^t} \left[-Ay'^2 f_r^m + \frac{(F_r^G + F_r^{VDW} + F_r^{EDL})}{6\pi\mu a_p} \right] \tag{5.122}$$

Tabulated values for the hydrodynamic correction factors (f terms) were presented by Rajagopalan (1974) and Rajagopalan and Tien (1979). Similarly, the balance of forces in the θ direction, with the use of eqn (5.120), yields the tangential particle velocity:

$$r\frac{d\theta}{dt} = u_\theta = \frac{1}{s_1} \left[s_2 B_1 a_p + s_3 B_2 y^+ a_p^2 + \frac{F_\theta^G}{6\pi\mu a_p} \right] \tag{5.123}$$

where $y^+ = y'/a_p$ and:

$$s_1(H) = \frac{f_\theta^t g_\phi^r - f_\theta^r g_\phi^t}{g_\phi^r} \tag{5.124}$$

$$s_2(H) = \frac{f_\theta^r g_{1\phi}^m + y^+ f_{1\theta}^m g_\phi^r}{g_\phi^r} \tag{5.125}$$

$$s_3(H) = \frac{f_\theta^{\mathrm{r}} g_{2\phi}^{\mathrm{m}} + y^+ f_{2\theta}^{\mathrm{m}} g_\phi^{\mathrm{r}}}{g_\phi^{\mathrm{r}}} \tag{5.126}$$

with $H = (y' - a_p)/a_p$; H is the dimensionless surface-to-surface separation distance. The g and f terms are correction factors for hydrodynamic interaction discussed previously in this section.

5.6.3 The trajectory equation

The components of the particle velocity in dimensionless form are given by:

$$\frac{1}{U}\frac{\mathrm{d}r}{\mathrm{d}t} = u_r^* = \frac{1}{f_{\mathrm{t}}^{\mathrm{r}}}\left[-A_1^+ y^{+2} f_{\mathrm{r}}^{\mathrm{m}} - N_{\mathrm{G}} \cos\theta + F^{*\mathrm{VDW}} + F^{*\mathrm{EDL}}\right] \tag{5.127}$$

$$\frac{r}{U}\frac{\mathrm{d}\theta}{\mathrm{d}t} = u_\theta^* = \frac{1}{s_1}\left[B_1^+ s_2 + B_2^+ y^+ s_3 + N_{\mathrm{G}} \sin\theta\right] \tag{5.128}$$

where $F^{*\mathrm{EDL}}$ and $F^{*\mathrm{VDW}}$ are the double layer and van der Waals forces in dimensionless form. The other dimensionless variables are defined in Table 5.2. Dividing eqn (5.127) by eqn (5.128) yields the final form of the trajectory equation:

$$\frac{\mathrm{d}r^+}{\mathrm{d}\theta} = r^+ \frac{\left[-A_1^+ y^{+2} f_r^{\mathrm{m}} - N_{\mathrm{G}} \cos\theta + F^{*\mathrm{VDW}} + F^{*\mathrm{EDL}}\right]}{\left[N_{\mathrm{D1}} B_1^+ y^+ + N_{\mathrm{D2}} B_2^+ y^{+2} + N_{\mathrm{D3}} N_{\mathrm{G}} \sin\theta\right]} \tag{5.129}$$

The various dimensionless groups in this equation are defined in Table 5.2.

The trajectory equation is a first-order ordinary differential equation and does not have a closed form solution. It can be integrated numerically once an initial

Table 5.2 *Dimensionless variables used in eqns (5.127)–(5.129)*

Parameter	*Definition*
A_1^+	$(A_1 a_p^2)/U$
B_1^+	$(B_1 a_p)/U$
B_2^+	$(B_2 a_p^2)/U$
$F^{*\mathrm{VDW}}$	$-N_{\mathrm{Lo}}/[4H^2(1 + 5.3HN_{\mathrm{Rtd}})]$
$F^{*\mathrm{EDL}}$	$F_{\mathrm{r}}^{\mathrm{EDL}}/(6\pi\mu a_p U)$
N_{G}	$(2a_p^2 \Delta\rho g)/(9\mu U)$
N_{Lo}	$A/(9\pi\mu a_p^2 U)$
N_{Rtd}	a_p/λ
N_{D1}	$(s_2 f_r^{\mathrm{t}})/(s_1 y^+)$
N_{D2}	$(s_3 f_r^{\mathrm{t}})/(s_1 y^+)$
N_{D3}	f_r^{t}/s_1

position for the suspended particle is specified. Since we are interested in the limiting trajectory, the integration is initialized at the rear of the collector at $r^+ = (1 + a_c/a_p)$ and $\theta = \pi$. The trajectory is followed upstream to the location where it intersects the outside of the Happel shell at $r = b$, $\theta = \theta_s$ and $\Psi = \Psi_L$. The limiting trajectory is illustrated in Figure 5.8. All trajectories of particles to the left of the illustrated limiting trajectory (or inside Ψ_L) are intercepted by the collector, while all on the right will pass by the collector. In practice, the numerical integration is initiated at a small separation distance from the collector surface (a few angstroms) as singularities exist in some of the force expressions at zero separation distance. In addition, particles cannot approach the collector at $H = 0$ due to the presence of Born repulsive forces at short distances (a few angstroms).

The single collector removal efficiency is calculated from the ratio of the flux of particles through the circular cross-section enclosed by the limiting trajectory to the total particle flux entering the Happel shell:

$$\eta = \frac{2\pi\Psi_L C_0}{\pi b^2 U C_0} \tag{5.130}$$

where C_0 is the concentration of particles at the entrance of the shell. This results in:

$$\eta = \sin^2 \theta_s \tag{5.131}$$

Hence, once the limiting trajectory is determined by numerical integration, one can determine θ_s, from which η can be calculated. Numerical techniques to solve the trajectory analysis are discussed by Rajagopalan and Tien (1979), Vaidyanathan (1986) and Tobiason (1987).

5.7 Representative model predictions of particle deposition

Representative model predictions for the effect of various physical and chemical-colloidal properties on the rate of deposition of colloidal particles will be presented in this section. These will be carried out for deposition onto a spherical collector in porous media, based on the theoretical approach described in Section 5.4 of this chapter. In these calculations, the convective-diffusion equation is solved numerically for the concentration distribution of particles around a spherical collector. The dimensionless deposition rate η (known as the single collector efficiency) is then calculated from eqn (5.93). Description of the numerical methods to solve the convective-diffusion equation is beyond the scope of this chapter. The reader is referred to the articles by Elimelech and Song (1992b) and Song and Elimelech (1993) for detailed description and analysis of the numerical procedures.

It should be emphasized that the general dependence of the deposition rate on the various variables examined is similar to that obtained when other collectors (as those discussed previously in this chapter) are considered. The model predictions presented here are for two distinct cases. The first is for deposition of colloidal particles onto similarly charged collectors in aqueous media. In this case, repulsive double layer interactions develop. The second case is for deposition of colloidal particles onto oppositely charged collectors, where attractive double layer interactions develop.

5.7.1 Deposition in the presence of repulsive double layer interactions

Representative model predictions for the effect of ionic strength, particle size, surface potentials of particles and collectors, Hamaker constant of the interacting media, and double layer interaction mode on the rate of deposition of Brownian particles are presented. In deposition in the presence of repulsive double layer interactions, it is often convenient to present the results in terms of the so-called collision efficiency α (also termed attachment or deposition efficiency). The collision efficiency is defined as the ratio of the actual deposition rate to the deposition rate in the absence of electrical double layer repulsion. For particle deposition onto spherical collectors, the collision efficiency is given by:

$$\alpha = \eta/\eta_0 \tag{5.132}$$

where η_0 is the single collector efficiency calculated from the convective-diffusion equation, without the inclusion of double layer interaction energy. In

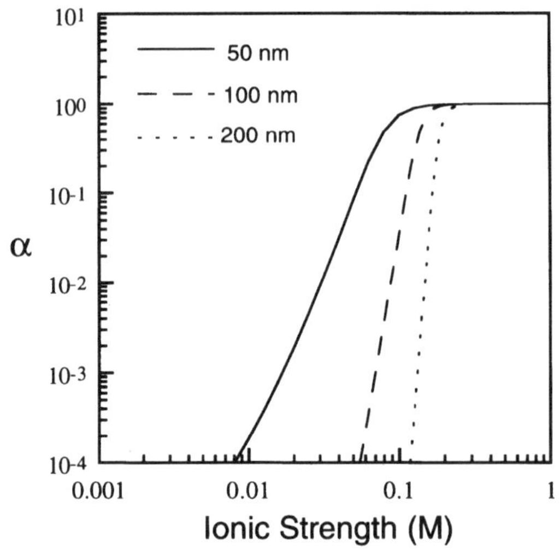

Figure 5.9 *Stability curves (collision efficiency as a function of ionic strength) in the presence of 1–1 electrolyte, for three colloidal suspensions of different particle diameter (d_p). The following parameters were used in the calculations: surface potentials of particles and collectors = –40 and –20 mV, respectively; collector diameter = 0.2 mm; approach velocity = 10^{-3} m s^{-1}; porosity = 0.4; temperature = 25°C; and Hamaker constant = 1×10^{-20} J*

the calculations of η presented here, the expression of Hogg *et al.* (1966) for the double layer interaction energy was employed. This expression is for interaction at constant surface potential. The van der Waals interaction energy was calculated from the expression of Gregory (1975) for retarded interaction. Expressions for these interaction energies, among others, were presented in Chapter 3.

Effect of ionic strength and particle size

The effect of ionic strength (1–1 electrolyte) on the collision efficiency of three colloidal suspensions is shown in Figure 5.9, for the specified physical and chemical conditions. These curves, describing the logarithm of the collision efficiency as a function of the logarithm of the electrolyte concentration, are usually referred to as *stability curves*. It is demonstrated that the collision efficiency increases with an increase in the ionic strength of the solution. This is caused by the decrease in the range and magnitude of the repulsive double layer interactions, and, consequently, the height of the energy barrier as the ionic strength increases. At high ionic strength, the energy barrier is eliminated and the collision efficiency approaches unity (i.e. $\eta = \eta_0$). The deposition under this condition is termed *favourable deposition* as opposed to *unfavourable deposition* when double layer repulsion predominates.

The results shown in Figure 5.9 also demonstrate the effect of particle size on the collision efficiency in particle deposition kinetics. It is observed that particle

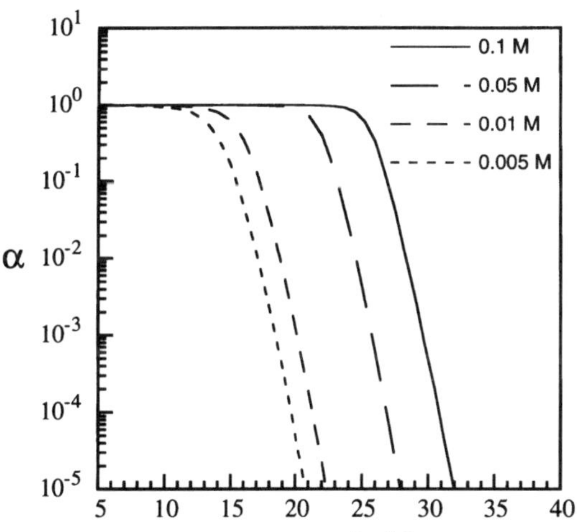

Figure 5.10 *Collision efficiency as a function of the surface potentials of particles (ψ_p) and collectors (ψ_c); the surface potentials are assumed to be similar. The calculations were carried out for different ionic strengths (1–1 electrolyte), as specified in the figure. The following parameters were used in the calculations: particle diameter = 100 nm; collector diameter = 0.2 mm; approach velocity = 10^{-3} m s^{-1}, porosity = 0.4; temperature = 25° C; and Hamaker constant = 1×10^{-20} J*

size has a marked effect on the collision efficiency when repulsive double layer interactions predominate. At a given ionic strength, smaller particles show a larger collision efficiency in the unfavourable deposition region. The slopes of the stability curves are found to be proportional to the size of the particles. The slopes are about 4, 8 and 16 for the 50-nm, 100-nm and 200-nm suspensions, respectively. In addition, it is illustrated that the critical deposition concentrations increase with an increase of particle size. The critical deposition concentration is the electrolyte concentration corresponding to the transition from unfavourable to favourable deposition (i.e. the bend in the stability curves).

The effects of particle size demonstrated in Figure 5.9 are a result of the linear dependence of the electrical double layer repulsion energy and, consequently, the height of the energy barrier on particle size. DLVO theory predicts higher energy barriers for larger particles, resulting in smaller collision efficiencies. A detailed analysis and discussion of the role of particle size in colloid deposition in the presence of repulsive double layer interactions is given by Elimelech and O'Melia (1990a).

Effect of surface potentials

Figure 5.10 illustrates the effect of surface potentials of particles and collectors on the collision efficiency at various ionic strengths. For convenience, it was assumed that the surface potentials of particles and collectors are identical in these model predictions. As observed, the collision efficiencies at a given ionic strength are dramatically reduced as the surface potential increases above a

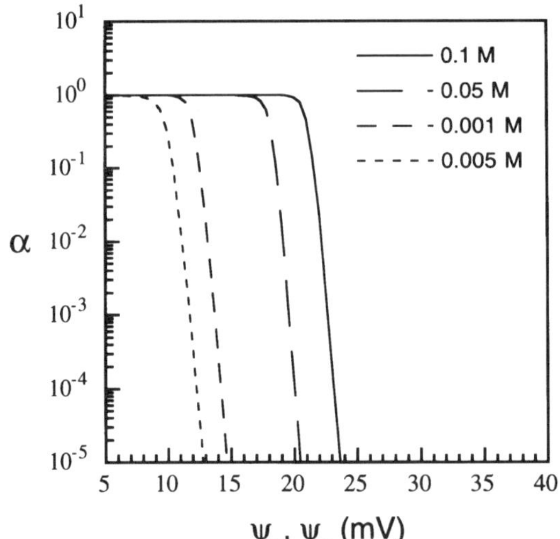

Figure 5.11 *Collision efficiency as a function of the surface potentials of particles (ψ_p) and collectors (ψ_c); the surface potentials are assumed to be similar. The calculations were carried out for different ionic strengths (1–1 electrolyte), as specified in the figure. The parameters that were used in the calculations are similar to those described in Figure 5.10, except that the particle diameter was 500 nm*

certain value. This is attributed to the increase of the energy barrier with an increase of the surface potentials. The energy barrier is very sensitive to the surface potentials, and it markedly increases above a critical surface potential. At high ionic strengths, the transition from favourable deposition ($\alpha = 1$) to unfavourable deposition ($\alpha < 1$) occurs at higher surface potentials, since the energy barrier decreases as the ionic strength increases.

Model predictions for larger particles under similar chemical conditions are shown in Figure 5.11. As observed, the drop in the collision efficiency curves is more pronounced (compared to the results for the smaller particles shown in Figure 5.10). In addition, the transition from favourable to unfavourable deposition corresponds to smaller surface potentials. These findings are attributed to the effect of particle size. As discussed earlier, higher energy barriers are predicted for larger particles.

Effect of Hamaker constant

Based on the DLVO theory, the total interaction energy between a particle and a collector surface is the sum of van der Waals and electrical double layer interactions. As shown in Chapter 3, the van der Waals attraction energy between a particle and a surface is proportional to the Hamaker constant of the interacting media. Hence, varying the Hamaker constant can cause changes in the van der Waals attraction energy and, therefore, the total interaction energy.

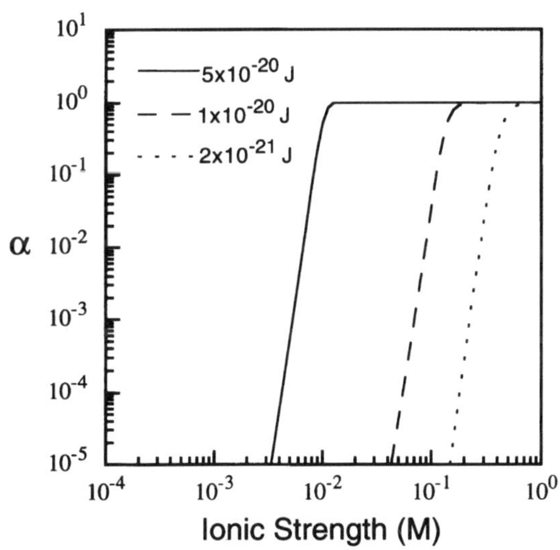

Figure 5.12 *Stability curves (collision efficiency as a function of ionic strength) in the presence of 1–1 electrolyte, calculated for three different Hamaker constants (specified in the figure). The following parameters were used in the calculations: surface potentials of particles and collectors = –40 and –20 mV, respectively; particle diameter = 100 nm; collector diameter = 0.2 mm; approach velocity =* 10^{-3} *m* s^{-1}, *porosity = 0.4; and temperature = 25°C*

The effect of the Hamaker constant of the interacting media on the collision efficiency is shown in Figure 5.12. The Hamaker constant has no effect on the slope of the stability curves since the universal van der Waals attraction is not affected by the ionic strength of the solution. However, the critical deposition concentrations are influenced by the Hamaker constant. Smaller critical deposition concentrations are observed for larger Hamaker constants. When the Hamaker constant of the interacting media is large, the energy barrier is eliminated at lower ionic strengths, resulting in $\alpha = 1$.

Effect of double layer interaction mode

As discussed in Chapter 3, several modes of double layer interaction in particle–particle or particle–collector interaction phenomena exist. Several expressions to calculate the double layer interaction energy were described in Chapter 3. Among these are: (1) the expression of Hogg *et al.* (1966) for interaction at constant surface potential; (2) the expression of Wiese and Healy (1970) for interaction at constant surface charge; and (3) the linear superposition approximation (Gregory, 1981) for intermediate interaction.

The significant effects of the double layer interaction mode on predicted stability curves are illustrated in Figure 5.13. When the expression for interaction

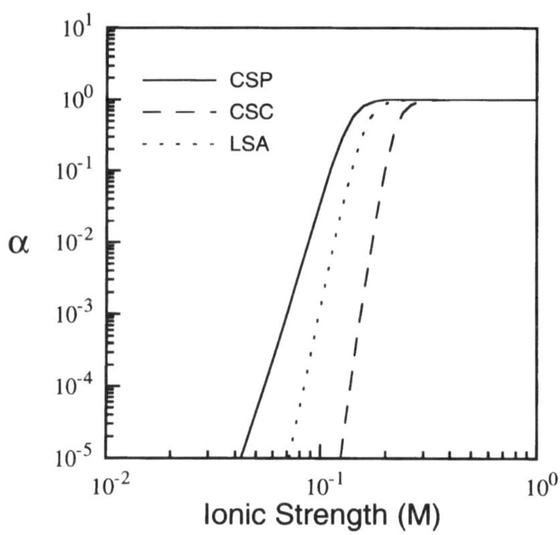

Figure 5.13 *Stability curves (collision efficiency as a function of ionic strength) in the presence of 1–1 electrolyte, calculated using various expressions for the electrical double layer interaction energy. The expressions that were used are for interaction at constant surface potential (CSP); constant surface charge (CSC); and the linear superposition approximation (LSA). The following parameters were used in the calculations: surface potentials of particles and collectors = –40 and –20 mV, respectively; particle diameter = 100 nm; collector diameter = 0.2 mm; approach velocity =* 10^{-3} *m* s^{-1}*; porosity = 0.4; temperature = 25°C; and Hamaker constant =* 1×10^{-20} *J*

at constant charge is used, the calculated electrical double layer repulsion and the energy barrier are higher. Consequently, both the critical deposition concentration and the slope of the corresponding stability curves are increased. On the other hand, using the expression for interaction at constant surface potential produces lower energy barriers. As a result, a smaller slope of the corresponding stability curve and a lower critical deposition concentration are obtained. Intermediate results are obtained with the use of the linear superposition approximation.

5.7.2 Deposition in the presence of attractive double layer interactions

Particle deposition rates can be enhanced when particles and collectors are oppositely charged. A detailed discussion of the mechanisms of deposition enhancement in the presence of attractive double layer interactions will be given in Chapter 11. Illustrative examples for the effect of some chemical-colloidal variables on particle deposition onto spherical collectors are described below. It is convenient to present the particle deposition results as the ratio of the dimensionless deposition rate in the presence of double layer attraction (η) to the

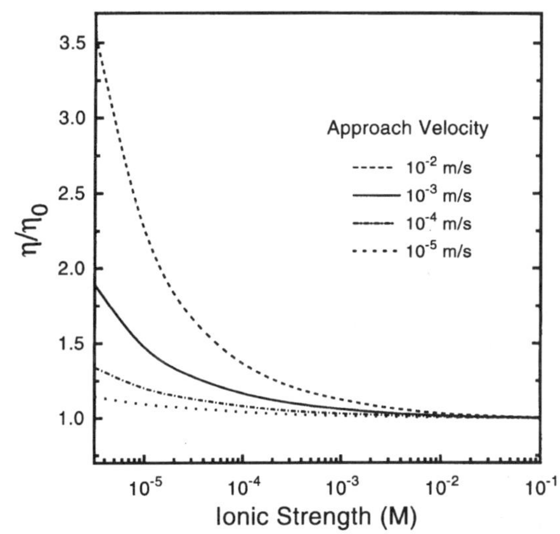

Figure 5.14 *The enhancement in particle deposition (expressed as η/η_0) as a function of the ionic strength (1–1 electrolyte) of the solution at various approach velocities. The curves from top to bottom are for approach velocities of 10^{-2}, 10^{-3}, 10^{-4} and 10^{-5} m s^{-1}, respectively. The following parameters were used in the calculations: surface potentials of particles and collectors = –40 and 20 mV, respectively; particle diameter = 100 nm; collector diameter = 0.2 mm; porosity = 0.38; temperature = 25°C; and Hamaker constant = 1×10^{-20} J*

deposition rate in the absence of double layer interaction (η_0). This ratio is larger than or equal to unity; it reflects the enhancement in colloid deposition rate induced by the attractive double layer interactions. In all the model predictions below (unless otherwise specified), the expression of Hogg *et al.* (1966) for the double layer interaction energy was employed. This expression is for interaction at constant surface potential. The van der Waals interaction energy was calculated from the expression of Gregory (1975) for retarded interaction. Expressions for these interactions were described in Chapter 3.

Effect of ionic strength and fluid velocity

The ratio η/η_0 as a function of ionic strength (1–1 electrolyte) for two colloidal suspensions of different size are presented in Figures 5.14 and 5.15. The curves were calculated for various approach velocities as indicated in these figures. It is shown that particle deposition rates for both suspensions increase as the ionic strength of the solution decreases. The increase is much more pronounced for larger particles (Figure 5.15) and higher fluid velocities. At ionic strengths larger than 10^{-3} M, particle deposition rates approach a constant value equal to the deposition rate in the absence of double layer interaction (η_0). At high fluid velocities, more particles are transported from the bulk solution to the region where attractive double layer interactions are important, thus enhancing the

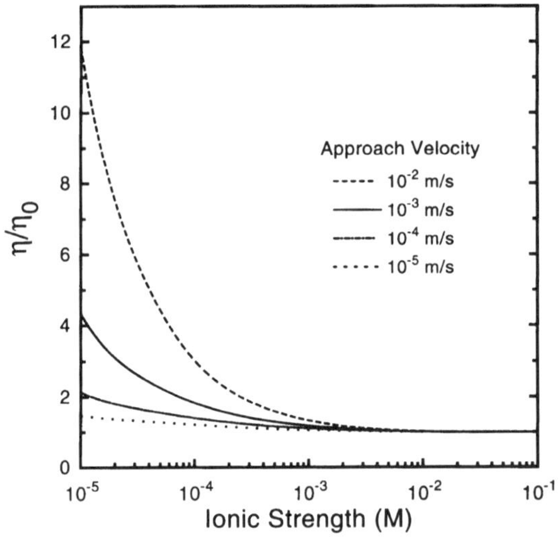

Figure 5.15 *The enhancement in particle deposition (expressed as η/η_0) as a function of the ionic strength (1–1 electrolyte) of the solution at various approach velocities. The curves from top to bottom are for approach velocities of 10^{-2}, 10^{-3}, 10^{-4} and 10^{-5} m s^{-1}, respectively. The following parameters were used in the calculations: surface potentials of particles and collectors = –40 and 20 mV, respectively; particle diameter = 800 nm; collector diameter = 0.2 mm; porosity = 0.38; temperature = 25°C; and Hamaker constant = 1×10^{-20} J*

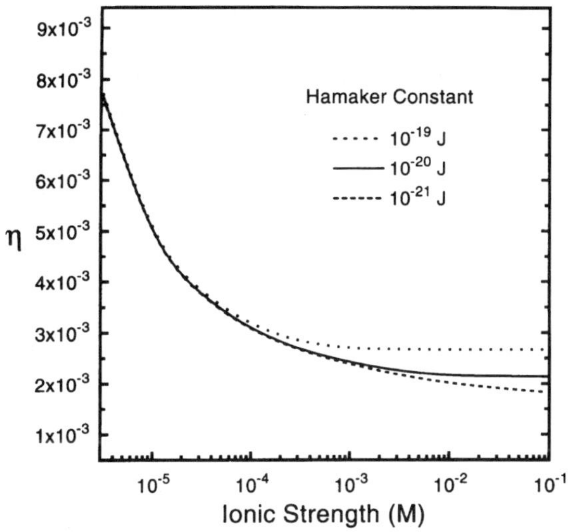

Figure 5.16 *Particle deposition (expressed as η) as a function of solution ionic strength (1–1 electrolyte) for various Hamaker constants of the interacting media. The following parameters were used in the calculations: surface potentials of particles and collectors = –40 and 20 mV, respectively; particle diameter = 500 nm; collector diameter = 0.3 mm; porosity = 0.38; temperature = 25°C; and approach velocity = 10^{-3} m s^{-1}*

transport rate of particles towards the collector. The role of particle size in particle deposition in the presence of attractive double layer interactions will be discussed in more detail in Chapter 11.

Effect of Hamaker constant

The effect of the Hamaker constant on the rate of particle deposition under attractive double layer interactions is shown in Figure 5.16. The Hamaker constant was varied over two orders of magnitude. For most interacting bodies in aqueous media, Hamaker constants fall into this range.

The results show that the particle deposition rates at low ionic strengths are not influenced by the value of the Hamaker constant. At ionic strengths larger than 10^{-4} M, however, deposition rates are affected by the Hamaker constant. These results are attributed to the relative range of the double layer and van der Waals forces. At low ionic strengths, the range of the double layer attractive interactions is much larger than that of the van der Waals attraction (Elimelech and Song, 1992a). As a result, particle deposition rates are controlled solely by double layer interactions. At higher ionic strengths, on the other hand, the range of the double layer and van der Waals interactions is comparable, and, as a result, deposition rates are affected by the Hamaker constant.

Effect of surface potentials of particles and collectors

It was shown previously (Figures 5.14 and 5.15) that, for a given colloidal suspension, the deposition rates are enhanced at low ionic strengths. This was attributed to the increase in the range of the electrical double layer interactions as the ionic strength decreases. These model predictions, however, were carried out for fixed surface potentials of particles and collectors. Surface potentials control the magnitude of the double layer interaction energy and, thus, may have an effect on the enhancement of particle deposition rates. The role of surface potentials of particles and collectors in the enhancement of particle deposition rates under attractive double layer interactions is described below.

The effect of surface potentials of particles and collectors on the enhancement of particle deposition (η/η_0) is shown in Figure 5.17. Each one of the curves was calculated at a different ionic strength (1–1 electrolyte). The absolute values of the surface potentials of particles and collectors were assumed to be equal (but of opposite sign) in these calculations. As observed, the magnitude of the surface potentials has a substantial effect on the deposition rate at low ionic strengths, whereas it has a negligible effect on the deposition rate at high ionic strength. These results demonstrate that the presence of long-range attractive double layer forces (which exist only at low ionic strengths) is a necessary condition for the enhancement of particle deposition. It is also shown that the magnitude of the double layer forces does affect the deposition rate under this condition.

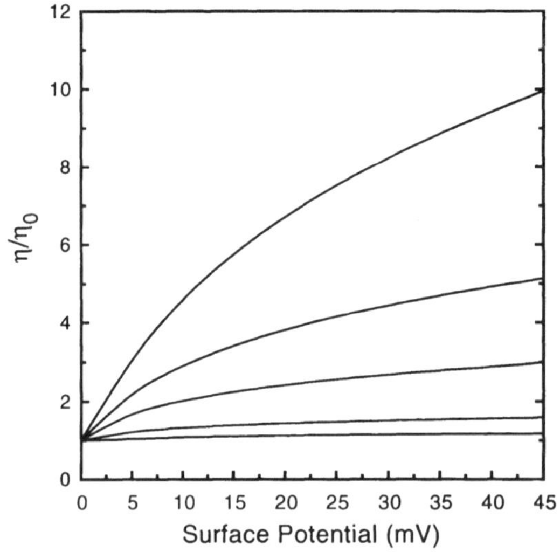

Figure 5.17 *The enhancement in particle deposition (expressed as η/η_0) as a function of the surface potentials of particles and collectors (assumed to have similar surface potentials). The curves were calculated for various ionic strengths of 1–1 electrolyte (the curves from top to bottom are for 10^{-6}, $10^{-5.5}$, 10^{-5}, 10^{-4} and 10^{-3} M, respectively). The following parameters were used in the calculations: approach velocity = 10^{-3} m s^{-1}; particle diameter = 400 nm; collector diameter = 0.2 mm; porosity = 0.38; temperature = 25°C; and Hamaker constant = 1×10^{-20} J*

Enhancement of particle deposition at high ionic strength is not feasible, even when the magnitude of the attractive double layer energy is substantial (i.e. at high surface potentials of particles and collectors). At high ionic strengths, the range of the electrical double layer forces is not sufficiently large to enhance the transport rate of particles towards the collector surfaces.

Effect of double layer interaction mode

In the previous model predictions (Figures 5.14–5.17), the electrical double layer interaction energy was calculated using the expression of Hogg *et al.* (1966) for interaction at constant surface potential. However, different interaction energy profiles are obtained when using other expressions, such as those for constant surface charge or intermediate interactions. It was shown previously (Figure 5.13) that calculated deposition rates are significantly affected by the mode of double layer interaction, when particles and collectors are similarly charged. This effect for oppositely charged particles and collectors is shown below.

The enhancements in deposition rates as a function of ionic strength (1–1 electrolyte), calculated with the expressions for constant surface potential and the linear superposition approximation, are compared to each other in Figure 5.18. It is shown that the deposition rates obtained with both expressions are comparable,

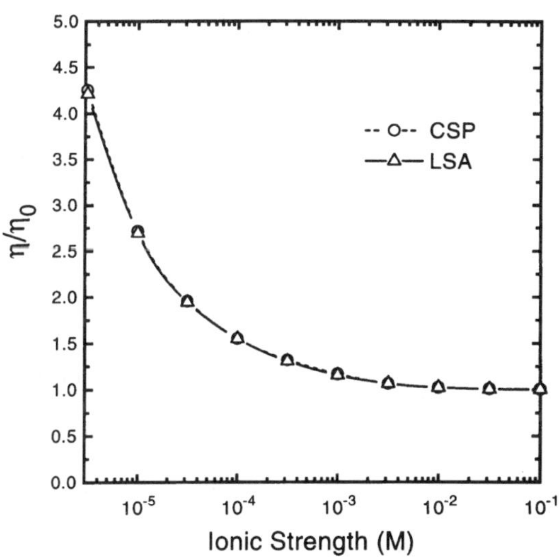

Figure 5.18 *The enhancement in particle deposition (expressed as* η/η_0*) as a function of ionic strength (1–1 electrolyte) calculated using two different expressions for the electrical double layer interaction energy. The expressions that were used are for interaction at constant surface potential (CSP) and the linear superposition approximation (LSA). The following parameters were used in the calculations: surface potentials of particles and collectors = –40 and 20 mV, respectively; approach velocity =* 2×10^{-3} *m* s^{-1}*; particle diameter = 600 nm; collector diameter = 0.4 mm; porosity = 0.38; temperature = 25°C; and Hamaker constant =* 1×10^{-20} *J*

in contrast to the case where particles and collectors are similarly charged. The similarity in the curves is attributed to the dependence of the interaction energy profiles at large distances. At large distances from the collector surface, the interaction energies calculated from both expressions (i.e. the constant surface potential and intermediate interactions) are comparable. The main difference in the interaction energy profiles is observed at short distances. The attraction at short distances is much larger when the expression for constant surface potential is used. However, since the enhancement in colloid deposition rates at low ionic strengths is attributed to the attractive interactions at large distances, these differences in the interaction energy at short distances have no effect on the overall deposition rates.

Bibliography

Churchill, S. W. (1988) *Viscous Flows: The Practical Use of Theory*, Butterworth Publishers, Stoneham, Massachusetts

Levich, V. G. (1962) *Physicochemical Hydrodynamics*, Prentice Hall, New Jersey

Russel, W. B., Saville, D. A. and Schowalter, W. R. (1989) *Colloidal Dispersions*, Cambridge University Press, Cambridge

Schlichting, H. (1960) *Boundary Layer Theory*, 7th edn, English Translation by J. Kestin, McGraw-Hill, New York

Tien, C. (1989) *Granular Filtration of Aerosols and Hydrosols*, Butterworth Publishers, Stoneham, Massachusetts

van de Ven, T. G. M. (1989) *Colloidal Hydrodynamics*, Academic Press, London

References

Albrecht, F. (1931) Deposition of dust from air currents and the theory of dust filter. *Physik Z.*, **32**, 48–56

Adamczyk, Z. and van de Ven, T. G. M. (1981) Deposition of particles under external forces in laminar flow through parallel-plate and cylindrical channels. *J. Colloid Interface Sci.*, **80**, 340–356

Adamczyk, Z., Czarnecki, J., Dabros, T. and van de Ven, T. G. M. (1983) Particle transfer to solid surfaces. *Adv. Colloid Interface Sci.*, **19**, 183–252

Adamczyk, Z., Zembala, M., Siwek, B. and Czarnecki, J. (1986) Kinetics of latex deposition from flowing suspensions. *J. Colloid Interface Sci.*, **110**, 188–200

Bowen, B. D. and Epstein, N. (1979) Fine particle deposition in smooth parallel-plate channels. *J. Colloid Interface Sci.*, **72**, 81–97

Bowen, B. D., Levine, S. and Epstein, N. (1976). Fine particle deposition in laminar flow through parallel-plate and cylindrical channels. *J. Colloid Interface Sci.*, **72**, 81–97

Brenner, H. (1961) The slow motion of a sphere through a viscous fluid towards a plane surface. *Chem. Eng. Sci.*, **16**, 242–251

Brinkman, H. C. (1947) A calculation of the viscous force exerted by a flowing fluid on a dense swarm of particles. *Appl. Sci. Res.*, **A1**, 27–34

Chari, K. and Rajagopalan, R. (1985a) Deposition of colloidal particles in stagnation-point flow. *J. Chem. Soc., Faraday Trans 2*, **81**, 1354–1366

Chari, K. and Rajagopalan, R. (1985b) Transport of colloidal particles over energy barriers. *J. Colloid Interface Sci.*, **107**, 278–282

Cochran, W. G. (1934) The flow due to a rotating disc. *Proc. Camb. Phil. Soc.*, **30**, 365–375

Dabros, T. and Adamczyk, Z. (1979) Noninertial particle transfer to the rotating disc under an external force field (laminar flow). *Chem. Eng. Sci.*, **14**, 1041–1049

Dabros, T. and van de Ven, T. G. M. (1983) A direct method for studying particle deposition onto solid surfaces. *Colloid & Polymer Sci.*, **261**, 694–707

Dabros, T. and van de Ven, T. G. M. (1987) Deposition of latex particles on glass surfaces in an impinging jet. *Phys Chem Hydrodynamics*, **8**, 171–172

Dahneke, B. (1974) Diffusional deposition of particles. *J. Colloid Interface Sci.*, **48**, 520–522

Elimelech, M. (1991) Kinetics of capture of colloidal particles in packed beds under attractive double layer interactions. *J. Colloid Interface Sci.*, **146**, 337–352

Elimelech, M. and O'Melia, C. R. (1990a) Effect of particle size on collision efficiency in the deposition of Brownian particles with electrostatic energy barriers. *Langmuir*, **6**, 1153–1163

Elimelech, M. and O'Melia, C. R. (1990b) Kinetics of deposition of colloidal particles in porous media. *Environ. Sci. Technol.*, **246**, 1153–1163

Elimelech, M. and Song, L. (1992a) Theoretical investigation of colloid separation from dilute aqueous suspensions by oppositely charged granular media. *Separations Technol.*, **2**, 2–12

Elimelech, M. and Song, L. (1992b) Deposition of colloids in porous media: theory and numerical solution. In *Transport and Remediation of Subsurface Contaminants: Colloidal, Interfacial, and Surfactant Phenomena*, (ed. Sabatini, D. A. and Knox, R. C.) *American Chemical Society Symposium Series 491*, 26–39

Goldman, A. J., Cox, R. G. and Brenner, H. (1967a) Slow viscous motion of a sphere parallel to a plane wall – I. Motion through a quiescent fluid. *Chem. Eng. Sci.*, **22**, 637–651

Goldman, A. J., Cox, R. G. and Brenner, H. (1967b) Slow viscous motion of a sphere parallel to a plane wall – II. Couette flow. *Chem. Eng. Sci.*, **22**, 653–660

Goren, S. L. and O'Neill, M. E. (1971) On the hydrodynamic resistance to a particle of a dilute suspension when in the neighborhood of a large obstacle. *Chem. Eng. Sci.*, **26**, 325–338

Gregory, J. (1975) Interaction of unequal double layers at constant charge. *J. Colloid Interface Sci.*, **61**, 44–51

Gregory, J. (1981) Approximate expressions for the retarded van der Waals interaction. *J. Colloid Interface Sci.*, **83**, 138–145

Happel, J. (1958) Viscous flow in multiparticle systems: slow motion of fluids relative to beds of spherical particles. *Am. Inst. Chem. Eng. J.*, **4**, 197–201

Hogg, R., Healy, T. W. and Fuerstenau, D. W. (1966) Mutual coagulation of colloidal dispersions. *Trans Faraday Soci.*, **66**, 1638–1651

Kuwabara, S. (1959) The forces experienced by randomly distributed parallel circular cylinders of spheres in a viscous flow at small Reynolds numbers. *J. Phys. Soc. Japan*, **14**, 527–532

Leveque, A. (1928) Laws of heat transmission by convection. *Ann Mines*, **13**, 201–299

Levich, V. G. (1962) *Physicochemical Hydrodynamics*, Prentice Hall, New Jersey

Marshal, J. K. and Kitchener, J. A. (1966) The deposition of colloidal particles on smooth solids. *J. Colloid Interface Sci.*, **22**,, 342–351

Matijevic, E. and Kallay, N. (1983) Kinetics of deposition of colloidal metal oxides particles on a steel surface. *Croatica Chem. Acta*, **56**, 649–661

O'Melia, C. R. and Stumm, W. (1967) Theory of water filtration. *J. Am. Water Works Assoc.*, **59**, 1393–1412

Payatakes, A. C., Rajagopalan, R. and Tien, C. (1974) On the use of Happel's model for filtration studies. *J. Colloid Interface Sci.*, **49**, 321–325

Prieve, D. C. and Lin, M. M. (1980) Adsorption of Brownian hydrosols onto a rotating disc aided by a uniform applied force. *J. Colloid Interface Sci.*, **76**, 32–47

Prieve, D. C. and Ruckenstein, E. (1974) Effect of London forces upon the rate of deposition of Brownian particles. *Am. Inst. Chem. Eng. J.*, **20**, 1178–1187

Prieve, D. C. and Ruckenstein, E. (1976) Rates of deposition of Brownian particles calculated by lumping interaction force into a boundary condition. *J. Colloid Interface Sci.*, **57**, 547–550

Rajagopalan, R. (1974) Stochastic modeling and experimental analysis of particle transport in water filtration. PhD Dissertation, Syracuse University, Syracuse, New York

Rajagopalan, R. and Kim, J. S. (1981) Absorption of Brownian particles in the presence of potential barriers: effect of different modes of double layer interaction. *J. Colloid Interface Sci.*, **83**, 428–448

Rajagopalan, R. and Tien, C. (1976) Trajectory analysis of deep bed filtration with the sphere-in-cell porous media model. *Am. Inst. Chem. Eng. J.*, **22**, 523–533

Rajagopalan, R. and Tien, C. (1979) The theory of deep bed filtration. In *Progress in Filtration and Separation* (ed. Wakeman, R. J.) vol. 1, Elsevier, Amsterdam, 179–269

Ruckenstein, E. and Prieve, D. C. (1973) Rate of deposition of Brownian particles under the action of London and double-layer forces. *J. Chem. Soc. Faraday Trans II*, **69**, 1522–1536

Schlichting, H. (1960) *Boundary Layer Theory*, 7th edn, English Translation by J. Kestin, McGraw-Hill, New York

Sell, W. (1931) Staubausscheidung en einfachen Koerpern und in Luftiltern. *Ver. Dent. Ing. Forschungsheft*, No. 347.

Sjollema, J. and Busscher, H. J. (1989) Deposition of polystyrene latex particles toward polymethylmethacrylate in a parallel plate flow cell. *J. Colloid Interface Sci.*, **132**, 382–394

Sjollema, J. and Busscher, H. J. (1990) Deposition of polystyrene particles in a parallel plate flow cell. 1. The influence of collector surface properties on the experimental deposition rate. *Colloids Surfaces*, **47**, 323–336

Song, L. and Elimelech, M. (1992) Deposition of Brownian particles in porous media: modified boundary conditions for the sphere-in-cell model. *J. Colloid Interface Sci.*, **153**, 294–297

Song, L. and Elimelech, M. (1993) Calculation of particle deposition rate under unfavourable particle–surface interactions. *J. Chem. Soc. Faraday Trans.* **89**, 3443–3542

Sparrow, F. M. and Gregg, J. L. (1959) Heat transfer from a rotating disc to fluids of any Prandtl number. *J. Heat Transfer, Tans. ASME Ser. C*, **81**, 249–252.

Spielman, L. A. and FitzPatrick, J. A. (1973) Theory for particle collection under London and gravity forces. *J. Colloid Interface Sci.*, **42**, 607–623

Spielman, L. A. and Friedlander, S. K. (1974) Role of electrical double layer in particle deposition by convective diffusion. *J. Colloid Interface Sci.*, **46**, 22–31

Tien, C. (1989) *Granular Filtration of Aerosols and Hydrosols*, Butterworth Publishers, Stoneham, Massachusetts

Tobiason, J. E. (1987) Physicochemical aspects of particle deposition in porous media. PhD Dissertation, The Johns Hopkins University, Baltimore, Maryland

Vaidyanathan, R. (1986) Hydrosol filtration in granular beds. MS Thesis, Syracuse University, Syracuse, NY

Von Kármán, Th. (1921) Laminare und turbulente reibung. *Z. Angew. Math. Mech.*, **1**, 233–252

Wiese, G. R. and Healy, T. W. (1970) Effect of particle size on colloid stability. *Trans Faraday Soc.*, **66**, 490–499

Yao, K. M. (1968) Influence of suspended particle size on the transport aspect of water filtration. PhD Dissertation, University of North Carolina, Chapel Hill, North Carolina

Yao, K. M., Habibian, M. T. and O'Melia, C. R. (1971) Water and wastewater filtration: concepts and applications. *Environ. Sci. Technol.*, **5**, 1105–1112

Chapter 6

Modelling of aggregation processes

Nomenclature

a_i	radius of particle i
A	Hamaker constant
B	constant in eqn (6.43)
c	electrolyte concentration
C	constant in eqn (6.47)
d	separation distance between particle surfaces
d_F	fractal dimension
d_{max}	limiting aggregate size
D_i	diffusion coefficient of particle i
D_{ij}	mutual diffusion coefficient for particles i and j
e	electronic charge (1.602×10^{-19} C)
f	fraction of added polymer necessary for destabilization
$f(x)$	frequency function
g	gravitational acceleration
G	shear rate
J_i	flux of i-particles to a sphere
J_{ij}	collision frequency for particles of type i and j
k	Boltzmann's constant (1.381×10^{-23} J K^{-1})
k'	aggregation number
$\bar{k}$	average aggregation number
k_a	aggregation rate constant
k_{ij}	collision rate constant for particles of type i and j
L	size of aggregate
M	mass of aggregate
n	exponent in eqn (6.47)
n_0	initial concentration of primary particles
n_i	number concentration of particles of type i
n_T	total particle concentration
R_{ij}	collision radius for particles i and j
t	time
t_{Ads}	characteristic adsorption time
T	absolute temperature
u	dimensionless separation distance – see eqn (6.32)
v	volume of an aggregate
v_g	number median particle volume
v_T	total volume of particles

w	degree of homogeneity – see eqn (6.53)
W	stability ratio
x	dimensionless aggregate size – see eqn (6.50)
y	exponent in eqn (6.43)
z	valence of ion
α	collision efficiency
α_0	limiting collision efficiency
$\beta(u)$	function of u, given in eqn (6.36)
γ	dimensionless function of surface potential – see eqn (6.35)
δ	separation of particles in primary minimum (m)
ϵ	power input per unit mass
ζ	electrokinetic (zeta) potential
η_K	Kolmogoroff microscale
θ	fractional surface coverage
κ	Debye–Hückel parameter
λ	wavelength
μ	dynamic viscosity
ν	kinematic viscosity
ρ	density of fluid
ρ_E	effective (buoyant) density of aggregate
ρ_s	density of particle
σ	geometrical standard deviation
τ	characteristic aggregation time
ϕ	volume fraction
ϕ_A	Van der Waals interaction
ϕ_{max}	height of energy barrier
ϕ_T	total interaction energy
ψ_δ	Stern potential

6.1 Collisions and aggregation: the Smoluchowski approach

Most discussions of the rate of aggregation start from the classic work of Smoluchowski (1917), which laid the foundations of the subject. It is convenient to think in terms of a dispersion of initially identical particles (primary particles), which, after a period of aggregation, contains aggregates of various sizes and different concentrations – n_i particles of size i, n_j particles of size j etc. Here, n_i etc. refer to the number concentrations of different aggregates and 'size' implies the number of primary particles comprising the aggregate; we can speak of 'i-fold' and 'j-fold' aggregates. A fundamental assumption is that aggregation is a second-order rate process, in which the rate of collision is proportional to the product of concentrations of two colliding species. (Three-body collisions are usually ignored in treatments of aggregation – they only become important at very high particle concentrations). Thus, the number of collisions occurring between i and j particles in unit time and unit volume, J_{ij}, is given by:

$$J_{ij} = k_{ij} n_i n_j \tag{6.1}$$

where k_{ij} is a second-order rate constant, which depends on a number of factors, such as particle size and transport mechanism (see below).

In considering the rate of aggregation, we must recognize that, because of interparticle forces, not all collisions may be successful in producing aggregates. The fraction of successful collisions is called the collision efficiency and given the symbol α. If there is strong repulsion between particles then practically no collision gives an aggregate and $\alpha \approx 0$. When there is no significant net repulsion or when there is an attraction between particles, then the collision efficiency can approach unity.

Although there are some theoretical difficulties, it is usual to assume that the collision rate is independent of colloid interactions and depends only on particle transport. This assumption can often be justified on the basis of the short-range nature of interparticle forces, which operate over a range which is usually much less than the particle size, so that particles are nearly in contact before these forces come into play. The 'decoupling' of transport and attachment steps greatly simplifies the analysis of aggregation kinetics and a similar assumption is common in simple treatments of particle deposition.

A very important case where this approach is not justified is that of hydrodynamic or viscous interaction, which involves much longer-range effects. This will be treated separately in Section 6.3.3.

For the present, we shall assume that every collision is effective in forming an aggregate (i.e. the collision efficiency, $\alpha = 1$), so that the aggregation rate constant is the same as the collision rate constant. It is then possible to write the following expression for the rate of change of concentration of k-fold aggregates, where $k = i + j$:

$$\frac{dn_k}{dt} = \frac{1}{2} \sum_{\substack{i+j \to k \\ i=1}}^{i=k-1} k_{ij} n_i n_j - n_k \sum_{k=1}^{\infty} k_{ik} n_i \tag{6.2}$$

The first term on the right-hand side represents the rate of formation of k-fold aggregates by collision of any pair of aggregates, i and j, such that $i + j = k$. Carrying out the summation by this method would mean counting each collision twice and hence the factor 1/2 is included. The second term accounts for the loss of k-fold aggregates by collision, and aggregation, with any other aggregates. The terms k_{ij} and k_{ik} are the appropriate rate constants. It is important to note that eqn (6.2) is for *irreversible* aggregation, since no allowance is made for break-up of aggregates.

For continuous particle size distributions, an integral version of eqn (6.2) can be written. In principle, it is then possible to derive the evolution of the aggregate size distribution with time (see Section 6.6.1), but there are formidable difficulties, especially in assigning values to the rate coefficients. These depend greatly on the nature of the particles and on the way in which collisions are brought about. The simplest assumption is that spherical particles *coalesce* on contact to form a larger sphere with the same total volume. This is physically unrealistic except for liquid (emulsion) droplets, but has often been assumed in earlier treatments of aggregation kinetics.

There are three important transport mechanisms in practice: (1) Brownian diffusion (giving *perikinetic* aggregation); (2) fluid motion (*orthokinetic* aggregation); and (3) differential settling. These will be considered in the next section. In all cases, we shall assume that the particles are spherical and that the collision efficiency is unity (every collision is effective in forming a permanent aggregate). Also, hydrodynamic interaction will be neglected for the time being. Although these assumptions are not realistic for practical systems, they enable simple results to be derived which can be used to illustrate the essential features of the various aggregation mechanisms.

Effects of colloidal and hydrodynamic interactions on aggregation rates will be considered in Section 6.3.

6.2 Collision mechanisms

6.2.1 Perikinetic aggregation

Small particles in suspension can be seen to undergo continuous random movements or Brownian motion. The diffusion coefficient of a spherical particle is given by the Stokes–Einstein equation:

$$D_i = \frac{kT}{6\pi a_i \mu} \tag{6.3}$$

where k is Boltzmann's constant, T the absolute temperature, a_i the particle radius and μ the viscosity of the suspending fluid.

Smoluchowski (1917) calculated the rate of diffusion of spherical particles of type i to a fixed sphere j. If each i particle is captured by the central sphere on contact, then the i particles are effectively removed from the suspension and a concentration gradient is established in the radial direction towards the sphere, j. After a very brief interval, steady-state conditions are established and it can easily be shown that the number of i particles contacting j in unit time is:

$$J_i = 4\pi R_{ij} D_i n_i \tag{6.4}$$

where D_i is the diffusion coefficient of particles of type i and n_i is their concentration in the bulk suspension. R_{ij} is the collision radius for particles i and j, which is the centre-to-centre distance at which they may be taken to be in contact. In practice, it can usually be assumed that this is simply the sum of the particle radii, i.e:

$$R_{ij} = a_i + a_j \tag{6.5}$$

When there is significant long-range attraction between particles they may be effectively 'captured' at greater distances, so that the collision radius is rather larger than that given by eqn (6.5). However, there is usually very little error in using this approximation.

Of course, in practice, the central sphere j is not fixed, but is itself subject to Brownian diffusion. It is only necessary to replace D_i in (6) by the *mutual diffusion coefficient*, D_{ij}, to account for the motion of the j particle, with:

$$D_{ij} = D_i + D_j \tag{6.6}$$

If the concentration of j particles is n_j, then the number of i–j collisions occurring in unit volume per unit time is simply:

$$J_{ij} = 4\pi R_{ij} D_{ij} n_i n_j \tag{6.7}$$

Comparing this with eqn (6.1) gives the rate constant for perikinetic collisions, which after substituting for R_{ij} and D_{ij}, using eqns (6.3), (6.5) and (6.6), gives:

$$k_{ij} = \frac{2kT}{3\mu} \frac{(a_i + a_j)^2}{a_i a_j} \tag{6.8}$$

This result has the very important feature that, for particles of approximately equal size, the collision rate constant becomes almost independent of particle size. The term $(a_i + a_j)^2/a_i a_j$ has a constant value of about 4 when $a_i \approx a_j$. Physically, this is because increasing particle size leads to a lower diffusion coefficient, but a larger collision radius and these two effects cancel each other out when the particles are of nearly the same size. Under these conditions, the rate constant becomes:

$$k_{ij} = \frac{8kT}{3\mu} \tag{6.9}$$

Inserting values appropriate to aqueous dispersions at 25°C gives $k_{ij} = 1.23 \times 10^{-17}\,\mathrm{m^3 s^{-1}}$.

For particles of different size, eqn (6.8) shows that the collision rate constant will always be *greater* than that for equal particles. However, the assumption of a constant value of k_{ij} is a reasonable approximation for particles which differ in size by a factor of about 2 or less.

We can now return to eqn (6.2) and insert the appropriate values for the rate constants, giving the rate of change of aggregate concentration. The simplest case to consider is the very early stages of the aggregation of a monodisperse suspension. The change in the concentration of *primary* particles, n_1, can be calculated from only the second term on the right-hand side of eqn (6.2) since only the loss of such particles needs to be considered. Furthermore, the loss of primary particles, in the early stages, is almost entirely caused by collisions with other primary particles, since there are very few aggregates. This leads to the result:

$$\left(\frac{\mathrm{d}n_1}{\mathrm{d}t}\right)_{t \to 0} = -k_{11} n_1^2 \tag{6.10}$$

where k_{11} is the rate constant for primary particle collisions and has the value given in eqn (6.9).

The initial rate of decrease of the *total* particle concentration, n_T can also be calculated, since each binary collision leads to the loss of *one* particle (two primary particles lost, one aggregate gained):

$$\left(\frac{dn_T}{dt}\right)_{t\to 0} = -\frac{k_{11}}{2} n_1^2 \tag{6.11}$$

The rate constant in this case, $k_{11}/2$, is sometimes called the *aggregation rate constant* and will here be given the symbol k_a. (In other texts, the same quantity may be called the *coagulation* or *flocculation* rate constant.) The factor ½ applied to k_{11} to give the aggregation rate constant has sometimes been a source of confusion, although there is a clear physical meaning. For aqueous dispersions at 25°, the aggregation rate constant has a value of $6.13 \times 10^{-18}\ m^3s^{-1}$.

These expressions could be integrated to give the concentration of primary particles and the total particle concentration as a function of time, but the restriction to $t \to 0$ would limit the utility of the results. In fact, Smoluchowski showed that this approach is not restricted to the very early stages of the aggregation process. If we use eqn (6.2) to write the rate of change of concentration of the various aggregates as:

$$\frac{dn_1}{dt} = -k_{11}n_1^2 - k_{12}n_1n_2 - k_{13}n_1n_3 \ldots$$

$$\frac{dn_2}{dt} = \tfrac{1}{2}k_{11}n_1^2 - k_{12}n_1n_2 - k_{23}n_2n_3 \ldots \tag{6.12}$$

$$\frac{dn_3}{dt} = \tfrac{1}{2}k_{12}n_1n_2 - k_{13}n_1n_3 - k_{23}n_2n_3 \ldots$$

then the rate of change of the *total* particle concentration can be obtained by summing the terms in eqn (6.12), provided that all of the rate constants, k_{11}, k_{12}, etc., are assumed to be equal. It was shown above that, for particles not too different in size, this is a reasonable assumption and the collision rate constant is then given by eqn (6.9). This approach leads to the very simple result:

$$\frac{dn_T}{dt} = -k_a n_T^2 \tag{6.13}$$

where $n_T = n_1 + n_2 + n_3 + \ldots$ and $k_a = 4kT/3\mu$, as explained above.

The only difference between eqns (6.11) and (6.13) is that the latter contains n_T on the right-hand side, rather than n_1, which enables eqn (6.13) to be integrated immediately. With the initial condition $n_T = n_0$ when $t = 0$ (n_0 is the initial concentration of primary particles), the result is:

$$n_T = \frac{n_0}{1 + k_a n_0 t} \tag{6.14}$$

It can be seen from eqn (6.14) that the total particle concentration is reduced to half of the initial value after a characteristic aggregation time, τ, given by:

$$\tau = \frac{1}{k_a n_0} \tag{6.15}$$

This characteristic time is sometimes called the *coagulation time* or the *half-life* of the aggregation. It can also be thought of as the average time that a particle in the original suspension spends before colliding with another particle. Since the value of k_a is known for given conditions (e.g. $6.13 \times 10^{-18}\,m^3\,s^{-1}$ in water at 25°C), it is possible to calculate typical values of τ, which then depend only on the initial particle concentration. For an aqueous dispersion with $n_0 = 10^9\,cm^{-3}$ (corresponding to a volume fraction of about 5×10^{-4}, for spherical particles of 1 μm diameter) the characteristic time would be about 163 s. For a 100-fold greater particle concentration (a volume fraction of about 5% for 1-μm particles), τ would be less than 2 s.

With the above definition of τ, eqn (6.14) can be rewritten as:

$$n_T = \frac{n_0}{1 + t/\tau} \tag{6.16}$$

It is worth stressing that this expression is based on the assumptions of constant k_{ij} values and that the colliding particles are spherical. The latter assumption implies that 'coalescence' of particles occurs after collision, which may be reasonable in the case of emulsion droplets, but cannot hold for solid particles, where various aggregate shapes can arise (see Section 6.4.1). However, provided that the decrease in diffusion coefficient is roughly balanced by the increase in collision radius as aggregates grow, then the previous treatment should still be reasonably applicable.

The ratio of the initial particle number to the total number after a period of aggregation, n_0/n_T, is a measure of the average aggregate size (the average number of primary particles in an aggregate) and hence a measure of the mean aggregate *volume*. Equation (6.16) shows that the average volume increases linearly with time, for $t \gg \tau$. For 'coalesced spheres' this means that the aggregate radius should increase linearly with the *cube root* of time.

In many practical applications of aggregation, such as the flocculation process in water treatment, it is required to produce large aggregates, containing thousands of primary particles, implying a very large reduction in total particle number. It follows from eqn (6.16) that a reduction of particle concentration, by a factor of, say, 1000 (giving an average aggregate of 1000 primary particles), would require a time of about 1000τ. For dilute suspensions this could represent many hours or days. Even for concentrated suspensions, several minutes may be needed to achieve a large degree of aggregation. For this reason practical flocculation processes can rarely be carried out relying only on Brownian motion to bring about the required collisions.

It is also possible from eqn (6.12) to derive the concentrations of individual aggregate types at different times:

$$n_1 = \frac{n_0}{(1 + t/\tau)^2}$$
$$n_2 = \frac{n_0(t/\tau)}{(1 + t/\tau)^3} \qquad (6.17)$$
$$n_k = \frac{n_0(t/\tau)^{k-1}}{(1 + t/\tau)^{k+1}}$$

Results from these expressions are plotted in Figure 6.1, for aggregates up to trimers, showing the dimensionless aggregate concentration n_k/n_0 against the reduced time t/τ. It is clear that, for all aggregates, the concentration passes through a maximum at a certain time and then declines slowly. At all times the concentration of singlets (primary, unaggregated particles) exceeds that of any other individual aggregate type.

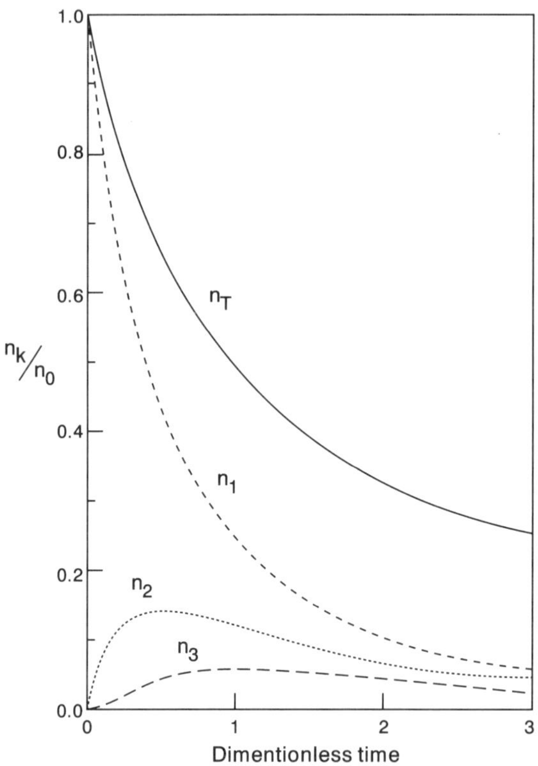

Figure 6.1 *Concentrations of primary particles* (n_1)*, doublets* (n_2)*, triplets* (n_3) *and total particles* (n_T) *as a function of reduced time* (t/τ)*, according to the Smoluchowski result, eqn (6.17). All concentrations are expressed as fractions of the initial concentration of primary particles,* n_0

Despite the simplifying assumptions made in deriving eqn (6.17), it gives reasonable agreement with measured aggregate size distributions in the early stages of the process, although with a value of k_a somewhat lower than that calculated above (Sonntag, 1993). The simple Smoluchowski approach cannot be expected to apply after substantial aggregation has occurred. In such cases, collisions between aggregates of very different sizes become significant, for which the collision rate constant will, according to eqn (6.8), be larger than that assumed here. Also, as aggregates grow, the assumption of spherical shape, implicit in the Smoluchowski treatment, becomes more doubtful. This topic will be deferred for discussion in Section 6.4.

6.2.2 Orthokinetic aggregation

We have seen that collisions brought about by Brownian motion do not generally lead to the rapid formation of very large aggregates, especially in dilute suspensions. In practice, aggregation (flocculation) processes are nearly always carried out under conditions where the suspension is subjected to some form of shear, either by stirring or by flow. Particle transport brought about by fluid motion can give an enormous increase in the rate of interparticle collisions, and aggregation brought about in this way is known as *orthokinetic aggregation*. The first treatment of the rate of orthokinetic aggregation was also due to Smoluchowski (1917), who considered only the case of uniform laminar shear. Such conditions are rarely, if ever, encountered in practice, but it is convenient to start from this simple case and then to modify the result for other conditions.

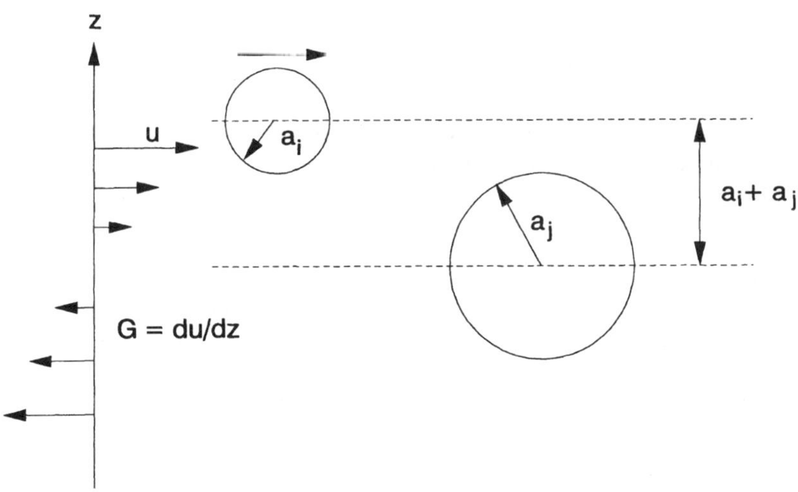

Figure 6.2 *Model for orthokinetic collision of spheres in a uniform shear field. The particles are on streamlines which are separated by a distance equal to the collision radius,* $a_i + a_j$, *and so will just collide*

A uniform laminar shear field is one in which the fluid velocity varies linearly in only one direction, perpendicular to the direction of flow. Smoluchowski assumed that particles would follow straight fluid streamlines and collide with particles moving on different streamlines, according to their relative position. The collision frequency depends on the sizes of the particles and on the velocity gradient or *shear rate*, G. By considering a fixed central sphere of radius a_j and flowing particles of radius a_i, it can be assumed that those moving particles on streamlines that bring their centres within a distance $a_i + a_j$ (the collision radius, R_{ij}) of the central particle will collide with it (Figure 6.2). The collision frequency can then be calculated by considering the flux of particles through a cylinder of radius R_{ij}, the axis of which passes through the centre of the fixed sphere j. For the conditions of Figure 6.2 it is clear that the particles in the upper half of the cylinder will move from left to right and vice versa. The total flux towards the central particle, J_i is just twice that in one half of the cylinder and is given by:

$$J_i = 4Gn_i \int_0^{R_{ij}} z \sqrt{(R_{ij}^2 - z^2)}\ \mathrm{d}z = 4Gn_i R_{ij}^3 \tag{6.18}$$

The total number of collisions occurring between i and j particles in unit volume and unit time is then simply:

$$J_{ij} = \frac{4}{3} n_i n_j G(a_i + a_j)^3 \tag{6.19}$$

It follows, by analogy with eqn (6.1), that the rate constant for orthokinetic collisions between i and j particles is:

$$k_{ij} = \frac{4}{3} G(a_i + a_j)^3 \tag{6.20}$$

The most important difference between this result and the corresponding rate constant for perikinetic collisions, eqn (6.8), is the dependence on the sizes of the colliding particles. As pointed out in Section 6.2.1, for particles of roughly equal size the perikinetic collision rate constant is nearly independent of particle size. This is most definitely not the case for orthokinetic collisions, where the rate is proportional to the *cube* of the collision radius, which has a major effect on aggregate growth rate. As aggregation proceeds and the aggregate size increases, the chance of capture becomes greater. Physically, it is clear that a large particle in a flowing or stirred suspension 'sweeps out' a larger volume than a smaller particle and has a greater opportunity of colliding with other particles. In the perikinetic case, the increased collision radius is largely compensated by the lower diffusion coefficient of larger particles.

The great dependence of the rate constant on particle size means that the assumption of a constant value of k_{ij} is not acceptable beyond the very early stages of aggregation, and aggregate concentrations, as in Figure 6.1, cannot be derived analytically. It is possible, making assumptions about the form of aggregates, to carry out numerical computations of aggregate size distributions under orthokinetic conditions (e.g. Wiesner, 1992), but this subject will be considered in Section 6.5.

For the sake of comparison with the perikinetic case, the rate of reduction in total particle concentration as a result of orthokinetic aggregation can be derived for the initial stages of the process where practically all collisions are between primary particles. By analogy with eqn (6.11), the corresponding orthokinetic rate is:

$$\frac{dn_T}{dt} = -\frac{16}{3} n_T^2 G a_1^3 \tag{6.21}$$

It has been assumed that the rate constant is given by eqn (6.20), with $a_i = a_j = a_1$, the primary particle radius, and that the total particle concentration can be taken as that of primary particles, which will only be the case in the very early stages of aggregation.

Despite its limitations, eqn (6.21) provides a useful comparison with the corresponding perikinetic expression, eqn (6.13). The ratio of the aggregation rate constants is:

$$\frac{k_a(\text{ortho})}{k_a(\text{peri})} = \frac{4G\mu a_1^3}{kT} \tag{6.22}$$

The aggregation rate constants are equal for particles of diameter 1 μm and a shear rate of about $10\,s^{-1}$ (corresponding to only mild agitation), for aqueous dispersions at room temperature. With higher shear rates and (especially) larger particles, the orthokinetic rate becomes very much greater. This explains, at least in a qualitative manner, why stirring of suspensions gives a very great improvement in the aggregation rate.

The form of eqn (6.21) is such that a simple transformation gives the result in terms of the *volume fraction*, ϕ, of the suspended particles ($=4\pi a_1^3 n_T/3$):

$$\frac{dn_T}{dt} = -\frac{4G\phi n_T}{\pi} \tag{6.23}$$

If ϕ is assumed to remain constant during the aggregation process in a closed system, then eqn (6.23) suggests a simple first-order rate law, which can be integrated to give:

$$\frac{n_T}{n_0} = \exp\left(\frac{-4G\phi t}{\pi}\right) \tag{6.24}$$

The particle concentration should thus decrease exponentially with time (and hence the average aggregate size, n_0/n_T, should increase exponentially with time). This type of behaviour has been found experimentally in some cases (Higashitani *et al.*, 1980). Although the assumption of constant volume fraction of the dispersed phase might appear reasonable from the standpoint of mass conservation, this is not the case. Growing aggregates usually adopt a rather open, *fractal* structure (see Section 6.4.1), which means that the effective occupied volume can be much larger than the total volume of the primary particles. The collision radius of an aggregate is greater than that calculated assuming 'coalesced spheres' and leads to more rapid aggregation than that predicted by eqn (6.24) (Jiang and Logan, 1991).

Nevertheless, the simplicity of eqn (6.24) has led to its widespread use for the design of practical flocculation units. At least it points out the importance of shear rate and solids concentration on the flocculation rate. For a fixed solids concentration, the extent of flocculation depends on the dimensionless number Gt, often called the 'Camp number' after Thomas R. Camp (Camp, 1953). In principle, the same degree of aggregation could be achieved by a short period of high shear or a long period of low shear, provided that the value of Gt was maintained constant. By increasing volume fraction, the same relative degree of aggregation (n_0/n_T) can be achieved with a smaller Gt value, a feature which is exploited in many practical flocculation units, where a region of high solids concentration is maintained, e.g. in upflow clarifiers or with solids recirculation. A more satisfactory dimensionless number related to the degree of aggregation is $G\phi t$ (Tambo and Watanabe, 1979), since this includes the particle concentration.

So far, it has been assumed that particles are in a uniform laminar shear field, but this is not a realistic assumption for real flocculation processes, which are carried out under turbulent conditions. One way of approaching this problem (Camp and Stein, 1943) is to calculate a mean velocity gradient, $\overline{G}$, from the power input per unit mass of fluid, ϵ:

$$\overline{G} = \sqrt{\frac{\epsilon}{\nu}} \tag{6.25}$$

where ν is the kinematic viscosity of the fluid ($= \mu/\rho$, where ρ is the density).

This average shear rate can then be inserted, in place of G, in the appropriate Smoluchowski expression, such as eqn (6.19):

$$J_{ij} = \left(\frac{4}{3}\right) n_i n_j \left(\frac{\epsilon}{\nu}\right)^{1/2} (a_i + a_j)^3 \tag{6.26}$$

The result is very like a more rigorous expression derived by Saffman and Turner (1956) for particle collisions in isotropic turbulence, but with a slightly different numerical factor (4/3 rather than 1.29). This apparent agreement is likely to be fortuitous, since the simple averaging procedure in the Camp and Stein approach is not well suited to the highly complex nature of turbulent flow (Spielman, 1978).

Turbulent flow is characterized by eddies of various sizes. The largest eddies are of comparable size to the vessel or impeller. The energy in large-scale eddies cascades through eddies of decreasing size, until, below a certain length scale, the energy is dissipated as heat. The well-known *Kolmogoroff microscale* separates the *inertial range*, where energy is transferred with very little dissipation, from the *viscous subrange*, where the energy is dissipated as heat. This length scale is given by:

$$\eta_K = \left(\frac{\nu^3}{\epsilon}\right)^{1/4} \tag{6.27}$$

For typical values of average shear rate ($\overline{G} \approx 50–100\,s^{-1}$) in aqueous dispersions, the Kolmogoroff microscale is of the order of 100–150 μm. The collision rate of particles smaller than η should be reasonably well approximated by an expression of the form of eqn (6.26), although with some uncertainty over the numerical factor. For larger particles, transport by eddies in the inertial range is important and it is likely that the collision rate shows a different dependence on power input – proportional to $\epsilon^{2/3}$ rather than $\epsilon^{1/2}$ (Cleasby, 1984). However, aggregates in this size range (>100 μm) are more likely to be broken in turbulent flow and so it is more difficult to draw conclusions about the rate of floc growth (see Section 6.5).

6.2.3 Differential sedimentation

Another important collision mechanism arises whenever particles of different size or density are settling from a suspension. Larger particles will sediment faster than smaller ones and can capture the latter as they fall. The appropriate rate can be easily calculated, assuming spherical particles and using Stokes' law for their sedimentation rate; see, for example Friedlander (1977). The resulting collision frequency, for particles of equal density, is:

$$J_{ij} = \left(\frac{2\pi g}{9\mu}\right)(\rho_s - \rho)n_i n_j(a_i + a_j)^3(a_i - a_j) \qquad (6.28)$$

where g is the acceleration due to gravity, ρ_s is the density of the particles and ρ is the density of the fluid.

Clearly, differential settling will be more important when the particles are large and dense and in such cases this collision mechanism can be very important in promoting aggregation. Even for an initially uniform suspension of equal particles, aggregates of different size will be formed which settle at different rates. It is often in the later stages of flocculation that floc growth by sedimentation becomes significant.

6.2.4 Comparison of rates

Since we have now covered the most important collision mechanisms, it is convenient to compare the rates for typical conditions. The simplest comparison is between the various collision rate constants, defined by eqn (6.1). These are, for the different mechanisms, repeating previous equations for convenience:

$$\text{Perikinetic: } k_{ij} = \frac{2kT}{3\mu}\frac{(a_i + a_j)^2}{a_i a_j}$$

$$\text{Orthokinetic: } k_{ij} = \frac{4}{3}G(a_i + a_j)^3 \qquad (6.29)$$

$$\text{Differential settling: } k_{ij} = \left(\frac{2\pi g}{9\mu}\right)(\rho_s - \rho)(a_i + a_j)^3(a_i - a_j)$$

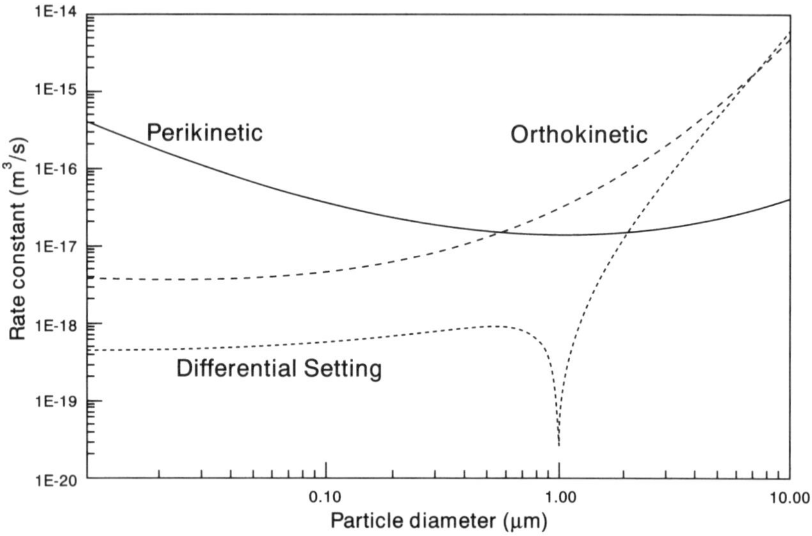

Figure 6.3 *Comparison of collision rate constants for different transport mechanisms (see text)*

The orthokinetic result is for the case of laminar shear, but should also be acceptable for turbulent collisions of particles smaller than the Kolmogoroff microscale. For comparison of these results, it is convenient to take one particle of fixed size and compute the various rate constants as a function of the size of the second particle.

The results of such computations are shown in Figure 6.3, in which one particle is taken to have a diameter of 1 μm and the other diameter varies between 0.01 and 10 μm. The shear rate is assumed to be 50 s^{-1} and the density of the particles 2 g cm^{-3}. All other values are appropriate for aqueous dispersions at 25°C. It is clear that the perikinetic mechanism gives the highest collision rates for particles less than about 0.6 μm in diameter, but that for larger particles orthokinetic collisions and differential settling become much more important. Note that the perikinetic rate passes through a minimum for equal-size particles, as expected from eqn (6.8). As the size of the second particle becomes greater than a few micrometres, the collision rate due to sedimentation increases sharply and becomes comparable to the shear-induced rate.

Of course, results like those in Figure 6.3 depend greatly on the assumed conditions. Choosing, for instance, a lower density of the particles would give a reduced effect of differential settling. Also, the various collision rate constants are affected in different ways by colloidal and hydrodynamic interactions (see Section 6.3). Nevertheless, the broad conclusions concerning the relative importance of the different mechanisms and the effect of particle size remain essentially correct.

6.3 Collision efficiencies

All of our previous treatment of aggregation rates has been based on the assumption that all particle collisions are successful in producing aggregates. In practice, this is very often not the case and allowance has to be made for the reduced *collision efficiency* (i.e. the fraction of successful collisions). Formally, all that is needed is to incorporate the collision efficiency, α, into the rate expressions discussed earlier. For example, the rate of change of the primary particle concentration, given by eqn (6.10), becomes:

$$\left(\frac{\mathrm{d}n_1}{\mathrm{d}t}\right)_{t\to 0} = -\alpha k_{11} n_1^2 \tag{6.30}$$

There remains the problem of assigning a value to α and this presents some difficulties. The collision efficiency can be very significantly reduced as a result of repulsive colloidal interactions, such as double layer repulsion or steric interaction (see Chapter 3). Another major effect is due to *hydrodynamic* or *viscous* interaction (Chapter 4), which tends to hinder the approach of colliding particles. Collisions brought about by diffusion or by induced particle motion are affected in different ways by these interactions and a comprehensive treatment is difficult. Initially, we will consider the effect of colloidal interactions on Brownian collisions and then go on to discuss hydrodynamic effects.

6.3.1 Stability ratio – the Fuchs approach

The effect of repulsive colloidal interactions on perikinetic aggregation is to give a reduction in the rate. It is conventional to speak of a *stability ratio*, W, rather than collision efficiency. The stability ratio is simply the ratio of the aggregation rate in the absence of colloidal interactions (i.e. the diffusion-controlled rate) to that found when there is repulsion between particles. It should be clear that W is just the reciprocal of the collision efficiency, i.e. $W = 1/\alpha$.

For cases where only van der Waals attraction and electrical repulsion need to be considered, and there is an energy barrier hindering the close approach of particles (see 3.5.1), the stability ratio can be calculated by treating the problem as one of diffusion in a force field (Fuchs, 1934), in a manner analogous to the treatment of particle deposition by convective diffusion (see 5.5). The result is:

$$W = 2\int_0^{\infty} \frac{\exp(\phi_\mathrm{T}/kT)}{(u+2)^2}\, du \tag{6.31}$$

where ϕ_T is the total interaction at a particle separation distance d, and u is a function of d and particle size. For spherical particles of different radii, a_i and a_j, u is given by:

$$u = \frac{2d}{(a_i + a_j)} \tag{6.32}$$

For equal particles this becomes simply $u = d/a$.

So, to evaluate W, the integral in eqn (6.31) has to be evaluated numerically, using appropriate expressions for the electrical and van der Waals interactions (see Chapter 3). Because of the exponential term, most of the contribution to the integral in eqn (6.31) comes from a region close to the maximum (see Figure 5.7) and a simple approximation due to Reerink and Overbeek (1954) can be used, which for unequal particles becomes:

$$W \approx \frac{1}{\kappa(a_i + a_j)} \exp\left(\frac{\phi_{max}}{kT}\right) \tag{6.33}$$

where κ is the Debye–Hückel parameter and ϕ_{max} is the height of the energy barrier.

Even for quite modest barrier heights, of the order of a few kT units, eqn (6.33) can yield very high stability ratios. Thus, for equal, 1-μm-diameter particles in a 100 mM solution of a 1–1 electrolyte, the term κa is about 500. For a barrier height of $10kT$, the stability ratio would be about 45, and for a $20kT$ barrier W is about 1×10^6 (i.e. only one in every million collisions would occur between particles having sufficient energy to overcome the barrier).

It is not surprising that quite small changes in electrolyte concentration or in surface potential of particles can have dramatic effects on aggregation rates. A quantitative prediction of the effect of (indifferent) salt concentration on stability ratio can be made on the basis of eqn (6.33), assuming simple DLVO-type interaction (Reerink and Overbeek, 1954). It is found that the rate of change of W with salt concentration, when plotted in log–log form, is given by:

$$\frac{d\log W}{d\log c} = -2.06 \times 10^9 \frac{a\gamma^2}{z^2} \tag{6.34}$$

where c is the concentration (mol dm^{-3}) of symmetrical (z–z) electrolyte, a is the particle radius (m) and the numerical constant is appropriate for aqueous dispersions at 25°C. γ is a dimensionless function of the surface potential of the particles, given by:

$$\gamma = \tanh \frac{ze\psi_\delta}{4kT} \tag{6.35}$$

where ψ_δ is the Stern potential and e the electronic charge.

In fact, the stability ratio does not decrease indefinitely as the salt concentration is increased, as implied by eqn (6.34). At a critical concentration, the potential energy barrier just disappears and aggregation should then occur at a transport-limited rate, determined by the collision frequency. At this point, the stability ratio and the collision efficiency should both become unity. This is often called the *critical coagulation concentration* (CCC). The form of the stability ratio versus electrolyte concentration plot is shown in Figure 6.4. Beyond the critical salt concentration there may be significant attraction between the particles, but this is rarely of sufficiently long range to give an appreciable increase in the collision radius of the particles and an enhanced collision frequency.

Experimental values of stability ratio are often found to give an approximately linear decrease with salt concentration on a log–log plot, but the slopes of the

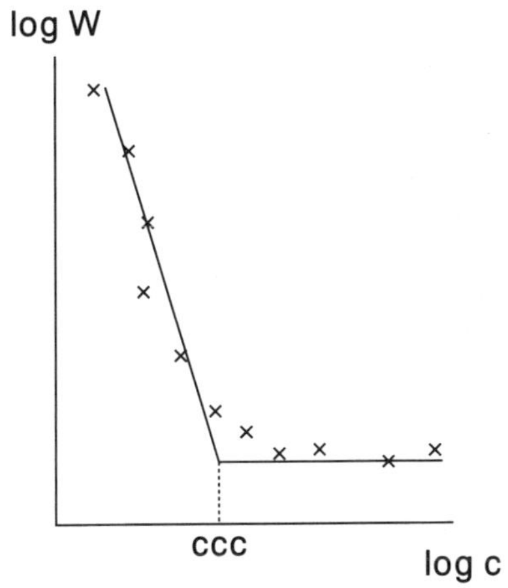

Figure 6.4 *Schematic diagram showing the effect of increasing salt concentration on stability ratio,* W. *On a log–log plot an approximately linear decrease is found until the critical coagulation concentration is reached, when the stability ratio becomes constant*

lines do not show the dependence on particle size and z predicted by eqn (6.34). Measurement of the aggregation rates of polystyrene latices with different particle sizes showed no systematic dependence on size (Ottewill and Shaw, 1966). Aggregation of negatively-charged silver iodide sols gave slopes which were not very different for La^{3+}, Ba^{2+} and K^+ salts (Frens and Heuts, 1988), despite the predicted dependence on $1/z^2$.

On closer inspection, linear behaviour of $\log W$–$\log c$ plots is difficult to justify on the basis of eqn (6.34), because of the γ^2 term. This means that a linear plot would be expected only if the Stern potential of the particles remained constant over a range of electrolyte concentrations. Judging by the effect of salt concentration on zeta potentials, this is a rather dubious assumption, especially close to the CCC, where there is often a large decrease in zeta potential over quite a small range of salt concentration. Nevertheless, more exact computations of stability ratio, by numerical integration of eqn (6.31) rather than using the approximation in eqn (6.33), give a predicted dependence of stability ratio on salt concentration which is rarely in accord with measured values.

This subject has received a great deal of attention in the colloid literature (Overbeek, 1980) and several suggestions have been made to reconcile measured stability ratios with those predicted on the basis of eqn (6.31). Aggregation in secondary minima, distributions of particle size and surface potential, hydrodynamic effects (see Section 6.3.3) and dynamic effects in double layer interaction (Dukhin and Lyklema, 1990) have all been invoked to account for the discrepancies. Although incorporation of such effects may improve the agreement with measured stability ratios, some uncertainties still remain. The position is rather like the problem of predicting rates of particle deposition in the presence of a repulsive interaction (see Section 11.1), where measured deposition rates can be very much higher than those computed from convective-diffusion theory.

From a practical standpoint, the lack of quantitative agreement between measured and computed 'slow' aggregation rates may not be of great significance, since the critical salt concentration for rapid (diffusion-limited) aggregation can be fairly well predicted from DLVO theory. Nevertheless, until a satisfactory modelling of 'slow' aggregation has been achieved, our understanding of the phenomena involved will remain incomplete.

6.3.2 Orthokinetic collision efficiencies

For collisions of non-Brownian particles (greater than a few micrometres in size), the Fuchs concept of diffusion in a force field is not appropriate and we have to consider the relative motion of particles induced by fluid shear or by external forces such as gravity. In such cases it may be possible for colliding particles to overcome potential energy barriers as a result of their relative motion. It has been observed that aggregation of otherwise stable colloids can sometimes be achieved by the application of sufficiently high shear. For instance, Zollars and Ali (1986) found that latex particles which were stable against Brownian aggregation for up to 4 years could be coagulated in a few minutes by the shearing at very high rates. The phenomenon of *shear flocculation* (Warren, 1981) is probably an example of such an effect. It should be clear that, for a given suspension, the collision efficiency for Brownian aggregation could be very different from that for orthokinetic collisions. In some cases similar values are found (Swift and Friedlander, 1964), but the agreement is probably fortuitous.

The question of collision efficiency for orthokinetic collisions cannot be adequately discussed without reference to hydrodynamic interaction. This is the subject of the next section.

6.3.3 Hydrodynamic interaction

The Smoluchowski approach to aggregation kinetics takes no account of the effect of the viscosity of the suspending medium on the collision of particles. In fact, *hydrodynamic* or *viscous* effects can have a great effect on aggregation rates. As particles approach very close it becomes increasingly difficult for the liquid between them to drain out of the gap and this tends to slow the aggregation process.

In perikinetic aggregation the effect is manifested as a reduced diffusion coefficient as particles approach close together which can be calculated exactly for the case of equal particles as a function of separation distance (Spielman, 1970). A useful empirical approximation was found by Honig *et al.* (1971):

$$\frac{D(u)}{D(\infty)} \approx \frac{6u^2 + 4u}{6u^2 + 13u + 2} = \frac{1}{\beta(u)} \tag{6.36}$$

where $D(u)$ and $D(\infty)$ are diffusion coefficients for particles separated by a dimensionless distance u $(=d/a)$ and at infinite separation respectively. $\beta(u)$ is

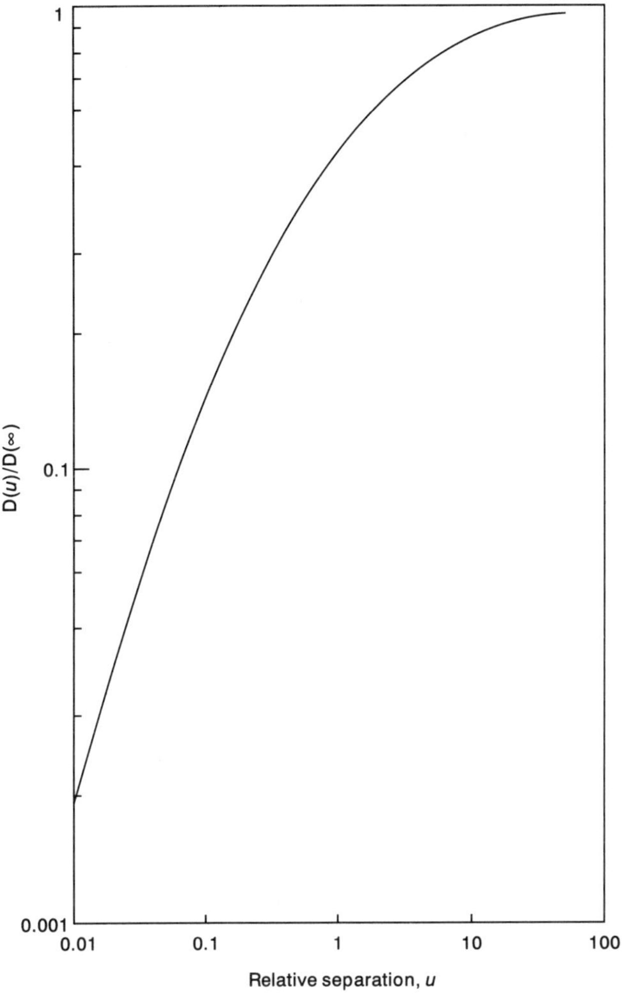

Figure 6.5 *Approximate values of* D(u)/D(∞), *from eqn (6.36), plotted against dimensionless separation distance,* u

the factor by which the diffusion coefficient is reduced at separation u, relative to the value for an isolated particle.

The function $1/\beta(u)$ is plotted against u in Figure 6.5 and it is apparent that the effect is of rather long range, giving an appreciable decrease in diffusion coefficient at separations up to at least 10 times the particle radius. Note also that the diffusion coefficient approaches zero as the particles come into contact, which means that, in the absence of any other interaction, aggregation could not occur. In reality, universal van der Waals attractive forces counteract the hydrodynamic resistance at close approach and particle contact can be achieved.

The factor $\beta(u)$ can be inserted in the Fuchs expression, eqn (6.31), to give a modified stability ratio. If we assume that the particles are fully destabilized, so that only van der Waals attraction needs to be considered, the following result for the *limiting stability ratio*, W_{lim}, is obtained (Honig *et al.*, 1971):

$$W_{\text{lim}} = 2 \int_0^{\infty} \beta(u) \frac{\exp(\phi_A/kT)}{(u+2)^2} \, du \tag{6.37}$$

where ϕ_A is the van der Waals interaction energy. Without the exponential term, the integral would diverge, giving an infinite value of the stability ratio.

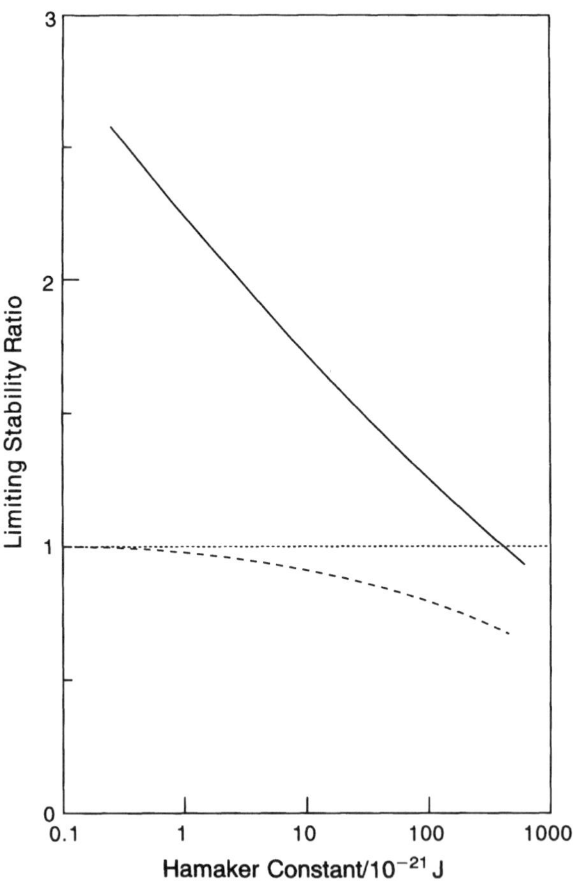

Figure 6.6 *Limiting stability ratio* W_{lim}, *from eqn (6.37), as a function of the Hamaker constant (——). For comparison the result in the absence of hydrodynamic interaction is also shown (---)*

The combined effect of van der Waals and hydrodynamic interactions on the limiting stability ratio of spherical particles in water is shown in Figure 6.6, where $W_{\lim}$ is plotted against the effective Hamaker constant. The procedure was to calculate the van der Waals attraction energy, for a given value of Hamaker constant and over a range of particle separations, and to evaluate the integral in eqn (6.37) numerically, using the value of $\beta(u)$ from eqn (6.36). For comparison, the value of $W_{\lim}$ in the absence of hydrodynamic interaction (calculated by setting $\beta(u)=1$) is also plotted in Figure 6.6 (dashed line).

These results show that over the likely range of Hamaker constants ($1\text{–}100 \times 10^{-21}$ J), the limiting stability ratio would vary from about 2.3 to 1.2. Only for the improbably high value of $A = 3 \times 10^{-19}$ J would van der Waals attraction entirely cancel out viscous resistance and give $W_{\lim} = 1$. Measurements of perikinetic aggregation rates for latex particles give apparent rate coefficients around half the calculated Smoluchowski value (Lichtenbelt *et al.*, 1974), indicating $W \approx 2$ and a Hamaker constant of about 5×10^{-21} J, in reasonable agreement with accepted values.

In the absence of hydrodynamic interaction, van der Waals attraction has little effect on the stability ratio over the usual range of values. This is because dispersion forces operate over a rather limited range, much less than the particle radius, and so give only a minor increase in the effective collision radius. The larger effect seen in Figure 6.6 when viscous interaction operates arises because van der Waals attraction becomes significant at close approach, where the hydrodynamic effect also becomes very large. The two effects counteract each other to a considerable extent.

Although it needs to be considered, the hydrodynamic effect does not have a dramatic effect on perikinetic aggregation, slowing the rate by a factor of no more than about 2. In the case of *orthokinetic* aggregation, viscous interaction can be much more significant.

The Smoluchowski treatment of orthokinetic collision rate assumes that particles in a uniform shear field move along straight streamlines until they collide with another particle (see Figure 6.2). In reality, streamlines curve around obstacles such as other particles, which makes collision less likely. The conventional 'no slip' assumption implies that, in the absence of any other effect, particles could not come into contact as a result of fluid motion, because the last layer of liquid between them could not be removed by viscous flow. Brownian motion and the presence of attractive interparticle forces allow aggregation to occur.

A full treatment of this problem involves computing trajectories of particles, incorporating hydrodynamic and colloidal interactions. It is convenient to consider a reference sphere at the centre of the coordinate system and compute the number of other particles colliding with this in unit time. By finding the 'limiting trajectory', which just brings a particle into contact with the reference particle, and tracing this trajectory back to a point far upstream, an orthokinetic collision efficiency can be calculated. This is the fraction of particles approaching within a cylinder defined by the collision radius (Figure 6.2) which collide with the reference particle. The method is similar to the trajectory analysis discussed in Section 5.6 for particle deposition on collectors.

Van de Ven and Mason (1977) used such an approach, incorporating van der Waals attraction and double layer repulsion. In the absence of repulsion, a *limiting collision efficiency*, α_0, was derived. This is defined in such a way that the collision frequency can be obtained simply by multiplying the appropriate Smoluchowski result by α_0. Thus, for equal particles of radius a and concentration n, the collision frequency, J, is:

$$J = \frac{32}{3}\alpha_0 n^2 G a^3 \tag{6.38}$$

This is obtained directly from eqn (6.19), by setting n_i and n_j equal to n, and a_i and a_j to a.

The computed results of van de Ven and Mason are well represented by a simple empirical expression:

$$\alpha_0 = f(\lambda/a)C_A^{0.18} \tag{6.39}$$

where $f(\lambda/a)$ is a function of the dispersion wavelength (since retardation effects were included) and the particle size, for which some values were given, and C_A is given by:

$$C_A = \frac{A}{36\pi\mu G a^3} \tag{6.40}$$

This expression shows that the collision efficiency decreases as the particle size and the shear rate increase. The dependence on shear rate means that the collision frequency varies as $G^{0.82}$, rather than being linearly dependent on shear rate, as in the Smoluchowski result.

Taking $A = 10^{-20}$ J, $\lambda = 100$ nm and $\mu = 10^{-3}$ Pa s as reasonable values for aqueous suspensions, values of α_0 can be calculated for different particle sizes and for different shear rates. In Figure 6.7 the limiting collision efficiency is plotted as a function of shear rate for three different particle sizes (1 μm, 2 μm and 4 μm diameter). The implication of these results is that, as aggregates grow, especially at high shear rate, there will be a corresponding decrease in α_0, which will tend to restrict further aggregation. However, the collision rate increases for larger particles, eqn (6.19), so that the aggregation rate may not be greatly affected.

The results just discussed are for colliding particles of equal size. For *unequal* particles, collision efficiencies can become very low (Adler, 1981). It is found that in some cases the trajectory of a smaller particle around a larger one is such that a large separation between the particles is maintained, with virtually no chance of capture. For instance, for particles with radii of 10 μm and 1 μm, the distance of closest approach has been calculated to be about 1.4 μm (van de Ven and Mason, 1981). This is far greater than the range of van der Waals attraction and it is difficult to see how any aggregation could occur in such cases. Brownian motion and the porous nature of aggregates may modify this conclusion. All real aggregates have a rather open structure and it is very likely that hydrodynamic resistance is greatly reduced when the approaching surfaces are permeable to the liquid.

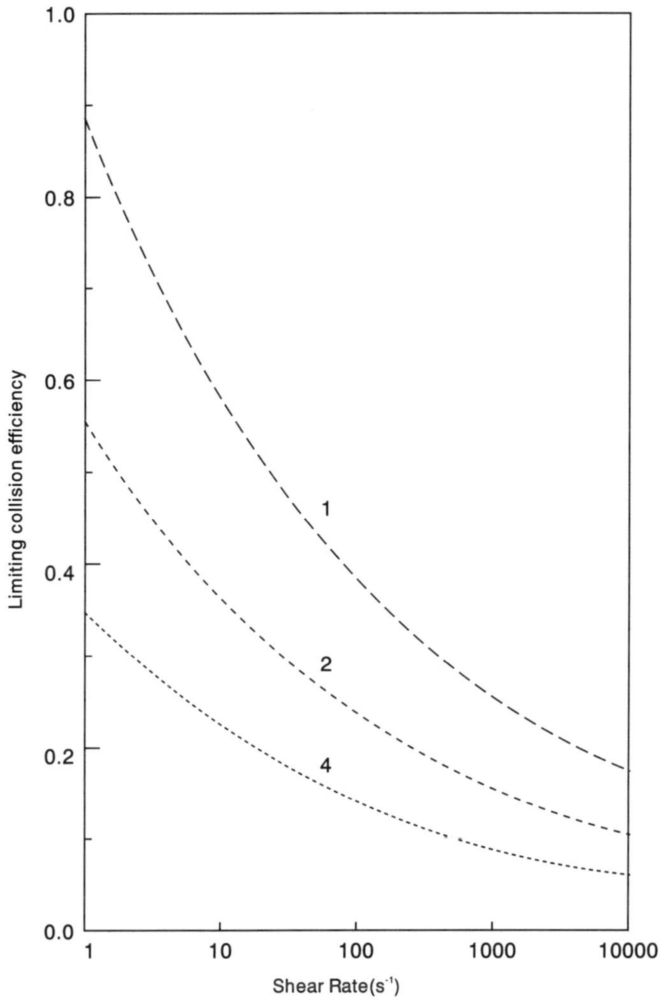

Figure 6.7 *Limiting collision efficiencies, for orthokinetic aggregation of fully destabilized particles, as a function of the applied shear rate. The colliding particles are assumed to be equal spheres with diameters (μm) shown on the curves*

In the presence of double layer repulsion the computations of van de Ven and Mason (1977), for equal spheres, show rather complex behaviour. Depending on conditions, particles may be stable or may aggregate into primary or secondary minima. It is possible that a suspension may be stable over a certain range of shear rates, but aggregate at higher and lower values. For aggregation into a secondary minimum, high shear rates would tend to pull aggregated particles apart. However, high shear rates can increase the chance of colliding particles overcoming a potential energy barrier and being captured in a primary minimum. Predictions of such behaviour in real systems are severely limited by the lack of detailed information on the relevant colloidal interactions and the appropriate parameters (zeta potentials, Hamaker constants, adsorbed species, etc.). Quite small changes in the assumed parameters can have very large effects on the computed results.

In the case of differential sedimentation, the effect of hydrodynamic interaction on aggregation rate has been analysed by Melik and Fogler (1984).

Hydrodynamic interaction affects the different aggregation mechanisms to different extents and will modify the previous discussion based on Figure 6.3, especially for particles larger than about 2 μm. Nevertheless, the qualitative conclusions concerning the relative importance of the different types of collision process remain broadly correct.

6.4 Form of aggregates

When solid particles aggregate, no coalescence can occur and the resulting clusters may adopt many different forms. In the simplest case of equal spheres, there is no doubt about the shape of a doublet, which must be in the form of a dumbbell. However, a third particle can attach in several different ways and with higher aggregates the number of possible structures rapidly increases, as indicated in Figure 6.8. In real aggregation processes, aggregates containing hundreds or thousands of primary particles can arise and it will never be possible

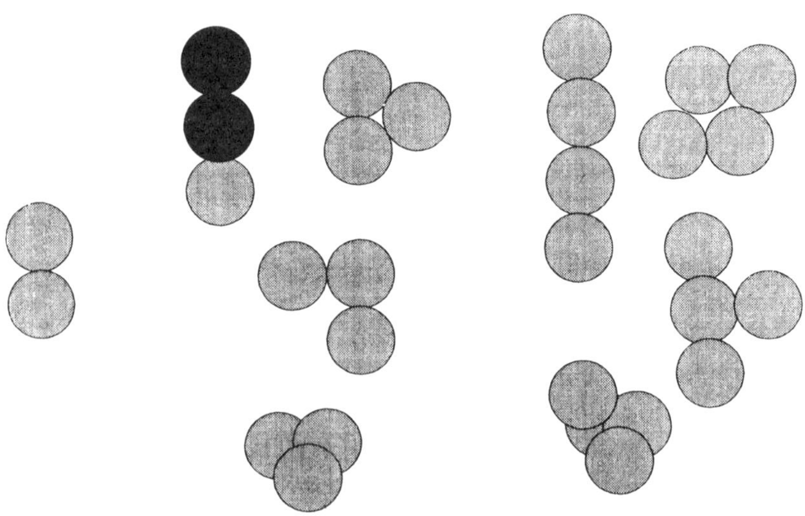

Figure 6.8 *Showing possible forms of aggregates of equal spheres:* (a) *doublets,* (b) *triplets, and* (c) *quadruplets*

to provide a detailed description of their structure. Some convenient method is needed which enables aggregate structure to be characterized in general terms, but still conveys useful information. A great deal of progress was made in this area during the 1980s, largely as a result of computer simulation of aggregate formation and the study of model aggregates.

6.4.1 Model studies: fractal clusters

Aggregates are now recognized as *fractal objects* (Meakin, 1988). If, for a large number of aggregates, the mass is plotted against aggregate size (diameter, for example), the plot may be linear, but with a non-integer slope. (For regular, three-dimensional objects the slope of such plots is 3.) For aggregates, the slope of the line, d_F, is called the *fractal dimension*, and can be considerably less than 3. There are several other ways of defining fractal dimension. Strictly, d_F should be called a *mass fractal dimension*, but the general term will be used here. The lower the fractal dimension, the more open (or 'stringy') is the aggregate structure. The relationship between aggregate mass, M, and size, L, is just:

$$M \propto L^{d_F} \tag{6.41}$$

The 'size' L can be defined in various ways and in fundamental studies it is often taken as the radius of gyration of the aggregate. However, the precise definition of L does not affect the form of eqn (6.41) and it may be convenient to use the largest diameter of an irregular aggregate as the measure of its size. If the relationship in eqn (6.41) applies over a wide range of aggregate sizes, then it implies that the aggregates have a *self-similar* structure, which is independent of the scale of observation (or the degree of magnification). This concept is illustrated schematically in Figure 6.9, where the fundamental unit is assumed to be a triplet of equal spheres. This simple two-dimensional illustration is not intended to represent real aggregates, but only to convey the idea of self-similarity. It is reminiscent of earlier ideas (e.g. Michaels and Bolger, 1962) on the 'hierarchical' nature of aggregation, in which small aggregates combine to form larger 'flocs'. Such models were normally restricted to just a few discrete levels of aggregation, rather like the simple picture in Figure 6.9. The essence of self-similar aggregates is that there is a continuum of 'levels' from large-scale structures down to individual primary particles.

Computer modelling of aggregation has given very useful insights into the process. Early studies were based only on the random addition of single particles to growing clusters (Vold, 1963). Later simulations (Meakin, 1988) of diffusion-controlled (i.e. 'rapid' perikinetic) aggregation, with single-particle addition (diffusion-limited aggregation), gave fairly dense structures, with d_F about 2.5. In many cases, single-particle addition is not a realistic model, since, throughout most of an aggregation process, growth occurs as a result of cluster–cluster encounters. In this case, computer simulations and experimental studies on a range of model colloids (Lin *et al.*, 1989) show much more open structures with a fractal dimension of around 1.8. It is important to realize that these simulations are based on the assumption that particles attach permanently to other particles

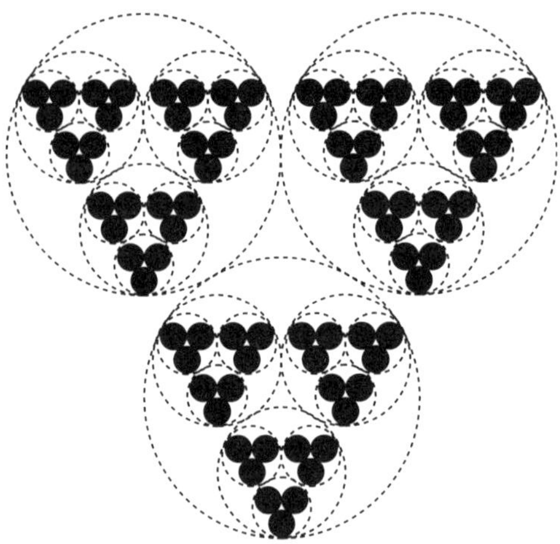

Figure 6.9 *Schematic illustration of self-similar aggregate structure*

at the first contact; the process is controlled entirely by diffusion – hence 'diffusion-limited aggregation' (DLA).

Intuitively, there is a simple reason for the difference in structure of aggregates produced by particle–cluster and cluster–cluster aggregation, illustrated in Figure 6.10. In the former case, a particle is able to penetrate some way into a cluster before encountering another particle and sticking. In an encounter of two clusters the first contact is likely to occur before the clusters have interpenetrated to a significant extent, which leads to a much more open structure.

When there is interparticle repulsion, so that the collision efficiency is reduced, aggregation is then said to be 'reaction-limited' and very different aggregate structures can be obtained under these conditions. It is found (Lin *et al.*, 1989) that reaction-limited aggregates are more compact than those produced by DLA, with a fractal dimension $d_F \approx 2.1$, for the cluster–cluster case. Again, it is not difficult to find an explanation for this effect. When the collision efficiency is low, particles (or clusters) need to collide many times before sticking occurs. This gives more opportunity to explore other configurations and to achieve some degree of interpenetration.

Less attention has been paid to aggregates produced by mechanisms other than diffusion. There is evidence that 'ballistic' aggregation (where encounters occur as a result of linear trajectories) gives rather more compact structures, especially in the particle–cluster case, where d_F can approach a value of 3. In the cluster–cluster case simulations of ballistic aggregation give $d_F \approx 1.9$ (Tence *et al.*, 1986). It is not clear how ballistic aggregation relates to shear-induced collisions and orthokinetic aggregation. Torres *et al.* (1991b) have simulated aggregation in viscous flows and found that aggregates produced by cluster–cluster encounters in shear and extensional flow are very like those formed by DLA, with $d_F \approx 1.8$. For particle–cluster aggregation, much more compact structures result with d_F

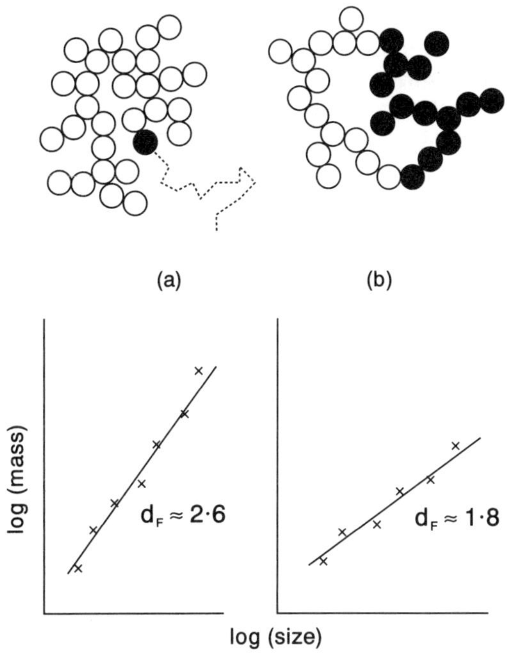

Figure 6.10 *Fractal aggregates produced by* (a) *single-particle addition to an aggregate and* (b) *cluster–cluster aggregation. In the latter case, aggregates are more open in structure and have a lower fractal dimension,* d_F

values up to about 2.9, although there is some dependence on the nature of the flow.

Measurements of fractal dimensions of aggregates formed from colloidal nickel hydroxycarbonate have been reported by Hoekstra *et al.* (1992), for both perikinetic and orthokinetic aggregation and for high and low colloid stability. Although their systems were far from 'ideal', the results were broadly in line with previous model studies and simulations. For rapid, diffusion-controlled aggregation, a fractal dimension of 1.7–1.8 was found. At low salt concentration, where aggregation was reaction limited, d_F was in the region of 2.0–2.1. For orthokinetic aggregation (in Couette flow) at high salt concentration, d_F was found to be shear-dependent, increasing from 1.7 at zero shear to about 2.2 at a shear rate of 200 s^{-1}. For orthokinetic collisions under reaction-limited (low-salt) conditions, fractal dimensions up to 2.7 at 200 s^{-1} were found.

The assumption so far that contacts between particles, once formed, are permanent, means that aggregates cannot undergo any subsequent restructuring. In fact, changes in aggregate structure have often been observed, always giving more compact forms (higher d_F). For example, Aubert and Cannell (1986) found that coagulation of silica particles under diffusion control initially gave aggregates with $d_F \approx 1.75$, but, after a few hours, these became more compact, with $d_F \approx 2.1$, which is the value achieved by reaction-limited aggregation.

It is very likely that agitation, as in stirred vessels, is effective in giving some compaction of flocs (higher d_F), probably by causing some deformation and rearrangement. This is a subject of great practical interest, but it has not received much fundamental attention.

6.4.2 Aggregate density

A very important consequence of the fractal, self-similar nature of aggregates is that their density decreases appreciably as the size increases. Actually, such behaviour had been empirically observed well before the concept of fractal aggregates was introduced. Lagvankar and Gemmell (1968) found that flocs produced under water treatment conditions had effective densities that decreased markedly with increasing size. Similar observations had been made in Japan from the 1960s by Tambo and coworkers (Tambo and Watanabe, 1979).

The effective, buoyant density of an aggregate in a liquid, ρ_E, is simply:

$$\rho_E = \rho_F - \rho_L = \phi_S(\rho_S - \rho_L) \qquad (6.42)$$

where ρ_S, ρ_L and ρ_F are the densities of the solid particles, the liquid and the aggregate (floc), respectively, and ϕ_S is the volume fraction of solid in the floc.

Since most measurements of floc density involve some form of sedimentation procedure, the buoyant density, ρ_E is the relevant property. When ρ_E is plotted against the floc diameter, in log–log form, a linear decrease is often found, implying a relationship of the form:

$$\rho_E = Ba^{-y} \qquad (6.43)$$

where B and y are constants.

Because of the proportionality between ρ_E and ϕ_s, it follows that the exponent y is related to the fractal dimension:

$$d_F = 3 - y \qquad (6.44)$$

Measurements of aggregate densities for many suspensions of practical interest, including mineral particles, sewage sludges and aluminium hydroxide flocs, give y values in the range 1–1.4, corresponding to fractal dimensions of 2–1.6, which are in line with those obtained by simulation and in model systems. This lends considerable support to the idea of 'universality' in particle aggregation (Lin *et al.*, 1989).

The values of y quoted above imply a very substantial decrease in density with increasing floc size. With $y = 1.2$ ($d_F = 1.8$), a 10-fold increase in size gives a 16-fold reduction in the effective floc density. Assuming that Stokes' law holds, the corresponding increase in sedimentation rate would be by a factor of less than 6, rather than the 100-fold increase that would be expected on the basis of constant floc density. So, the fractal nature of flocs has very important implications for gravitational or centrifugal separation processes.

Other solid–liquid separation processes are also affected by the density of flocs. Dewatering operations proceed more rapidly when flocs are more compact since, effectively, there is less solid surface in contact with water and hence less drag. Also, more open, low-density flocs are more likely to be compressed during cake filtration and this can lead to blocking of pores and a reduced filtration rate. It is very significant that substantial improvements in filterability of flocs can be achieved by recirculation of flocculated solids (Knocke and Kelley, 1987). In this way newly-introduced, unflocculated particles are brought into contact with preformed flocs and aggregation of the particle–cluster type would be expected. From simulation studies, this is known to produce more compact aggregates than cluster–cluster processes. The higher density of flocs formed in upflow clarifiers, where incoming particles flow through a layer of preformed floc, can probably be explained by essentially the same reasoning.

In some cases, lower density flocs might be preferred. For instance, the capture of particles in granular filters and in orthokinetic flocculation is greatly dependent on the effective collision radius, which, for a given degree of aggregation, will become greater as the floc density decreases. The effect of the fractal nature of aggregates on aggregation kinetics will be briefly discussed in the next section.

6.4.3 Collision rates of fractal aggregates

The Smoluchowski treatment of aggregation kinetics is based on the assumption that the colliding particles are spheres. Even for spherical primary particles, aggregation quickly leads to shapes like those in Figure 6.8, and their collision rates cannot be calculated exactly. Only in the case of coalescing liquid droplets could the assumption of spherical particles be justified.

For perikinetic aggregation, the growth of aggregates gives an increasing collision radius and a reduced diffusion coefficient and these effects tend to cancel out, giving a collision rate constant which is not greatly dependent on aggregate size. For fractal aggregates, the hydrodynamic radius (which determines the drag and hence the diffusion coefficient) is likely to be somewhat less than the outer 'capture radius', corresponding to the physical extent of the aggregate. For high degrees of aggregation, the ratio of these two radii has been calculated to be about 0.6 (Torres *et al.*, 1991a). This means that Brownian collisions will occur rather more rapidly than predicted from the rate constant in eqn (6.9). However, for aggregates greater than a few micometres in size, perikinetic aggregation is negligible and collisions induced by shear become far more important.

In the orthokinetic case, it is the effective capture radius of a fractal aggregate that is of paramount importance and this is greatly dependent on the fractal dimension (e.g. Wiesner, 1992). Instead of eqn (6.20), the collision rate constant for orthokinetic collisions between i and j particles can be written:

$$k_{ij} = \frac{4Ga_0^3}{3}(i^{1/d_F} + j^{1/d_F})^3 \tag{6.45}$$

where a_0 is the radius of the primary particles and it has been assumed that the radius of an i-fold aggregate is given by:

$$a_i = a_0 i^{1/d_F} \tag{6.46}$$

by analogy with the definition of fractal dimension in eqn (6.41).

For the 'coalesced sphere' assumption, $d_F = 3$, and the increase in aggregate size is relatively slow (a 10-fold increase in capture radius for 1000-fold aggregates). For lower values of d_F, the aggregate size increases more rapidly, which can give a dramatic increase in aggregation rate (Jiang and Logan, 1991). The implications for the evolution of aggregate size distribution have been discussed by Wiesner (1992). An obvious consequence of the fractal nature of aggregates is that the effective floc volume will not be conserved, as assumed in the derivation of eqn (6.23). There will be a substantial increase in floc volume for typical values of d_F and this is the reason for the increased collision frequency.

Another very important outcome of the fractal nature of aggregates is that hydrodynamic interaction is much less significant than for solid particles and the very large effects expected for unequal-size particles (Section 6.3.3) are not found for aggregates. Torres *et al.* (1991a) calculated stability ratios for aggregates in shear flow and found values only slightly in excess of 1. It should be clear from the simple pictures in Figure 6.10 that particle–cluster and cluster–cluster collisions will be hindered much less by hydrodynamic effects than would similar encounters involving solid particles equivalent in size to the clusters.

6.5 Aggregate strength and break-up

It was stated at the beginning of our discussion of aggregation kinetics that aggregation would be regarded as irreversible, and this assumption is implicit in expressions such as eqn (6.2). This is a convenient assumption, since the break-up of aggregates is very difficult to model. However, since nearly all aggregation processes are carried out with some form of agitation, the break-up process cannot be ignored. In practice, it is often found that aggregates (flocs) reach a certain, limiting size, which depends on the applied shear or energy dissipation and on the *floc strength*. Empirically, the size may depend on the energy dissipation according to (e.g. Mühle and Domasch, 1991):

$$d_{max} = C\epsilon^{-n} \tag{6.47}$$

where C and n are constants.

There are several theoretical approaches for floc break-up in turbulent flow which lead to expressions of the form of eqn (6.47) (Tambo and François, 1991; Mühle, 1993). The exponent depends on the size of the floc relative to the turbulence microscale; for instance, for flocs large compared to the microscale an exponent of around –0.4 may be found, whereas for much smaller flocs the dependence on energy input is not so great and $n \approx 0.3$. However, these values are difficult to check experimentally and may be highly system-specific. As a convenient rule of thumb, it is sometimes assumed that the limiting floc size in a turbulent flow field is of the same order as the Kolmogoroff microscale, given by eqn (6.27).

Even in laminar shear, it is not easy to predict maximum aggregate size. Torres *et al.* (1991a) used the following expression to model break-up of large flocs in simple shear, derived by balancing the van der Waals force between two particles with the hydrodynamic force acting to separate two aggregates:

$$(R_{Hi} + R_{Hj}) = \left(\frac{A}{18\pi\mu Ga\delta^2}\right)^{1/2} \tag{6.48}$$

where R_{Hi} and R_{Hj} are the dimensionless hydrodynamic radii of two colliding aggregates (scaled by the primary particle radius, a), A is the Hamaker constant and δ is the separation of particles in the primary minimum.

The parameter δ is subject to considerable uncertainty and is often treated as a fitting parameter, with a value of the order of 1 nm or less. Although eqn (6.48) applies to simple shear and is based on simplifying assumptions, it does highlight some important factors governing aggregate strength. For instance, the maximum floc size is predicted to vary as $G^{-0.5}$, which is equivalent to a dependence on $\epsilon^{-0.25}$, and this exponent is of the same order as those found experimentally in some cases (Tambo and François, 1991).

The criterion for limiting floc size in eqn (6.48) is based on the assumption that a collision between two aggregates can only lead to attachment if the sum of their hydrodynamic radii does not exceed a certain critical value. For larger aggregates, the shear force tending to separate them is greater than the binding force. This is equivalent to finding the size of colliding aggregates, under given shear conditions, for which the collision efficiency becomes zero. Such an approach was adopted by Brakalov (1987) for turbulent conditions and the predictions of his model agree quite well with measurements on aggregates of metal hydroxides.

However, there are many cases where the concept of a limiting size based on a vanishing collision efficiency is not appropriate. For instance, aggregates formed under low-shear conditions may break when subjected to higher shear. In that case, floc breakage may occur in several ways, not necessarily into the aggregates from which the floc was formed at the last collision. Experimental observations (e.g. Glasgow and Liu, 1991) indicate that floc breakage is a complex phenomenon, with large-scale fragmentation as well as surface erosion of small components occurring simultaneously. At present, there is no satisfactory model to account for the observed effects, and progress is hampered by the lack of an accepted experimental method for studying floc breakage. Also there is no widely accepted definition of common terms such as 'floc strength'.

Intuitively, the strength of an aggregate must depend on the attractive forces between component particles and the number of particle–particle contacts. The latter must depend on the density of the aggregate, which determines the effective 'coordination number' of the component particles. Because of the fractal nature of aggregates, an increase in size means a decrease in density and a reduced number of particle–particle contacts per unit volume of aggregate. Since the disruptive force increases with size, the limiting size may be reached when the aggregation *number* is still quite small. The same number of primary particles in a smaller, more compact aggregate could be said to be 'stronger' in that it resists a shearing force which would disrupt the larger, lower density aggregate. For this reason, assessment of 'floc strength' on the basis of the limiting hydrodynamic size achieved under given shear conditions may be misleading. Information on the *mass* of the aggregate would also be relevant.

Inclusion of floc break-up in modelling of aggregation processes usually involves an assumption of the limiting aggregate size under given conditions. Aggregates exceeding this size are assumed to break into two or more 'daughter' aggregates and the precise form of breakage assumed can greatly influence the computed aggregate size distribution.

6.6 Aggregate size distributions

6.6.1 Analytical approaches

There has been a great deal of fundamental interest in solving the Smoluchowski equation, eqn (6.2), to obtain an aggregate size distribution as a function of time. The form of the expression in eqn (6.2) is for *discrete* particle sizes, but an equivalent form can be written for *continuous* particle size distributions (e.g. Rosen, 1984):

$$\frac{\partial n(i,t)}{\partial t} = \frac{1}{2}\int_0^i k(i-j,j)n(j,t)n(i-j,t)\mathrm{d}j - n(i,t)\int_0^\infty k(i,j)n(j,t)\,\mathrm{d}j \qquad (6.49)$$

where the left-hand side represents the rate of change of concentration of particles of size (volume) i and the coefficients $k(j,i-j)$ are equivalent to the collision rate constants k_{ij} etc. in eqn (6.2). In integral expressions these are known as collision *kernels*.

Solution of these expressions is made very difficult by the fact that the collision kernels depend on the size of the colliding aggregates, and the dependence on size is not generally known, except in very simple cases.

The simplest possible assumption is that the collision kernels are constant, independent of size, and in this case eqn (6.49) leads to a very simple form of aggregate size distribution (Meesters and Ernst, 1987). This is most conveniently expressed in terms of a dimensionless aggregate size, x, normalized by the average size, such that, for an aggregate of volume v, the dimensionless size is given by:

$$x = \frac{v}{\bar{v}} = \frac{k}{\bar{k}} \qquad (6.50)$$

where $\bar{v}$ is the average aggregate volume and $\bar{k}$ is the number of primary particles in an aggregate of average volume. As before, the volume of an aggregate is determined by the number of primary particles it contains, so that the average, $\bar{k}$, is given by:

$$\bar{k} = \frac{n_0}{n_T} \qquad (6.51)$$

where, as before, n_0 and n_T are the concentrations of primary particles and *total* particles, respectively.

With this definition, and appropriate initial conditions, the solution of eqn (6.49) turns out to have a simple exponential form. In terms of dimensionless size, x, the size distribution for the constant-kernel case can be written (Meesters and Ernst, 1987):

$$f(x) = \exp(-x) \tag{6.52}$$

where $f(x)$ is the frequency function, such that the fraction of aggregates with (reduced) sizes in the range x to $x + dx$ is $f(x)\,dx$.

The assumption of constant collision kernel is essentially that made by Smoluchowski in deriving the aggregate size distribution in eqn (6.17) for the discrete, perikinetic case. In fact, for long times ($t \gg \tau$), the Smoluchowski result, expressed in dimensionless form, approaches the exponential distribution given by eqn (6.52), as shown in Figure 6.11. In this figure, the points represent aggregate size distributions derived from eqn (6.17), for different reduced times (t/τ) and expressed in terms of the average aggregate size given by eqn (6.51). (For reduced times of 1, 5 and 10, values of $\bar{k}$ are 2, 6 and 11, respectively.) For reduced times greater than 5 and for aggregates larger than the average size ($x > 1$), the points lie close to the exponential result.

Of course, in many practical cases, especially for orthokinetic aggregation, the collision kernels depend greatly on aggregate size and cannot be treated as constant. In order to solve eqn (6.49) for non-constant kernels, it is convenient to

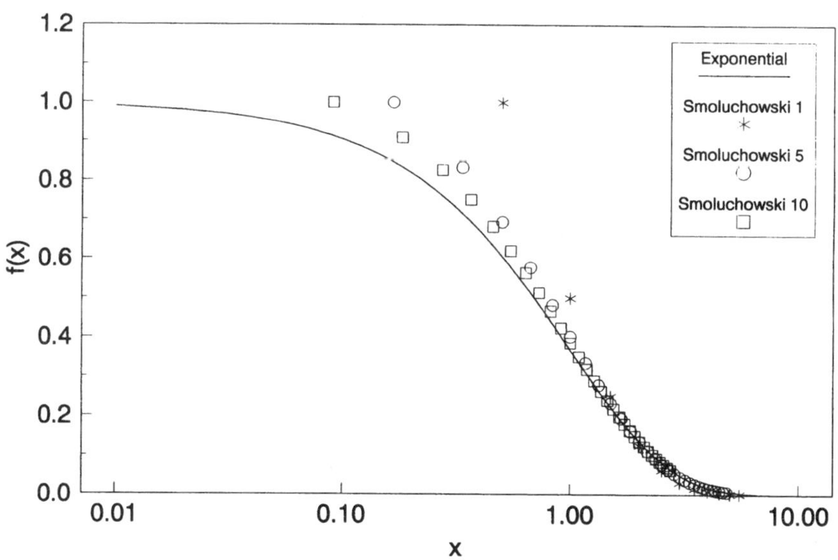

Figure 6.11 *The aggregate size distribution,* f(x), *in terms of the reduced size,* x. *The points are derived from the Smoluchowski expression, eqn (6.17), at different values of the reduced time* (t/τ): *1, 5 and 10. The drawn line is the exponential distribution, eqn (6.52)*

assume that the kernels are homogeneous functions of aggregate size, so that, if the aggregates are increased in size by a certain factor, m, then the kernel increases by a factor m^w, where w is the *degree of homogeneity*:

$$k(mi,mj) = m^w k(i,j) \tag{6.53}$$

When $w=0$ (the constant-kernel case), the exponential form, eqn (6.52), is recovered. For $w>0$, different forms are obtained, typically bell-shaped curves as shown by numerical computations of Meesters and Ernst (1987). The reason is that, for $w>0$, collisions between large and small aggregates (or primary particles) occur more rapidly than in the constant-kernel case, so that small aggregates are less likely to remain. This extra depletion of the smaller aggregates gives bell-shaped, rather than exponential, distributions.

6.6.2 'Self-preserving' distributions

Solution of eqn (6.49) is made much easier if some 'universal' or 'scaling' form is assumed. The first example of such an approach was the 'self-preserving' distribution of Swift and Friedlander (1964), discussed further by Friedlander (1977). This is based on the assumption that the fraction of aggregates in a size range dv is a function only of the dimensionless size, x, defined in eqn (6.50). Thus:

$$\frac{n\,\mathrm{d}v}{n_\mathrm{T}} = f\left(\frac{v}{\bar{v}}\right)\mathrm{d}\left(\frac{v}{\bar{v}}\right) = f(x)\,\mathrm{d}x \tag{6.54}$$

This can be rearranged, using the fact that the average aggregate volume, $\bar{v}$, is simply given by $\bar{v}=v_\mathrm{T}/n_\mathrm{T}$, to give:

$$n(v,t) = \frac{n_\mathrm{T}^2}{v_\mathrm{T}} f(x) \tag{6.55}$$

There are also the integral relations which give the total number of particles, n_T and the total volume of particles, v_T, per unit volume:

$$n_\mathrm{T} = \int_0^\infty n\,\mathrm{d}v \tag{6.56}$$

and:

$$v_\mathrm{T} = \int_0^\infty nv\,\mathrm{d}v \tag{6.57}$$

For a closed system, v_T remains constant and n_T decreases with time.

The implication of these expressions is that, if a form of the aggregate size distribution is known for particular values of n_T and v_T, then the distribution for another value of n_T (i.e. at another time) can be determined from eqn (6.54). When plotted in suitable form, the shapes of the distributions at different times are similar and hence the distribution is said to be 'self-preserving'.

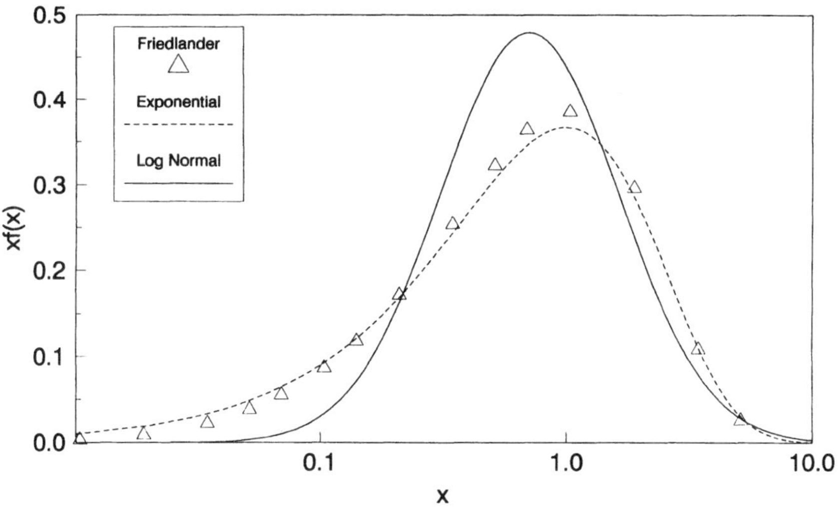

Figure 6.12 *Comparison of 'self-preserving' distributions, in terms of the reduced aggregate size,* x, *as* x f(x) *versus* x. *The results are from the computations of Friedlander and Wang (1966), the exponential form and the log-normal distribution of Lee (1985), eqn (6.59)*

Friedlander and Wang (1966) showed how the integral form of the Smoluchowski expression, eqn (6.49), together with the scaling relation eqn (6.55) and the integral relations above, could be used to derive a numerical solution for Brownian aggregation. The kernels used were from eqn (6.8), so that the dependence on aggregate size is included. The self-preserving distribution obtained in this way is plotted in Figure 6.12, as $x f(x)$ versus x, and appears to be approximately exponential in form.

An alternative approach to Brownian aggregation proposed by Lee (1983, 1985) is to *assume* a log-normal form for the aggregate size distribution and to compute the parameters of the distribution using kernels of the form given by eqn (6.8). Lee showed that the distribution of aggregate sizes tended to a universal form at long times, depending on the breadth of the initial distribution but typically of the order of 10–200 τ (the characteristic aggregation time defined in eqn (6.15)).

The log-normal distribution, expressed in terms of aggregate volume, is written by Lee as:

$$\frac{n(v,t)}{n_T} = \frac{1}{3\sqrt{2\pi}\, v \ln \sigma} \exp\left[\frac{-\ln^2 (v/v_g)}{18 \ln^2 \sigma}\right] \tag{6.58}$$

where $v_g(t)$ is the number median particle volume and $\sigma(t)$ is the geometrical standard deviation based on the aggregate *radius*. This is the reason for the presence of the factors 3 and 18 on the right-hand side of eqn (6.58), instead of the more familiar values 1 and 2, which appear when the entire distribution is given in terms of particle radius. Note that there is an implicit assumption that $d_F = 3$ and so the aggregates do not have fractal character.

Lee (1985) showed that the value of σ approached a limiting value of 1.32 after long times and that the volume distribution in reduced form could be written:

$$x\,f(x) = \frac{1}{\sqrt{2\pi \ln 2}} \exp\left[\frac{-(\ln \sqrt{2}\,x)^2}{\ln 4}\right] \tag{6.59}$$

This result is also plotted in Figure 6.12, for ease of comparison with the Friedlander 'self-preserving' distribution. The two curves are seen to be broadly similar, but with peaks at rather different sizes. There is no *a priori* reason to suppose that the distribution will follow the log-normal form and Lee's result has less fundamental justification than that of Friedlander. Nevertheless, log-normal distributions are widely employed and have convenient mathematical properties, so that eqn (6.59) has some attraction.

An interesting consequence of the log-normal result is that the total particle concentration can be shown to vary with time as:

$$\frac{n_T}{n_0} = \frac{1}{1 + 1.04 k_a n_0 t} \tag{6.60}$$

where $k_a = 4kT/3\mu$ (see eqn (6.13)). This differs very little from the Smoluchowski result, eqn (6.14).

Our discussion of limiting or 'self-preserving' distributions has only been concerned with closed systems, where the total number of primary particles remains constant. Steady-state size distributions of environmental particles (e.g. in lakes) are affected by input of particles and loss by sedimentation (e.g. Jiang and Logan, 1991). This aspect will not be considered here.

6.6.3 The 'maximum entropy' approach

Another way of looking at aggregate size distributions is to consider the *most probable* way in which primary particles are distributed among aggregates of different size. For a fixed number of primary particles, n_0, (i.e. for a closed system) and a *total* number of particles n_T in unit volume of suspension, the average aggregate size is n_0/n_T, as in eqn (6.51), and the reduced aggregate size, x, is given by eqn (6.50). If we consider the number of ways in which n_0 primary particles can be distributed among n_T objects, it is possible to find the distribution which represents the maximum number of possible arrangements of particles among aggregates. This is essentially a statistical-mechanical approach and is equivalent to finding the *maximum entropy* of the distribution

(Rosen, 1984). The entropy (or uncertainty), S, of the distribution $f(x)$ is given by:

$$S = \int_0^\infty f(x) \ln f(x) \, dx \tag{6.61}$$

When this function is maximized, using the constraints in eqns (6.56) and (6.57), it turns out that the most probable distribution is given simply by (Rosen, 1984):

$$f(x) = \exp(-x) \tag{6.62}$$

which is just the same as the expression derived analytically for the constant-kernel case, as discussed in Section (6.6.1). For comparison with the Friedlander and log-normal 'self-preserving' distributions, the exponential distribution is also plotted in Figure 6.12 as $x\,f(x)$ versus x.

It is worth remembering that no assumption about the collision process is made in deriving eqn (6.62) and yet a distribution very like the Smoluchowski result emerges.

Another approach to the most probable distribution (Botet and Jullien, 1984) uses weighting factors based on collision kernels and leads to the result:

$$f(x) = \frac{(1 - w)^{1-w}}{\Gamma(1 - w)} x^{-w} \exp[-(1 - w)x] \tag{6.63}$$

where Γ is the gamma function. For the constant-kernel case, where $w = 0$, eqn (6.63) reduces to the simple exponential form, as expected. For higher values of w, eqn (6.63) gives rather higher concentrations of large aggregates, but the overall form of the distribution is not greatly changed.

Since the simplest 'maximum entropy' method (Rosen, 1984) makes no assumptions about collision mechanisms or aggregate break-up, it is interesting to speculate on its applicability to aggregation in agitated systems. Here, orthokinetic collision processes and break-up of aggregates by shear also lead to a steady-state floc size distribution. In real systems it is not possible to analyse these processes in detail, but they are essentially stochastic in nature and a probabilistic approach might be fruitful.

Cohen (1991) has investigated steady-state size distributions in stirred suspensions, using an entropy approach and assuming that aggregation and break-up processes occur entirely at random. However, there is no way in which aggregate strength could be introduced into such a procedure. In the absence of detailed information, it seems reasonable to take the mean aggregate size as a variable to be determined empirically. By analogy with limiting floc sizes found in stirred suspensions, the mean size should depend on the stirring conditions and on the 'floc strength', which is affected by the nature of the particle interaction and hence by the additives used. Given an empirical value of the mean aggregate size, a 'scaling' relationship, perhaps the very simple exponential form of eqn (6.62), might then describe the aggregate size distribution reasonably well.

6.7 Flocculation by polymers

6.7.1 Introduction

In practical applications, suspensions are often flocculated by polymers and the mode of action of polymeric flocculants is broadly understood (Gregory, 1987). In many cases high molecular weight polymers are effective by virtue of a *bridging* mechanism, in which segments of a single polymer chain are adsorbed on more than one particle. With polyelectrolytes and oppositely charged particles, there is also the possibility of charge effects; either simple charge neutralization or some form of 'electrostatic patch' interaction (Gregory, 1976). In some texts the term 'flocculation' is used to imply aggregation by a polymer bridging mechanism, as distinct from 'coagulation' caused by the addition of salts and the consequent effect on double layer interaction (see Chapter 3). However, there are some difficulties in maintaining this distinction, e.g. in the case of polyelectrolytes, where the relative importance of polymer bridging and charge effects may not be known.

At the moment of addition of a polymer solution to a suspension of particles, usually under conditions of agitation, several processes are initiated, which are illustrated schematically in Figure 6.13:

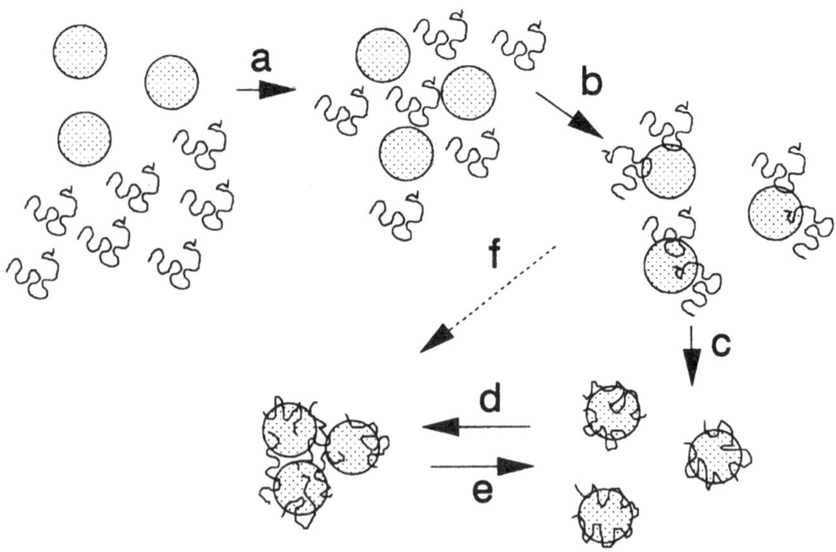

Figure 6.13 *Schematic illustration of the stages involved in the flocculation of a suspension by added polymer:* (a) *mixing of polymer chains among the particles;* (b) *adsorption of polymer;* (c) *re-conformation of adsorbed chains;* (d) *flocculation; and* (e) *break-up of flocs. The dashed arrow,* (f) *shows the possibility of bridging flocculation before the adsorbed chains have reached their equilibrium conformation*

(a) mixing of polymer solution throughout the suspension
(b) adsorption of polymer molecules on the particles, to give some degree of destabilization (either by charge neutralization or by giving opportunity for 'bridging' between particles)
(c) rearrangement of adsorbed polymer molecules from an initially extended state to a flatter, equilibrium conformation
(d) collisions between destabilized particles to give aggregates (flocs)
(e) break-up of flocs under the influence of applied shear.

These processes occur at different rates, which depend on a number of factors. Furthermore, they do not occur sequentially, but, to a large extent, concurrently, which makes the whole system difficult to analyse, but can help to explain the importance of mixing and dosing effects.

6.7.2 Mixing

In many cases, effective polymeric flocculants are of quite high molecular weight and are added as rather viscous solutions. For this reason the thorough distribution of polymer throughout the suspension may take some time and small droplets of polymer solution may persist for an appreciable fraction of the mixing time. These could possibly act as 'nuclei' for floc formation, although there is little evidence for such an effect (Stewart and Sutton, 1986). If the time needed for a thorough distribution of added polymer among the particles is long relative to the adsorption and flocculation rates, there may be significant practical consequences. For instance, excess polymer may adsorb on a fraction of the particles, which could then become restabilized. This effect is thought to be responsible for the residual haze and poor filterability found for some flocculated suspensions (Slater and Kitchener, 1966).

There is a good deal of evidence, largely empirical, that a high degree of turbulence at the point of addition of polymer solution can have a beneficial effect on the flocculation process. This is especially the case for fairly concentrated suspensions, where adsorption and particle collision rates are rapid. For instance, Warden (1984) found that the polymer-assisted thickening and dewatering of waterworks sludge could be greatly improved by careful attention to the polymer dosing point. It was shown that poor thickening was obtained at low and high values of applied shear, with intermediate values giving optimum results. There are many other applications, such as papermaking, where mixing conditions are known to be important.

Many laboratory flocculation trials are conducted in stirred vessels and the stirring rate can have a marked effect on the results. Such systems are complicated by the fact that there is a wide variation of effective shear rates (or energy dissipation) in a stirred vessel (Glasgow and Kim, 1986), so that floc formation and break-up may occur in different regions. It is typically found that flocs pass through a maximum size at some time after polymer dosing, and can show a marked reduction in size with continued stirring. These changes are reflected in dewatering rate (Kayode and Gregory, 1988) and settling rate (Li

Guibai and Gregory, 1991), which also reach maximum values after a certain stirring time.

By dosing polymer solution slowly or intermittently to a stirred suspension, considerable improvement in flocculation has been observed by Hogg (1989). In fact, Hogg showed that, in some cases, floc growth occurred only during polymer addition. Clearly, these effects must be closely related to mixing, since the more slowly polymer is added, the less likely is local overdosing and flocs are not allowed to grow very rapidly during the very early stages of the process.

6.7.3 Adsorption

It seems reasonable to suppose that the adsorption of polymer molecules on particles is a process that occurs at a transport-limited rate, at least in the early stages, when the fractional coverage is still quite low. In that case, the rate of adsorption depends on the rate of arrival of polymer molecules at a particle surface, and the number of encounters in unit time and unit volume can be written, by analogy with eqn (6.1):

$$J_{12} = k_{12} n_1 n_2 \tag{6.64}$$

where n_1 and n_2 are the number concentrations of particles and polymer molecules respectively and k_{12} is the appropriate rate constant (see below).

In practice, a certain fraction of added polymer needs to be adsorbed on particles to give adequate destabilization and hence allow flocculation to occur under the given shear conditions. If this fraction is f, then the time required for adsorption can be derived simply, assuming that the particle concentration remains constant, giving:

$$t_{\mathrm{Ads}} = -\frac{\ln(1-f)}{k_{12} n_1} \tag{6.65}$$

A very important feature of this result is the inverse dependence on particle number concentration – the adsorption time will be lower at higher particle concentrations. The required time is also independent of the polymer concentration, provided that the same fraction of added polymer needs to be adsorbed. This last condition implies that the amount of adsorbed polymer needed to destabilize particles is proportional to particle concentration.

The adsorption rate constant can be estimated if it is assumed that both particles and polymer molecules can be treated as spheres. This assumption may seem unreasonable in the case of long-chain polymers, but such chains often adopt an approximately spherical, random coil conformation in solution and can be treated, hydrodynamically, as equivalent spheres. The transport of polymer molecules to particles may be by diffusion (perikinetic) or under the influence of fluid motion (orthokinetic). In each case, the rate constant can be derived from the appropriate Smoluchowski expression (see Gregory, 1988). For the perikinetic case, the rate constant is given by eqn (6.8) and the orthokinetic value is given by eqn (6.20).

The high shear forces present at a particle surface may reduce the adsorption rate and may even cause desorption of adsorbed polymer (Lee and Fuller, 1985). Although a quantitative analysis of this aspect is not yet possible, it may be one reason to expect an 'optimum' mixing intensity for flocculation by polymers.

Rate constants calculated from eqns (6.8) and (6.20) show that orthokinetic transport will often be much more significant than diffusion in promoting particle–polymer contacts (Gregory, 1988). This is the case for particles of about 1 μm in size or greater and polymer molecules larger than about 0.1 μm, with effective shear rates greater than about $50\,s^{-1}$ (see the numerical example in Section 6.7.5). Since typical polymer mixing conditions can correspond to shear rates of several hundred per second or greater, orthokinetic transport is likely to be important in most cases involving polymers of moderately high molecular weight. In such cases, the rate of polymer adsorption will depend strongly on the effective shear rate.

6.7.4 Reconformation

At equilibrium, an adsorbed polymer chain adopts a conformation consisting of trains, loops and tails, and a great deal of effort has been expended in both theoretical and experimental determinations of the segment density distribution as a function of distance from the adsorbing surface, (see, for example, Cosgrove, 1990). However, this equilibrium arrangement may be very different from the conformation of a polymer molecule in free solution and at the moment of initial contact with the particle.

The time over which the equilibrium conformation is attained can have a great influence on flocculation, since it is very likely that bridging contacts will occur more readily when the adsorbed polymer is in an extended, non-equilibrium state. When there is strong interaction between adsorbing polymer segments and the particle surface, e.g. between ionic groups of opposite sign, it is likely that a polymer chain will eventually lie flat on the surface, thus presenting no opportunity for bridging contacts. It could be argued that bridging contacts formed by polymer chains very soon after adsorption, or by simultaneous adsorption on two particles, would be weaker than those formed by adsorbed chains in a more 'flat' configuration. In the latter case, more segments of the chain would be attached to the particles and this should result in stronger binding between particles. Also, the more extended chains would occupy a smaller fraction of the particle surface and so the collision efficiency might be lower (see below).

Pelssers *et al.* (1990) assumed that adsorbed polymer chains would either be in an 'active' or 'inactive' state, the former corresponding to the extended, non-equilibrium condition in which bridging contacts are possible. After some time, relaxation to a flatter, inactive configuration occurs. The rate of this relaxation process depends on the nature of the polymer and on the particle–polymer interaction, but characteristic times of a few seconds for polymers with molecular weights of several million have been estimated (Cohen Stuart and Tamai, 1988). These times fit quite well the flocculation data of Pelssers *et al.* (1990). However, very much longer relaxation times (many minutes or even hours) are indicated from experiments of Pefferkorn and Elaissari (1990), with quaternized polyvinyl pyridine of lower molecular weight (3.6×10^5).

At present, there is little other information which is of help in modelling the reconformation process under practical flocculation conditions. The effect of shear on the process seems to have been neglected.

6.7.5 Flocculation

The rate of aggregation (flocculation) depends on the collision rate of particles, also given by eqn (6.1). The value for the collision rate constant of primary particles, k_{11}, or the aggregation rate constant, $k_a (= k_{11}/2)$ can be used to calculate the characteristic aggregation time, τ, given by eqn (6.15).

It has been shown (Gregory, 1988) that this approach, applied to stirred suspensions and fairly high molecular weight polymers, predicts flocculation times considerably lower than adsorption times. A simple numerical example may be helpful.

Consider an aqueous suspension at 25°C, containing particles of 2 μm diameter at a number concentration of $10^{15}\,m^{-3}$. The suspension is stirred to give an effective shear rate of $50\,s^{-1}$, and 80% of the added polymer needs to be adsorbed to give effective destabilization. The polymer molecules are assumed to have an effective diameter of 200 nm (a reasonable value for a random coil polymer with a molecular weight of several million). It is then possible to calculate the appropriate rate constants from eqns (6.8) and (6.20) and then the times to give the required degree of adsorption from eqn (6.65). The characteristic aggregation times can be obtained from eqn (6.15). The following results are found:

	Perikinetic	*Orthokinetic*
t_{Ads}	43 s	7.8 s
τ	163 s	3.7 s

If the adsorption and aggregation are controlled only by diffusion (the perikinetic case), then the effective amount of polymer is adsorbed in a time which is appreciably less than the aggregation time. In this case the aggregation process would not be limited by the rate of adsorption. Under the assumed shear conditions, orthokinetic collisions increase the rates of both polymer adsorption and particle aggregation. However, the aggregation rate is more strongly influenced and the characteristic aggregation time becomes less than the adsorption time. In this case, the aggregation process would be limited by the relatively slow adsorption – particles would undergo several collisions before sufficient polymer had been adsorbed to give adequate destabilization.

Even when the added polymer is fully adsorbed, the collision efficiency may be less than unity, so that every collision does not result in permanent attachment. If the fraction of particle surface with adsorbed polymer is θ, and it is assumed that the only successful collisions are those between coated and uncoated surfaces, then it follows that the flocculation rate will depend on the product $\theta(1-\theta)$. This leads to the well-known 'half surface coverage' condition for optimum flocculation (Healy and La Mer, 1964). If the amount of polymer added to the suspension is not sufficient to give $\theta > 0.5$, then the collision efficiency should increase throughout the adsorption phase. It follows that there will be a period after polymer dosing during which particles are not yet sufficiently destabilized for flocculation to occur. For dilute suspensions this 'lag time' can be appreciable (several minutes). It can be reduced by increasing the shear rate, which increases both adsorption and flocculation rates. Such effects have been verified experimentally (Gregory and Lee, 1990).

For more concentrated suspensions, polymer adsorption is sufficiently rapid for flocculation to begin almost immediately after dosing. In such cases, mixing and reconformation effects can be very important.

6.7.6 Floc break-up

This topic has been considered briefly in Section 6.5. Although no adequate quantitative models are available, the break-up of flocs produced by polymers can be qualitatively very different from that of other aggregates. Long-chain polymers giving bridging flocculation lead to much stronger flocs (higher C values) than simple salts. Polyelectrolytes acting by charge neutralization give flocs of intermediate strength. Typically, flocs formed by polymer bridging can have diameters approaching 1 mm, which is larger than the microscale of turbulence (of the order of 100 μm in most cases).

A very important point concerning floc breakage is that it is often irreversible in the case of polymer bridging. Although flocs formed by long-chain polymers are more resistant to breakage, they tend not to re-form if breakage does occur. When charge neutralization is the mode of action, flocs are generally weaker, but can regrow after breakage if the shear rate is reduced (Ditter *et al*, 1982). Irreversible floc breakage has important implications for the choice of the duration and intensity of mixing during polymer dosing.

Bibliography

Chandrasekhar, S., Kac, M. and Smoluchowski, R. (1986) *Marian Smoluchowski. His Life and Scientific Work*, Polish Scientific Publishers, Warsaw

Dobiáš, B. (ed.) (1993) *Coagulation and Flocculation, Theory and Applications*, Marcel Dekker, New York

Sonntag, H. and Strenge, K. (1987) *Coagulation Kinetics and Structure Formation*, VEB Deutscher Verlag der Wissenschaften, Berlin

References

Adler, P. M. (1981) Heterocoagulation in shear flow. *J. Colloid Interface Sci.*, **83**, 106–115

Aubert, C. and Cannell, D. S. (1986) Restructuring of colloidal silica aggregates. *Phys. Rev. Lett.*, **56**, 738–741

Botet, R. and Jullien, R. (1984) Size distribution of clusters in irreversible kinetic aggregation. *J. Phys. A.*, **17**, 2517–2530

Brakalov, L. B. (1987) A connection between the orthokinetic coagulation capture efficiency of aggregates and their maximum size. *Chem. Eng. Sci.*, **42**, 2373–2383

Camp, T. R. (1953) Flocculation and flocculation basins. *Proc. ASCE*, **79**, 1–18

Camp, T. R. and Stein, P. C. (1943) Velocity gradients and internal work in fluid motion. *J. Boston Soc. Civ. Eng.*, **30**, 219–238

Cleasby, J. L. (1984) Is velocity gradient a valid flocculation parameter? *J. Env. Eng. ASCE.*, **110**, 875–897

Cohen R. D. (1991) Evolution of the cluster-size distribution in stirred suspensions. *J. Chem. Soc. Faraday Trans.*, **87**, 1163–1168

Cohen-Stuart, M. A. and Tamai, M. (1988) Dynamics of adsorbed polymers. I. Thickness relaxation of poly(vinyl pyrrolidone) on glass. *Macromolecules*, **21**, 1863–1866

Cosgrove, T. (1990) Volume fraction profiles of adsorbed polymers. *J. Chem. Soc. Faraday Trans.*, **86**, 1323–1332

Ditter, W., Eisenlauer, J. and Horn, D. (1982) Laser optical method for dynamic flocculation testing in flowing dispersions. In *The Effect of Polymers on Dispersion Properties*, (ed. Tadros, Th.F.) Academic Press, London, 323–342

Dukhin, S. S. and Lyklema, J. (1990) Dynamics of colloid particle interaction. Incomplete desorption relaxation. *Faraday Discuss. Chem. Soc.*, **90**, 261–269

Frens, G. and Heuts, J. J. F. G. (1988) The double layer potential ϕ_δ as a rate determining factor in the coagulation of electrocratic colloids. *Colloids Surfaces*, **30**, 295–305

Friedlander, S. K. (1977) *Smoke, Dust and Haze*, John Wiley, New York

Friedlander, S. K. and Wang, C. S. (1966) The self-preserving particle size distribution for coagulation by Brownian motion. *J. Colloid Interface Sci.*, **22**, 126–132

Fuchs, N. (1934) Über die Stabilität und Aufladung der Aerosole. *Z. Physik*, **89**, 736–743

Glasgow. L. A. and Kim, Y. H. (1986) Characterization of agitation intensity in flocculation processes. *J. Env. Eng. ASCE*, **112**, 1158–1163

Glasgow, L. A. and Liu, X. (1991) Response of aggregate structures to hydrodynamic stress. *AIChE J.*, **37**, 1411–1414

Gregory, J. (1976) The effect of cationic polymers on the colloidal stability of latex particles. *J. Colloid Interface Sci.*, **55**, 35–44

Gregory, J. (1987) Flocculation by polymers and polyelectrolytes. In: *Solid–Liquid Dispersions* (ed. Tadros, Th.F.) Academic Press, London, 163–181

Gregory, J. (1988) Polymer adsorption and flocculation in sheared suspensions. *Colloids Surfaces*, **31**, 231–253

Gregory, J. and Lee, S. Y. (1990) The effect of charge density and molecular mass of cationic polymers on flocculation kinetics in aqueous solution. *J. Water SRT-Aqua*, **39**, 265–274

Healy, T. W. and La Mer, V. K. (1964) Energetics of flocculation and redispersion by polymers. *J. Colloid Sci.*, **19**, 323–332

Higashitani, K., Miyafusa, S., Matsuda, T. and Matsuno, Y. (1980) Axial change of total particle concentration in Poiseuille flow. *J. Colloid Interface Sci.*, **77**, 21–28

Hoekstra, L. L., Vreeker, R. and Agterhof, W. G. M. (1992) Aggregation of nickel hydroxycarbonate studied by light scattering. *J. Colloid Interface Sci.*, **151**, 17–25

Hogg, R. (1989) The dynamics of polymer-induced flocculation of fine-particle suspensions. In *Flocculation and Dewatering*, (ed. Moudgil, B. M. and Scheiner, B. J.) Engineering Foundation,

New York, 143–151

Honig, E. P., Roebersen, G. J. and Wiersema, P. H. (1971) Effect of hydrodynamic interaction on the coagulation rate of hydrophobic colloids. *J. Colloid Interface Sci.*, **36**, 97–109

Jiang, Q. and Logan, B. E. (1991) Fractal dimensions of aggregates determined from steady-state size distributions. *Environ. Sci. Technol.*, **25**, 2031–2038

Kayode, T. O. and Gregory, J. (1988) A new technique for monitoring alum sludge conditioning. *Water Res.*, **22**, 85–90

Knocke, W. R. and Kelley, R. T. (1987) Improving heavy metal sludge dewatering characteristics by recycling preformed sludge solids. *J. Water Pollut. Control Fed.*, **59**, 86–91

Lagvankar, A. L. and Gemmell, R. S. (1968) A size–density relationship for flocs. *J. Am. Water Wks Assoc.*, **60**, 1040–1046

Lee, J-J and Fuller, G. G. (1985) Adsorption and desorption of flexible polymer chains in flowing systems. *J. Colloid Interface Sci.*, **103**, 569–577

Lee, K. W. (1983) Change of particle size distribution during Brownian coagulation. *J. Colloid Interface Sci.*, **92**, 315–325

Lee, K. W. (1985) Conservation of particle size distribution parameters during Brownian coagulation. *J. Colloid Interface Sci.*, **108**, 199–206

Li Guibai and Gregory, J. (1991) Flocculation and sedimentation of high-turbidity waters. *Water Res.*, **25**, 1137–1143

Lichtenbelt, J. W. Th., Pathmamanoharan, C. and Wiersema, P. H. (1974) Rapid coagulation of polystyrene latex in a stopped flow spectrophotometer. *J. Colloid Interface Sci.*, **49**, 281–285

Lin, M. Y., Lindsay, H. M., Weitz, D. A., Ball, R. C., Klein, R. and Meakin, P. (1989) Universality in colloid aggregation. *Nature*, **339**, 360–362

Meakin, P. (1988) Fractal aggregates. *Adv. Colloid Interface Sci.*, **28**, 249–331

Meesters, A. and Ernst, M. H. (1987) Numerical evaluation of self-preserving spectra in Smoluchowski coagulation theory. *J. Colloid Interface Sci.*, **119**, 576–587

Melik, D. H. and Fogler, H. S. (1984) Gravity-induced flocculation. *J. Colloid Interface Sci.*, **101**, 72–83

Michaels, A. S. and Bolger, J. C. (1962) The plastic flow behavior of flocculated colloidal sediments. *Ind. Eng. Chem. Fund.*, **1**, 153–162

Mühle, K. (1993) Floc stability in laminar and turbulent flow. In *Coagulation and Flocculation, Theory and Applications* (ed. Dobiáš, B.) Marcel Dekker, New York, 355–390

Mühle, K. and Domasch, K. (1991) Stability of particle aggregates in flocculation with polymers. *Chem. Eng. Process.*, **29**, 1–8

Ottewill, R. H. and Shaw, J. N. (1966) Stability of monodisperse polystyrene latex dispersions of various sizes. *Discuss. Faraday Soc.*, **42**, 154–163

Overbeek, J. Th. G. (1980) The rule of Schulze and Hardy. *Pure Appl. Chem.*, **52**, 1151–1161

Pefferkorn, E. and Elaissari, H. (1990) Adsorption–desorption processes in charged polymer/colloid systems; structural relaxation of adsorbed macromolecules. *J. Colloid Interface Sci.*, **138**, 187–194

Pelssers, E. G. M., Cohen Stuart, M. A. and Fleer, G. J. (1990) Kinetics of bridging flocculation. Role of relaxations in the polymer layer. *J. Chem. Soc. Faraday Trans.*, **86**, 1355–1361

Reerink, H. and Overbeek, J. Th. G. (1954) The rate of coagulation as a measure of the stability of silver iodide sols. *Discuss. Faraday Soc.*, **18**, 74–84

Rosen, J. M. (1984) A statistical description of coagulation. *J. Colloid Interface Sci.*, **99**, 9–19

Saffman, P. G. and Turner, J. S. (1956). On the collision of drops in turbulent clouds. *J. Fluid Mech.*, **1**, 16–30

Slater, R. W. and Kitchener, J. A. (1966) Characteristics of flocculation of mineral suspensions by polymers. *Discuss. Faraday Soc.*, **42**, 267–275

Smoluchowski, M. (1917) Versuch einer mathematischen Theorie der Koagulationskinetic kolloider Lösungen. *Z. Phys. Chem.*, **92**, 129–168

Sonntag, H. (1993) Coagulation kinetics. In *Coagulation and Flocculation, Theory and Applications* (ed. Dobiáš, B.) Marcel Dekker, New York, 57–99
Spielman, L. A. (1970) Viscous interactions in Brownian coagulation. *J. Colloid Interface Sci.*, **33**, 562–571
Spielman, L. A. (1978) Hydrodynamic aspects of flocculation. In *The Scientific Basis of Flocculation* (ed. Ives, K. J.) Sijthoff and Noordhoff, Alphen aan den Rijn, 63–88
Stewart, R. F. and Sutton, D. (1986) Characterization of the structure of concentrated flocculated suspensions. *Part. Sci. Technol.*, **4**, 251–264
Swift, D. L. and Friedlander, S. K. (1964) The coagulation of hydrosols by Brownian motion and laminar shear flow. *J. Colloid Sci.*, **19**, 621–647
Tambo, N. and François, R. J. (1991) Mixing, breakup and floc characteristics. In *Mixing in Coagulation and Flocculation.* (ed. Amirtharajah, A., Clark, M. M. and Trussell, R. R.) American Water Works Association Research Foundation, Denver, 256–281
Tambo, N. and Watanabe, Y. (1979) Physical aspects of flocculation. I. The floc density function and aluminium floc. *Water Res.*, **13**, 409–419
Tence, M., Chevalier, J. P. and Jullien, R. (1986) On the measurement of the fractal dimension of aggregated particles by electron microscopy: experimental method, corrections and comparison with numerical models. *J. Phys.*, **47**, 1989–1998
Torres, F. E., Russel, W. B. and Schowalter, W. R. (1991a) Floc structure and growth kinetics for rapid shear coagulation of polystyrene colloids. *J. Colloid Interface Sci.*, **142**, 554–574
Torres, F. E., Russel, W. B. and Schowalter, W. R. (1991b) Simulations of coagulation in viscous flows. *J. Colloid Interface Sci.*, **145**, 51–73
van de Ven, T. G. M. and Mason, S. G. (1977) The microrheology of colloidal suspensions. VII. Orthokinetic doublet formation of spheres. *Colloid Polymer Sci.*, **255**, 468–479
van de Ven, T. G. M. and Mason, S. G. (1981) Comparison of hydrodynamic and colloid forces in paper machine headboxes. *Tappi*, **64**, 171–175
Vold. M. J. (1963) Computer simulation of floc formation in a colloidal sediment. *J. Colloid Sci.*, **18**, 684–695
Warden, J. H. (1984) Thickening and dewatering of hydroxide sludges. In *Solid–Liquid Separation* (ed. Gregory, J.) Ellis Horwood, Chichester, 89–99
Warren, L. J. (1981) Shear flocculation. *Chemtech*, **11**, 180–185
Wiesner, M. R. (1992) Kinetics of aggregate formation in rapid mix. *Water Res.*, **26**, 379–387
Zollars, R. I. and Ali, S. I. (1986) Shear coagulation in the presence of repulsive interparticle forces. *J. Colloid Interface Sci.*, **114**, 149–166

Chapter 7

Selection of a simulation method

Nomenclature

a	particle radius
$\boldsymbol{a}_i$	acceleration vector of particle i
A	Helmholtz free energy
A_S	parameter used in Smart MC algorithm
A, B	dynamic variables, quantities or functions
$\boldsymbol{A}_{ij}$,$\boldsymbol{B}_{ij}$	pair diffusion tensors
$c_0, c_1, c_2, \ldots$	numerical constants
c_{AA}	autocorrelation coefficient
c_{AB}	correlation coefficient for variables A and B
D	diffusion constant
$\boldsymbol{D}$	grand diffusion tensor
$\boldsymbol{D}_i^0$	diffusion tensor for a single particle i
D_{ii}	self-diffusion coefficient
$\boldsymbol{D}_{ij}$	diffusion tensor for particle pair i and j
$\boldsymbol{e}_{ij}$	unit vector pointing from particle i to particle j
E	internal energy
E_0	initial value of internal energy
f	normalizing factor
$\boldsymbol{F}$	force vector
G	Gibbs free energy
G	shear rate
$\hbar$	Planck's constant
H	Hamiltonian value
i	index
$\boldsymbol{I}$	unit tensor
iL	Liouville operator
$g(r)$	pair distribution function
k	Boltzmann constant
K	kinetic energy
m_i	mass of particle i
n	integer
N	number of particles

$\mathbf{p}$	generalized conjugate momenta
$\mathbf{p}^{\mathrm{G}}$	random momentum term from Gaussian distribution
p_i	momentum of particle $i=1,2,\ldots,N$
p_m,p_n	elements of a limiting distribution vector of state space
P	pressure
$\mathbf{P}$	limiting distribution vector of state space
$\mathbf{q}$	generalized coordinates
q_i	generalized coordinates of particle i
q_{mn}	transition probability from state m to state n
Q	partition function
$\mathbf{Q}$	transition matrix
r	separation distance
$\mathbf{r}$	generalized position in Cartesian coordinate system
$\mathbf{r}^{\mathrm{G}}$	random displacement term from Gaussian distribution
$\mathbf{r}_i$	position vector of particle i
$\Delta\mathbf{r}_i$	displacement vector of particle i
Δr_{max}	maximum displacement
$\mathbf{R}_i^{\mathrm{G}}$	Gaussian random displacement for particle i
$\mathbf{R}_i^{\mathrm{U}}$	uniform random displacement for particle i
$\boldsymbol{R}$	rate-of-strain tensor
S	entropy
$\boldsymbol{S}_i$	shear tensor
t	time
Δt	time interval
T	absolute temperature
v	interparticle potential energy
$\mathbf{v}$	velocity
$\mathbf{v}^{\mathrm{G}}$	random velocity term from Gaussian distribution
V	volume
V_{n}	normalizing volume
$w(r)$	pair virial function
$W(r)$	weighting factor
α_{mn}	element of the underlying matrix
β	$1/kT$
δ	Kronecker delta or Dirac delta function
Ψ	thermodynamic potential
γ	flow-type parameter
Γ	thermodynamic state or phase point
λ	constant
λ_{B}	Broglie wavelength
μ	chemical potential
ρ	probability density or distribution function
ρ_{W}	weighted probability density or distribution function
σ	standard deviation
τ_{c}	correlation time
ξ	friction coefficient

Subscript

i	particle i
j	particle j
ij	particle pair
m	state m
n	state n
mn	transition from state m to n
nm	transition from state n to m

Superscript

c	correction stage
G	Gaussian distribution
m	state m
n	state n
(n)	n-particle
p	prediction stage
U	uniform distribution

7.1 Overview of simulation protocol

The modelling techniques employed in Chapters 4–6 to describe particle aggregation and deposition phenomena were essentially of two types, depending upon whether the problem under consideration was defined at a microscopic or at a macroscopic level.

At the macroscopic level of observation, Eulerian methods were used to describe the system properties in terms of particle concentration in space and time, either for a steady-state condition or to reveal time-dependent behaviour. Alternatively, at the microscopic level of observation the equation of motion, or trajectories, of individual particles in the colloidal system were solved to predict steady-state or time-dependent positions of every particle in the dispersion. Both these calculation procedures contain inherent limitations, many of which can, at least in principle, be overcome by using a *simulation method* geared to the appropriate degree of rigour and complexity that is required to solve the real practical problem of interest. Computer simulation methods adopt a microscopic level of observation and a variety of techniques can be employed to abstract the equilibrium properties and sometimes the dynamic or kinetic properties of the dispersion. These various simulation methods can be further classified into two types, as shown in Figure 7.1.

The figure shows the essential features of the 'stochastic' and 'dynamic' methods and, for each type, two further divisions according to the theoretical basis of the method, which is then further exemplified by examples of the principal named methods, their algorithms, and typical simulation applications.

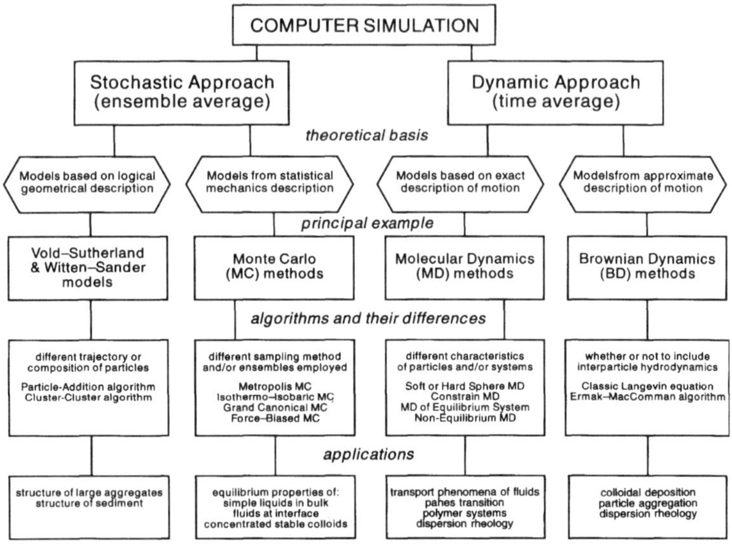

Figure 7.1 *Overview of common methods used to stimulate particle dispersions*

Moving across the figure from the left- to the right-hand side also corresponds to the chronological order in which the techniques were derived, as will be described later in this chapter.

Stochastic methods can be used to identify the positions and velocities of the dispersed phase, normally in the limiting case of equilibrium conditions, since the calculations cannot be scaled to a meaningful time dimension. As the name implies, dynamic methods can reveal dynamic properties from which it is possible to abstract information on the kinetics of the phenomena under investigation.

Since the 1960s computer simulations have played an increasingly important role in understanding the behaviour of simple liquid physics. The success of simulation techniques in this area has inspired its application in many other fields of science and technology involving 'particles' which may be as small as electrons or as large as planets. The significant advantage of computer simulation lies in the possibility of providing essentially exact results for problems which would otherwise only be soluble by approximate methods or, sometimes, may well be totally intractable. Furthermore simulation procedures may be able to offer details and insights of real (rather than 'model') processes and so assist in interpreting experimental results. Indeed, in some instances where experimental measurements are practically difficult or impossible to perform, simulation methodologies may even be able to replace experimentation, with due caution.

Since the first computer simulation was accomplished in the 1950s the science has undergone some dramatic development. Some of the most significant advances include the evolution:

(1) from using hard sphere to soft sphere models
(2) from equilibrium to non-equilibrium systems
(3) from one- or two-dimensional to three-dimensional problems
(4) from monoatomic to complex molecular liquids, or even 'liquids' of macroscopic grains
(5) from the basic canonical to many more sophisticated ensembles (considered in detail below)
(6) from smooth spheres to rough and/or non-spherical particles
(7) from classical mechanics to modern quantum descriptions.

Most of these developments have been made in the course of liquid simulation, and are well documented elsewhere (e.g. Hansen and McDonald, 1986; Allen and Tildesley, 1987).

This chapter aims to describe the fundamental concepts and principles underlying relevant simulation methods. Description of individual techniques and examples of particular methods will be focused upon those that are, or may be, applicable to deposition and aggregation in colloidal dispersions.

Widely used computer simulation methods include the Monte Carlo method (MC), molecular dynamics method (MD) and Brownian dynamics method (BD) (Figure 7.1). In general, the MC method is easier to implement but can only provide information on systems in their equilibrium state. The MD method, on the other hand, is more complex but capable of simulating non-equilibrium as well as equilibrium systems. Both methods are accurate in the sense that no simplifying assumptions need to be made in their formulation. In contrast, the BD method is only an approximate realization of, for example, Newton's equation of motion. However, in some cases, it may be the only feasible option.

These methods will be described in detail later, but a brief overview is appropriate here. The MC method, which derives its name from the extensive use of random numbers, relies on sampling a given statistical ensemble by generating a succession of configurations in accordance with the probability density for that ensemble. Different ensembles require different recipes of sampling; hence the so-called canonical MC, isothermal–isobaric MC, grand canonical MC, etc. The most widely used method is the Metropolis MC procedure, which samples from the canonical ensemble. In this procedure, each particle is picked up at random or in sequence and given a uniform random displacement, as mathematically stated in the following algorithm:

$$\Delta r_i = \Delta r_{\max} R_i^{\mathrm{U}} \tag{7.1}$$

where $\Delta r_{\max}$ is an adjustable parameter which controls the maximum displacement of the particles during each trial move, $\mathbf{R}_i^{\mathrm{U}}$ represents a random vector whose components are uniform random numbers on (–1, 1). At each step the total potential energy is calculated and compared with that of the preceding step. If the trial move renders a decrease in the potential energy, the move is accepted and the particle position is updated; otherwise, the trial move is accepted with a probability $\exp(-\Delta V/kT)$, where ΔV is the energy difference. After every particle has been moved (whether accepted or not), a MC cycle is said to be completed. To obtain reliable and stable results, at least 10^5–10^6 cycles are usually required. Compared with other simulation methods, the relative ease of use makes the MC method the obvious first choice for evaluating time-independent properties. It must be stressed that MC is only suitable for calculation of equilibrium properties.

In contrast to the stochastic nature of MC methods, MD methods are deterministic in nature because in the MD procedure configurations are generated by solving a set of Newton's equations of motion for each of the interacting particles. In practice, these equations of motion are replaced by a set of finite difference equations and are solved in a step-by-step manner. Perhaps the most commonly used method for integrating the equations of motion is the so-called Verlet algorithm:

$$r_i(t + \Delta t) = 2r_i(t) - r_i(t - \Delta t) + \Delta t^2 a_i(t) \qquad (7.2)$$

Since acceleration $\mathbf{a}(t)$ is involved, calculation of the net forces on each particle is required and this usually determines the overall execution (or CPU) time of the MD simulation program. The amount of computation (and hence the CPU time on a scalar machine) used by a MD time step is generally comparable to that of a MC cycle. Because there is a huge difference between the timescales used to describe the motion of solvent molecules or ions and the motion of colloidal particles, simulation of colloidal systems by the MD procedure can be prohibitively demanding of computing time. Also the Brownian motion, to which colloidal particles are subjected, cannot be readily allowed for by the Newton's deterministic equation of motion employed in the MD procedure. Therefore, the MD method is generally not suitable for simulation of colloidal dispersions. Hence, in these situations the BD method becomes useful.

In the BD method, Newton's equation of motion is replaced by some type of Langevin equation, and the effects of solvent particles are represented by a combination of a random force term and a frictional term, as shown in the following algorithm:

$$\Delta r_i(t) = \Delta t \sum_j \frac{D_{ij} \bullet F_j}{kT} + \Delta t \nabla_{r_j} \bullet D_{ij} + R_i^{G} \qquad (7.3)$$

where $\boldsymbol{D}_{ij}$ is the diffusion tensor in which hydrodynamic effects are incorporated, $\mathbf{F}_j$ is the net systematic force and $\mathbf{R}_i^{G}$ represents the random motion whose components are random variables selected from the $3N$-variate Gaussian distribution with zero mean and covariance matrix $\langle \Delta\mathbf{r}_i \Delta\mathbf{r}_i \rangle = 2\boldsymbol{D}_{ij}\Delta t$, $I, j = 1, 2, \ldots, N$. In MC or MD procedure the energy/force loop dictates the overall speed of a simulation, but in a BD procedure calculation of the diffusion tensors and generating multivariate correlated random numbers can be much more time-consuming than force evaluation. Therefore, a BD simulation with hydrodynamic interactions being taken into account usually requires more CPU time than MC or MD. Roughly speaking, on a scalar machine, the CPU time used by MC or MD simulations is proportional to N^2, whereas, for BD simulations with complex matrix manipulations, it is proportional to N^3.

Finally it should be mentioned that since all of the simulation methods described above have been developed from studies of the liquid state, modifications to these methods are usually needed to suit specific requirements in the investigation of colloidal dispersions. Examples of such modifications have been described in detail elsewhere (Jia, 1992) and some of the important principles concerning each simulation strategy will be discussed in detail below.

7.2 Useful concepts in statistical mechanics

Statistical mechanics is the theory used to analyse the behaviour of natural or spontaneous fluctuations. It provides a link between the detailed information at the microscopic level (e.g. positions, velocities, etc. of the constituent particles) and the observed and sometimes measurable macroscopic properties (e.g. pressure, internal energy, viscosity). A computer simulation essentially provides only the microscopic information. Statistical mechanics is used to convert these details into the desired information at the macroscopic level. Therefore it is necessary to review some basic concepts in statistical mechanics.

Consider an isolated macroscopic system consisting of N particles; the Hamiltonian H of the system may then be written as the sum of kinetic energy $K(\mathbf{p})$ and potential energy $E(\mathbf{q})$:

$$H(\mathbf{q},\mathbf{p}) = K(\mathbf{p}) + E(\mathbf{q}) \tag{7.4}$$

where $\mathbf{q} \equiv (q_1, q_2, \ldots, q_N)$ stands for the generalized coordinates of the particles and $\mathbf{p} \equiv (p_1, p_2, \ldots, p_N)$ the conjugate momenta. The values of these variables define a phase point, Γ, in a multidimensional space called phase space. In spherically isotropic systems consisting of simple atoms or spheres, $\mathbf{q}$ is simply a set of Cartesian coordinates, $\mathbf{r} \equiv (r_1, r_2, \ldots, r_N)$ and the phase space has $6N$ dimensions. Usually the kinetic energy K takes the form:

$$K = \sum_{i=1}^{N} |p_i|^2/2m_i \tag{7.5}$$

where m_i is the mass of particle i. The form of the potential energy E can be much more complex (as can be seen from Chapter 3) but it is often crucial to computer simulation since the amount of calculation involved in the simulation, the (structural) properties and the evolution of the systems to be simulated all depend on the potential energy.

A macroscopic system can have a large or even infinite number of microscopic states. An *ensemble* is the assembly of all possible microscopic states that are consistent with the constraints imposed to characterize the system macroscopically. The distribution of phase point Γ of an ensemble is described by a *phase space probability density*, or *phase space distribution function*, denoted by $\rho(\Gamma,t)$. The quantity $\rho(\Gamma,t)\,d\Gamma$ is the probability that at time t the physical system is in a microscopic state represented by a phase point Γ lying in the infinitesimal phase space element $d\Gamma$. For the probability density, there is a conservation law known as Liouville's theorem:

$$\frac{\partial\rho(\Gamma,t)}{\partial t} = -iL\rho(\Gamma,t) \tag{7.6}$$

which describe the fact that systems, or phase points of the ensemble, are neither created nor destroyed as time evolves. In eqn (7.6), L is the Liouville operator defined as:

$$iL \equiv \left(\sum_i \dot{q}_i \bullet \nabla_{q_i} + \sum \dot{p}_i \bullet \nabla_{p_i} \right) \equiv \left(\dot{q} \bullet \nabla_q + \dot{p} \bullet \nabla_p \right) \tag{7.7}$$

so that (7.6) has a formal solution which may be written as:

$$\rho(\Gamma,t) = \exp(-iLt)\rho(\Gamma,0) \tag{7.8}$$

where

$$\exp(-iLt) = 1 - iLt - \frac{1}{2} L^2 t^2 + \ldots .$$

If A denotes the instantaneous value of some dynamical variable or property that can be assumed as a function of the phase point, i.e. $A(\Gamma)$, then the equation of motion that describes the time dependence of A may be written as:

$$\dot{A}[\Gamma(t)] = iLA[\Gamma(t)] \tag{7.9}$$

$$A[\Gamma(t)] = \exp(iLt)A[\Gamma(0)] \tag{7.10}$$

A principal assumption of statistical mechanics is that an experimentally observable property A_{obs} is really the time average of $A(\Gamma)$ taken over a long time interval t_{obs}, i.e.

$$A_{\text{obs}} = \langle A[\Gamma(t)] \rangle_t = \lim_{t_{obs} \to \infty} \frac{1}{t_{\text{obs}}} \int_0^{t_{\text{obs}}} A[\Gamma(t)]\, dt \tag{7.11}$$

The integral here implies that phase space is treated as a continuum (the classical viewpoint). The alternative to the time average is an *ensemble average*, taken over all the members of the ensemble at a particular time:

$$A_{\text{obs}} = \langle A(\rho(\Gamma,t) \rangle_{\text{ens}} = \sum_{\Gamma} A(\Gamma)\rho(\Gamma) \tag{7.12}$$

The time average and the ensemble average gives the same result in the thermodynamic limit if the system is *ergodic*. An ergodic (or irreducible) system is a system in which a single closed trajectory can pass through all the non-zero $\rho(\Gamma)$ points in phase space.

Some common statistical ensembles are listed in Table 7.1. Here the symbol Q denotes the *partition function*, defined as the sum over states. For a given ensemble, it is has a definite value which may be used to normalize the phase space probability density and is related to the corresponding thermodynamic potential Ψ by:

$$\Psi = -\ln Q \tag{7.13}$$

The function Ψ has the property that at thermodynamic equilibrium it reaches a minimum value. There are different forms of Ψ, corresponding to different ensembles.

Table 7.1 *The commonly used classical ensembles*

Ensemble	*Probability density*	*Partition function proportional to*	*Thermodynamic relations*
Microcanonical (constant NVE)	$\delta[H(\Gamma) - E]$	$\sum_{\Gamma} \delta[H(\Gamma) - E]$ $\frac{1}{N!}\frac{1}{h^{3N}}\int dr\, dp\,\delta(H - E)$	$\Psi = S$ (S = entropy) $\beta S = \ln Q$
Canonical (constant NVT)	$\exp[-\beta H(\Gamma)]$	$\sum_{\Gamma} \exp[-\beta H(\Gamma)]$ $\frac{1}{N!}\frac{1}{h^{3N}}\int dr\, dp\, \exp(-\beta H)$	$\Psi = A$ (A = Helmholtz free energy) $A = E - TS$ $\beta A = -\ln Q$
Isothermal–isobaric (constant NPT)	$\exp[-\beta(H + PV)]$	$\sum_{\Gamma}\sum_{V} \exp[-\beta(H + PV)]$ $\frac{\int dV \int dr\, dp \exp[-\beta(H + PV)]}{V_n h^{3N} N!}$	$\Psi = G$ (G = Gibbs free energy) $\beta G = -\ln Q$ $G = A + PV$
Grand canonical (constant μVT)	$\exp[-\beta(H - \mu N)]$	$\sum_{\Gamma}\sum_{N} \exp[-\beta(H - \mu N)]$ $\sum_{N} \frac{\exp(\mu N \beta) \int dr\, dp \exp(-\beta H)}{N! h^{3N}}$	$\Psi = PV$ $\beta PV = \ln Q$

Table 7.1 does not exhaust all the ensembles that may be of interest in computer simulation. Sometimes, it may be more convenient to use an ensemble with extra constraint(s) imposed on the ensemble (e.g. the canonical ensemble with constant total momentum as used in MD). Also, non-equilibrium ensembles may be set up. In fact, the possibilities of constructing the ensemble are endless, provided that the corresponding phase space probability density $\rho(\Gamma)$ can be written down and a proper method for generating state points can be devised.

In computer simulations, instead of the full phase space distribution function, the configurational distribution function is often used. In a classical system, the *configurational distribution function*, $\rho(\mathbf{r})$, is defined as:

$$\rho(r) = \int_{p} \rho(r,p)\, dp \tag{7.14}$$

Similarly, we have the *momentum distribution function* $\rho(\mathbf{p})$:

$$\rho(p) = \int_{r} \rho(r,p)\, dr \tag{7.15}$$

In this way, the phase space distribution function breaks into two parts:

$$\rho(\Gamma) = \rho(r)\rho(p) \tag{7.16}$$

By integrating eqn (7.14) over all coordinates except those pertaining to the particles of interest, the reduced configurational distribution function is obtained:

$$\rho^{(n)}(r_1, r_2, \ldots, r_n) = \int \mathrm{d}r^{N-n} \rho(r) \tag{7.17}$$

Here the superscript (n) is used to distinguish from the *full* distribution function $\rho(r)$, and $\mathrm{d}r^{N-n}$ is a shorthand for $\mathrm{d}r_{n+1}\mathrm{d}r_{n+1} \ldots \mathrm{d}r_N$. $\rho^{(n)}$ means the joint distribution function that in an N-particle system particle 1 will be found at $\mathbf{r}_1$, particle 2 at $\mathbf{r}_2$, ... and particle n at $\mathbf{r}_n$. If the N particles are indistinguishable, the corresponding distribution function $\rho^{(1)}$ is related to $\rho^{(n)}$ $(\mathbf{r}_1, \mathbf{r}_2, \ldots, \mathbf{r}_n)$ by:

$$\rho^{(n)} = \frac{N!}{(N-n)!} \rho^{(1)} \tag{7.18}$$

For a homogeneous system:

$$\rho^{(1)} = \frac{N}{V} = \rho \tag{7.19}$$

where ρ is what is commonly called the number density. For an 'ideal gas' in which particles are uncorrelated, $\rho^{(2)}(\mathbf{r}_1, \mathbf{r}_2)$ factors as $\rho^{(1)}(\mathbf{r}_1)\rho^{(1)}(\mathbf{r}_2)$, i.e.

$$\rho^{(2)}(r_1, r_2) = \frac{N(N-1)}{V^2} = \rho^2 \left(1 - \frac{1}{N}\right) \approx \rho^2 \tag{7.20}$$

At this point, it is convenient to use the radial distribution function $g(r)$, which is defined as:

$$g(r) \equiv \rho^{(2)}(r_1, r_2)/\rho^2 \tag{7.21}$$

This function gives the probability of finding a pair of particles a distance r apart, relative to the probability expected for a completely random distribution at the same density. The radial distribution function is useful for an isotropic and homogeneous system, not only because it provides insight into the structure of the dispersion (liquid), but also because many thermodynamic properties of the system can be written as integrals over $g(r)$ provided that the particles interact through pairwise-additive forces. For example, the internal energy may be written as:

$$E = \tfrac{3}{2}NkT + 2\pi N\rho \int_0^\infty r^2 v(r) g(r)\, \mathrm{d}r \tag{7.22}$$

where $v(r)$ is the interparticle pair potential, and the pressure:

$$PV = NkT - \tfrac{2}{3}\pi N\rho \int_0^\infty r^2 w(r) g(r)\, \mathrm{d}R \tag{7.23}$$

where $w(r) = r\,\mathrm{d}v(r)/\mathrm{d}r$ is called the interparticle *pair virial function.*

The final group of concepts in this introductory section is concerned with the so-called correlation functions, which are encountered in other areas of engineering and measurement science. Correlations between two different quantities A and B are measured via the correlation coefficient c_{AB}:

$$c_{AB} = <\delta A \delta B>/\sigma(A)\sigma(B) \tag{7.24}$$

where $\sigma(A)$ and $\sigma(B)$ are the root-mean-square deviations, e.g.

$$\sigma^2(A) = <\delta A^2> = <A^2> - <A>^2 \tag{7.25}$$

$$\delta A = A - <A>$$

The absolute value of c_{AB} lies between 0 and 1. Higher values of c_{AB} indicate a high degree of correlation. Useful examples of the correlation functions include:

- *time correlation function* $c_{AB}(t)$, obtained when A and B are evaluated at two different times
- *autocorrelation function* $c_{AA}(t)$, if A and B are identical. The time integral (from $t=0$ to $t=\infty$) gives a correlation time τ_c.

These functions are important quantities in that they provide a clear picture of the dynamics in a fluid; the correlation time τ_c may be directly related to macroscopic transport coefficients (e.g. the diffusion coefficient), and their Fourier transform may be related to experimental spectra.

7.3 Monte Carlo methods

The MC method aims to generate a trajectory in phase space which samples from a given ensemble so that the ensemble average can be evaluated. As described in Section 7.1, the name 'Monte Carlo' originates from the extensive use of random numbers in the sampling procedures. The development of modern MC methods was instigated by Metropolis and coworkers (Metropolis and Ulam, 1949) at the end of World War II to study diffusion of neutrons in fissionable material. Although the idea of using random sampling procedures in finding summation or integration is not new, the novel contribution of these authors was that, for the first time on a digital electronic computer, they demonstrated the possibility of treating determinate mathematical problems by finding a probabilistic analogue which can then be solved by a stochastic sampling experiment. In the authors' original words (Metropolis *et al.*,1953), the difference between the traditional scheme and their novel method is that 'instead of choosing configurations randomly and then weighting them with $\exp(-E/kT)$, we choose configurations with a probability $\exp(-E/kT)$ and weight them evenly'.

As implied in the above quotation, the method was based on sampling the canonical ensemble. To distinguish it from other ensemble samplings that have appeared subsequently, their method is often referred to as the Metropolis MC (MMC) method, while others are usually named according to the ensembles to be sampled.

7.3.1 Metropolis Monte Carlo method

Ideally, an ensemble average should be calculated by sampling the phase space thoroughly. This is, however, practically impossible and usually unnecessary because many of the samples will turn out to make little contribution to the average. Therefore, it is desirable for the phase space to be explored by a finite number of samplings, and each of the sampled (or, strictly speaking, generated) configurations makes an important contribution to the average. Such a sampling procedure is termed *importance sampling*. With importance sampling, the desired ensemble average of a quantity A(Γ):

$$\langle A \rangle_{\text{ens}} = \int A(\Gamma)\rho_{\text{ens}}(\Gamma)\,\text{d}\Gamma \tag{7.26}$$

is approximated by:

$$\langle A \rangle_{\text{ens}} = \langle A\rho_{\text{ens}}/\rho_{\text{tri}} \rangle_{\text{tri}} \tag{7.27}$$

where ρ_{ens} is phase space probability density for the given ensemble, and ρ_{tri} corresponds to a distribution function from which the configuration is sampled. In the Metropolis MC method (MMC), ρ_{tri} is chosen to be equal to ρ_{ens}, or ρ_{NVT}, since their calculations are carried out in the canonical ensemble, so that the ensemble average is simply:

$$\langle A \rangle_{\text{ens}} = \langle A \rangle_{\text{tri}} \tag{7.28}$$

Such a choice, $\rho_{\text{tri}} = \rho_{\text{NVT}}$, implies that the quantity $A\rho_{\text{NVT}}$ will be statistically significant where ρ_{NVT} is significant. This assumption can be justified for most functions A of practical interest, but not always.

Now the remaining question is how to accomplish the importance sampling so that the ensemble average can be evaluated by eqn (7.28). One way is to set up a so-called *Markov chain*, with the limiting distribution of the chain equal to ρ_{ens}. A Markov chain is a finite sequence of trial states $\{\Gamma_1, \Gamma_2, \ldots, \Gamma_m, \Gamma_n, \ldots\}$ that constitute a *state space* in which each state is related only to the one that immediately precedes it. Denoting **Q** as the transition matrix for the Markov chain, consisting of the transition probability q_{mn} (i.e. the probability of going from Γ_m to Γ_n), and **P** the limiting distribution vector for the state space, with its elements p_n given by $\rho_{\text{ens}}(\Gamma_n)$, the property of the Markov chain is formally described by the eigenvalue equation:

$$PQ = P \tag{7.29}$$

or

$$\sum_m p_m\, q_{mn} = \text{f}_n \tag{7.30}$$

with an eigenvalue of unity. The transition matrix **Q** is often called the stochastic matrix since its rows must add to one:

$$\sum_n q_{mn} = 1 \tag{7.31}$$

In a MC simulation, the limiting distribution vector **P** is known once the ensemble is chosen, but the transition matrix **Q** is not and must be determined. Since for a given **P**, the solution to eqn (7.29) satisfying eqn (7.31) is not unique, there is often a great deal of flexibility in finding an appropriate transition matrix **Q**. The only constraint is that the elements of the matrix q_{mn} should be independent of the partition function Q (the reason will become clear below).

A useful short-cut in searching for a solution of eqn (7.29) is to utilize the microscopic reversibility condition:

$$p_m q_{mn} = p_n q_{nm} \tag{7.32}$$

which considerably eases the task and still guarantees that eqn (7.29) is satisfied. It was based on this microscopic condition and eqn (7.31) that Metropolis *et al.* (1953) suggested a sampling procedure:

$$q_{mn} = \begin{cases} \alpha_{mn} & m \neq n, \quad p_n \geq p_m \\ \alpha_{mn}(p_n/p_m) & m \neq n, \quad p_n < p_m \\ 1 - \sum\limits_{n \neq m} q_{mn} & m = n \end{cases} \tag{7.33}$$

often known as the asymmetrical solution. Note that this solution only involves the ratio $p_n/p_m = \rho(\Gamma_n)/\rho(\Gamma_m)$, so that the partition function is cancelled out. Therefore, it is independent of the partition function Q.

In the asymmetrical solution, $\alpha_{mn} = \alpha_{nm}$. Therefore, α_{mn} constitutes a symmetrical stochastic matrix, usually called the underlying matrix of the Markov chain. The underlying matrix is not unique and there is considerable freedom in choosing it.

The flexibility in choosing the stochastic matrix and the underlying matrix has made it possible for a variety of sampling schemes to be proposed (e.g. Barker, 1965). However, the Metropolis scheme is by far the most commonly used because evidently it leads to a faster convergence of the Markov chain (Valleau and Whittington, 1977).

7.3.2 Monte Carlo methods for various ensembles

Although the importance sampling established by Metropolis *et al.* (1953) is based on the canonical ensemble, it is a rather general sampling strategy and can be readily applied to other ensembles. On the other hand, the difficulty inherent in the Metropolis MC is providing a direct route to statistical properties, such as the entropy S and Helmholtz free energy A that require evaluation of the partition function Q, also raises the need to exploit other ensembles.

The extension of the method to the isothermal–isobaric (i.e. constant-NPT) ensemble was first made by Wood (1968, 1970), with the Markov chain of state having a limiting distribution proportional to $\exp[-\beta(PV+E)+N\ln V]$. Like the MMC, the isothermal–isobaric MC cannot calculate properties such as the Gibbs free energy since the method only samples importance regions in phase space.

To calculate the Gibbs free energy, the grand canonical MC method (GCMC) may be used. In this method the Markov chain of states is generated such that the limiting distribution is proportional to $\exp[-\beta(E-N\mu)-\ln N!-3N\ln\lambda_B + N\ln V]$, λ_B being the thermal de Broglie wavelength defined as $\lambda_B = (\hbar^2/2\pi mkT)^{1/2}$. A number of techniques for generating this chain have been proposed, e.g. Salsburg *et al.* (1959), Chesnut (1963), Norman and Filinov (1969) and Rowley *et al.* (1975). Compared with the MMC, the grand canonical MC is much more complicated and difficult to handle and will not be presented here.

7.3.3 Advanced Monte Carlo methods

Umbrella sampling

The conventional MC methods, described in the last section, are based on the importance sampling strategy. This procedure creates a problem for the MMC and its extensions when calculating some statistically important properties (such as free energy) because their calculation demands a substantial sampling over the higher-energy states. Unfortunately, these higher energy states are mostly rejected in the importance sampling procedure. The solution, therefore, lies in finding some non-Boltzmann-type sampling method designed to encourage the system to explore those regions alien to (or rarely encountered by) the MMC.

One of the methods proposed to calculate free energy in the canonical ensemble is so-called umbrella sampling (Torrie and Valleau, 1974, 1977). The umbrella sampling generates the Markov chain from a distribution of the following form:

$$\rho_w = \frac{W(r)\exp[-\beta E_0(r)]}{\int dr W(r)\exp[-\beta E_0(r)]} \tag{7.34}$$

For instance, in the case of a molecular fluid, this method actually calculates the free energy difference between the fluid of interest and a reference fluid characterized by potentials E and E_0, with $\Delta E = E - E_0$. The positive-valued weighting factor $W(r) = W[\Delta E(r)]$ is specified in the beginning of a simulation run, often taking the form $W = \exp(-\beta\Delta E/2)$. The final ensemble average is then corrected by the free energy of the reference fluid to give the desired absolute value. This method is easier and more flexible than the purpose-built direct grand canonical MC method.

Other relevant techniques include the *acceptance ratio* method (Bennett, 1976), which is an extension of the umbrella sampling, and the particle insertion method (Widom, 1963), which simply inserts a non-disturbing test particle in the course of a conventional MMC simulation and calculates its interaction with others in order to estimate the chemical potential. For a comparison of these methods, the reader should consult Allen and Tildesley (1987).

Preferential sampling

The methods described in the preceding section enable the conventional MMC method to be extended so that the statistical properties such as the free energy or chemical potential of a liquid can be calculated. There are also methods to enhance the MMC in other ways, e.g. with the objective of improving its efficiency. This feature is important, and provides the focus of the discussion below.

In the conventional MC methods, every particle is given an equal weight in its selection, and an equal probability governing their direction of motion. This 'evenness' makes the methods easy to implement but often inefficient to perform. In view of the nature of many real systems it may be desirable to move some particles more often than others, or to bias the moves in some preferred directions. This strategy gives rise to *preferential sampling methods*.

The use of a preferential sampling procedure means that a non-symmetrical underlying matrix has to be constructed. To incorporate this sampling technique into the MMC method, it is necessary to modify the Metropolis solution, eqn (7.33), so that:

$$q_{mn} = \begin{cases} \alpha_{mn} & m \neq n, \quad \alpha_{nm}p_n \geq \alpha_{mn}p_m \\ \alpha_{mn}(\alpha_{nm}p_n/\alpha_{mn}p_m) & m \neq n, \quad \alpha_{nm}p_n < \alpha_{mn}p_m \\ 1 - \sum\limits_{n \neq m} q_{mn} & m = n \end{cases} \tag{7.35}$$

It is easy to show that the microscopic reversibility condition is satisfied, although now $\alpha_{mn} \neq \alpha_{nm}$. As such, the weighting factor changes from p_n/p_m to $\alpha_{nm}p_n/\alpha_{mn}p_m$, and the remaining task is just to find a suitable underlying matrix for the problem in hand. As in the case of importance sampling in the conventional MC methods, the calculation does not have to be carried out in the canonical ensemble. Examples of the MC methods based on this prescription are given below.

For instance, in order to make simulation of dilute ion solutions more efficient, two methods have been suggested (Owicki and Scheraga, 1977; Owicki, 1978). In the first, a solvation shell is constructed around each ion. Solvent molecules inside the shell are picked and moved as usual but those outside the shell are given a prescribed probability to be selected so that they are moved less frequently. In the second method, the clear boundary imposed by the solvation shell is removed. Instead, all solvent particles are subjected to a non-uniform weighting in selection. The weighting function monotonically decays with the distance from the solute ion, so again prejudice is successfully introduced into the species selection process.

Force-bias methods

Improvement of simulation efficiency may also be achieved through the direction biasing in systems consisting of interacting particles. Since in real liquids particles tend to move in directions according to the forces acting on them, it is possible to make the Markov chain converge to its limiting distribution more quickly by building the force bias into the underlying stochastic matrix. Based on this idea, Pangali *et al.* (1978) proposed the following formulation for the underlying Markov chain:

$$\alpha_{mn} = \frac{\exp\left[\lambda\beta(F_i^m \bullet \Delta r_i^{nm})\right]}{f(F_i^m, \lambda, \Delta r_{\max})} \tag{7.36}$$

where F_i^m is the force acting on particle i in state Γ_m, $\Delta \mathbf{r}_i^{nm} = \mathbf{r}_i^n - \mathbf{r}_i^m$ is the displacement in a trial move to state Γ_n, λ is a constant, and f a normalizing factor. The constant λ typically lies on $[-1,1]$. When $\lambda = 0$, the conventional MC is recovered; when $0 < \lambda \leq 1$, the method is called the force-bias MC. If $-1 \leq \lambda < 0$, it is termed the anti-force-bias MC (useful in helping particles overcome large energy barriers). Both have been shown to produce considerably faster convergence over the conventional MC (Rao *et al.*, 1979; Mehrotra *et al.*, 1983; Cao and Berne, 1990). Since in essence the MD character has been introduced in the force-bias MC, it ought to give better agreement with MD results than the conventional MC.

Smart Monte Carlo method

If particles are also subjected to Brownian motion, the combination of the systematic force $\mathbf{F}_i^m$ and a random force term leads to another scheme, called the *smart MC method*, devised by Rossky *et al.* (1978). The trial displacement in this method is written as:

$$\Delta r_i^{nm} = \beta A_S F_i^m + R_i^{\mathrm{G}} \tag{7.37}$$

where $\mathbf{R}_i^{\mathrm{G}}$ corresponds to the displacement due to Brownian motion, whose components are chosen from a Gaussian distribution with zero mean and variance $2A_S$. Equation (7.37) is actually the same as the algorithm used in the BD simulation without hydrodynamic corrections. The difference is that, in MC simulation, the parameter A_S can be given a much larger value than normally allowed in BD and still yields the correct results. This is because the MC method is not a dynamic simulation in which the time evolution of the system must be faithfully followed in order to obtain the correct time averages. Rather it is a stochastic simulation and relies on statistical criteria to guarantee the correct ensemble averages. The underlying stochastic matrix for this procedure is:

$$\alpha_{nm} = (4\pi A_S)^{-3/2} \exp\left[\frac{-\left(\Delta r_i^{nm} - \beta A_S F_i^m\right)^2}{4A_S}\right] \tag{7.38}$$

Both the force-bias MC and the smart MC offer a substantial improvement over the Metropolis MC method, and can be used in other ensembles as well as the canonical ensemble. For instance, Mezei (1983) has applied similar techniques based on these methods to the constant-NPT ensemble. Ultimately the choice of method depends on the specific system under investigation.

7.4 Molecular dynamics methods

The MD method provides a way of solving a set of coupled equations of motion, usually ordinary differential equations, for many-body systems, using the finite difference techniques. It was applied by Alder and Wainwright (1957, 1959) to systems of mechanically hard cores including hard discs, hard spheres and square-well particles. These early models of MD are characterized by the use of discontinuous or purely impulsive forces, and the technical details differ qualitatively from those of soft-sphere models with continuous interparticle potentials. Hard-sphere models rely upon a collision-by-collision basis, whereas soft-sphere models take a regular step-by-step integration over time.

The MD method has been widely applied to homogeneous systems such as monoatomic liquids to produce the correct time ordering of the generated configurations. However, for mixtures like colloidal dispersions where hydrodynamic and stochastic forces must be correctly included as well as the usual interparticle interactions (Chapter 3), this method is hardly practical at present because of the complexity and computational 'cost' involved. Nevertheless, it merits a general description here for the sake of completeness and since certain underlying concepts will be utilized in another context later (e.g. the idea of constraint dynamics developed for the MD simulation can be adopted to BD simulations of colloids).

7.4.1 Equations of motion and finite difference methods

The classical equations of motion for a particle in the many-body system can have several forms, including the Lagrangian form and the Hamiltonian form. For convenience, the more familiar Newton's form will be used so that the motion of each particle may be described by either of the following equations:

$$m_i \ddot{r}_i = F_i \tag{7.39}$$

$$\dot{r}_i = p_i / m_i \tag{7.40}$$

where the force $F_i = \dot{p}_i$ is usually expressed as the spatial derivative of the interparticle interaction potential E (hence the significant difference between soft core MD algorithms and hard core MD algorithms mentioned above):

$$\dot{p}_i = -\nabla_{r_i} E = F_i \tag{7.41}$$

In principle, solution of these equations will automatically satisfy the conservation laws of energy and/or total momentum (depending on the system under consideration) and particle trajectories generated using these equations are strictly reversible in time (i.e. changing the signs of all the velocities or momenta will cause the particles to retrace their trajectories). In practice, slight deviations from these observations or requirements are often inevitable because of the finite accuracy inherent in both the MD algorithms and the computers.

In soft sphere simulations, the equations of motion are replaced by a set of finite difference equations which, in turn, are solved on a step-by-step basis. Given particle positions, velocities and other dynamical properties at time t, the updated positions, velocities, etc. at a later time, $t + \Delta t$, can be obtained quite easily and to a sufficient degree of accuracy. The equations are solved on a step-by-step basis. The choice of the time interval Δt will depend on the individual methods but, in any case, Δt will be considerably smaller than the typical time taken by a particle to travel its own length or the mean interparticle spacing.

Many algorithms utilize this general finite difference approach. Among the criteria for selecting the appropriate algorithm for MD simulations, perhaps the most important desirable features are those of simplicity and the ability to maintain an acceptable level of energy conservation. Two of the most frequently used algorithms are described briefly below.

Verlet-type algorithms

The Verlet algorithm was devised for integrating the second-order equations, such as eqn (7.39) (Verlet, 1967). The equation for advancing the particle positions is:

$$r(t + \Delta t) = 2r(t) - r(t - \Delta t) + \Delta t^2 a(t) \tag{7.42}$$

where **a** represents the acceleration of the particle. The velocity **v** is not included but may be obtained using:

$$v(t) = \frac{r(t + \Delta t) - r(t - \Delta t)}{2\Delta t} \tag{7.43}$$

Compared with other MD algorithms, the Verlet algorithm requires less computer storage and is easier to program. In addition, it is exactly reversible in time and, given conservative forces, is guaranteed to conserve linear momentum. It has been shown (Verlet, 1967; Fincham and Heyes, 1982; Heyes and Singer, 1982) to have excellent energy-conserving properties even with long time steps. However, in dealing with velocities, the Verlet algorithm is rather awkward and results are not sufficiently accurate.

Modifications to this basic algorithm have been proposed to address these deficiencies. One of them is a so-called half-step 'leap-frog' scheme (Hockney, 1970; Potter, 1972):

$$v(t + \tfrac{1}{2}\Delta t) = v(t - \tfrac{1}{2}t) + \Delta t a(t) \tag{7.44}$$

$$r(t + \Delta t) = r(t) + \Delta t v(t + \tfrac{1}{2}\Delta t) \tag{7.45}$$

$$v(t) = \tfrac{1}{2}[v(t + \tfrac{1}{2}\Delta t) + v(t - \tfrac{1}{2}\Delta t)] \tag{7.46}$$

Although it is equivalent to the Verlet algorithm, the explicit inclusion of velocity terms in the above equations makes programming simpler and, sometimes, the overall requirement of word storage is smaller. The elimination of subtraction in calculating the velocity in eqn (7.43) minimizes the loss of numerical precision.

Another equivalent to the Verlet algorithm which handles velocities in a rather more satisfactory manner has been used by Swope *et al.* (1982). This 'velocity Verlet' algorithm takes the form:

$$r(t + \Delta t) = r(t) + \Delta t v(t) + \tfrac{1}{2}\Delta t^2 a(t) \tag{7.47}$$

$$v(t + \Delta t) = v(t) + \tfrac{1}{2}\Delta t[a(t) + a(t + \Delta t)] \tag{7.48}$$

This algorithm, in fact, involves two stages to compute the velocity since both $\mathbf{a}(\Delta t)$ and $\mathbf{a}(t+\Delta t)$ are required in eqn (7.48):

$$v(t + \tfrac{1}{2}\Delta t) = v(t) + \tfrac{1}{2}\Delta t a(t) \tag{7.49}$$

$$v(t + \Delta t) = v(t + \tfrac{1}{2}\Delta t) + \tfrac{1}{2}\Delta t a(t + \Delta t) \tag{7.50}$$

This algorithm uses approximately the same computer storage as the original Verlet algorithm. Its numerical stability and simplicity make it a practical and attractive proposition (Allen and Tildesley, 1987).

Gear-type algorithms

Another widely used type of algorithm is the Gear prediction–correction algorithm. As is implied by the name, these algorithms involve essentially two stages: prediction and correction. At the first stage, the positions, velocities, accelerations, etc. are predicted using the current values of these quantities. At the second stage, the forces and accelerations are evaluated from the new position, and then the evaluated new accelerations are used to correct the predicted values. The corrected values are stored as the new values and may be used to calculate any variables of interest. For illustration purpose, a so-called four-value Gear algorithm is described below.

An estimation of the positions, velocities, etc. at time $t+\Delta t$ may be obtained by Taylor expansion about time t as follows:

$$r^{p}(t + \Delta t) = v(t) + \Delta t v(t) + \ldots \tag{7.51}$$

$$v^{p}(t + \Delta t) = v(t) + \Delta t a(t) + \tfrac{1}{2}\Delta t^{2} b(t) + \ldots \tag{7.52}$$

$$a^{p}(t + \Delta t) = a(t) + \Delta t b(t) + \ldots \tag{7.53}$$

$$b^{p}(t + \Delta t) = b(t) + \ldots$$

where **b** denotes all the third time derivatives of **r**, and the superscript ‘p’ is used as a reminder that the quantities are for the prediction stage. Since these equations take no account of real motion of particles, they must be modified in order to generate correct trajectories. This may be performed as follows. First, from the predicted position $\mathbf{r}^{p}$, the forces and accelerations at time $t+\Delta t$ are recalculated. Next, the corrected value of acceleration $\mathbf{a}^{c}(t+\Delta t)$ is compared with the predicted $\mathbf{a}^{p}(t+\Delta t)$ to give an estimate of the error in the prediction stage:

$$\Delta a(t + \Delta t) = a^{c}(t + \Delta t) - a^{p}(t + \Delta t) \tag{7.54}$$

Then the estimated error is used to correct the predicted values:

$$r^{c}(t + \Delta t) = r^{p}(t + \Delta t) + c_1 \Delta a(t + \Delta t) \tag{7.55}$$

$$v^{c}(t + \Delta t) = v^{p}(t + \Delta t) + c_2 \Delta a(t + \Delta t) \tag{7.56}$$

$$a^{c}(t + \Delta t) = a^{p}(t + \Delta t) + c_3 \Delta a(t + \Delta t) \tag{7.57}$$

$$b^{c}(t + \Delta t) = b^{p}(t + \Delta t) + c_4 \Delta a(t + \Delta t) \tag{7.58}$$

The four correction coefficients c_1 to c_4 have been given and discussed in details by Gear (1966, 1971) and will not be given here. However, some remarks regarding the Gear algorithms should be noted (Allen and Tildesley, 1987). First, the correction step may be iterated so as to further refine the positions and velocities. In many applications this iteration is the key to obtaining an accurate solution. However, since evaluation of accelerations from particle positions requires calculation of forces (thereby making it the most time-consuming part of the simulation) normally only one, or occasionally two, corrector steps are carried out. Secondly, since the fewer the corrector steps the better, a Taylor expansion is used to ensure accurate estimation, although in principle arbitrary guesses are allowable. Thirdly, algorithms of different orders (e.g four-value and five-value) are available, each utilizing a slightly different predictor equation. However, since the prediction procedure is a purely mathematical manipulation, merely increasing the order of the equation does not greatly improve the accuracy of the MD solution as a whole.

In general, for short time intervals Δt, the higher-order Gear methods are more accurate; but with long time intervals Δt (which are of more interest in most simulations), the Verlet algorithm becomes more attractive. An extensive discussion of the merits of various MD algorithms have been given by Berendsen and van Gunsteren (1986).

7.4.2 Application of molecular dynamics to particulate systems

Hard sphere systems

The MD method was first adopted to study many-body systems (Alder and Wainwright, 1957, 1959) considering hard spheres. The motion of hard spheres is characterized by a straight trajectory and a constant velocity, which only changes as a consequence of collisions. The time evolution of a hard sphere system can be treated as a sequence of strictly binary elastic collisions.

The recipe for the calculation of the trajectories consists of, first, advancing the coordinates of all particles until a collision occurs somewhere in the system, and then calculating velocity changes of the colliding particles using the energy and linear conservation laws. Such a procedure is repeated as many times as are necessary to yield results of statistical reliability for the problem of interest. In hard sphere MD simulations, calculation of the collision time (i.e. the time interval between collisions for each pair of spheres) is the computationally intensive part of the program since, in principle, all possible collisions between distinct pairs must be considered.

The generalization of this scheme to the case of the square-well or more complicated 'multistep' potentials is straightforward (Chapela *et al.*, 1984). In this case, each pair of particles has more than one distance at which 'collisions' occur. At each 'collision' the velocity changes are determined in the same way as described above. The hard-sphere MD method has also been used to include particle surface roughness with the spherical symmetry still preserved (Subramanian and Davis, 1975; O'Dell and Berne, 1975). For rough spheres, the total momentum consists of angular as well as linear momenta, but the usual conservation laws are still satisfied and can be used to calculate the velocity changes.

Soft sphere systems

Extension of the MD method to cope with soft sphere systems was first reported by Rahman (1964) and developed by Verlet and coworkers (Verlet, 1967, 1968; Levesque *et al.*, 1973). The simulation is initiated by placing the particles at arbitrarily chosen sites $\mathbf{r}_i(0)$. A set of corresponding coordinates at an earlier time step $\mathbf{r}_i(-\Delta t)$ is typically set at random with a Maxwell–Boltzmann distribution appropriate to the temperature of interest and chosen such that the net linear momentum is a constant (usually zero). The subsequent calculation is organized as a loop over time.

In the early stages of the simulation, it is normal for the temperature to drift away from its original value, and an occasional rescaling of particle velocities is therefore needed. Once equilibrium is reached, a further run of order 10^3 time steps is usually sufficient. The equations of motion used to generate particle trajectories and the methods for solving them have been discussed in Section 7.4.1.

As can be seen, unlike the hard sphere MD model, where the generated trajectories are exact but the time intervals are variable, the soft sphere MD model generates approximate trajectories and normally uses a constant time interval for all particles. The time step used for the soft sphere MD simulations is often chosen on the basis of energy conservation, and is usually smaller than the corresponding collision time interval.

Molecular dynamics simulation in different ensembles

The use of Newtonian equations of motion in the conventional MD method implies that the simulation is only applicable to systems characterized by fixed values of *N*, *V* and *E* (and often the total momentum as well). In certain applications, it may be desirable to have the temperature and/or pressure fixed. This calls for MD methods which utilize other forms of equations of motion.

Several schemes have been developed for this purpose, many of which were inspired by Andersen *et al.* (1984). These schemes include those developed by Andersen *et al.* (1984), Andrea *et al.* (1983), Nose (1984) and Woodcock (1971) for constant-temperature MD methods; and those by Haile and Graben (1980), Parrinello and Rahman (1980, 1981), Hoover *et al.* (1982) and Evans and Morriss (1983, 1984) for constant-pressure MD methods.

Non-equilibrium systems

One of the advantages of MD over MC is its ability to provide dynamic or time-dependent properties i.e. the coefficients used to describe energy, momentum or mass transport of a fluid (e.g. viscosity and diffusion constant). This requires the MD to be applied to non-equilibrium ensembles. Although these properties can sometimes be calculated with alternative methods such as the linear response or perturbation theory in statistical mechanics, the non-equilibrium MD offers improved efficiency and enhanced versatility. Early attempts to induce momentum or energy flow in a MD simulation (Hoover and Ashurst, 1975) involved introducing an external momentum or energy reservoir to interact, at boundaries, with particles contained in the simulation cell. Since such treatments are incompatible with the periodic boundary conditions commonly used in computer simulations (see Chapter 8), modern non-equilibrium MD simulations have been developed to make use of such perturbations, imposed on the usual equations of motion in one way or another, that are easily adapted to preserve translational invariance and periodicity.

Examples of this type of approach have been reported for a mixture of soft spheres (Evans and Hanley, 1979; Bernu *et al.*, 1987). One of the most significant developments in this area has been the use of non-equilibrium MD to investigate the physics of non-Newtonian flow in relation to the rheology of concentrated colloidal dispersions, including comparisons of theoretical behaviour with experimental measurements (Heyes, 1986; Barnes *et al.*, 1987, 1989; Johma *et al.*, 1991).

Other reviews of these methods have been provided by Ciccotti *et al.* (1979), Hoover (1983), Evans and Morriss (1984) and Allen and Tildesley (1987).

7.5 Brownian dynamics methods

In mixtures containing different particulate species such as dispersions of macromolecules or colloids, the timescale characterizing the motion of each species can differ by several orders of magnitude. To simulate such systems using the standard MD methods would require extremely short time intervals to handle the fast motion of some species and exceedingly long runs to allow the slow evolution of the others. Thus, where the fast motions are not of great interest MD methods are simply too demanding of computing resources to be practical. Fortunately, approximate approaches such as a Brownian dynamics (BD) method can be used in such cases. In Brownian dynamics, the fast motions of solvent particles are omitted from the simulation and their impact on the solute is represented by a combination of random and frictional force terms. In other words, the Newtonian equations of motion are now replaced by Langevin-type equations of motion.

Traditionally, Brownian motion theory was rather phenomenological, i.e. no attempt was made to base the stochastic theory on the exact equations of motion (Wang and Uhlenbeck, 1945). Later theories of Brownian motion, largely attributed to Zwanzig (1960), have now been formally established, entirely on the basis of microscopic first principles. Zwanzig (1960) first used a projection operator method to derive an exact generalized Fokker–Planck equation. Similar methods were employed by Mori (1965) to obtain a linear generalized Langevin equation. Later, Nordholm and Zwanzig (1975) presented a unified derivation of the generalized theory, and demonstrated complementarity with the earlier results of Zwanzig (1960) and Mori (1965). Despite their equivalence, the Langevin-type equations are usually more convenient for constructing the BD algorithms. These theories will not be described in great detail here; instead, appropriate results for BD simulations will be considered below.

7.5.1 Brownian dynamics for hydrodynamically independent particles

Consider a set of non-hydrodynamically interacting Brownian particles suspended in a host fluid, for which the equation of motion from the traditional *ad hoc* treatment may be more convenient to use:

$$\dot{p}(t) = -\xi p(t) + F(t) + p^{G}(t) \tag{7.59}$$

where $\mathbf{p}$ is the momentum vector, $\mathbf{F}$ the interparticle force vector, ξ the friction coefficient (treated as a constant, independent of particle position and momentum), and $\mathbf{p}^G$ the random term to represent the effect of Brownian motion.

A BD simulation method based on eqn (7.59) has been developed by Ermak (Ermak, 1976; Ermak and Buckholtz, 1980). In this method, the equation of motion is integrated, in the usual MD fashion, over a succession of time intervals Δt during which the force term $\mathbf{F}$ can be considered to be constant. The resulting algorithm is similar to those of MD methods. For one-component dispersions of N particles, it yields:

$$r(t + \Delta t) = r(t) + c_1 \Delta t v(t) + c_2 \Delta t^2 a(t) + \Delta r^G \tag{7.60}$$

$$v(t + \Delta t) = c_0 v(t) + c_1 \Delta t a(t) + \Delta v^G \tag{7.61}$$

where the numerical coefficients are given by:

$$c_0 = \exp(-\xi \Delta t) \tag{7.62}$$

$$c_1 = (\xi \Delta t)^{-1}(1 - c_0) \tag{7.63}$$

$$c_2 = (\xi \Delta t)^{-1}(1 - c_1) \tag{7.64}$$

and the random variables are defined as stochastic integrals:

$$\Delta r^G = \int_t^{t+\Delta t} \frac{1 - \exp[-\xi(t + \Delta t - \tau)]}{m\xi} p^G(\tau)\, d\tau \tag{7.65}$$

$$\Delta v^G = \int_t^{t+\Delta t} \frac{1 - \exp[-\xi(t + \Delta t - \tau)]}{m} p^G(\tau)\, d\tau \tag{7.66}$$

whose vectorial components of $\Delta\mathbf{r}^G$ and $\Delta\mathbf{v}^G$ are sampled independently from a bivariate Gaussian distribution:

$$\rho(\Delta r^G, \Delta v^G) = \frac{\exp\left\{-\dfrac{1}{2(1 - c_{rv}^2)}\left[\left(\dfrac{\Delta r^G}{\sigma_r}\right)^2 + \left(\dfrac{\Delta v^G}{\sigma_v}\right)^2 - 2c_{rv}\left(\dfrac{\Delta r^G}{\sigma_r}\right)\left(\dfrac{\Delta v^G}{\sigma_v}\right)\right]\right\}}{2\pi\sigma_r\sigma_v(1 - c_{rv}^2)^{1/2}} \tag{7.67}$$

with zero mean, and variances σ_r, σ_v and correlation coefficient c_{rv}:

$$\sigma_r^2 = \Delta t^2 \frac{kT}{m\xi\Delta t}\left\{2 - \frac{3 - 4\exp(-\xi\Delta t) + \exp(-2\xi\Delta t)}{\xi\Delta t}\right\} \tag{7.68}$$

$$\sigma_v^2 = \frac{kT[1 - \exp(-2\xi\Delta t)]}{m} \tag{7.69}$$

$$\sigma_r\sigma_v\sigma_{rv} = \frac{kT[1 - \exp(-\xi\Delta t)]^2}{m\xi} \tag{7.70}$$

As ξ approaches zero, the instantaneous systematic force term in eqn (7.59) dominates, so Newtonian mechanics are recovered and eqns (7.60) and (7.61) become a Taylor series predictor algorithm, which, as noted in Section 7.3.1, is not a particularly accurate method. Therefore, for low values of ξ, it is necessary to replace eqns (7.60) and (7.61) with some generalized versions of the predictor–corrector or Verlet algorithms with the stochastic and frictional effects being taken into account. One such generalized Verlet-like algorithm is given by:

$$v(t + \Delta t) = c_0 v(t) + (c_1 - c_2)\Delta t a(t) + c_2 \Delta t a(t + \Delta t) + \Delta v^{\mathrm{G}} \qquad (7.71)$$

where the corresponding algorithm for displacement has the same form as eqn (7.60).

If only the configurational dynamics over a long period of time are of interest, a more appropriate and simpler equation of motion (often referred to as the *position Langevin equation*) may be used (Lax, 1966; Zwanzig, 1969):

$$\dot{\mathbf{r}}(t) = \frac{D}{kT}\mathbf{F}(t) + \mathring{\mathbf{r}}^{\mathrm{G}} t) \qquad (7.72)$$

where D is the diffusion constant, related to the friction coefficient by $\xi = kT/mD$, $\mathring{\mathbf{r}}^{\mathrm{G}}(t)$ is a random term which may be taken to have a delta function correlation for each particle:

$$\langle r_i^{\mathrm{G}}(t) r_j^{\mathrm{G}}(t)\rangle = 2D\Delta t\delta_{ij} \qquad i, j = 1, 2, \ldots, 3N \qquad (7.73)$$

An algorithm based on this position Langevin equation (Ermak and Yeh, 1974; Ermak, 1975) may be written as:

$$r(t + \Delta t) = r(t) + (D/kT)F(t)\Delta t + \Delta \mathbf{r}^{\mathrm{G}} \qquad (7.74)$$

where each component of $\Delta\mathbf{r}^{\mathrm{G}}$ is chosen from a Gaussian distribution with zero mean and variance $<(\Delta\mathbf{r}^{\mathrm{G}})^2> = 2D\Delta t$.

It should be borne in mind that due to the nature of the stochastic variables involved, the equations of motion and their corresponding BD algorithms described above cannot yield a unique, determinate trajectory for each of the particles – only representative trajectories can be generated. Consequently, the calculated velocity has no well-defined meaning and the short-time dynamic behaviour generated by these equations may be far from realistic. This is especially true for the position Langevin equation of motion. At longer times, the latter method resembles the smart MC more than it does the MD methods.

7.5.2 Brownian dynamics of hydrodynamically interacting particles

When the effects of hydrodynamic interactions are taken into account, it is clear from Chapter 4 that the friction coefficient ξ is no longer a constant independent of the positions and momenta of the particles. Physically, this is because the solvent flow, induced by one particle, must have an effect through frictional forces on others. Under creeping-flow conditions (normally satisfied for small particles such as colloids and macromolecules in liquids), the diffusion tensor becomes a configuration-dependent matrix. The corresponding position Langevin-type equations can have several forms depending on the convention adopted for stochastic differentials; however, they are practically all the same (Tough *et al.*, 1986).

Algorithms for quiescent dispersions

The associated BD algorithm has been given by Ermak and McCammon (1978), and is written as:

$$r_i(t + \Delta t) = r_i(t) + \sum_j \frac{D_{ij}(t) \bullet F_j(t)}{kT} \Delta t + \nabla_{ij} \bullet D_{ij}(t)\Delta t + \Delta r^{\mathrm{G}} \tag{7.75}$$

where $\boldsymbol{D}_{ij}$ is the pair diffusion tensor for particles i and j (see Section 4.4.1), and the components of $\Delta \mathbf{r}^{\mathrm{G}}$ are random variables selected from the $3N$-variate Gaussian distribution with zero mean and covariance matrix:

$$\langle \Delta r^{\mathrm{G}} \Delta r^{\mathrm{G}} \rangle = 2\Delta t D_{ij} \tag{7.76}$$

Several features of this BD algorithm can be highlighted:

- the inclusion of a configuration-dependent diffusion tensor means that the hydrodynamic effects are considered at the position Langevin equation rather than the more general momentum Langevin equation level. Although approximate (since the momentum dependence is dropped), such a treatment is the only feasible solution of the complex problem
- in deriving the algorithm, it has been assumed that during the time interval Δt is much larger than the momentum relaxation time, i.e. $\Delta t > m_i D_{ii}(t)/kT$, where m_i is the mass of particle i and D_{ii} its self-diffusion coefficient. Meanwhile, for the sake of numerical accuracy, the time step Δt should be sufficiently short so that the force on the particles and the gradient of the diffusion tensor are essentially constant
- in contrast to the simple non-interacting particle BD algorithm, for simulations with hydrodynamic interactions the generation of random numbers from the multivariate Gaussian distribution becomes the most computationally 'expensive' part of the calculations, since it involves some complex manipulation of the $3N \times 3N$ covariance matrix
- the simple diffusion tensors such as Oseen and Rotne–Prager tensors (described in Chapter 4) are usually favoured in BD simulations instead of the higher-order ones. This is because both tensors have the property $\nabla \bullet D = 0$, which enables the gradient term to be dropped from the algorithm
- however, as has been noted in Section 4.4.1, the Oseen and Rotne–Prager tensors are only applicable for dilute dispersions. Their adoption in concentrated systems can lead to physically absurd predictions (e.g. negative diffusion coefficient) (Glendinning and Russel, 1982; Bacon *et al.*, 1983). Whilst higher-order tensors are too computationally expensive to use in practice, an effective screened pair diffusion tensor (Snook *et al.*, 1983) might be the way around the problem but such an approach has been criticized as lacking in vigour (Beenakker, 1983). It should also be acknowledged that some controversy remains as to whether or not the use of the pair diffusion tensors is consistent with the periodic boundary conditions.

Algorithms for flowing dispersions

In the presence of an external flow, eqn (7.75) should be extended to include the effects of the flow. For instance, Ansell *et al.* (1985) have proposed a BD algorithm for a case of laminar flow:

$$r_i(t+\Delta t) = r_i(t) + \sum_j \frac{D_{ij}(t)\bullet F_j(t)}{kT}\Delta t + \nabla_{r_j}\bullet D_{ij}(t)\Delta t + S_i\bullet R\Delta t + v(r_i)\Delta t + \Delta r^{\mathrm{G}} \tag{7.77}$$

where $\mathbf{v}(\mathbf{r}_i)$ is the velocity of the undisturbed flow at $\mathbf{r}_i$, $\boldsymbol{S}_i$ is the shear tensor and $\boldsymbol{R}$ the rate-of-strain tensor. $\boldsymbol{S}_i \bullet \boldsymbol{R}$ represents the effect of hydrodynamic interactions on the shear-induced motion. For a two-dimensional shear flow, v is given as:

$$v(r_i) = \begin{pmatrix} v_x \\ v_y \\ v_z \end{pmatrix} = \begin{pmatrix} 0 & 1 & 0 \\ \gamma & 0 & 0 \\ 0 & 0 & 0 \end{pmatrix} \begin{pmatrix} r_x \\ r_y \\ r_z \end{pmatrix} G \tag{7.78}$$

where G is the shear rate and γ the flow type parameter (see Section 4.4.2). In a simple shear flow, $\gamma=0$ and $v=\hat{x}Gr_y$. The form of the shear tensor, up to order r^{-5}, has been given by Mazur and van Saarloos (1982):

$$\begin{aligned} S_i = {} & \tfrac{5}{2}\sum_{i\neq j} a_j^3 r_{ij}^{-2} \boldsymbol{e}_{ij}\boldsymbol{e}_{ij}\boldsymbol{e}_{ij} - \tfrac{5}{6}\sum_{i\neq j} a_j^3 r_{ij}^{-4}(5a_i^2 + 3a_j^2)(\boldsymbol{e}_{ij}\boldsymbol{e}_{ij}\boldsymbol{e}_{ij} - \tfrac{2}{5}\mathbf{1}\boldsymbol{e}_{ij}) \\ & - \tfrac{25}{4}\sum_{k\neq i}\sum_{j\neq k} a_j^3 a_k^3 r_{ik}^{-2} r_{kj}^{-3} \boldsymbol{e}_{ik}\left\{[5(\boldsymbol{e}_{ik}\bullet\boldsymbol{e}_{kj})^2 - 1]\boldsymbol{e}_{kj}\boldsymbol{e}_{kj} - 2(\boldsymbol{e}_{ik}\bullet\boldsymbol{e}_{kj})\boldsymbol{e}_{ik}\boldsymbol{e}_{kj}\right\} \end{aligned} \tag{7.79}$$

This algorithm was designed to be applicable to the situation where aggregated particles rotate due to an external shear field.

Another BD algorithm, proposed by Peters and Gupta (1989), emphasizes the situation usually encountered in particle deposition systems. It allows for particle motion in the vicinity of a fixed macroscopic (collector) surface, which may, or may not, be subjected to an external flow:

$$r_i(t+\Delta t) = r_i(t) + \sum_j \frac{D_{ij}(t)\bullet F_j(t)}{kT}\Delta t + \nabla_{r_j}\bullet D_{ij}(t)\Delta t + v(r_i)\Delta t + \Delta r^{\mathrm{G}} \tag{7.80}$$

The algorithm has the same form as the Ermak–McCarmmon algorithm, eqn (7.75), except for the additional velocity term to account for fluid convection effects and the modifications to the diffusion tensor due to the presence of the collector surface. According to Felderhof (1978), the grand diffusion tensor:

$$D = \sum_{i=1}^{N}\sum_{j=1}^{N} D_{ij} \tag{7.81}$$

can be approximated as a sum of single and pair particle interactions:

$$\begin{aligned} D_{ij} &= D_i^0 + \sum_{k=1}^{N} A_{ik} \quad && i \neq k \\ D_{ij} &= B_{ij} && i \neq j \end{aligned} \tag{7.82}$$

where $\boldsymbol{D}_i^0$ is the diffusion tensor for single particles, and $\boldsymbol{A}_{ij}$ and $\boldsymbol{B}_{ij}$ are diffusion tensors for pair interactions. Expressions for these tensors have been given in Chapter 4.

It may be noted that both algorithms described above have utilized the additivity assumption in dealing with the hydrodynamic interactions, but in different ways. In the first algorithm (Ansell *et al.*, 1985), the effect of shear is separated from the diffusion tensor, whereas in the second algorithm (Peters and Gupta, 1989), the wall effect and, presumably, the shear-induced effect are both incorporated directly into the diffusion tensor.

The two algorithms have been derived following more or less the same route as adopted by Ermak and McCammon (1978), although each has a different primary objective. Therefore, they should be regarded as being complementary in purpose but not exclusive of each other in application. A combination of the two could be more useful for use in the simulation of deposition of aggregating particles from a flowing suspension.

7.5.3 Constraint Brownian dynamics

In systems of polyatomic molecules or polycolloidal flocs, the motion of each molecule or floc is characterized by two distinct modes, i.e. the fast vibration of atoms or primary colloids and the slower motion of the molecule or floc as a whole. Clearly, in dynamic simulation of molecules or flocs it would be quite unrealistic to assume total rigidity of these molecules or flocs. On the other hand, eliminating *some* of the fast-mode (e.g. the highest-frequency) components whilst retaining a reasonable degree of flexibility would allow a larger time step to be used in the simulation and thus improve the computational efficiency.

Although it is possible to construct a set of generalized coordinates obeying some purpose-built equations of motion in which the high-frequency components to be eliminated only appear implicitly, such an approach has seldom been used (Ryckaert and Bellemans, 1975) because of its complexity. A simpler way of handling the problem is to treat these components as constraints imposed on the usual equations of motion. To this end several techniques are available (Ciccotti and Ryckaert, 1986), of which the most widely used is the so-called SHAKE procedure (Ryckaert *et al.*, 1977). The procedure operates, at each time step of the simulation, as follows:

(1) evaluate new positions, etc. of each primary particle in the absence of any constraints
(2) adjust the positions resulting from the constraint-free run, one by one, so as to satisfy the constraints
(3) calculate the constraint forces, etc. required to keep the desired bond lengths constant
(4) correct the final results using the constraint forces together with the real interparticle or intraparticle interactions.

The SHAKE is a very efficient and versatile procedure which applies equally well to totally rigid and highly flexible molecules or flocs. For MD simulations, it is most easily incorporated into the Verlet-type algorithms, whereas for BD simulations the most convenient algorithms are those based on position Langevin equations of motion. It should be stressed that although for flexible molecules (or flocs) there is ample choice as to where to apply constraints, inclusion of bond angle constraint(s) is generally not recommended (van Gunsteren and Berendsen, 1977; van Gunsteren, 1980) because it tends to reduce the program efficiency and is hardly physically justified.

Bibliography

Allen, M. P. and Tildesley, D. J. (1987) *Computer Simulation of Liquids*, Clarendon Press, Oxford

Ciccotti, G., Frenkel, D. and McDonald, I. R. (1987) *Simulation of Liquids and Solids: Molecular Dynamics and Monte Carlo Methods in Statistical Mechanics*, North-Holland, Amsterdam

Hansen, J. P. and McDonald, I. R. (1986) *Theory of Simple Liquids*, 2nd edn, Academic Press, New York

Nicholson, D. and Parsonage, N. (1982) *Computer Simulation and the Statistical Mechanics of Adsorption, Academic Press, New York*

References

Alder, B. J. and Wainwright, T. E. (1957) Phase transition for a hard sphere system. *J. Chem. Phys.*, **27**, 1208–1209

Alder, B. J. and Wainwright, T. E. (1959) Studies in molecular dynamics. I. General method *J. Chem. Phys.*, **31**, 459–466

Allen, M. P. and Tildesley, D. J. (1987) *Computer Simulation of Liquids*, Clarendon Press, Oxford

Andersen, H. C., Allen, M. P., Bellemans, A., Board, J., Clarke, J. H. R., Ferrario, M., Haile, J. M., Nose, S., Opheusden, J. V. and Ryckaert, J. P. (1984) New molecular dynamics methods for various ensembles. *Rapport d'activité scientifique du CECAM*, 82–115

Andrea, T. A., Swope, W. C. and Andersen, H. C. (1983) The role of long ranged forces in determining the structure and properties of liquid water. *J. Chem. Phys.* **79**, 4576–4584

Ansell, G. C., Dickinson, E. and Ludvigsen, M. (1985) Brownian dynamics of colloidal-aggregate rotation and dissociation in shear flow. *J. Chem. Soc. Faraday Trans. 2*, **81**, 1269–1284

Bacon, J., Dickinson, E., Parker, R., Anastasiou, N. and Lal, M. (1983) Motion of flocs of two or three interacting colloidal particles in a hydrodynamic medium. *J. Chem. Soc. Faraday Trans. 2*, **79**, 91–109

Barker, A. A. (1965) Monte Carlo calculations of the radial distribution functions for a proton-electron plasma. *Aust. J. Phys.*, **18**, 119–133

Barnes, H. A., Edwards, M. F. and Woodcock, L. V. (1987) Applications of computer simulations to dense suspension rheology. *Chem. Eng. Sci.*, **42**, 591–608

Barnes, H. A., Hutton, J. F. and Walters, K. (1989) *An Introduction to Rheology*, Elsevier, Amsterdam

Beenakker, C. W. J. (1983) Comments in general discussion. *Faraday Disc. Chem. Soc.*, **76**, 240

Bennett, C. H. (1976) Efficient estimation of free energy difference from Monte Carlo data. *J. Comp. Phys.*, **22**, 245–268

Berendsen, H. J. C. and van Gunsteren, W. F. (1986) Practical algorithms for dynamic simulations. *Molecular Dynamic Simulations of Statistical Mechanical Systems*. Proceedings of the Enrico Fermi Summer School, Varenna, 1985, 43–65 Soc. Italiana di Fisica, Bologna

Bernu, B., Hansen, J. P., Howatari, Y., and Pastore, G. (1987) Soft-sphere model for glass-transition in binary-alloys – pair structure and self-diffusion. *Phys. Rev.*, **A**, 36, 4891–4903

Cao, J. and Berne, B. J. (1990) Monte Carlo methods for accelerating barrier crossing: anti-force-bias and variable step algorithms. *J. Chem. Phys.*, **92**, 1980–1985

Chapela, G. A., Martinez-Casas, S. E. and Alejandre, J. (1984) Molecular dynamics for discontinuous potentials. I. General method and simulation of hard polyatomic molecules. *Mol. Phys.*, **53**, 139–159

Chesnut, D. A. (1963) Monte Carlo calculations for the two-dimensional triangular lattice gas: supercritical region. *J. Chem. Phys.*, **39**, 2081–2084

Ciccotti, G., Jacucci, G. and McDonald, I. R. (1979) 'Thought experiments' by molecular dynamics. *J. Stat. Phys.*, **21**, 1–22

Ciccotti, G. and Rycaert, J. P. (1986) Molecular dynamics simulation of rigid molecules. *Comput. Phys. Rep.* **4**, 345–392

Ermak, D. L. (1975) A computer simulation of charged particles in solution I. Technique and equilibrium properties. *J. Chem. Phys.*, **62**, 4189–4196

Ermak, D. L. (1976) Brownian dynamics techniques and their application to dilute solutions. *Rapport d'activité scientifique du CECAM*

Ermak, D. L. and Buckholtz, H. (1980) Numerical integration of the Langevin equation: Monte Carlo simulation. *J. Comp. Phys.*, **35**, 169–182

Ermak, D. L. and McCammon, J. A. (1978) Brownian dynamics with hydrodynamic interactions. *J. Chem. Phys.*, **69**, 1352–1360

Ermak, D. L. and Yeh, Y. (1974) Equilibrium electrostatic effects on behaviour of polyions in solution: polyion-mobile ion interaction. *Chem. Phys. Lett.*, **24**, 243–248

Evans, D. J. and Hanley, H. J. M. (1979) Viscosity of a mixture of soft spheres. *Phys. Rev. A*, **20**, 1648–1654

Evans, D. J. and Morriss, G. P. (1983) Isothermal isobaric molecular dynamics. *Chem. Phys.* **77**, 63–66

Evans, D. J. and Morriss, G. P. (1984) Non-Newtonian molecular dynamics. *Comp. Phys. Rep.*, **1**, 297–343

Felderhof, B. U. (1978) The diffusion of interacting Brownian particles *J. Phys.* **A11**, 929–937

Fincham, D. and Heyes, D. M. (1982) Integration algorithms in molecular dynamics. *CCP5 Quarterly*, **6**, 4

Gear, C. W. (1966) The numerical integration of ordinary differential equations of various orders. *Report ANL 7126*, Argonne National Laboratory

Gear, C. W. (1971) *Numerical Initial Value Problems in Ordinary Differential Equations*, Prentice-Hall, Englewood Cliffs, NJ

Glendinning, A. B. and Russel, W. B. (1982) A pairwise additive description of sedimentation and diffusion in concentrated suspensions of hard spheres. *J. Colloid Interface Sci.*, **89**, 124–143

Haile, J. M. and Graben, H. W. (1980) Molecular dynamics simulations extended to various ensembles. I. Equilibrium properties in the isoenthalpic-isobaric ensemble. *J. Chem. Phys.* **73**, 2412–2419

Hansen, J. P. and McDonald, I. R. (1986) *Theory of Simple Liquids*, 2nd edn, Academic Press, New York

Heyes, D. M. (1986) Non-Newtonian behaviour of simple liquids. *J. Non-Newtonian Fluid Mech.*, **21**, 137–155

Heyes, D. M. and Singer, K. (1982) A very accurate molecular dynamics algorithm. *CCP5 Quarterly*, **6**, 11

Heyes, D. M., McKenzie, D. J. and Buscall, R. (1991) Rheology of weakly flocculated suspensions: experiment and Brownian dynamics simulation. *J Colloid Interface Sci.*, **142**, 303–316

Hockney, R. W. (1970) The potential calculation and some applications. *Methods. Comp. Phys.*, **9**, 135–211

Hoover, W. G. (1983) Nonequilibrium molecular dynamics. *Annu. Rev. Phys. Chem.*, **34**, 103–127

Hoover, W. G. and Ashurst, W. T. (1975) Nonequilibrium molecular dynamics. In *Theoretical Chemistry: Advances And Perspectives* (ed. Eyring, H. and Henderson, D.), Vol. 1, Academic

Press, New York, 1–51

Hoover, W. G., Ladd, A. J. C. and Moran, B. (1982) High strain rate plastic flow studied via nonequilibrium molecular dynamics. *Phys. Rev. Lett.* **48**, 1818–1820

Jia, X. (1992) Simulation of colloidal aggregation and deposition. PhD Thesis, University of Manchester Institute of Science and Technology

Johma, A. I., Merrington, A., Woodcock, L. V., Barnes, H. A. and Lips, A. (1991) Recent developments in dense suspension rheology. *Powder Technol.*, **65**, 343–370

Lax, M. (1966) Classical noise. IV. Langevin methods. *Rev. Mod. Phys.*, **38**, 541–566

Levesque, D., Verlet, L. and Kurkijarvi, J. (1973) Computer 'experiments' on classical fluids. IV. Transport properties and time correlation functions of the Lennard-Jones liquid near its triple point. *Phys. Rev. A*, **7**, 1690–1700

Mazur, P. and van Saarloos, W. (1982) Many-sphere hydrodynamic interactions and mobilities in a suspension. *Physica A*, **115**, 21–57

Mehrotra, P. K., Mezei, M. and Beveridge, D. L. (1983) Convergence acceleration in Monte Carlo simulation on water and aqueous solutions. *J. Chem. Phys.*, **78**, 3156–3166

Metropolis, N. and Ulam, S. (1949) The Monte Carlo method. *J. Am. Stat. Assoc.*, **44**, 335–341

Metropolis, N, Rosenbluth, A. W., Rosenbluth, M. N., Teller, A. H. and Teller, E. (1953) Equation of state by fast computing machines. *J. Chem. Phys.*, **21**, 1087–1092

Mezei, M. (1983) Virial bias Monte Carlo methods: efficient sampling in the (T,P,N) ensemble. *Mol. Phys.*, **48**, 1075–1082

Mori, H. (1965) Transport, collective motion, and Brownian motion. *Prog. Theor. Phys.*, **33**, 423–455

Nordholm, S. and Zwanzig, R. (1975) A systematic derivation of exact generalised Brownian motion theory. *J. Stat. Phys.*, **13**, 347

Norman, G. E. and Filinov, V. S. (1969) Investigations of phase transitions by a Monte Carlo method. *High Temp (USSR)*, **7**, 216

Nose, S. (1984) A molecular dynamics method for simulation in the canonical ensemble. *Mol. Phys.* **52**, 255–268

O'Dell, J. and Berne, B. J. (1975) Molecular dynamics of the rough sphere fluid. I. Rotational relaxation. *J. Chem. Phys.*, **63**, 2376–2394

Owicki, J. C. (1978) Optimization of sampling algorithm in Monte Carlo calculations on fluids. In *Computer Modelling of Matter* (ed. Lykos, P.) ACS Symposium Series Vol. 86, 159–171, American Chemical Society, Washington

Owicki, J. C. and Scheraga, H. A. (1977) Preferential sampling near solutes in Monte Carlo calculations on dilute solutions. *Chem. Phys. Lett.*, **67**, 600

Pangali, C., Rao, M. and Berne, B. J. (1978) On a novel Monte Carlo scheme for simulating water and aqueous solutions. *Chem. Phys. Lett.*, **55**, 413–417

Parrinello, M. and Rahman, A. (1980) Crystal structure and pair potentials: a molecular dynamics study. *Phys. Rev. Lett.* **45**, 1196–1199

Parrinello, M. and Rahman, A. (1981) Polymorphic transitions in single crystals: a new molecular dynamics method. *J. Appl. Phys.* **52**, 7182–7190

Peters, M. H. and Gupta, D. (1989) Brownian dynamics simulation of convective transport and adsorption of hydrodynamically interacting particles onto surfaces. In *Studies in Physical and Theoretical Chemistry*, Vol. 63, *Math/Chem/Comp 1988* (ed. Graovac, A.) Elsevier, Amsterdam, 387–410

Potter, D. (1972) *Computational Physics*, Wiley, New York

Rahman, A. (1964) Correlations in the motion of atoms in liquid argon. *Phys. Rev.*, **136A**, 405–411

Rao, M., Pangali, C. and Berne, B. J. (1979) On the force bias Monte Carlo simulation of water: methodology, optimisation and comparison with molecular dynamics. *Mol. Phys.*, **37**, 1773–1798

Rossky, P. J., Doll, J. D. and Friedman, H. L. (1978) Brownian dynamics as smart Monte Carlo simulation. *J. Chem. Phys.*, **69**, 4628–4633

Rowley, L. A., Nicholson, D. and Parsonage, N. G. (1975) Monte Carlo grand canonical ensemble calculation in a gas–liquid transition region for 12–6 argon. *J. Comp. Phys.*, **17**, 401–414

Ryckaert, J. P. and Bellemans, A. (1975) Molecular dynamics of liquid *n*-butane near its boiling point. *Chem. Phys. Lett.* **30**, 123–125

Ryckaert, J. P., Ciccotti, G. and Berendsen, H. J. C. (1977) Numerical integration of the cartesian equations of motion of a system with constraints: molecular dynamics of *n*-alkanes. *J. Comput. Phys.* **23**, 327–341

Salsburg, Z. W., Jacobson, J. D., Fickett, W. and Wood, W. W. (1959) Application of the Monte Carlo method to the lattice gas model I. Two dimension traiangular lattice. *J. Chem. Phys.*, **30**, 65–72

Snook, I., van Megen, W. and Tough, R. J. A. (1983) Diffusion in concentrated hard sphere dispersions: effective two-particle mobility tensors. *J. Chem. Phys.*, **78**, 5825–5836

Subramanian, G. and Davis, H. T. (1975) Molecular dynamical studies of rough sphere fluids. *Phys. Rev.*, **A11**, 1430–1439

Swope, W. C., Andersen, H. C., Berens, P. H. and Wilson, K. R. (1982) A computer simulation method for the calculation of equilibrium constants for the formation of physical clusters of molecules: application to small water clusters. *J. Chem. Phys.*, **76**, 637–649

Torrie, G. M. and Valleau, J. P. (1974) Monte Carlo free energy estimates using non-Boltzmann sampling: application to the sub-critical Lennard-Jones fluid. *Chem. Phys. Lett.*, **28**, 578–581

Torrie, G. M. and Valleau, J. P. (1977a) Nonphysical sampling distributions in Monte Carlo free energy estimation: umbrella sampling. *J. Comp. Phys.*, **23**, 187

Torrie, G. M. and Valleau, J. P. (1977b) Monte Carlo study of a phase separating liquid mixture by umbrella sampling. *J. Chem. Phys.*, **66**, 1402–1408

Tough, R. J. A., Pusey, P. N., Lekkerkerker, H. N. W. and van den Broeck, C. (1986) Stochastic descriptions of dynamics of interacting Brownian particles. *Mol. Phys.*, **59**, 595–619

Valleau, J. P. and Whittington, S. G. (1977) Monte Carlo in statistical mechanics: choosing between alternative transition matrices. *J. Comp. Phys.*, **24**, 150–157

van Gunsteren, W. F. (1980) Constraint dynamics of flexible molecules. *Mol. Phys.*, **40**, 1015–1019

van Gunsteren, W. F. and Berendsen, H. J. C. (1977) Algorithms for macro-molecular dynamics and constraint dynamics. *Mol. Phys.*, **34**, 1311–1327

Verlet, L. (1967) Computer 'experiments' on classical fluids I. Thermodynamical properties of Lennard-Jones molecules. *Phys. Rev.*, **159**, 98–103

Verlet, L. (1968) Computer 'experiments' on classical fluids II. Equilibrium correlation functions. *Phys. Rev.*, **165**, 201–214

Widom, B. (1963) Some topics in the theory of fluids. *J. Chem. Phys.*, **39**, 2808–2812

Wang, M. C. and Uhlenbeck, G. E. (1945) On the theory of Brownian motion II. *Rev. Mod. Phys.*, **17**, 323–342

Wood, W. W. (1968) Monte Carlo calculations for hard disks in the isothermal–isobaric ensemble. *J. Chem. Phys.*, **48**, 415–434

Wood, W. W. (1970) NPT-ensemble Monte Carlo calculations for the hard disk fluid. *J. Chem. Phys.*, **52**, 729–741

Woodcock, L. V. (1971) Isothermal molecular dynamics calculations for liquid salts. *Chem. Phys. Lett.* **10**, 257–261

Zwanzig, R. (1960) Ensemble method in the theory of irreversibility. *J. Chem. Phys.*, **33**, 1338–1341

Zwanzig, R. (1969) Langevin theory of polymer dynamics in dilute solution. *Adv. Chem .Phys.*, **15**, 325–331

Chapter 8

Implementation of computer simulations

Nomenclature

a	positive integer
$a_1, a_2, \ldots a_{12}$	numerical constants
c	constant
c	positive integer
c_{ij}	element of a convariance matrix
$\boldsymbol{C}$	convariance matrix
$\boldsymbol{D}$	diffusion tensor
$F_{\mathrm{p_i}}$	fraction of code amounting to the ith task performed on m processors
F_{p}	sum of F_{p}
F_{s}	fraction of operations performed on a single processor
F_{v}	fraction of operations performed in vector mode
l_{ij}	element of lower triangle matrix
$\boldsymbol{L}$	lower triangle matrix
m_i	number of processors used by a particular task
M	modulus
n	integer constant
N	number of particles
N_{P}	number of processors in a parallel computer
N_{R}	number of discrete position of simulation space R
OH	computational overhead
P	measure of computing performance
r	centre-to-centre separation distance
$\mathbf{r}$	position vector
$\mathbf{r}_i^m$	position of particle i in state m
$\mathbf{r}_i^n$	position of particle i in state n
$\Delta \mathrm{r}_{\mathrm{max}}$	maximum possible displacement
v	pair interaction potential energy
v_n	n-body interaction potential energy, $n=1,2,3,\ldots$
V	configurational energy
V_{S}	vector speedup ratio
ΔV_{nm}	change of potential energy from state m to state n
x	variable
X_i	integer
$\mathbf{X}$	vector

α_{nm}	element of underlying stochastic matrix
β	$1/kT$
ϵ_E	depth of energy well
γ	(non-uniform) random number
λ	intermediate variable
σ	standard deviation of a Gaussian distribution
σ_1	inner limit of potential energy well
σ_2	outer limit of potential energy well
σ_c	collision diameter
ρ	distribution function
ρ_m	distribution function for state m
ρ_n	distribution function for state n
ξ	uniform deviate
ζ	random number from a Gaussian distribution

This chapter considers the practical implementation, at the software and hardware level, of the principles underlying the computer simulation methods outlined in the preceding chapter. Broadly speaking, the implementation of a computer simulation study comprises three stages, involving modelling of the process, generating appropriate coding and executing the problem on a computing device. Although the modelling stage precedes the other two stages, in practice much of it will have to be dealt with in the context of the particular problem in hand, since simulation models are usually purpose-built and problem-dependent. Nevertheless, some general techniques relevant to simulation models will be described here. The remaining part of the chapter provides a brief description of the computer hardware which can be used for the purpose of colloidal simulation. Some specific results obtained using some of the machines described will be presented later in Part IV of the book. Finally, a statement of some general guidelines for efficient programming of advanced computer architectures is listed in an Appendix.

8.1 Pair potential models

In computer simulations, the instantaneous, systematic forces acting on each of the particles are usually expressed in the form of interaction potentials. For monocomponent systems, the configurational energy may be calculated as the sum over individual particles, pairs, triplets, etc.

$$V = \sum_i v_1(r_i) + \sum_i \sum_{j>i} v_2(r_i,r_j) + \sum_i \sum_{j>i} \sum_{k>j>i} v_3(r_i,r_j,r_k) + \ldots \tag{8.1}$$

where $v_1(r_i)$ represents the effect of an external force field on particle i, and $v_2(r_i,r_j)$, $v_3(r_i,r_j,r_k)$, etc. are the pair, triplet and multibody potentials respectively. Since the calculation of any quantity involving a sum over triplets or multiple particles is very complicated and time-consuming, a common practice in computer simulations is to omit the multibody forces and take only the two leading terms in eqn (8.1). To compensate for multibody effects, an effective pair potential is usually used.

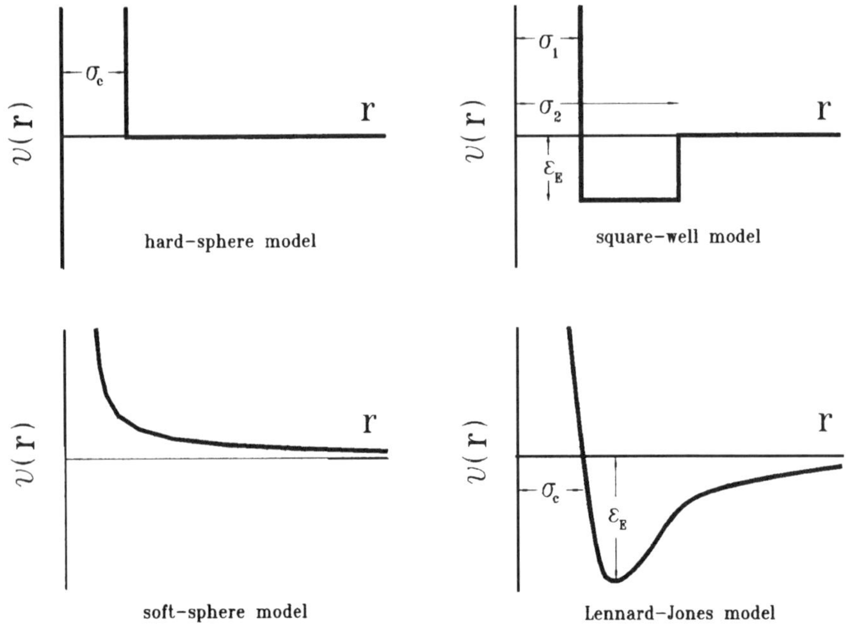

Figure 8.1 *Schematic representation of various pair potential models*

Although less precise, the pairwise approximation has been proved to give a remarkably good description of liquid properties. The same is expected to be true in colloidal systems where the effective pair interactions are the familiar DLVO-type (e.g. van der Waals and electrical double layer) potentials.

Due to the complexity and difficulty involved in accurately representing the effective pair interactions, some highly idealized models for these pair potentials often have to be used. The most commonly used functions are given in Figure 8.1, and include the following:

Hard sphere model

$$v(r) = \begin{cases} \infty & r < \sigma_c \\ 0 & r \geq \sigma_c \end{cases} \tag{8.2}$$

where σ_c denotes the closest approach distance.

The square-well model

$$v(r) = \begin{cases} \infty & r < \sigma_1 \\ -\epsilon_E & \sigma_1 \leq r < \sigma_2 \\ 0 & r \geq \sigma_2 \end{cases} \tag{8.3}$$

where σ_1 and σ_2 are the inner and outer limits of the energy well, respectively.

The soft sphere model

$$v(r) = cr^{-n} \tag{8.4}$$

where c is a constant and n is a parameter which controls the 'hardness' of the potential and is often chosen to be an integer.

The Lennard-Jones model for atomic particles

$$v(r) = 4\epsilon_E \left[\left(\frac{\sigma_c}{r} \right)^{12} - \left(\frac{\sigma_c}{r} \right)^{6} \right] \tag{8.5}$$

where ϵ_E is the depth of the energy well and σ_c the collision diameter (i.e. the distance at which $v = 0$).

8.2 Periodic boundary conditions

Since any computing facility has a limited amount of storage and finite execution speed, the number of the particles that can be accommodated in a computer simulation is not infinite. Usually, it ranges from 10 to 10^4, and rarely beyond 10^5. A given quantity of particles is generally confined within some kind of control volume, called the simulation cell, or box, so that their motion can be easily followed.

Since, in most cases, the number of particles within the simulation cell represents a tiny portion of the real system, an important question arises as to how to represent properties of the real system with the calculated averages based on such a small sample system. A major obstacle here comprises 'surface effects' associated with the finite size of the simulation cell, since for any small sample there will be a large fraction of particles lying on the surface of the simulation cell (e.g. for 1000 particles arranged in a $10 \times 10 \times 10$ cube, no less than 488 particles appear on the cube faces). These particles will experience quite different forces from those in the bulk, and hence the results obtained from such sample systems will almost certainly differ from the properties of the real systems they are meant to represent.

To eliminate such problems, a simple yet effective method proposed by Born and von Karman (1912) and generally referred to as the periodic boundary conditions can be employed. In this method the simulation cell is replicated throughout space to form an infinite lattice. As a result, as a particle moves in the original cell, its images in the replica will move in exactly the same way. If a particle leaves the original cell, one of its images will move in through the opposite face (Figure 8.2). The periodic boundary condition also implies that each particle in the original cell interacts not only with other particles inside the cell but also with their images in the replica. In this manner, the surface effects are eliminated and the number density of the cell (and hence the entire system) is conserved. Although, as particles move around, the members in the simulation cell may continually change at each simulation step, it is only necessary to consider those particles currently resident in the control volume.

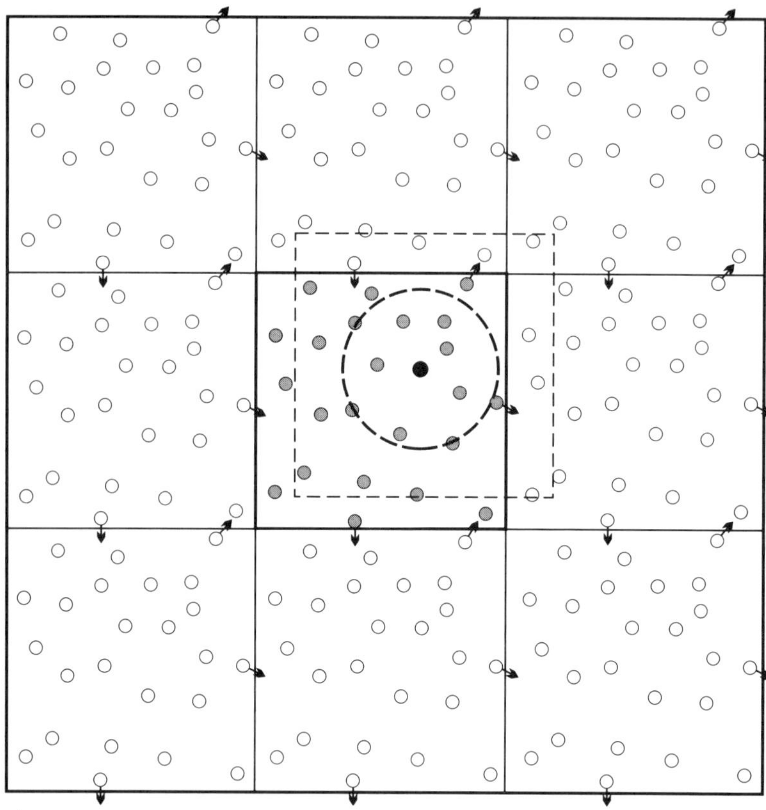

Figure 8.2 *Schematic representation of a two-dimensional simulation box (the central box) with periodic boundary conditions on all sides. Under the minimum image convention only particles enclosed by the box delineated by the dashed line are used to calculate the potential energy associated with the central particle. If potential truncation is also applied, only particles inside the zone delineated by the dashed circle are considered*

To minimize the error introduced through the symmetry of the periodic lattice, straightforward use of periodic boundary conditions is generally restricted to systems where the interparticle forces are short-ranged in comparison with the dimension of the cell. To handle long-range interactions, either the size of the simulation box must be increased (which is obvious, but not always feasible), or the Eward summation should be used. A second limitation of the periodicity is that any attempt to obtain quantities or simulate processes (e.g. phase transition) whose characteristic length scales are larger than the cell dimension is bound to be futile. Thirdly, periodic boundary conditions are not universally applicable. For example, if an external force field is present, either the corresponding potential must have the same periodicity, or the periodic boundary conditions must be abandoned. The presence of a collector surface in deposition systems will render the latter case to be true.

When calculating the interparticle forces the concept of minimum image convention (Metropolis *et al.*, 1953) is often adopted to save computation time. This short-cut is generally used in conjunction with the periodic boundary condition. With the periodic boundary condition alone, the total force acting on a particle should be calculated as a sum of all the interactions between a given (central) particle and every other particle in the simulation box as well as their images in the replicate cells (of infinite number). In practice this is impossible to perform, and hence with the minimum image convention the summation is only carried out over its interactions with the $N-1$ closest neighbouring particles (i.e. the particles enclosed in the dashed box in Figure 8.2, with the centre of the box being the particle of interest).

Even with such simplification, this may still be a substantial calculation for a system of modest scale. Further reduction in computing time may be achieved by potential truncation. The truncation of potential energy is justifiable on the grounds that the largest contribution to the force or the potential energy of a given particle normally comes from its closest neighbours. The potential truncation is accomplished by specifying a *cut-off distance*, so that only those particles lying in the cut-off sphere (the dashed circle in Figure 8.2) are counted, whereas those outside are simply assumed to be zero. Where necessary, the loss of accuracy due to the potential truncation may sometimes be recovered by applying some long-range corrections.

8.3 Generating random numbers

Random numbers play a central role in both MC and BD simulations, so the various means of generating random numbers are of critical importance. A number of practical computer random number generators are in use. To distinguish these computer-generated, repeatable and completely deterministic random numbers from naturally occurring random events, the term pseudo-random numbers is sometimes used. Some statistical criteria are available for judging whether or not a sequence of irregular numbers is sufficiently random (Knuth, 1981).

8.3.1 Uniform distribution

Uniform deviates or the random numbers uniformly distributed on, for example, (0,1), are the most important and widely used types of random number. Machine-supplied random number generators are almost always linear congruential generators which produce a sequence of numbers $X_i, X_2, \ldots$, each between 0 and $M-1$, by the recurrence relation:

$$X_{i+1} = \mathrm{MOD}[(aX_i + c), M] \tag{8.6}$$

where a and c are positive integers called the multiplier and increment respectively, M is a large positive integer called the modulus, and MOD denotes the modulo operation. The initial value used for X is called the seed. Once the seed is given, a series of random numbers, with a period of up to $M-1$, can be generated using eqn (8.6). A random real number ξ on (0,1) can be obtained as REAL(X_{i+1}/M).

For generators based on the linear congruential algorithm to be well-behaved, some important rules of thumb have to be adhered to, as noted by Knuth (1981):

- M should be as large an integer as possible, and so it is often chosen to be 2^L, L being the word-length of the computer
- if M is a power of 2, choose a so that MOD$(a,8)=5$ or 3
- give c a prime value close to $\left(\frac{1}{2}-\frac{\sqrt{3}}{6}\right)M$
- avoid using low-order (insignificant) bits because they tend to be less random.

Generally, the standard random number generators (such as the NAG library (NAG, 1984)) are recommended, but to obtain a different sequence from these routines care must be taken in choosing the seed because the actual period depends on the value of the initial seed. Normally, the seed should be chosen to be relatively prime to M. Some simple routines for further refining the machine-supplied generators have been given by Press *et al.* (1989).

8.3.2 Gaussian distribution

Once the uniform deviates are generated, it is possible to use them to construct random numbers from other distributions (Rubinstein, 1981). For instance, random numbers from the normal distribution

$$\rho(x) = \frac{1}{\sigma\sqrt{2\pi}} \exp\left(\frac{-(x-\langle x\rangle)^2}{2\sigma^2}\right) \tag{8.7}$$

may be constructed in the following way (Box and Muller, 1958):

(1) generate two uniform deviates ξ_1 and ξ_2 on (0,1)
(2) calculate $\zeta_1=(-2\ln\xi_1)^{1/2}\cos(2\pi\xi_2)$ and $\zeta_2=(-2\ln\xi_1)^{1/2}\sin(2\pi\xi_2)$; ζ_1 and ζ_2 are two independent normally distributed random numbers with zero mean and unit variance
(3) use $\gamma=\langle x\rangle+\sigma\zeta$ to obtain the desired Gaussian random numbers.

Alternative methods are also available, one of which is due to Knuth (1981); at least for vector computers (described later), this should be faster than the Box–Muller method. The procedure is:

(1) generate two uniform deviates ξ_1 and ξ_2 on (0,1)
(2) calculate $\lambda = \frac{1}{4}\left(\sum_{i=1}^{12}\xi_i - 6\right)$

(3) use the polynomial $\zeta=((((a_1\lambda^2+a_2)\lambda^2+a_3)\lambda^2+a_4)\lambda^2+a_5)\lambda$ to obtain the desired unit variance Gaussian random number, where the coefficients are:

$a_1=0.029899776$, $a_2=0.008355968$, $a_3=0.076542912$,
$a_4=0.252408784$, $a_5=3.949846138$.

8.3.3 Multivariate Gaussian distribution

Given two independent Gaussian random numbers, ζ_1 and ζ_2, generated as above with zero mean and unit variance, a pair of random numbers from a bivariate Gaussian distribution, with zero mean and variances σ_1 and σ_2, can be calculated as $\gamma_1=\sigma_1\zeta_1$ and $\gamma_2=\sigma_2[c_{12}\zeta_1+(1-c_{12}^2)^{1/2}\zeta_2]$ where c_{12} is a correlation coefficient (such as that specified in Section 7.5). To sample n correlated normal deviates γ_i with zero mean and covariance matrix $c_{ij}=<\gamma_i\ \gamma_2>$ from the multivariate Gaussian distribution:

$$\rho(X) = \frac{1}{[C^{-1}(2\pi)^n]^{1/2}} \exp(-X\bullet C^{-1}\bullet X) \tag{8.8}$$

Ermak and McCammon (1978) have proposed a method which relies on finding a lower triangle matrix $\boldsymbol{L}$ to satisfy $\boldsymbol{L}^{\mathrm{T}}\ \boldsymbol{L}=\boldsymbol{C}$. The procedure is:

(1) generate n independent Gaussian random numbers, ζ_1 to ζ_n, with zero mean and unit variances
(2) determine the elements of $\boldsymbol{L}$ using the following expressions:

$$l_{11} = c_{12}^{1/2}$$

$$l_{i1} = c_{i1}/l_{11}$$

$$l_{ii} = \left[c_{ii} - \sum_{k=1}^{i-1} l_{ik}^2\right]^{1/2} \qquad i > 1$$

$$l_{ij} = \left[c_{ij} - \sum_{k=1}^{j-1} l_{ik}l_{jk}\right] l_{jj} \qquad i > \mathrm{j} > 1$$

(3) calculate the desired random variables as:

$$\gamma_i = \sum_{j=1}^{i} l_{ij}\zeta_j$$

In hydrodynamic BD simulations, $n=3N$ and $\boldsymbol{C}=<\Delta\mathbf{r}\Delta\mathbf{r}>=2\mathbf{D}\Delta t$, where N is the number of the particles, $\mathbf{r}$ is a $3N$ position vector, and $\mathbf{D}$ a $3N\times 3N$ diffusion tensor.

8.3.4 Exponential distribution

Random numbers from an exponential distribution

$$\rho(x) = <x>^{-1} \exp(-x/<x>) \tag{8.9}$$

can be generated by first generating a uniform random number x on $(0, 1)$, and the desired random number is $\gamma=-<x>\ln\xi$.

8.3.5 Arbitrary distribution

Finally, it is useful to describe a general method for sampling random numbers from an arbitrary distribution $\rho(x)$. For any distribution on $[x_1, x_2]$, the following equation always holds (Rubinstein, 1981):

$$\rho(x) = \Lambda\rho(x)\delta(x,x_1,x_2)/\Lambda \tag{8.10}$$

$$\delta(x,x_1,x_2) = \begin{cases} 1 & x_1 \leq x \leq x_2 \\ 0 & \textit{otherwise} \end{cases}$$

where $\Lambda \geq 1$ is a normalizing constant used to ensure that ρ/Λ is always less than 1. The procedure for sampling from $r(x)$ is:

(1) generate two uniform deviates ξ_1 and ξ_2,
(2) calculate $\gamma = x_1 + (x_2 - x_1)\xi_2$ and $\eta = \rho(\gamma)/\Lambda$
(3) accept γ as the desired random number on $\rho(x)$, if $\xi_1 \leq \eta$.

8.4 Example: implementation of Metropolis MC simulation

Since the Metropolis method is the most widely used MC technique (and generally referred to as the 'standard' procedure), its implementation warrants a detailed description. A particular emphasis will be placed on some practicalities not covered in previous sections. Some of the techniques described below have now become common practice in all the MC methods mentioned in Chapter 7, since many of the 'advanced' MC methods are direct or indirect extensions of the Metropolis MC method.

The general procedure is shown in the form of a flow chart in Figure 8.3. Most computer simulations are carried out in a cubic box, for ease of implementing periodic boundary conditions, although cells of other shapes may sometimes be preferred. The chart shows 'equilibration' and 'production' stages. Unless the simulation starts with an initial configuration which is in an equilibrium state, the system must be allowed to 'evolve' to equilibrium prior to accumulating information on the equilibrium properties themselves.

In MC simulations all particles are subjected to a maximum possible displacement Δr_{max}. When a particle is selected, a displacement sphere of radius Δr_{max}, or, more conveniently, a displacement cube of side $2\Delta r_{max}$ is constructed around the particle at r_i^m. Given the finite accuracy on the computer, there is a large but finite number N_R of new positions r_i^n for the selected particle to choose, and hence the underlying stochastic matrix is given by:

$$\alpha_{mn} = \frac{1}{N_R} \tag{8.11}$$

Particles are normally moved one by one. In the original MC simulation reported by Metropolis *et al.* (1953), particles were chosen randomly. Later MC simulations tended to use an equivalent but simpler procedure, i.e. to pick each

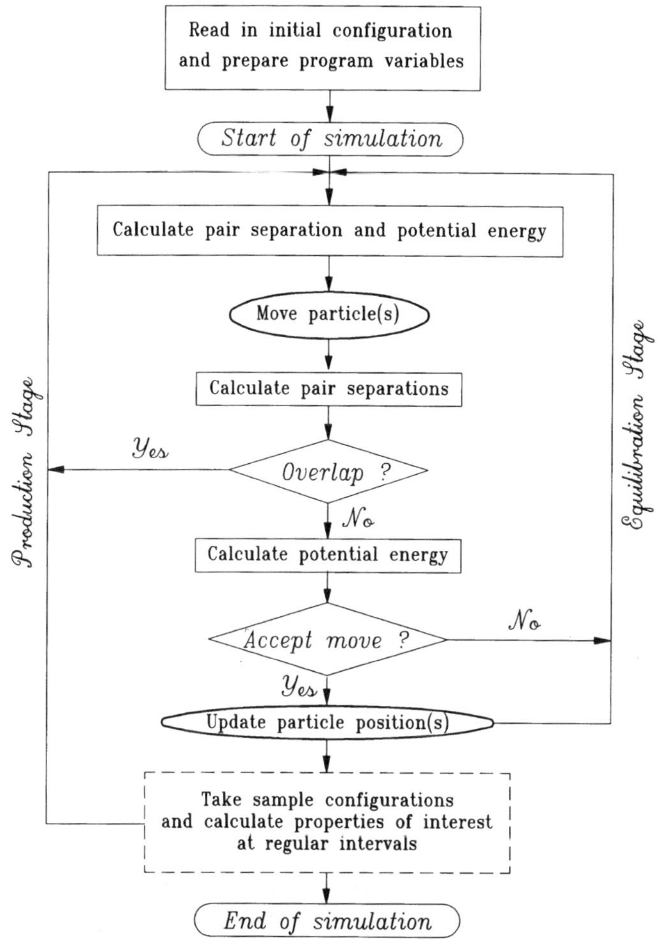

Figure 8.3 *Flow diagram showing the general calculation procedures involved in performing a Monte Carlo simulation*

particle in the same sequence as it is stored in the computer. When every particle has been moved once, a MC cycle is said to be completed. Since at each step only one particle moves, it is only necessary to calculate the potentials associated with this particle because the other part of the configurational energy potential will cancel out in the ratio $\rho_n/\rho_m = exp(-\beta\Delta V_{nm})$. Trial moves are accepted with the probability MIN(1, ρ_n/ρ_m). If $\rho_n/\rho_m \leq 0$, the move is 'downhill' in terms of energy change. Otherwise, it is accepted only when $\rho_n/\rho_m > \xi$, ξ being a uniform deviate generated at each step. This procedure is analogous to sampling random numbers from a non-uniform (e.g. from exponential) distribution, as outlined in Section 8.3.4, and is central to MC simulation.

In principle, under periodic boundary conditions, there is no limit on the value of Δr_{max}. However, if Δr_{max} is too small, most of the trials will be accepted because the value of ΔV_{nm} will be very small and hence the probability will be very close to 1. This situation corresponds to a Markov chain of highly correlated states which explores phase space very slowly. If Δr_{max} is too large, few trial moves will be accepted, which again results in a slow exploration of phase space. Therefore, it is necessary to adjust Δr_{max} continually during a MC simulation to render an efficient exploration. There is no standard criterion for an optimum Δr_{max}, but it is generally chosen so that the acceptance ratio (number of accepted moves/number of total moves) is approximately 0.5. The best choice may be problem-dependent.

By convention, MC methods are based on one-by-one trial moves, and this suits the conventional (scalar) computers very well. However, there is no reason why several or even all particles cannot be moved at once. Multi-particle-move MC simulations are particularly appealing to modern parallel computers, although they are more difficult to handle and vulnerable to the pitfall of having a very low acceptance ratio (since the chance of moving to an improbable configuration increases rapidly with the number of particles). Several examples of multi-particle-move have been reported in the literature MC simulations (Ree, 1970; Ceperly *et al.*, 1977; Rossky *et al.*, 1978; Rao *et al.*, 1979; Chapman and Quirke, 1985). There is limited evidence that multi-particle-move MC methods can offer a significant improvement of efficiency over the conventional Metropolis MC method (Rossky *et al.*, 1978; Chapman and Quirke, 1985).

This example has sought to demonstrate the computation principle for a MC simulation. However, for comparison, the general procedures employed in BD methods (see Section 7.5) are summarized in Figure 8.4. In these procedures 'preparation' and 'production' are required, since for the simulation results to be representative of conditions defined, the system should be allowed to lose any 'memory' of a particular initial configuration. This is important if the goal is to obtain 'steady-state' aggregation or deposition rates. Hence, accumulation of information about the system under investigation should start after the system begins to exhibit dynamic equilibrium. Examples of simulations performed using this procedure are given in Section 15.1.

8.5 Computer hardware

The past three decades, particularly the late 1980s, have seen remarkable advances in computer technology, with a roughly 10-fold increase in arithmetic computing speed in every five years (Hockney and Jesshope, 1990). This dramatic increase in computer speed has been made possible partly through technological improvements in the performance of hardware components and, perhaps more crucially, by the introduction of ever greater parallelism at all levels of the computer architecture. In addition there has been a substantial increase in the size of computer memories over the last few years. These advances in computer technology have had a significant impact in the field of numerical simulations. They have allowed ever more complex phenomena in

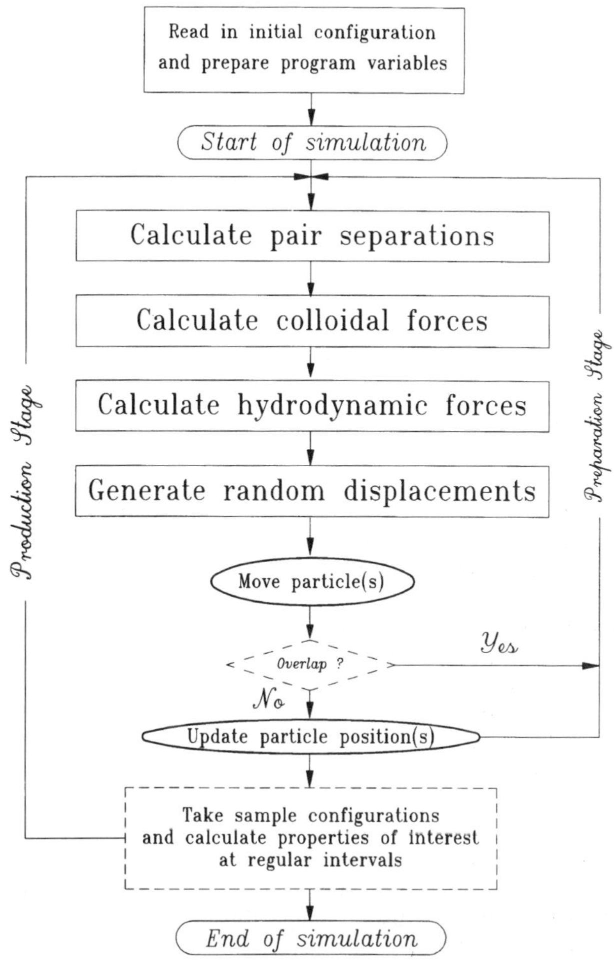

Figure 8.4 *Flow diagram showing the general calculation procedures involved in performing a Brownian dynamics simulation*

many disciplines to become amenable to study using computational methods and have prompted new programming strategies in order to make efficient use of modern architectures.

Even though most simulation codes are written in high-level computer languages, from which all technical details about the hardware are hidden, a clear understanding of the basic concepts concerning how the hardware works will prove to be beneficial to those engaged in computer simulation. For this purpose, a selection of some important concepts relating to computer hardware and consideration of their influence on program execution will be presented in this section.

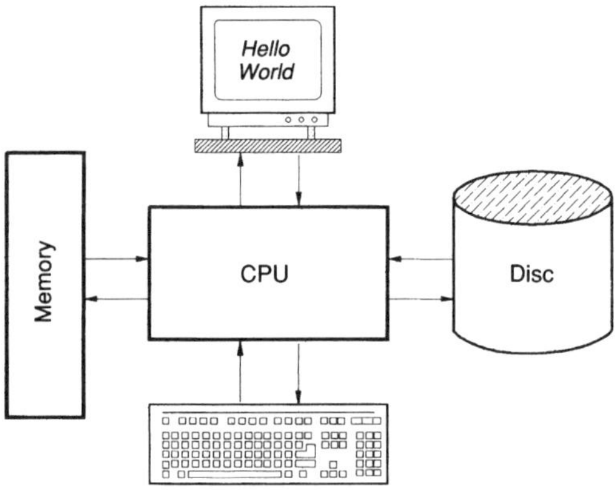

Figure 8.5 *Principal components of a computing machine*

From a programmer's point of view, the computer can be regarded as a group of connected electronic devices or units (Figure 8.5) including the central processing unit(s) (CPU), storage (memory and disc), and input/output consoles (I/O). Of these basic units, the CPU is obviously the most important. However, good design of other components, especially those responsible for data transfer, are crucial if the computer is to work efficiently.

Whatever the nature of the simulation, it must be verifiable (to some acceptable extent), and run sufficiently fast. The verification qualification includes considerations of the assumptions and simplifications embedded in the simulation model – which may be a function of the computer power available as well as the state of knowledge of the physical results – and of the actual results obtained. In many cases it is advantageous to *visualize* the results rather than attempt to interpret numerical data directly. Hence, in computational terms, in selecting a computing machine we require one which is large enough, fast enough and has sufficient graphical display capability for the particular simulation to be attempted. In addition it is essential to have the necessary software tools to facilitate program development. The development of vector and parallel computer architectures (described below) has contributed considerably to increased performance. However, it should be borne in mind that these performance gains can only be obtained to the full by careful matching of the problem to the architecture. It is not difficult to write code that performs very poorly. This is true for 'conventional' scalar machines as well. It is mentioned here as a reminder that improved simulations cannot be achieved merely by porting existing programs onto a more powerful computer.

8.5.1 Classification of computers

The speed of the CPU is related to the clock cycle or 'click', which is the basic unit of time for a computer – in less than one clock cycle nothing happens as far as the programmer is concerned. In theory, at each click the CPU can issue an instruction for the computer to perform certain operation(s) in order to produce a desired result. Therefore, the inverse of the clock cycle, usually quoted as millions of instructions per second or Mip(s), provides a convenient measure of the raw power of the computer. However, this is the maximum theoretical instruction issue rate and the actual rate is generally only about one-tenth to one-half of the maximum because of the resource conflicts during the execution of a program. From a programmer's point of view, the difference between a 'conventional' computer and a supercomputer is that, on the issuance of one instruction, the conventional computer can, at most, deliver one operation (hence no more than one result) whereas all supercomputers have features for generating many operations (hence more than one result). Thus, for a conventional computer the Mip rating may be a sensible measure, but for supercomputers it is of less significance.

Since many practical applications involve floating-point operations, the performance measure for supercomputers is often in terms of millions of floating-point operations per second or Mflop(s) in preference of Mip(s). Usually, both addition and multiplication are included in counting the Mflops and the reference to an operation is often assumed to be on 64-bit operands. As with conventional computers, the rating of supercomputers is referred to their peak performance (e.g. some popular high-performance computers are rated as 420 Mflops for a dual processor Cray X-MP, 533 Mflops for a uniprocessor Amdahl 1200, 432 Mflops for a four-processor IBM 3090/VF-400, and 400 Mflops for a uniprocessor CDC CYBER 205). In fact, the performance of a computer is dependent, in a very complex way, on many interrelated aspects such as the nature of the application, the size of the problem, the high-level language, the human effort in program optimization, the quality of the compiler, the operating system, the hardware characteristics, etc.. Thus, the rating of a computer, whether in Mip or Mflop, can only be regarded as a reference point for further evaluations.

Reducing the clock cycle has always been a goal of computer design but the introduction of concurrency in various forms and at various levels of system architecture is a more important reason for success of the present supercomputers. The principal methods for achieving concurrency may be summarized as (Hockney and Jesshop, 1990):

- Pipelining – applying 'assembly-line' techniques to improve the performance of an arithmetic or control unit
- Functional units – partitioning a single CPU into independent functional units, such as logic, addition or multiplication, and allowing them to perform different tasks simultaneously on different data
- Arrays – using an array of identical arithmetic logic units (or processing elements) under common control, all performing the same operation simultaneously but on different data streams
- Multiprocessing – using a number of processors, each obeying its own instructions, and either communicating via a common memory, or connected in a network.

Since individual designs may combine some or all of the above features, these features are not suitable for distinguishing different types of computers. Instead, the classification (or 'taxonomy') proposed by Flynn (1972) is often used, according to which computers are grouped into four categories, namely

- single-instruction single-data (SISD) machines
- single-instruction multiple-data (SIMD) machines
- multiple-instruction multiple-data (MIMD) machines
- multiple-instruction single-data (MISD) machines.

Of the four categories, SISD is characteristic of the simplest conventional computers; the SIMD and MIMD types cover all current supercomputers; and the last simply does not exist in reality. Vector processors fall naturally into the third (SIMD) category. Parallel computers fall into the third (MIMD) category. Of all the supercomputers designed to date, the single-CPU vector processor is the most successful and has received the most attention from FORTRAN compiler writers. Nevertheless, massively parallel or array processors are the subject of extensive research and are likely to be of significant use in the simulation of colloidal processes.

Every computer system has a prescribed set of machine instructions inbuilt at the hardware level, based on which all application programs are compiled. Different computers have different instruction sets. Based on the instruction set, computers can be broadly divided into two groups. One is the complex instruction set computer (CISC), and the other is the reduced instruction set computer (RISC). The RISC processors have captured considerable attention in recent years, since the use of a smaller number of instructions allows them to be executed more rapidly on a RISC processor than the corresponding instructions on a CISC processor. The RISC architecture is also easier to manufacture (and hence more economical). Since any complex instructions omitted from the reduced instruction set are simulated at the software level by the compiler, the performance of the RISC computer depends critically on the quality of the compiler.

It is useful to follow some rules of coding when writing programs – this assists the compiler in optimizing the program for different types of computers. These are discussed in some detail in Appendix 8.7.2, together with a description of some typical computer hardware (Appendix 8.7.1).

8.5.2 Amdahl's law

In analysing the performance of vector or parallel computers, Amdahl's 'law' (Amdahl, 1967) is often employed. This states that, in any computer system having two or more processing modes of different speeds, the performance of the system will be dominated by the slowest mode. For computer systems with scalar as well as vector processing capacities this means that the overall performance is dominated by scalar performance. This can be seen from the following relationship:

$$p = \frac{1}{\left[1 - \frac{V_s - 1}{V_s} F_v\right]} \tag{8.12}$$

where P is a measure of the overall performance defined as the total number of operations per unit time, V_s is the vector speedup defined as the ratio of the time required to perform an operation in scalar mode and that in vector mode, and F_v is the fraction of operations performed in vector mode.

Since $(V_s - 1)/V_s$ is usually close to unity, P is strongly dependent on the fraction of operations performed in scalar mode, $(1 - F_v)$, and hence the scalar performance. Figure 8.6 shows the performance versus vectorization curves for $V_s = 10$ (solid line) and $V_s = \infty$ (dashed line), respectively. When $V_s = 10$ (the typical order of magnitude for current supercomputers), full vectorization will lead to a factor of 10 performance improvement but 55% vectorization only gives a factor of 2 improvement. On the other hand, for a given scalar speed, merely increasing the speedup ratio V_s does not assist greatly unless the vectorization F_v exceeds 90% (and such a high vectorization can rarely be achieved).

It should be noted that in practice most programs follow an 80–20 rule: 80% of the time is spent in 20% of the code. The precise values may vary with different codes, but the important feature to note is that most codes have only a few very CPU-intensive routines. Hence the key issue to ensure optimum speed

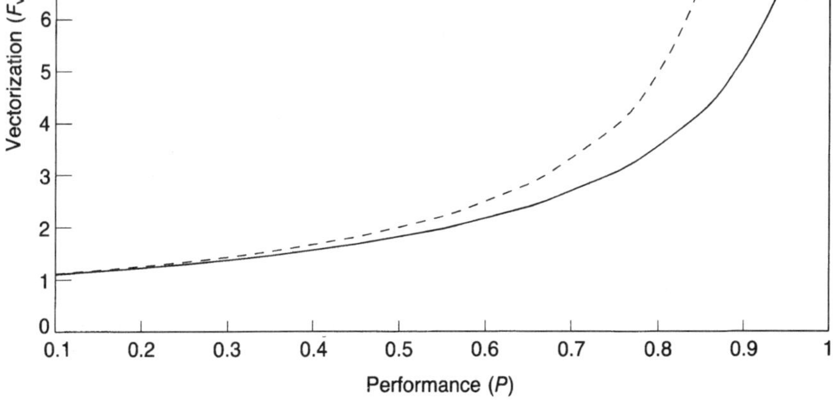

Figure 8.6 *Performance of a vector computer for various utilizations (see eqn (8.12))*

of vector computers lies in the identification and vectorization of these CPU-intensive, or heavily utilized, portions of the code.

Applying Amdahl's law to computers which achieve the speedup over the single-processor performance by spreading the computation over M processors yields a more complex relationship than in the case of vector computers. This is because, apart from parallelizable part(s) of the program, consideration must also be given to the overhead necessary for proper initiation and synchronization of each parallel processor, as well as the non-parallelizable part(s) of the program (that have to be executed on a single processor). Levesque and Williamson (1989) showed that:

$$P = \frac{1}{\left[F_{s} + \sum_{i=1}^{N} \left(\frac{F_{p_i}}{m_i} + OH_i\right)\right]} \tag{8.13}$$

$$F_{s} + \sum_{i=1}^{N} F_{p_i} = 1$$

where F_s is the fraction of operations performed on a single processor, N the number of disjoint parallel tasks, F_{p_i} *the fraction of code that amounts to the* ith task performed on m_i $(m_i \leq M)$ processors, and OH_i the corresponding overhead. From eqn (8.13), several important observations can be made.

First, if the overhead is negligible and all parallel tasks are executable on N_p processors, then the performance can be expressed as:

$$P = \frac{1}{\left(1 - \frac{N_P - 1}{N_P} F_p\right)} \tag{8.14}$$

where $F_p = \Sigma F_{p_i}$. This corresponds to the upper limit of the performance for a parallel computer, and has essentially the same form as for the vector computer, eqn (8.12).

Secondly, if the overhead is a significant fraction of the time taken to complete each parallel task itself, then the performance gain may well be lost. For instance, suppose a parallelizable task of 1.0 s duration is to be spread over 32 processors, then without any overhead the speedup factor would be 32, but if the overhead is 0.1 s the speedup factor is only about 8. Merely using more processors does not always result in the desired speedup – identifying and parallelizing as much code as possible is much more important.

The time spent on execution of the non-parallelizable code (on a single processor) can have a dramatic effect on the overall run time. When $F_s = 1\%$ the maximum speedup P is 100, even if there is no overhead and an infinite number of processors are used. Several P versus M plots are given in Figure 8.7, from which it can be seen that, unless full parallelization is achieved and the overhead is negligibly small, the overall performance improvement is not proportional to the number of processors used. For example, a task with 60% utilization of parallel capacity cannot effectively use more than four processors.

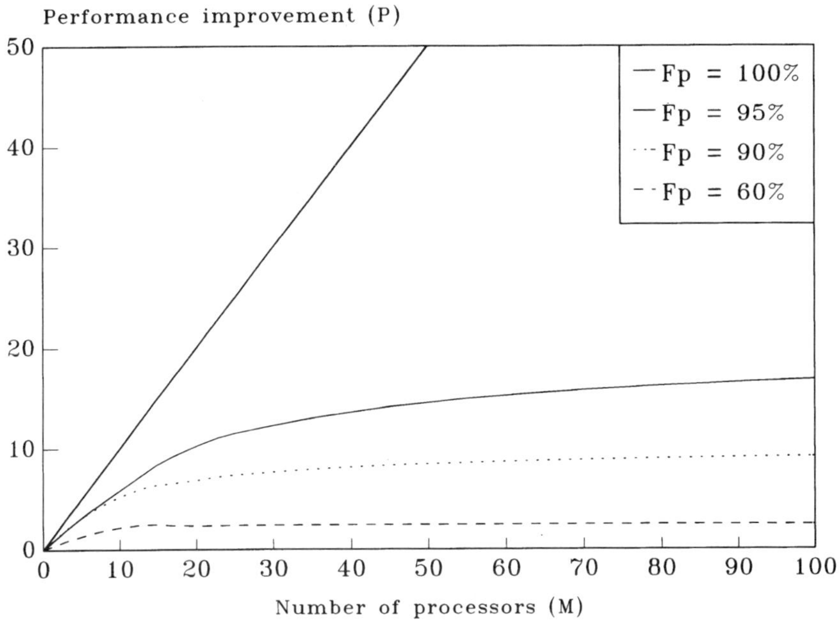

Figure 8.7 *Performance of parallel computer for various utilizations (see eqns (8.13) and (8.14))*

8.5.3 Registers and memories

Registers are a form of very high-speed memory built in the CPU to hold the most heavily used information during the execution of a program. They can provide their information to the CPU in one clock cycle, which is typically an order of magnitude faster than the ordinary memory. There are generally three types of registers:

- address registers used to hold the location of data within the memory
- operand registers used to hold data currently being manipulated, and
- instruction registers used to hold a copy of prefetched instructions in the part of the program currently being executed.

It is quite common for a computer to have eight registers for each of the first two types, with each type having as many (or more) backup registers. However, the size of the instruction stack and the scheme for prefetching instructions from the memory are often different on different computers. In vector computers, each operand register can hold many (usually the power of 2, e.g. 32 or 64) operands, thereby making it possible to perform vector operations. Although it is normally the task of the compiler to optimize the use of the registers, additional assistance from the programmer can often prove to be valuable.

Since registers are very expensive to manufacture, there are not many of them in a computer (usually, no more than a few hundred) – the bulk of the information (data and instructions) has to be held in the less expensive, and also much slower, ordinary memory system, which typically exceeds millions or even billions of bytes in size. The wide disparity between processor and memory speeds undermines the power of the CPU. In order to improve memory performance, several special techniques (such as interleaving, paging and caching) have been developed (Ibbett and Topham, 1989).

An interleaved memory system consists of a number of independent memory units or 'banks'. Each unit has its own specially built access circuits, capable of dealing with each access request within a period of a few clock cycles (called the bank cycle time). Although slightly improved, the access rate to a memory bank (dictated by the bank cycle time) is still much less than the request issue rate. The delay due to the long bank cycle time can be avoided, if consecutive memory locations are assigned onto the memory banks in a 'round-robin' manner so that each subsequent request is serviced by a different memory bank. Clearly, in order to be effective, the memory system must have at least as many banks as the number of clock cycles in the bank cycle time. Bank cycle times on supercomputers are typically four or eight clock cycles, so a memory system of eight banks is generally employed.

A memory cache is small, fast, expensive memory placed between the very fast CPU registers and the large slow main memory. Instead of directly interfacing with the main memory, the CPU will first check whether or not a requested item is resident in the cache. If so (a 'hit' is said to occur), the item will be fetched from the cache, typically in two clock cycles. Otherwise (a 'miss' is said to have taken place), the CPU will request a transfer of a packet of data from the main memory containing the item of interest plus a few more items as well, on the assumption that the data will be referenced contiguously. For computers with memory caches, the performance depends crucially on the hit/miss ratio – a high ratio can lead to a significant speedup whereas a low ratio may actually slow down the process.

The motivation for the development of a paging system stems not from the requirement of speeding up memory access but from the consideration of making more effective use of the actual memory space. The idea is to provide the users with a virtual address space through a page-mapping scheme, so as to simulate a much larger memory than is accessible physically on the computer. In other words, the virtual address space is considered to be split up into a series of contiguous pages of some convenient size (e.g. 512 bytes) so that, although the entire virtual memory is beyond the capacity of the real memory, individual pages of the virtual memory can be fitted in. During the execution of a program, a call for a virtual address in a certain page is translated into a real address by a table look-up. If that page currently resides in the memory it can be accessed immediately, otherwise it is first copied into the memory from a secondary storage device, e.g. a hard disc. Such indirect-addressing schemes have been widely used in computer designs.

8.6 Visualization of simulation results

Computer-generated high-resolution (photo-realistic) colour graphics has obvious attractions and advantages over the conventional methods of presentation. It has been widely used in a variety of disciplines in science and technology since the 1980s. The advent of and developments in personal computers, graphics workstations and other supercomputers (Sun, Meiko, Stardent, Silicon Graphics, etc.) means that direct visualization is becoming commonplace. Dynamic animations and visualizations of journeys through simulation cells can be produced. In the latter case, this is a so-called *virtual reality* recreation under the control of the user or programmer. The ability to visualize the behaviour of colloidal dispersions is of importance in the science of colloid formulation. An example is given in Section 15.1.

In this section, a graphics workstation with Dynamic Object Rendering Environment (Dore) graphics software will be used to illustrate some basic concepts of computer visualization.

To visualize a computer-generated image, the display device is of crucial importance. The major display devices that have so far been developed include the cathode ray tube (CRT), electroluminescent display (ELD), liquid crystal display (LCD) and plasma panel display systems, of which the CRT is by far the most widely used. Until the mid-1980s, the vector displays were common in CRT devices. In such systems, an image is effectively drawn on the screen – the electron beam is controlled to scan out the actual lines or points constituting the image in an arbitrary order dictated by the display commands; hence the term vector (or random) scan. Ever since the introduction of Apple Macintosh and IBM PC with their built-in raster displays as standard, the use of bitmap graphics for user–computer interaction has been increasingly popular. In raster display systems, the raster scan technique adopted from television technology is employed so that the electron beam scans sequentially, one raster line at a time, from top to bottom, on the screen to produce the entire image. A bitmap is a 'ones' and 'zeros' representation of the rectangular array of pixels (short for picture elements) that make up the screen. For monochrome bitmap graphics, each pixel requires one bit of memory to represent either 1 (bright) or 0 (black). At each pixel during the raster scanning, the intensity of the electron beam is set to reflect the brightness of the pixel. For colour graphics, more than one bit is used for each pixel and three beams – one each for red, green and blue primary colours – are controlled to make up a composite colour at each pixel (the RGB colour model).

Bitmap graphics has the advantage over vector graphics that the actual display of the image is handled by the regular, repetitive raster scan, which is far easier and less onerous to implement than the random scan of vector systems. The cost-effectiveness is the major reason for the popularity of raster displays in modern computers. Compared to the vector system, the raster system suffers from two major disadvantages.

First, since primitives such as lines and polygons are specified in terms of their endpoints or vertices, they must be scan-converted into their component pixels in the frame buffer (i.e. memory) before the image can be displayed. This puts a substantial burden on the CPU or, if present, graphics coprocessor, and limits the rate of image production, especially in real-time or dynamic animations. For instance, a Stardent 1500 workstation uses a raster colour display: the screen is made up from 1280 × 1024 pixels, refreshed at a rate of 70 Hz (or 70 frames per second), and the system has 24 bits per pixel (allowing a choice of any of 16 million colours simultaneously). A simple calculation reveals that each (screenful) frame requires approximately 4 MB storage, and the computer must be able to process up to 280 MB of data per second (in the worst case)! This is why graphics computers are necessarily of high performance.

The second drawback of the raster system is that mathematically smooth lines, polygons, and curves can only be approximated with finite-sized pixels on the raster grid, giving rise to a jagged image.

In parallel with the steady advances in graphics hardware technology, the software has gradually evolved from low-level device-dependent packages supplied by vendors for their particular display devices to high-level device-independent packages that can drive a broad range of display devices. The use of device-independent package in conjunction with a high-level programming language (such as FORTRAN, Pascal or C) promotes the portability of application programs and isolates the programmers from most machine peculiarities. The need for standards in such device-independent graphics packages has helped to bring about a series of standardized graphics specifications, such as GKS (the Graphical Kernel System) and PHIGS (the Programmer's Hierarchical Interactive Graphics System), both issued by the American National Standards Institute (ANSI). The Dore library is an example of a high-level device-independent graphics package.

Appendix 8.7.3 describes the Stardent 1500 graphics workstation and the attributes of the Dore graphics software, to illustrate some of the basic concepts of computer visualization. One feature of this type of system is that the computed images can be placed inside a three-dimensional box and used to display both static and dynamic processes. The images can be manipulated interactively, e.g. allowing rotation, selective magnification and translation.

8.7 Appendices

8.7.1 Hardware specifications relevant to computer simulations

Table 8.1 lists characteristics of some high-performance computers that may be used for performing computer simulations. All the machines shown in Table 8.1 support Fortran 77 and those with UNIX operating systems also support C Language. This is by no means an exhaustive list, but represents some of the systems employed in generating some of the results shown in Section IV.

The Amdahl VP1200 is vector mainframe and has a clock cycle of 7.0 ns, with a scalar speed of about 143 Mips. The memory system on the VP1200 machine is of the interleaved type with 128 memory banks, and supports a large vector register memory (64 KB for 1200) which is dynamically configurable from eight

Table 8.1 *Characteristics of some high-performance computers*

Type of computer	*Number of processors*	*Total memory*	*Performance per processor*
Amdahl VP1200 (vector mainframe)	1	256 MB (shared)	533 Mflops
Cray X-PM/4 (multi-processor mainframe)	4	128 MB (shared)	210 Mflops
Convex C220 (vector/parallel minicomputer)	2	256 MB (shared)	50 Mflops
iPSC/860 (multi-processor hypercube)	32	512 MB (distributed)	60 Mflops
Stardent 1500 (graphics superworkstation)	2	32 MB (shared)	16 Mflops
Meiko MK20 (transputer)	20	160 MB (distributed)	1.5 Mflops

registers of length 512–1024 elements to 256 registers of length 16–32 elements, with each element containing eight bytes.

The iPSC Hypercubes have 32 nodes each. A node may be considered as an independent, fully functional computing unit, with its own instruction set and memory, and capable of operating on its own data. An iPSC/2 node consists of an Intel 386 processor (CPU), a scalar acceleration board (SX) and a vector board (VX). An iPSC/860 node contains an Intel 386 processor (CPU) and an i860 coprocessor (RX). Each iPSC/2 node has a 4-MB scalar memory (on SX board) and a 1-MB vector memory (on VX board). An iPSC/860 node has an 8-MB total memory (on RX board) accessible for both scalar and vector modes. The main memory listed in Table 8.1 for each hypercube is the maximum collective memory. There is no common data store, but each node can communicate with any other node(s) in the hypercube by message passing via some special routing facilities or channels. The number of nodes is always of the form 2^n, where n is an integer known as the dimension of the hypercube. The hypercube is configured so that if 2^n is the number of nodes, then each node is connected to n of its neighbours. In practice, the hypercube may be divided into a number of cubes, so that its resource can be shared by a number of users or applications. A cube is a software-controlled collection of nodes. As with the hypercube, a cube always contains 2^n nodes.

Historically, the Stardent 1500 (Titan Computing System) was the first of a new generation of high-performance three-dimensional graphics computers. Unlike most advanced graphics workstations (e.g. Silicon Graphics), which depend on additional hardware, such as the specialized three-dimensional graphics transformation/clipping/ shading pipeline, in order to boost graphics performance, the Titan relies on its main CPU and the efficiency of vector processing. Each processor comprises an integer unit and a vector unit. The integer unit is a 16-Mip integer processing unit, with a 10-Mip sustained integer performance relative to the VAX-11/780 (with a nominal 1-Mip rating). The vector unit is rated as 16 Mflops, with a benchmark performance of 6 Mflops (when executing the 64-bit 100×100 double precision Linpack benchmark). Up to four memory boards can be installed in the machine. Each board has up to 32 MB memory with eight interleaves. In floating-point operations, 32 vector registers of length 64 elements are used (although other configurations are also possible).

8.7.2 Guidelines for efficient programming

More care is necessary in programming for supercomputers than for scalar computers, because current standard Fortran is inherently a sequential language demanding an approach that suits scalar computers more than vector/parallel supercomputers. Although most Fortran compilers supplied on supercomputers can perform fairly well in optimizing programs written in standard Fortran, a programmer's effort is always valuable in assisting and guiding the compiler to achieve a better optimized code (Jia, 1992). Adherence to sound *software* engineering practices is essential. This section summarizes a number of techniques that may be of use in optimizing Fortran programs for running on vector/parallel supercomputers.

Vector computers

In general, to accomplish a given task on a scalar computer, the fewer the machine operations involved, the faster the execution. However, on a vector machine, the speedup depends largely on how much code can be executed in vector mode rather than on how many machine operations are involved. Vectorizable codes are almost exclusively those involving operations on arrays in DO loops. To achieve high efficiency, as much computation as possible should be coded using arrays in DO loops. Consequently, much of the concern in program optimization is the optimal handling of array references to and from memory; and execution of a vectorized code usually requires more memory space than its scalar counterpart. Some general points that need to be borne in mind when writing programs for a computer with vector processing facilities are as follows.

(1) An inner loop with either or both of the following features is considered as a candidate for vectorization by most compilers:
   - involving a store into at least one array, with a loop variant subscript expression
   - containing at least one recognized reduction function

   However, before any vector instructions can be generated by the compiler, it must check that none of the following 'inhibitors' are present in the candidate loop:
   - subroutine call(s)
   - referencing user-defined external functions
   - any I/O statements

- recursion of certain forms
- certain nested IF blocks
- GO TO statements that exit the loop
- backward transfers within the loop.

When any of the above are encountered, the compiler's ability to optimize the loop will be compromised or completely paralysed. Thus, whenever possible, they have to be split out from the rest of the code in the loop.

(2) Due to the manner in which multi-column arrays are stored in memory, whenever possible:
- contiguous array references should be used instead of non-contiguous ones; only the left-most subscript in inner loops should be varied,
- the longest column of an array should be the left-most.

(3) In the case where use of non-contiguous array references is inevitable, care has to be taken to avoid memory-bank conflict. For example, if M is the number of memory banks, a stride of $M \times$ n, where n is an integer, should not be used if at all possible.

(4) Generally, use of too many scalar variables instead of array elements will undermine the compiler's ability to optimize the code, because the number of vector registers is usually very small. However, when used moderately in a loop, these scalar variables can be very useful in helping the compiler to recognize and eliminate common subexpressions.

(5) In DO loops, IF statements can play havoc with optimization but are often inevitable. In such cases, structured IF blocks should be used. Often a complex, nested IF block can be accomplished more efficiently by breaking it into a number of simple ones and executing them in several different loops tailored to each of the special conditions.

(6) Although a programmer cannot force the compiler to vectorize a non-vectorizable loop, most compilers do allow the programmer to suppress optimization of vectorizable loops by using compilation directives. This is useful when the vectorized loop turns out to run slower than its scalar version. For instance, if a loop contains an IF statement which, to the programmer's knowledge, has a very low true/false ratio, it is better left to execute in scalar mode.

(7) Since subroutine calls prohibit vectorization but optimization is possible within the subroutines, if they must be called in DO loops, all of the vectorizable code in the loops should be moved into the subroutines.

(8) The use of GO TO and EQUIVALENCE statements should be avoided, even if it means extra programming effort or more memory usage will be required.

Sample codes concerning the above points can be found in most user guides for supercomputers and compilers, and techniques for better vectorization have been illustrated by Levesque and Williamson (1989). Besides the general points listed above, attention must also be given to those machine-specific requirements. To give some specific examples, whilst the Amdahl 1200 machine cited in Table 8.1 favours the use of COMMON data blocks for large arrays to exploit its dynamic memory facility, the Stardent 1500 cannot vectorize loops involving arrays that appear in COMMON blocks. All intrinsic functions can be vectorized on the Amdahl vector computers, but only some are vectorizable on the iPSC systems.

Parallel computers

Although each processor in a parallel computer (e.g. iPSC Hypercube or transputers) may be capable of vector processing, the major attraction of the system is its explicit (or high-level) parallelism. Thus, for a given problem, the efficiency with which the system is utilized depends largely on the extent to which the problem is decomposed to process in parallel.

Since most applications executed on such systems involve a mix of computation and interprocessor communication, the goals of parallel design are to maximize the time a processor spends computing and minimize the time it spends communicating. The list below summarizes some general design guidelines that may be useful in writing concurrent applications on, in this instance, the iPSC Hypercube parallel computers.

(1) Separate the user interface from the core computation (the user interface is most likely to end up in a host program).
(2) Distribute or decompose the core computation evenly among the nodes. Either, or a combination, of the following two methods may be used:
   - domain decomposition, in which the data to be processed are partitioned and the partitions assigned to different nodes
   - controlled decomposition, in which the tasks (not the actual data) are divided up and performed on different nodes.
(3) Design a communication strategy so that the nodes spend as little time in communication as possible. This may involve:
   - decreasing the number of messages by increasing the size of each message
   - using global operations, whenever possible, in preference to individual node operations for communication
   - employing a particular node topology for certain types of application.
(4) Generalize the number of nodes required by the application (to eliminate the effort of recoding when the number of nodes available changes).

A program designed for vector computers is still logically sequential (as for the conventional scalar computers), since optimizations are carried out only on individual DO loops and the order in which the tasks to be performed in a program is unchanged. By contrast, a program designed for parallel computers is often logically non-sequential and involves a different programming philosophy. Hence, concurrent programs tend to be more difficult to design and debug than the vector programs.

Until now, the only programming paradigm available for users of highly parallel computers has been one in which the programmer writes code for each processor, including explicitly the coding of data passing from processor to processor as required. This has turned out to be much harder than writing serial code because, first, it is the programmer's responsibility to decide how data and the operations on the data should be distributed over the available processors and second, the programmer has to write code to deal with every *bit* of the data on *each* processor and has to keep track of all the bits as they are being processed on, and moved about, the processors. Hopefully, with the emerging new programming languages or standards, such as the High Performance Fortran (HPF) or Fortran 90, which contains a rich variety of array facilities, this difficult and tedious task can be considerably eased. Moreover, programs written in HPF are portable across SIMD, MIMD shared memory and MIMD distributed memory architectures.

8.7.3 Examples of visualization tools

One example of versatile visualization software is the Dore system, which was described briefly in Section 8.6. The principal characteristics of the Dynamic Object Rendering Environment (Dore) may be summarized as follows (Jia, 1992):

(1) high-quality, realistic images – full-colour, high-resolution, three-dimensional graphics can be generated. Besides the standard ray tracing, it also has some of the most advanced features currently available, including transparency, true shadows and environmental reflection. Images can be rendered in a variety of styles: points, wireframe, faceted, smooth-shaded surface, and any combination of these
(2) dynamic and ray traced images – both highly realistic static images and real-time dynamic images can be generated, and in both rendering modes the same database is used. It can interactively manipulate three-dimensional images, e.g. rotating, zooming and translating objects
(3) ease of use – the library does not require expert knowledge of computer graphics, although an understanding of some fundamental concepts is desirable. Only one basic description of an object's geometry and its attributes (how it is coloured, lit and modified) needs to be supplied, the rest of the chores being performed automatically
(4) device independence – there is no need to recode every time a piece of graphics hardware changes.

The basic steps for programming with the Dore library are analogous to the procedures of photography:

- Step 1: create the objects being modelled using primitives for points, lines, polygons, patches, and surfaces
- Step 2: choose attributes to describe the appearance of the objects, i.e. their colour, reflectivity, surface shadow effects, etc.
- Step 3: position and scale scene objects relative to one another
- Step 4: position various kinds of lights in relation to scene
- Step 5: select a camera lens and position the camera in relation to the scene
- Step 6: ask for the scene to be rendered (produced).

Although not essential, some proficiency in the C programming language and the concept of object-oriented programming is often helpful in developing visualizations. This is an area under active development, as evidenced by a growing number of periodical journals devoted to computer visualization, animation and virtual reality simulation.

Bibliography

Hockney, R. W. and Jesshope, C. R. (1990) *Parallel Computers 2: Architecture, Programming and Algorithms*, Adam Hilger, Bristol and Philadelphia

Ibbett, R. N. and Topham, N. P. (1989) *Architecture of High Performance Computers*, Macmillan Education Ltd, London

Levesque, J. M. and Williamson, J. W. (1989) *A Guidebook to Fortran on Supercomputers*, Academic Press, San Diego

References

Amdahl, G. M. (1967) The validity of the single processor approach to achieving large scale computing capabilities. *AFIPS Conf. Proc.*, **30**, 483

Born, M. and von Karman, Th. (1912) Uber Schwingungen in Raumgittern. *Physik. Z.*, **13**, 297–309

Box, G. E. P. and Muller, M. E. (1958) A note on the generation of random normal deviates. *Ann. Math. Stat.*, **29**, 610–611

Ceperley, D., Chester, G. V. and Kalos, M. H. (1977) Monte Carlo simulation of a many-fermion study. *Phys. Rev. B.*, **16**, 3081–3099

Chapman, W. and Quirke, N. (1985) Metropolis Monte Carlo simulation of fluid with multiparticle moves. *Physica*, **131B**, 34–40

Ermak, D. L. and McCammon, J. A. (1978) Brownian dynamics with hydrodynamic interactions. *J. Chem. Phys.*, **69**, 1352–1360

Flynn, M. J. (1972) Some computer organizations and their effectiveness. *IEEE Transactions on Computers,* C-21, 948–960

Hockney, R. W. and Jesshope, C. R. (1990) *Parallel Computers 2: Architecture, Programming and Algorithms*, Adam Hilger, Bristol and Philadelphia

Ibbett, R. N. and Topham, N. P. (1989) *Architecture of High Performance Computers*, Macmillan Education Ltd, London

Jia, X. (1992) Computer simulation of colloidal aggregation and deposition, PhD Thesis, University of Manchester Institute of Science and Technology.

Knuth, D. E. (1981) *The Art of Computer Programming*, 2nd edn, Addison-Wesley, Reading, Mass

Levesque, J. M. and Williamson, J. W. (1989) *A Guidebook to Fortran on Supercomputers*, Academic Press, San Diego

Metropolis, N., Rosenbluth, A. W., Rosenbluth, M. N., Teller, A. H. and Teller, E. (1953) Equation of state by fast computing machines. *J. Chem. Phys.*, **21**, 1087–1092

NAG (1984) *NAG Fortran Library Manual*, Vol. 4, Mark 9, NAG Central Office, 7 Banbury Road, Oxford, UK OX2 6NN

Press, W. H., Flannery, B. P., Teukosky, S. A. and Vetterling, W. T. (1989) *Numerical Recipes: The Art of Scientific Computing*, Cambridge University Press, New York

Rao, M., Pangali, C. and Berne, B. J. (1979) On the force bias Monte Carlo simulation of water: methodology, optimisation and comparison with molecular dynamics. *Mol Phys.*, **37**, 1773–1798

Ree, F. H. (1970) Statistical mechanics of single-occupancy systems of spheres, disks and rods. J. Chem. Phys., **53**, 920–931

Rossky, P. J., Doll, J. D. and Friedman, H. L. (1978) Brownian dynamics as smart Monte Carlo simulation. *J. Chem. Phys.*, **69**, 4628–4633

Rubinstein, R. Y. (1981) *Simulation and Monte Carlo Methods*, Wiley, New York

Part III

Experimental methods and model validation techniques

Chapter 9

Experimental techniques for aggregation studies

Nomenclature

a	particle radius
a_0	radius of primary particles
a_G	radius of gyration
A	cross-sectional area of light beam
b	$=2/d_F$
C	light-scattering cross-section
C_k	light-scattering cross-section of k-fold aggregate
d_1	diameter of primary particles
d_F	fractal dimension
D	diffusion coefficient
$f(k)$	aggregate size distribution function
$g(\Delta t)$	correlation function
I	transmitted light intensity
$I(0)$	initial scattered light intensity
I_θ	intensity of scattered light at an angle θ to the incident beam
I_0	incident light intensity
J_1	first-order Bessel function
k	aggregation number
$\bar{k}$	mean aggregation number ($=n_0/n_T$)
k_a	aggregation rate constant
k_C	Carman–Kozeny constant, eqn (9.35)
K	permeability
l	depth of packed bed
L	optical path length
m	relative refractive index
n	particle number concentration
n_0	initial particle number concentration
n_2	number concentration of doublets
n_k	number concentration of k-fold aggregates
n_T	total particle number concentration
$P(\theta)$	form factor, eqn (9.10)
q	scattering vector, eqn (9.12)
Q	scattering coefficient
Q_0	scattering coefficient of primary particles

Q_k	scattering coefficient for k-fold aggregate
r	distance from centre of particle
r_{ij}	centre–centre distance of two particles i and j
R	ratio of RMS to DC value, eqn (9.23)
R_θ	Rayleigh ratio, eqn (9.9)
S	specific surface
$S(\theta)$	structure factor, eqn (9.16)
t_C	relaxation time
u	$=qa$
v	filtration velocity
$\overline{V}$	mean voltage
V_{RMS}	root mean square value of voltage fluctuations
α	dimensionless particle size, eqn (9.5)
α_E	electro-optic parameter, eqn (9.22)
Γ	gamma function
ΔP	pressure difference
Δt	delay time
ϵ	porosity
λ	light wavelength
μ	viscosity
ρ	dimensionless function of m and α, eqn (9.7)
τ	turbidity
ϕ	volume fraction of dispersed particles

9.1 General – choice of technique

Experimental studies of particle aggregation may be undertaken for various reasons, ranging from fundamental to highly practical. In fundamental research, there may be interest in the absolute rate of aggregation of model suspensions in order to check theories of aggregation kinetics or to derive information on colloidal interactions. In many cases, *relative* rates of aggregation are sufficient (e.g. in deriving collision efficiencies or stability ratios – see Section 6.3.1). More practical investigations could be concerned with the properties of aggregates (flocs) produced under different conditions and with the determination of the optimum dosage of additives needed to give effective flocculation.

For fundamental studies, the most satisfactory means of following aggregation processes is to determine the particle size distribution of the aggregating suspension at frequent intervals (continuously if possible). Ideally, the method chosen should give information on the number concentrations of primary particles and of all aggregates. The only way of gaining such information unambiguously is by using a technique which directly counts particles and can discriminate between different particle sizes. Various methods of counting and sizing particles have been used in aggregation studies and have given valuable information. Such methods are generally limited to rather dilute suspensions and to particles which are larger than about 0.2 μm. These techniques give absolute rates of aggregation.

Slightly less direct methods are available, based on the light-scattering properties of aggregating suspensions. Light-scattering measurements can give information on the average aggregate size and also on the form of aggregates (e.g. the fractal dimension). It is possible to derive aggregate size distributions, but some assumptions are necessary. These and other methods can easily give *relative* rates of aggregation, by comparing the rate of change of, say, scattered light intensity under given conditions with that for a fully destabilized suspension. Under certain conditions light-scattering techniques can give absolute aggregation rates.

For practical purposes there are many methods which may be used to give aggregate properties, such as settling rate, floc size, density and permeability. By following the change of these properties with the amounts of additives used, it is possible to establish 'optimum dosages', which are frequently needed in industrial applications.

A brief survey of some available methods will be given in the following sections, paying more attention to principles than to practical details.

9.2 Particle counting and sizing

9.2.1 Microscopic methods

The earliest method used to follow the aggregation of suspensions was direct microscopic observation. Samples of the suspension are withdrawn at intervals, diluted if necessary and examined by optical microscopy. The number of particles in a given volume can be determined simply by counting. Ordinary optical microscopes cannot easily resolve particles much below about 1 μm in size, although the *ultramicroscope*, using dark-field illumination, allows much smaller particles to be detected (see, for example, Mysels (1959). However, it is not generally possible to determine the size of aggregates by optical microscopy (unless they grow to about 20 μm or more), and the method is normally used to give just the *total* number of particles. Nevertheless, such measurements can be used to determine the aggregation rate constant, k_a, from eqn (6.13). Many early estimates of stability ratios for slowly-coagulating sols were made in this way (Freundlich, 1926).

A recent development in microscopy is the confocal laser scanning microscope (Pluta, 1989), in which a very small sample (about 0.5 μm diameter) is illuminated by a laser beam. The laser scans the sample point by point and the scattered light is measured by a photodetector. In this way an image can be built up with a much higher resolution than a conventional light microscope, allowing smaller particles to be resolved. Another advantage is that rather concentrated suspensions can be examined. This device may prove useful in the analysis of aggregation kinetics and in studies of floc structure.

Electron microscopy provides much higher resolution than the light microscope, but the elaborate sample preparation, including a drying stage, makes it impractical for kinetic measurements. Aggregate structure can be examined by electron microscopy, provided that care is taken to avoid significant structural changes during sample preparation. The *freeze-fracture* technique has proved useful in this respect (Stewart and Sutton, 1986).

9.2.2 Sensing zone techniques

Automated particle counting can be achieved by allowing particles to pass singly through a zone in which their presence can be detected by a suitable sensor. Particles passing through the zone produce a response from the detector and can be counted as a series of pulses. If the sensor response depends on particle size, then the pulse height is size-dependent and this provides a means of discriminating between particles of different sizes.

It is important to ensure that particles do not pass simultaneously through the sensing zone, as otherwise two or more particles would produce just one pulse and be counted as one (larger) particle. This is known as the *coincidence* effect and, although statistical correction for this effect is possible, the most satisfactory procedure is to ensure that the sample is sufficiently dilute to avoid the problem. For many suspensions of interest, high degrees of dilutions are necessary, which may cause some change in the particle size distribution, especially for aggregating particles.

There are two commonly used detection techniques in particle counting, electrical and optical, giving *electrical sensing zone* (*electrozone*) and *optical sensing zone* (*light-scattering*) counters. The two types have their own advantages and disadvantages.

Electrozone counters are based on a principle developed by Coulter (1956) and made available commercially as the *Coulter counter*. The basic concept (often called the *Coulter principle*) is that particles in an electrolyte solution passing through an orifice cause a momentary change in electrical resistance and hence a voltage pulse if the current between two electrodes is maintained constant. The electrodes are located on either side of the orifice and a particle passing through displaces a volume of electrolyte equal to the particle volume. Most particles can be assumed to have infinite resistance, in comparison to the electrolyte solution, and so the voltage pulse produced is proportional to the particle volume.

The Coulter technique has been used very widely, with thousands of references in the literature (Miller and Lines, 1988). It is capable of quite high-speed counting (5000 or more particles per second) and can resolve particles only slightly different in size.

A unique feature of the electrozone method, in comparison with other (especially optical) techniques, is that it is virtually independent of the shape or composition of the particles. A particle passing through the orifice gives a pulse which depends only on the volume of electrolyte displaced and hence on the volume of the particle. For an aggregate, the pulse height is proportional to the total volume of the constituent particles, without the included fluid (since this is electrolyte solution it does not contribute to the extra resistance). Early studies of particle aggregation using the Coulter counter were reported by Matthews and

Rhodes (1970) and the technique has been used many times since (e.g. Pefferkorn *et al.*, 1988).

An often-quoted difficulty in using the electrozone technique to study aggregation phenomena is the possibility of aggregate breakage during passage through the orifice, where there may be very high shear rates. However, this question is not straightforward, since breakage of an aggregate gives fragments of the same total *volume* and the size registered by the counter should be the same as that of the original aggregate. Thus, breakage within the orifice should give one pulse, corresponding to the aggregate size. Breakage in the elongational flow field just ahead of the orifice could give several smaller pulses.

Another possible problem has to do with the need to mix the suspension with a rather concentrated (typically about 2%) salt solution, which would destabilize many suspensions. However, the low particle concentration (typically around 10^6 particles/ml) means that aggregation would be very slow and in the short time needed for a measurement this should not cause any difficulty.

The electrozone technique cannot conveniently be used over a very wide range of aggregate size. In order to detect small, primary particles, an orifice is needed with a diameter not more than about 50 times the primary particle size. (Particles smaller than about 0.5 μm are very difficult to monitor, which is a serious limitation for colloids.) As aggregates grow larger, they can reach a size where orifice blockage becomes a problem.

Particle counting and sizing can also be achieved using light-scattering techniques. In this case, particles are made to pass singly through a focused light beam (usually a laser beam) and either the transmitted light or scattered light intensity is measured. Each particle passing through the beam causes a reduction in transmitted light or an increase in the scattered light (see Section 9.3). The decrease in transmitted light is caused by scattering at all angles and should be very much greater than the increase in scattered light at any particular angle. Nevertheless, the scattering technique is the more sensitive and can be used for smaller particles than the transmission (or *light blockage*) method. The reason is that detecting particles by light transmission depends on reliably measuring a small difference between large quantities (the transmitted light intensities with and without a particle in the beam). In the case of scattered light, the value in the absence of a particle should be essentially zero, so that even a very small increase can be detected. Lower particle size limits are typically about 1 μm for light blockage counters and around 0.2 μm for light-scattering instruments. In both cases, the size of particles can be derived using standard light-scattering theory (see Section 9.3.2).

There have been several detailed studies of aggregation kinetics using optical particle counters (Cahill *et al.*,1987; Sonntag *et al.*, 1986; Pelssers, 1988) which have provided useful checks on the predictions of Smoluchowski theory (see Section 6.2.1). This subject has been reviewed by Sonntag (1993). Under optimum conditions, aggregates up to about five-fold can be distinguished (although highly uniform primary particles are needed). The very high dilutions necessary for such measurements and the experimental limitations confine this technique to fundamental studies. In some cases breakage of aggregates in passing through the sensing zone of optical particle counters may occur (Gibbs, 1982).

9.3 Light-scattering methods

A suspension of particles illuminated by a light beam causes some of the light to be scattered. In general, scattering by particles depends on their size, shape and refractive index, the light wavelength and the angle of observation. A complete theory (*Mie theory*) is available for spherical particles (Kerker, 1969) and there are several approximate approaches possible. For studies of particle aggregation it is usually necessary to make simplifying assumptions and even so it may not be possible to derive a detailed aggregate size distribution. Often, some form of the size distribution is assumed and appropriate parameters (e.g. moments of the distribution) are derived from light-scattering measurements.

A consequence of light scattering by particles is that there is a reduction in the intensity of light transmitted *through* the suspension. Measurement of *extinction* or *turbidity* can be a very simple method of deriving information on particle size or the state of aggregation of suspensions. More detailed information can be obtained by measuring the angular distribution of scattered light, although the technique is more elaborate. *Static light scattering* involves measurement of average scattered light intensities and can give information over a wide range of particle sizes. *Dynamic light scattering* gives values of the diffusion coefficients of particles, from which size information can be derived. This method is only applicable to rather small particles (below about 2 μm) and so is of limited use in aggregation studies. All of these light-scattering methods will be discussed briefly in the following sections.

9.3.1 Turbidity

Turbidity is conventionally defined in terms of the reduction in intensity (extinction) of a beam of light passing through a length L of suspension (Figure 9.1):

$$I = I_0 \exp(-\tau L) \qquad (9.1)$$

where I_0 is the incident light intensity, I is the transmitted light intensity and τ is the turbidity.

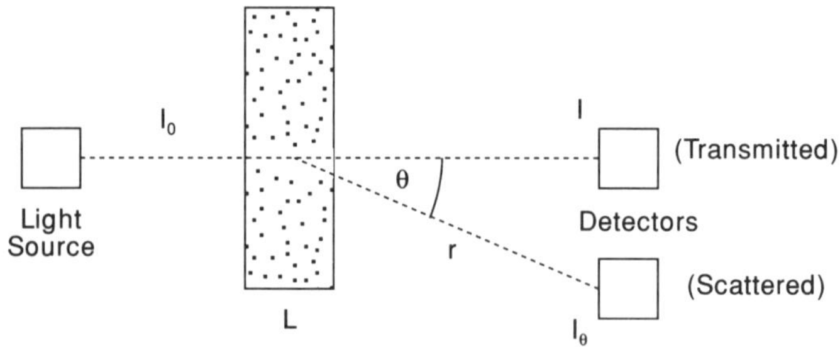

Figure 9.1 *Measurement of transmitted and scattered light (schematic). The incident light, intensity* I_0*, passes through a length,* L *of suspension. Scattering by particles causes a reduction of transmitted light intensity to* I *and a scattered light intensity* I_θ *at an angle* θ *to the light beam and a distance* r *from the sample*

(Note: extinction can also be due to *absorption* of light by the particles and an equation of similar form applies, but with an *extinction coefficient* instead of turbidity. For simplicity, we shall assume that the particles are non-absorbing, so that the extinction is entirely due to scattering.)

If the suspension is of uniform particles, with a concentration n per unit volume, then the turbidity is defined simply as:

$$\tau = nC \tag{9.2}$$

where C is the *light-scattering cross-section* of the particles.

C represents the effective cross-sectional area of the beam from which light is removed as a result of scattering by the particle. It is related to the geometrical cross-sectional area of the particle by a *scattering coefficient*, Q, which, for spherical particles, is simply:

$$Q = \frac{C}{\pi a^2} \tag{9.3}$$

where a is the particle radius.

The scattering coefficient depends greatly on the particle size, refractive index and the light wavelength. For very small particles (size less than about 5% of the wavelength, λ) Rayleigh theory applies (see Section 9.3.2) and the scattering coefficient is given by:

$$Q = \frac{8}{3} \alpha^4 \left(\frac{m^2 - 1}{m^2 + 2} \right)^2 \tag{9.4}$$

where m is the refractive index of the particles relative to that of the suspending medium and α is a dimensionless particle size given by:

$$\alpha = \frac{2\pi a}{\lambda} \tag{9.5}$$

For a typical value of relative refractive index (say 1.2, corresponding to polystyrene particles in water) and bearing in mind that α cannot exceed about 0.3 for Rayleigh theory to apply, eqn (9.4) is only valid up to Q values of about 10^{-3}, i.e. for particles which scatter less than 1/1000 of the light incident upon them.

For larger particles, Q can be computed exactly from Mie theory (see, for example, Kerker (1969)) and, when plotted against α, shows a regular series of maxima and minima, as in Figure 9.2. For very large particles, the scattering coefficient approaches a limiting value of 2, i.e. the amount of light scattered by a large particle is *twice* the light incident upon it (because of *diffraction* of light at the edge of the particle). The particle size at which Q becomes effectively constant depends on the wavelength and the refractive index of particles. For latex particles in water, the limiting value is not reached until the particle diameter is about 30 times the light wavelength.

When the particle refractive index is not greatly different from that of the suspension medium, and the particles are quite large, a very convenient

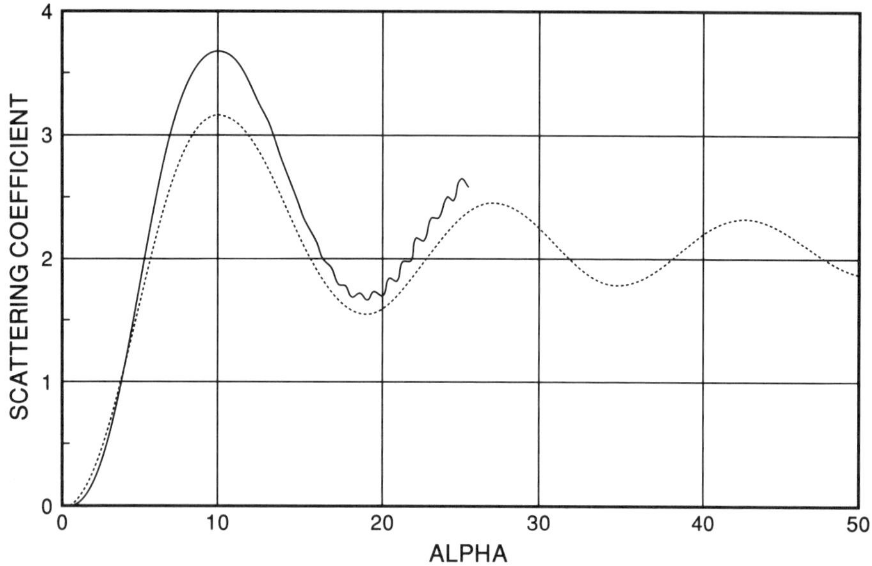

Figure 9.2 *Calculated scattering coefficient,* Q, *as a function of dimensionless size,* α. *Full line: exact computations based on Mie theory. Broken line: calculations from anomalous diffraction approximation, eqn (9.6)*

approximate form for Q may be used. This is the *anomalous diffraction approximation* (van de Hulst, 1957):

$$Q = 2 - \left(\frac{4}{\rho}\right) \sin \rho + \left(\frac{4}{\rho^2}\right)(1 - \cos \rho) \tag{9.6}$$

where ρ is a function of the dimensionless size, α, and the relative refractive index, m:

$$\rho = 2\alpha(m - 1) \tag{9.7}$$

This expression may be applied when α is large (greater than about 10) and m small (less than about 1.2). Some results from the anomalous diffraction approximation are shown in Figure 9.2, together with the corresponding 'exact' results from Mie theory. It can be seen that the approximate result follows very well the form of the exact computations, especially with regard to the positions of the maxima and minima. The approximate value of the scattering coefficient is, at worst, about 14% too low when $m = 1.2$. For smaller values of refractive index, the agreement between Mie theory and the anomalous diffraction approximation becomes even better.

For a suspension of uniform spherical particles, with radius a and number concentration n, the *volume fraction*, ϕ, is simply $4\pi na^3/3$. The *specific turbidity*, τ/ϕ, then follows from eqns (9.2) and (9.3):

$$\frac{\tau}{\phi} = \frac{3Q}{4a} \tag{9.8}$$

Since the scattering coefficient depends on particle size, it is possible, in principle, to derive particle size information from turbidity measurements on a suspension of known concentration. In an aggregating suspension, the volume fraction should remain constant and the change in turbidity might be a useful measure of the degree of aggregation.

For mono-sized particles the specific turbidity, when plotted against particle size, typically shows a pronounced maximum followed by an undulating region where there is a decreasing trend. For very large particles, where the scattering coefficient becomes nearly constant, eqn (9.8) implies that specific turbidity varies inversely with particle size.

Making use of the anomalous diffraction approximation, eqn (9.6), the specific turbidity can be calculated as a function of particle size and refractive index. Examples are shown in Figure 9.3, assuming a light wavelength of 700 nm. It is clear that, for small particles, there is a marked increase in turbidity as particle size increases, but that eventually turbidity declines at larger sizes. For this reason a single value of specific turbidity would not give an unambiguous value

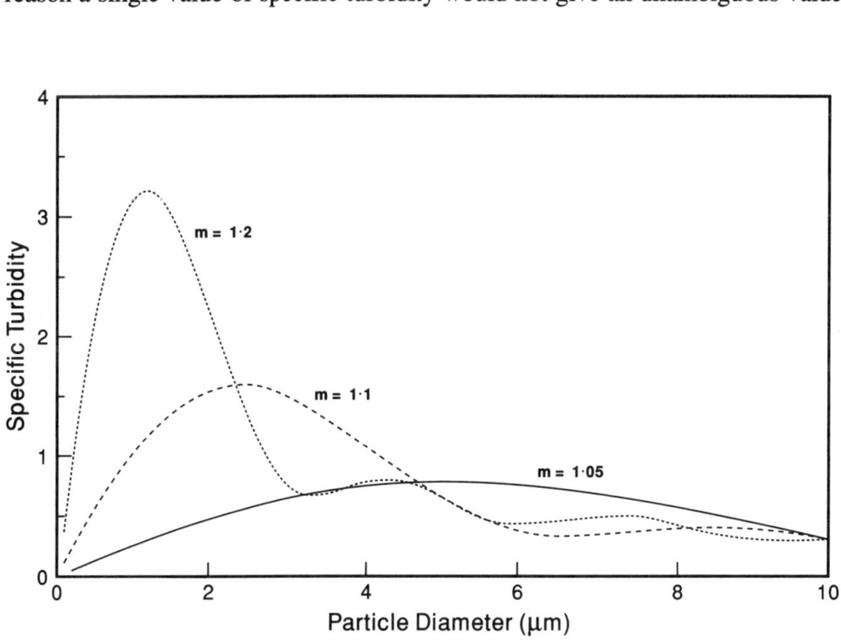

Figure 9.3 *Specific turbidity as a function of particle diameter, for suspensions of uniform, spherical particles, with different values of refractive index,* m. *Derived from eqn (9.8) with* Q *calculated from eqn (9.6). Wavelength assumed to be 700 nm*

of particle size. However, measurements at two or more wavelengths can be used to derive particle size; see, for example, Kerker (1969). For non-uniform particles, turbidity measurements can only give size information if some form of particle size distribution is assumed. For instance, in the case of the log-normal distribution (see Section 6.6.2), just two parameters are needed and these can be derived from specific turbidity measurements at two wavelengths (Zollars, 1980).

For aggregating suspensions, although the average particle size must increase, the applicability of specific turbidity measurements is not straightforward. The main problem is that, except in the case where particles 'coalesce' to form spherical aggregates, there is uncertainty over the light-scattering properties of aggregates. Nevertheless, for very small particles, much less than the light wavelength, there is a substantial increase in turbidity as aggregation occurs and this has long been used as a measure of relative coagulation rates and stability ratios (Reerink and Overbeek, 1954).

The turbidity of many suspensions increases visibly as aggregation occurs, but once the aggregates have reached a certain size (typically of the order of 1 μm) turbidity levels off and eventually declines. Provided only the initial coagulation rate is required, then turbidity can be a very convenient experimental approach. Absolute rates can only be derived in the case of very small particles, where Rayleigh theory applies, but it is often acceptable to derive rates relative to the 'rapid' coagulation of fully destabilized sols. In this way, stability ratios (Section 6.3.1) can be easily calculated.

9.3.2 Static light scattering

Mie theory

When a particle is illuminated by a light beam, its constituent molecules become polarized by the oscillating electric field of the light wave and radiate light in all directions. If the radiated (scattered) light is of the same wavelength as the incident light, the phenomenon is known as *elastic scattering*, since no energy is lost from the beam. For particles of arbitrary size, the scattering pattern is complicated by the fact that scattered radiation from different parts of the same particle is subject to phase differences and hence interference effects. An exact theory was developed by Mie early in the twentieth century (see Kerker, 1969). Mie theory gives the complete angular distribution of scattered light intensity from a homogeneous spherical particle as a function of its refractive index, size and the light wavelength. For larger particles the scattering pattern becomes highly asymmetrical, with most of the light being scattered in the forward direction.

Although measurements of the angular distribution of scattered light, or the intensity at just one angle, can give information on particle size, the full Mie theory is of very limited use for aggregating suspensions, because of the nature of real aggregates. Usually, an approximate expression has to be adopted, in order to interpret light-scattering data in terms of aggregate properties.

Rayleigh scattering

The simplest case is where the particles are much smaller than the light wavelength, so that all of the scattered radiation from one particle can be assumed to be in phase. For Rayleigh theory to apply, the particle diameter should be less than 10% of the light wavelength (i.e. well below 0.1 μm for visible light). In this case the intensity of unpolarized light scattered at an angle θ to the incident beam and at a distance r from the particle (see Figure 9.1) is given by:

$$\frac{I_\theta}{I_0} = \frac{1}{r^2}\left[\frac{8\pi^4 a^6}{\lambda^4}\left(\frac{m^2-1}{m^2+2}\right)^2 (1+\cos^2\theta)\right] = \frac{R_\theta}{r^2} \tag{9.9}$$

where a is the particle radius and I_0 is the incident light intensity. The term in square brackets, R_e, is often called the *Rayleigh ratio*.

Rayleigh scattering gives a symmetrical pattern about $\theta = 90°$, where the scattered intensity is minimum. (The horizontally polarized component of the scattered light, represented by the cos θ term in eqn (9.9), is zero at 90°.)

When the scattered light is integrated over all angles, the total amount of scattering by the particle is obtained, which can be expressed as a scattering coefficient. With this procedure, making use of the definitions of Q in eqn (9.3) and α in eqn (9.5), the result given previously for scattering coefficient, eqn (9.4), is obtained.

The Rayleigh expression predicts that light scattered by a particle varies as the *sixth* power of particle size and hence as the square of particle volume. So, for an aggregating suspension, where the total volume of particles remains constant, the total amount of light scattered should increase in proportion to the average aggregate volume. Unfortunately, the limitation to very small particle sizes severely restricts the application of this simple result.

Rayleigh–Gans–Debye scattering

A frequently used expression, which has a wider range of applicability, is the Rayleigh–Gans–Debye (RGD) approximation. The physical basis of RGD scattering is that a particle (of arbitrary shape) is assumed to consist of elements which behave as independent Rayleigh scatterers. This is appropriate when the parameter ρ (eqn 9.7) is very low, which implies small particles and a low value of refractive index. Although the RGD result can be used for particles considerably larger than in the case of the Rayleigh expression, the condition $\rho \ll 1$ means that the scattering coefficient Q must be very small (van de Hulst, 1957).

The RGD approximation can be written as a modified form of the Rayleigh result:

$$\frac{I_\theta}{I_0} = \frac{1}{r^2} R_\theta P(\theta) \tag{9.10}$$

The term $P(\theta)$ is known as a *form factor*, and represents a correction to the Rayleigh expression which accounts for effects due to the size and shape of the particle. This term can be evaluated for many simple geometrical shapes and for particle aggregates in terms of the mass distribution.

For homogeneous spheres, of radius a, the form factor becomes:

$$P(\theta) = \left[\frac{3(\sin u - u\cos u)}{u^3}\right]^2 \tag{9.11}$$

where $u = qa$ and q is the *scattering vector*, given by:

$$q = \frac{4\pi}{\lambda}\sin\frac{\theta}{2} \tag{9.12}$$

For very low scattering angles, series expansion of eqn (9.11) leads to:

$$P(\theta) = 1 - \frac{u^2}{5} + \ldots \tag{9.13}$$

For homogeneous particles of arbitrary shape, with radius of gyration a_G, $P(\theta)$ can be expressed as a series expansion:

$$P(\theta) = 1 - \frac{(qa_G)^2}{3} + \ldots \tag{9.14}$$

which becomes equivalent to eqn (9.13) in the limit of low scattering angle, since the radius of gyration of a homogeneous sphere is $\sqrt{(3/5)}a$.

In the limit of zero scattering angle, the form factor becomes unity and the scattered intensity is just that given by the Rayleigh expression. For this reason, light scattering at very low forward angles from an aggregating suspension of constant volume fraction should be proportional to the average aggregate volume and such measurements have been used to measure aggregation rates (Young and Prieve, 1991). In particular, it can be shown (e.g. Zeichner and Schowalter, 1979) that the scattered light should increase with time according to:

$$\left(\frac{1}{I(0)}\frac{\mathrm{d}I}{\mathrm{d}t}\right)_{t\to 0} = 2k_a n_0 \tag{9.15}$$

where k_a is the aggregation rate constant, eqn (6.13), and n_0 is the initial number concentration of primary particles. $I(0)$ is the measured light-scattering intensity for the initial suspension.

This result shows that absolute values of the aggregation rate constant, at least in the early stages, can be derived in a straightforward manner.

Light-scattering measurements at very low angles became much more feasible with the advent of lasers, because of the narrow parallel beam that can be generated. However, there are some practical difficulties associated with the technique, notably the very large scattering produced by stray dust particles. Measurements at larger angles are easier to carry out but more difficult to interpret, because of the need to consider the structure of small aggregates.

The RGD approach allows treatment of particle aggregates through a *structure factor*, $S(\theta)$, which is given by:

$$S(\theta) = \sum_i^k \sum_j^k \frac{\sin qr_{ij}}{qr_{ij}} \tag{9.16}$$

where r_{ij} is the centre-to-centre distance of a pair of particles in a k-fold aggregate and the summation is over all such pairs.

For aggregates of hard spheres it is, in principle, possible to calculate the structure factor for any assumed structure. However, for three-fold and higher aggregates the number of possible structures increases very rapidly (see Figure 6.8). Only in the case of a doublet is there a unique structure factor. It can be shown (van Zanten and Elimelech, 1992) that the scattered light intensity from a suspension of equal spheres undergoing perikinetic aggregation and following Smoluchowski kinetics increases according to:

$$I(\theta,t) = I(\theta,0)\left(1 + 2\frac{n_2}{n_0}\frac{\sin qd_1}{qd_1} + 2\sum_{k=3}^{\infty}\frac{n_k}{n_0}S_k(\theta)\right) \tag{9.17}$$

where d_1 is the diameter of primary particles (i.e. the centre-to-centre distance of two touching particles), and n_0, n_2 and n_k are the number concentrations of primary particles (initially), doublets and k-fold aggregates at time t. These concentrations are given by eqn (6.17). $S_k(\theta)$ is the structure factor for a k-fold aggregate for a scattering angle θ.

It follows that the initial rate of increase of scattered light from an aggregating dispersion is given by:

$$\left(\frac{1}{I(\theta,0)}\frac{\mathrm{d}I(\theta,t)}{\mathrm{d}t}\right)_{t\to 0} = 2k_{\mathrm{a}}n_0\frac{\sin qd_1}{qd_1} \tag{9.18}$$

which reduces to eqn (9.15) for very low scattering angles (when $q \to 0$).

By carrying out measurements at different angles simultaneously, van Zanten and Elimelech (1992) were able to derive absolute coagulation rate constants for

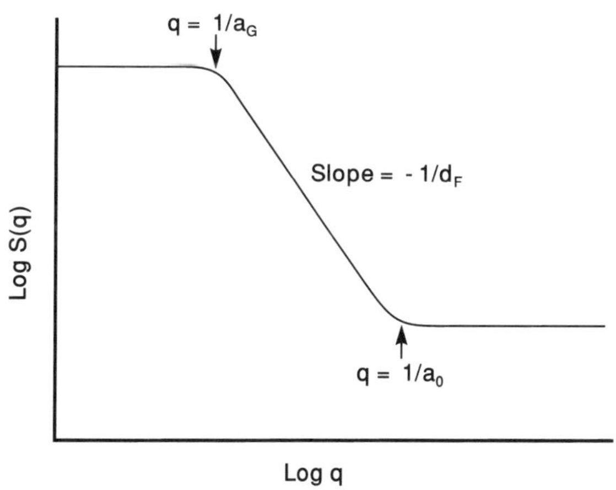

Figure 9.4 *Form of log–log plot of normalized light-scattering intensity, or the structure factor* S(q), *against the scattering vector* q, *for an aggregated suspension. When* 1/q *is greater than the radius of gyration of the aggregates,* a_G, *or less than the primary particle radius,* a_0, S(q) *is independent of* q. *For intermediate* q *values, the slope of the line is* $-1/d_F$

latex suspensions. This procedure should give more reliable values than those from measurements only at very low angles.

For a suspension containing aggregates, a log–log plot of normalized light-scattering intensity (or, effectively, the structure factor) against the scattering vector q can give information on the structure of the aggregates. Such a plot is shown schematically in Figure 9.4. The main features are constant scattering intensities at low and high scattering vectors with a linear decrease in the intermediate region. This behaviour arises because q represents a characteristic (inverse) length scale probed by the scattering experiment. The summand in eqn (9.16) is most sensitive to values of q such that $qr_{ij} \approx 1$.

At low q values (low scattering angles), $1/q$ is much greater than the aggregate size and $qr_{ij} \ll 1$ for all particle pairs in the aggregate, so that $\sin(qr_{ij})/qr_{ij} \approx 1$ for all particle pairs and the sum becomes simply k^2. So, for an aggregating suspension the total light scattered increases as the weight-averaged aggregate size. This is just the behaviour found in the low angle limit, mentioned above. The light scattering is insensitive to the structure of an aggregate and depends only on the total volume of the constituent particles.

When q is very high, such that $qa \gg 1$, the length scale probed is much less than the size of the primary particles and the only significant contribution to the sum in eqn (9.16) comes from terms where $i=j$ and hence $r_{ij}=0$. This gives $S(\theta)=k$ for a k-fold aggregate, so that the scattering is just the same as that from k isolated particles. Under these conditions aggregation of a suspension would give no change in scattered light.

In the intermediate region of q it can be shown (e.g. Schmidt, 1989) that the slope of the linear region in Figure 9.4 is $-d_F$, where d_F is the mass fractal dimension given by eqn (6.41). This result is obtained from an integral form of eqn (9.16), in terms of the radial distribution function of particles in an aggregate. No further details will be given here.

The range of q values accessible by scattering of visible light, bearing in mind that measurable scattering angles are typically 10–160°, is of the order of 2–30 μm^{-1}, so that only rather small aggregates (<1 μm) can be investigated. Nevertheless, this approach can give very useful information on fractal aggregates formed from primary particles with sizes less than, say, 0.1 μm. Chen and Russel (1991) used this method for studying the flocculation of silica dispersions and Hoekstra *et al.* (1992) were able to investigate fractal aggregates of nickel hydroxycarbonate in a similar manner.

By using radiation with shorter wavelengths (such as X-rays or neutrons), aggregate structure can be probed over shorter length scales. In this way information on, for instance, interparticle distances in flocs can be derived (Wong *et al.*, 1988).

The Rayleigh–Gans–Debye approximation cannot be applied to the kind of aggregates formed in many practical cases, which may range in size up to 1 mm or more. For studies of large aggregates, other approaches have to be used.

Fraunhofer diffraction

For particle sizes very much greater than the light wavelength, scattering can be treated as a problem in geometrical optics. A large spherical particle in a light beam can be treated as a circular disc, with the same diameter. At the edge of the disc, light is diffracted and gives a characteristic pattern of light and dark rings at a plane far from the particle. These rings represent maxima and minima in the intensity of the diffracted light and their positions are dependent only on the light wavelength and the particle diameter, not on the particle properties. Although diffraction by particles has been known since the early nineteenth century and was used to study the size of blood cells in 1918 (see Azzopardi, 1992), it only achieved widespread use from about 1980.

The relative diffracted light intensity, as a function of angle, is given by:

$$I(\theta) = (\text{constant}) \bullet \frac{\alpha^2 J_1^2 (\alpha \sin \theta)}{\sin^2 \theta} \tag{9.19}$$

where α is the dimensionless size parameter given in eqn (9.5) and J_1 is a first-order Bessel function.

Zero values of the Bessel function (i.e. dark bands in the diffraction pattern) occur when the argument ($\alpha \sin \theta$) takes values of 3.83, 7.02, 10.17 etc. For particles larger than about 10 μm and for optical wavelengths these bands occur at quite low angles. Nevertheless, using laser illumination, and good-quality optics, including a Fourier transform lens and an array of concentric detectors, it is possible to derive detailed information (e.g. Azzopardi, 1992). Several commercial instruments using diffraction methods are available and these find very wide application for routine particle size analysis. It is now recognized that simple Fraunhofer theory becomes unsuitable for particles not much larger than the light wavelength and the commercial software usually includes computations based on Mie theory. This means that information on the nature of the particles (such as refractive index) is needed.

For heterodisperse suspensions it is necessary to invert the diffraction data to give a particle size distribution. The only feasible way of doing this is to assume a form of the size distribution, such as the log-normal or Roslin–Rammler forms, and to derive the appropriate parameters by an iterative procedure (Zhang and Xu, 1992). Using such methods, available instruments are claimed to give particle size over a very wide range (typically 1–1000 μm).

Although diffraction methods have been used for aggregation studies, such as the flocculation of bacteria (Whittington and George, 1992), there is little information on diffraction by large aggregates, which can have very low density. The self-similar, fractal nature of aggregates means that the fraction of aggregate volume occupied by particles can be very low (less than 1% for large structures). An obvious consequence is that the effective refractive index could be very low and treating the aggregate as a 'solid' opaque particle may not be appropriate. A comparison between aggregate sizes from diffraction measurements with those from a more direct method, such as microscopy, would be worthwhile.

9.3.3 Dynamic light scattering

The basis of dynamic light scattering is that light scattered from a moving particle will have a slightly different frequency from the incident light, just as in the well-known *Doppler effect*. In colloidal dispersions, random Brownian motion of the particles causes scattered light to vary randomly in frequency, and interference between light scattered from different particles causes random fluctuations in the light intensity measured by a stationary detector. A characteristic 'speckle pattern' is visible when a suspension of Brownian particles is illuminated by a laser beam. Analysis of this effect involves autocorrelation of the scattered light intensity, measured as a train of pulses from a photomultiplier. The nature of the technique is such that it may be called *photon correlation spectroscopy* (*PCS*) or *quasi-elastic light scattering* (*QELS*) as well as dynamic light scattering. A review of the method, and its development since 1964, has been given by Weiner (1992). Practical implementation was made possible when coherent laser light and dedicated correlators became available. Commercial instruments were introduced in the early 1980s and are widely used for sizing of particles in the sub-micrometre range. (Because the effect is dependent on Brownian motion, a practical upper limit of particle size is about 3 μm.)

From the autocorrelation function of the photon pulse train, it is possible to derive a function $g(\Delta t)$ where Δt is the correlation delay time. For a monodisperse suspension of colloidal particles, it can be shown (e.g. Pusey, 1981) that:

$$g(\Delta t) = \exp\left(-\frac{\Delta t}{t_c}\right) \tag{9.20}$$

where t_C is the relaxation time of the fluctuations, the characteristic time of the speckle pattern at the detector. Its inverse, $1/t_C$, can be regarded as a linewidth representing the Doppler broadening of the frequency of the scattered light.

There is a very simple relation between t_C and the particle diffusion coefficient D:

$$t_c = \frac{1}{Dq^2} \tag{9.21}$$

where q is the scattering vector given in eqn (9.12).

Since the diffusion coefficient of spherical particles is related to their size by the Stokes–Einstein relation, eqn (6.3), dynamic light scattering provides a measure of particle size, at least for monodisperse, spherical particles.

For polydisperse samples, an intensity-weighted sum of exponential terms has to be used instead of eqn (9.20). Inversion of this expression to give a particle size distribution is an ill-conditioned Laplace transform problem, which is subject to considerable uncertainty. Commercial instruments usually give a mean size, together with a 'polydispersity index', which gives an indication of the breadth of the size distribution.

While dynamic light scattering is undoubtedly an extremely convenient means of estimating the size of colloidal particles, its use in aggregation studies is limited by the restriction to small sizes, where Brownian motion is appreciable. Nevertheless, for coagulation of sub-micrometre particles, at least in the early stages of the process, the method could be a useful alternative to static light scattering, since the assumptions involved are not as restrictive as the RGD approximation.

Rates of rapid coagulation of latex suspensions have been studied by Bolle *et al.* (1987) under conditions of high and low salt concentration, using dynamic light scattering. Their results were interpreted in terms of fractal dimensions of the aggregates and gave results as expected for diffusion-limited (high salt) and reaction-limited (low salt) coagulation (see Section 6.4.1). Herrington and Midmore (1989), also using a PCS technique, investigated the rapid aggregation of latex suspensions. They avoided the inversion (deconvolution) problem by assuming Smoluchowski kinetics and generating a theoretical autocorrelation function. Fitting this to the experimental data gave a measure of the degree of coagulation at various times and hence an estimate of the coagulation rate constant. Rate constants derived in this way, using latex samples of different particle size over a range of concentrations, were quite consistent and in agreement with values determined by other means.

Dynamic light scattering is not restricted to uniform spherical particles. Aggregation of particles such as silica (Zhou, *et al.*, 1991), nickel hydroxy-carbonate (Hoekstra *et al.*, 1992) and clay (Herrington and Midmore, 1993) has been studied by the same technique.

9.4 Other optical methods

9.4.1 Electro-optic techniques

Particles which are charged and have some degree of asymmetry may be orientated by an applied electric field. Because light scattering by asymmetrical particles may depend on their alignment relative to the incident light direction, changes in particle orientation can give measurable changes in scattered or transmitted light. The *electro-optic effect* (Stoylov, 1991) can be quantified by a parameter α_E, which is defined as the relative change in light scattering intensity (under given conditions) as a result of applying an electric field:

$$\alpha_E = \frac{I_E - I}{I} \tag{9.22}$$

where I_E and I are the intensities in the presence and absence of the electric field.

Rather than deriving information from the absolute value of α_E, a more common approach is to observe the change in this parameter after the field is switched on or off. For instance, when the field is removed particles adopt random orientations as a result of *rotary Brownian motion* and the time of this relaxation process is very dependent on particle size. The *rotary diffusion coefficient*, D_r, of a spherical particle is inversely proportional to the *cube* of particle size, and D_r can be derived from the initial slope of the α_E versus time plot when the field has been removed. The rate of decay decreases dramatically as particle size is increased and this could, in principle, be used as a measure of aggregation.

There are complications in using electro-optic techniques for aggregation studies, the most important of which is that aggregation not only gives an increase in size (and hence a decrease in the rotary diffusion coefficient) but may also cause a significant change in asymmetry, leading to a different electro-optic effect. In some cases, particles which are initially asymmetrical, may acquire a greater degree of symmetry as a result of aggregation. An important example comprises plate-like particles of a clay such as kaolinite. Aggregates of these platelets approach spherical symmetry and give a much lower α_E value than the original platelets. This effect has been used successfully to monitor the aggregation of clay particles (Trimm *et al.*, 1984).

By contrast, nearly spherical particles, such as latex, show virtually no electro-optic effect in the unaggregated state, but small aggregates are highly asymmetrical and give a much higher α_E. Nevertheless, the relative rate of decay of the effect can still give a very sensitive indication of aggregate size and this technique has been used in a number of studies (e.g. Starchev *et al.*, 1993).

9.4.2 Turbidity fluctuations

When a flowing suspension is illuminated by a narrow light beam, the transmitted light intensity is found to vary randomly about a mean value (Figure 9.5). The variation is due to random fluctuations in the number of particles in the probed volume, giving corresponding changes in the total light-scattering cross-section (see Section 9.3.1) and hence in the transmitted light.

Number fluctuations are found in a given volume of any suspension, e.g. as a result of Brownian diffusion into and out of the defined volume. Such

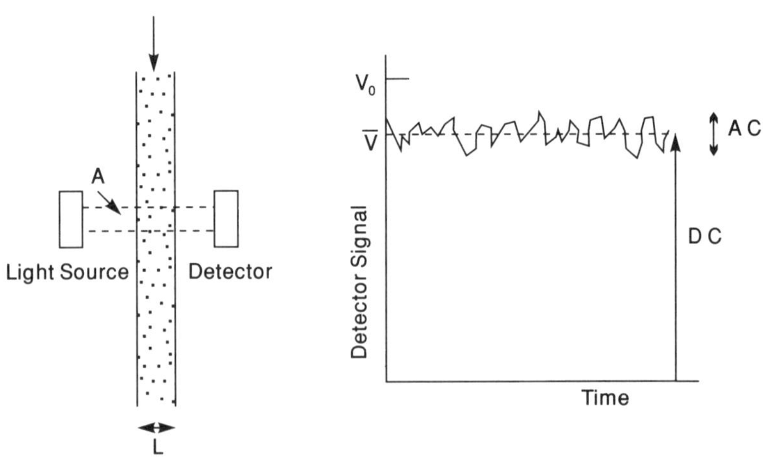

Figure 9.5 *Schematic illustration of the turbidity fluctuation technique. Light transmitted through a flowing suspension shows random variations in intensity as a result of fluctuations in the number of particles in the illuminated volume*

fluctuations follow the Poisson distribution, so that the variance about the mean is equal to the mean number in the sample volume. A suspension of large, non-Brownian particles also shows Poisson-distributed number fluctuations, as a result of random distribution during mixing. The light beam illuminating a flowing suspension (Figure 9.5) effectively probes a sequence of small samples in rapid succession, giving a transmitted light intensity which fluctuates randomly at a frequency which depends on the flow rate.

This effect was used by Gregory and Nelson (1984, 1986) for monitoring flocculation processes and the theoretical basis was given by Gregory (1985). The procedure is to derive the root mean square (RMS) value of the fluctuating (AC) signal from a photodiode and to divide this by the mean (DC) value to give a *ratio* value $R = \mathrm{RMS/DC}$. In practice, R may vary from less than 10^{-4} to 0.1 (i.e. for a 10-V DC value, the RMS value of the fluctuating signal may vary from 1 mV to 1 V). For a monodisperse suspension of particles with light-scattering cross-section C and number concentration n, it can be shown that:

$$R = \frac{V_{\mathrm{RMS}}}{\overline{V}} = \sqrt{\frac{nL}{A}}\, C \tag{9.23}$$

where L is the optical path length and A is the cross-sectional area of the light beam (i.e. the illuminated sample volume is LA). The dependence of R on the square root of the particle concentration is a consequence of the Poisson distribution of number fluctuations.

If the concentration is expressed as a volume fraction and the scattering coefficient, eqn (9.3) is introduced, we obtain:

$$\frac{R}{\sqrt{\phi}} = \sqrt{\frac{3\pi L a_0}{4A}}\, Q_0 \tag{9.24}$$

where the radius and scattering coefficient of the primary particles are designated a_0 and Q_0, to distinguish them from the corresponding values for aggregates (see below).

This expression shows that, for constant values of scattering coefficient and volume fraction, R should increase as the square root of particle size. For similar conditions, the turbidity, given by eqn (9.8), *decreases* as the particle size increases.

In an aggregating suspension, it is convenient to express aggregate size in terms of the aggregation number, k (see Section 6.6.1). For a fractal k-fold aggregate, the aggregate radius, a_k, can be written in terms of the primary particle radius and the appropriate fractal dimension, d_{F}:

$$a_k = a_0 k^{1/d_{\mathrm{F}}} \tag{9.25}$$

The corresponding scattering cross-section is then:

$$C_k = \pi a_0^2 k^{2/d_{\mathrm{F}}} Q_k \tag{9.26}$$

where Q_k is the scattering coefficient for the aggregate.

For a distribution of aggregate sizes $f(k)$ and a mean aggregate size $\bar{k}(=n_0/n_T$, eqn (6.51), it can be shown (Gregory and Chung, 1994):

$$\frac{R^2}{\phi} = \left(\frac{3\pi L a_0}{4A}\right)\frac{1}{\bar{k}}\int_0^\infty k^{4/d_F}\, Q_k^2\, f(k)\, dk \tag{9.27}$$

The corresponding expression for specific turbidity is:

$$\frac{\tau}{\phi} = \frac{3}{4\bar{k}a_0}\int_0^\infty k^{2/d_F}\, Q_k\, f(k)\, dk \tag{9.28}$$

To proceed further we need to consider the size distribution and scattering coefficients of aggregates. It was shown in Section 6.6 that aggregate size distributions may adopt a 'self-preserving' form and that a very simple exponential form, eqn (6.52), may be appropriate in some cases. In terms of aggregation number this can be written:

$$f(k) = \frac{1}{\bar{k}}\exp\left(-\frac{k}{\bar{k}}\right) \tag{9.29}$$

For fairly large aggregates, it may be acceptable to treat the scattering coefficient as constant, although this assumption may not hold for aggregates of rather open structure (low d_F). With the assumptions of an exponential size distribution and a constant Q_k, the integrals in eqns (9.28) and (9.29) have closed form solutions, giving:

$$\frac{R^2}{\phi} = \frac{3\pi L a_0 Q_k^2}{4A}(\bar{k})^{2b-1}\,\Gamma(2b+1) \tag{9.30}$$

and:

$$\frac{\tau}{\phi} = \frac{3Q_k}{4a_0}(\bar{k})^{b-1}\,\Gamma(b+1) \tag{9.31}$$

where $b = 2/d_F$ and Γ is the gamma function.

These two expressions may be combined to give:

$$\left(\frac{R}{\tau}\right)^2 = \frac{4\pi L a_0^3}{3A\phi}\bullet\frac{\Gamma(2b+1)}{[\Gamma(b+1)]^2}\bullet\bar{k} \tag{9.32}$$

If the suspension concentration is fixed and the scattering coefficient and fractal dimension of the aggregates can also be assumed constant, then all of the terms on the right-hand side of eqn (9.32) are constant apart from the mean aggregation number, $\bar{k}$, which would then be proportional to the term $(R/\tau)^2$. Since both R and τ are directly measurable, the simple set-up shown in Figure 9.5 provides a convenient means of estimating the relative change in aggregation number. It is noteworthy that the scattering coefficient has been cancelled out in deriving eqn (9.32), which implies that the assumption of constant Q_k may not be a serious limitation. A similar cancellation of Q_k occurs if a log-normal form of the aggregate size distribution is used, and a result not very different from eqn (9.32) is obtained.

Another important conclusion follows from a rearrangement of eqn (9.30):

$$R = (\text{Constant}) \bullet (\bar{k})^{(4-d_F)/2d_F} \tag{9.33}$$

Since, in a three-dimensional world, the fractal dimension cannot exceed 3, the exponent of $\bar{k}$ in eqn (9.33) must always be positive, so that R will always increase as aggregation proceeds. When $d_F = 2$, R increases as the square root of aggregation number. For $d_F = 3$, R increases only as $\bar{k}^{1/6}$. These cases correspond to those of 'extended aggregates' and 'coalescence' discussed by Gregory and Nelson (1984), who found the same dependence on aggregation number by a simpler method, assuming aggregates of uniform size.

Although quantitative conclusions can be drawn from turbidity fluctuation measurements, the main use of the technique has been as an empirical means of monitoring flocculation processes. It has been shown experimentally that R increases markedly as aggregation occurs, for a wide variety of suspensions, and this is borne out by the above analysis. There are several reports of flocculation studies using this method (e.g. Whittington and George, 1992; Alince and van de Ven, 1993; Ching *et al.*, 1994).

9.5 Aggregate properties

In practice, aggregation of suspensions is usually carried out in order to enhance solid–liquid separation processes, such as sedimentation, flotation, thickening and dewatering. An aggregate of fine particles effectively exposes less particle surface to the suspending fluid than do the original particles and hence there is less fluid drag. This results in enhanced sedimentation rates and greater permeability in, say, cake filtration of aggregated suspensions. Fundamental analysis of sedimentation rates and permeability in terms of aggregate properties is beyond the scope of this chapter. A very important aspect is the hydrodynamic behaviour of fractal aggregates (Chellam and Wiesner, 1993).

From an empirical standpoint, sedimentation and permeability methods have been widely used in flocculation studies, partly because of their relative simplicity, but also since they can be related to practical separation processes. We shall briefly discuss some of the experimental approaches.

9.5.1 Sedimentation methods

The sedimentation behaviour of aggregated suspensions depends very much on concentration. In dilute suspensions particles and aggregates settle independently and their motion is unaffected by the motion of other particles. In more concentrated suspensions, sedimentation is influenced by other particles and *hindered settling* (or *zone settling*) is observed. In hindered settling, the motion of a particle is affected by hydrodynamic interaction with other moving particles (Han and Lawler, 1991). Small particles tend to be pulled down by larger ones, whereas the sedimentation of large particles can be retarded by smaller ones. Another effect is that, in a closed system, sedimentation of particles causes an upward displacement of fluid, which gives an overall reduction in settling rate. The net effect is that sedimentation is slower in concentrated suspensions than for the corresponding isolated particles – hence the term 'hindered settling'. A natural consequence is that a fairly sharp interface develops between the sedimenting particles and a clear particle-free layer. Particles above this boundary, being in a region of lower concentration, tend to settle more rapidly and hence 'catch up with' their slower companions in the more concentrated region. This is the reason why hindered settling may also be called 'zone settling'.

The transition from free sedimentation to hindered settling occurs at a concentration which depends on the degree of aggregation. For single particles, free settling can occur at higher concentrations than for aggregates, since the latter effectively occupy more volume than single particles and have a greater influence on their neighbours. The conventional way of illustrating these effects is by a diagram like that in Figure 9.6, in which regions are mapped according to both particle (solids) concentration and degree of aggregation. Free sedimentation is seen to occur in regions of low solids concentration and/or low degrees of aggregation. Zone settling may be found at quite low solids concentrations if the particles are aggregated to a large extent. For high concentrations of aggregated particles, compression becomes significant, and this is typically observed at the bottom of a sedimentation vessel, as a result of the weight of overlying solids.

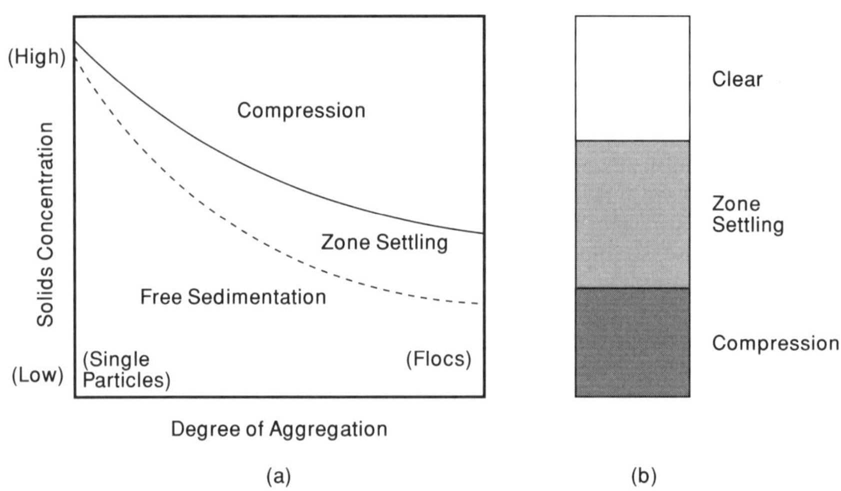

Figure 9.6 (a) *The different forms of sedimentation that may occur, depending on solids concentration and degree of aggregation.* (b) *Intermediate stage in the settling of a flocculated suspension. The upper region is essentially free of particles. Below this is a region of hindered sedimentation or zone settling. At the bottom, settled particles (flocs) are subject to compression*

Sedimentation of an isolated particle can be treated in fairly simple terms, by Stokes' law for instance, and, for solid particles, an effective diameter (the Stokes equivalent diameter) can be derived if the density is known. This is the basis of very well-established particle-sizing methods, which are still widely used (Allen, 1992). All of these methods involve following the particle concentration in a sedimenting suspension as a function of time. This may be achieved by gravimetric means, monitoring by light (photosedimentation) or X-ray scattering, or, more recently, by capacitance techniques (Williams *et al.*, 1991).

For aggregates, sedimentation rate cannot give their size unless the aggregate density is known. The fractal nature of most aggregates means that there is a very significant decrease of buoyant density with increasing size (Section 6.4.3). Sedimentation rates, together with aggregate size determination (usually by direct microscopic observation), have been used to determine floc density in a number of studies (e.g. Farrow and Warren, 1993).

For concentrated suspensions, where zone settling occurs, sedimentation rate can be readily observed by following the interface between the sedimenting solids and clear supernatant. The rate of fall of this sharp demarcation line can be measured using only a stopwatch and this method is still widely used, especially in empirical studies of flocculation of slurries and sludges. Initially the rate of fall is constant and provides a useful index of the effectiveness of a particular flocculant. Eventually the rate declines as the line approaches the height of the settled solids. In many cases, especially with flocculated suspensions, there follows a fairly slow *compression* or *consolidation* phase until the sediment reaches an equilibrium height. The distinction between zone settling and consolidation is that, in the latter case, the settled solids have formed an interconnected network which has a certain compressibility. Quantitative treatments of sedimentation and consolidation of aggregated suspensions are available (e.g. Buscall, 1990), but this aspect will not be considered here. Quite detailed information can be obtained by measuring solids concentration profiles in sedimenting suspensions. Such measurements can be made in various ways, including ultrasonic (Howe and Robbins, 1990), capacitance (Williams *et al.*, 1991) or gamma ray (Bergström, 1992) techniques.

The volume of sediment from an aggregated suspension can provide a sensitive indication of the state of aggregation – the greater the degree of aggregation, the larger the sediment volume. This is mostly used in an entirely empirical manner (e.g. Dollimore and Horridge, 1974), but Zrinyi *et al.* (1988) adopted a fractal approach to sediment volume. They showed that the density of a sediment composed of fractal aggregates will be a decreasing function of aggregate size. Experimental measurements of sediment volume for various systems were in qualitative agreement with their predictions.

9.5.2 Permeability methods

The flow of fluid through a packed bed of particles depends greatly on the size of the particles. The permeability of a packed bed, K, is defined in terms of the pressure difference across the bed, ΔP, the bed depth, l, and the velocity, v, and viscosity, μ, of the fluid:

$$v = \frac{K\Delta P}{\mu l} \tag{9.34}$$

The permeability depends on the structure of the packed bed and is often given in the form of the *Carman–Kozeny* equation:

$$K = \frac{\epsilon^3}{k_c S^2 (1 - \epsilon)^2} \tag{9.35}$$

where ϵ is the porosity of the bed, S is the *specific surface* of the particles (surface area per unit volume) and k_C is a constant, which is found to be about 5 in many practical cases.

Equation (9.35) shows that the permeability will increase as the porosity increases and also as the specific surface decreases (because of the reduced drag). For spherical particles, radius a, $S = 3/a$, and the specific surface decreases with increasing particle size. The same trend is found for irregular particles.

When a suspension of particles is filtered (e.g. through a membrane with pores smaller than the particles) a *filter cake* is built up which gives an increasing resistance to filtration. For very small particles the permeability is low and the filtration may be very slow. It is a common observation that aggregation of particles can greatly increase permeability and hence filtration rates. This can be qualitatively understood as being a result of the increased aggregate size, but quantitative analysis is very difficult because of the fractal character and intrinsic permeability of aggregates (Chellam and Wiesner, 1993), as well as their compressibility. Nevertheless, filterability has been commonly employed as a sensitive measure of the degree of aggregation of suspensions.

Filterability methods are widely used to evaluate polymeric flocculants in an empirical manner, not least because such additives are often intended to enhance the dewatering of suspensions and filtration rate is directly relevant to such operations. One of the earliest methods is the *refiltration rate* technique of La Mer (1966), in which a flocculated suspension is filtered through a suitable medium and then the filtrate is passed once more through the filter cake under constant pressure. The time for a certain volume to flow through the filter cake is measured and this gives the refiltration rate. This can give a very sensitive indication of 'optimum' flocculation conditions, although there are problems associated with the disproportionate effect of small amounts of unflocculated 'fines' (Slater and Kitchener, 1966).

A convenient measure of filterability of concentrated suspensions is the *capillary suction time* (CST) (Gale and Baskerville, 1967). The CST device measures the time for a certain volume of liquid to be drawn from the suspension by capillary action using a standard filter paper. The time is determined automatically as the advancing liquid front passes between electrical contacts. Although the CST method is used almost entirely for empirical studies, there has been some theoretical analysis (Tiller *et al.*, 1990).

Bibliography

Dobiáš, B. (ed) (1993) *Coagulation and Flocculation, Theory and Applications*, Marcel Dekker, New York
Kerker, M. (1969) *The Scattering of Light and Other Electromagnetic Radiation*, Academic Press, New York
Stanley-Wood, N. G and Lines, R. W. (eds) (1992) *Particle Size Analysis*, Royal Society of Chemistry, Cambridge
van de Hulst, H. C. (1981) *Light Scattering by Small Particles*, Dover Publications Inc., New York
Williams, R. A. and de Jaeger, N. C. (eds) (1991) *Advances in Measurement and Control of Colloidal Processes*, Butterworth-Heinemann, Oxford

References

Alince, B. and van de Ven, T. G. M. (1993) Stability of clay suspensions – effect of pH and polyethylenimine. *J. Colloid Interface Sci.*, **155**, 465–470
Allen, T. (1992) A review of sedimentation methods for particle size analysis. In *Particle Size Analysis*, (ed. Stanley-Wood, N. G. and Lines, R. W.) Royal Society of Chemistry, Cambridge, 454–476
Azzopardi, B. J. (1992) Instrumentation for particle size analysis by far field diffraction: accuracy, limitations and future. In *Particle Size Analysis*. (ed. Stanley-Wood, N. G. and Lines, R. W.) Royal Society of Chemistry, Cambridge, 108–132
Bergström, L. (1992) Sedimentation of flocculated alumina suspensions: γ-ray measurements and comparisons with model predictions. *J. Chem. Soc. Faraday Trans*, **88**, 3201–3211
Bolle, G., Cametti, C., Codastefano, P. and Tartaglia, P. (1987) Kinetics of salt-induced aggregation in polystyrene latices studied by quasi-elastic light scattering. *Phys. Rev. A*, **35**, 837–841
Buscall, R. (1990) The sedimentation of concentrated colloidal suspensions. *Colloids Surfaces*, **43**, 33–53
Cahill, J., Cummins, P. G., Staples, E. J. and Thompson, L. (1987) Size distribution of aggregates in flocculating dispersions. *J. Colloid Interface Sci.*, **117**, 406–414
Chellam, S. and Wiesner, M. R. (1993) Fluid mechanics and fractal aggregates. *Water Res.*, **27**, 1493–1496
Chen, M. and Russel, W. B. (1991) Characteristics of flocculated silica dispersions. *J. Colloid Interface Sci.*, **141**, 564–577
Ching, H-W., Elimelech, M. and Hering, J. (1994) Dynamics of coagulation of clay particles with aluminum sulfate. *J. Env. Eng. ASCE*, **120**, 169–189
Coulter, W. H. (1956) High speed automatic blood cell counter and cell size analyzer. *Proc. Natl Electron. Conf.*, **12**, 1034
Dollimore, D. and Horridge, T. A. (1974) The dependence of the flocculation behaviour of china clay–polyacrylamide suspensions on the suspension pH. *J.Colloid Interface Sci.*, **42**, 581–588
Farrow, J. B. and Warren, L. J. (1993) Measurement of the size of aggregates in suspension. In *Coagulation and Flocculation, Theory and Applications*. (ed. Dobiáš, B.) Marcel Dekker, New York, 391–426
Freundlich, H. (1926) *Colloid and Capillary Chemistry*, Methuen, London
Gale, R. S. and Baskerville, R. C. (1967) Capillary suction time method for determination of the filtration properties of a solid–liquid suspension. *Chem. Ind.*, 355–356
Gibbs, R. J. (1982) Floc breakage during HIAC light-blocking analysis. *Env. Sci. Technol.*, **16**, 298–299

Gregory, J. (1985) Turbidity fluctuations in flowing suspensions. *J. Colloid Interface Sci.*, **105**, 357–371

Gregory, J. and Chung, H-J. (1994) On-line measurements of floc properties by an improved turbidity fluctuation technique. In *Optimal Dosing of Coagulants and Flocculants*, Proc. IWSA-IAWQ Workshop, Mülheim, 12–13 January 1994, IWW, Mülheim, 101–115

Gregory, J. and Nelson, D. W. (1984) A new optical method for flocculation monitoring. In *Solid–Liquid Separation*, (ed. Gregory, J.) Ellis Horwood, Chichester, 172–184

Gregory, J. and Nelson, D. W. (1986) Monitoring of aggregates in flowing suspensions. *Colloids Surfaces*, **18**, 175–186

Han, M. and Lawler, D. F. (1991) Interactions of two settling spheres: settling rates and collision efficiency. *J. Hydraulic Eng. ASCE*, **117**, 1269–1289

Herrington, T. M. and Midmore, B. R. (1989) Determination of rate constants for the rapid coagulation of polystyrene microspheres using photon correlation spectroscopy. *J. Chem. Soc. Faraday Trans I*, **85**, 3529–3536

Herrington, T. M. and Midmore, B. R. (1993) Investigation of scaling effects in the aggregation of dilute kaolinite suspensions by quasi-elastic light scattering. *Colloids Surfaces A*, **70**, 199–202

Hoekstra, L. L., Vreeker, R. and Agterhof, W. G. M. (1992) Aggregation of colloidal nickel hydroxycarbonate studied by light scattering. *J. Colloid Interface Sci.*, **151**, 17–25

Howe, A. M. and Robbins, M. M. (1990) Determination of gravitational separation in dispersions from concentration profiles. *Colloids Surfaces*, **43**, 83–94

Kerker, M. (1969) *The Scattering of Light and Other Electromagnetic Radiation*, Academic Press, New York

La Mer, V. K. (1966) Filtration of colloidal dispersions flocculated by anionic and cationic polyelectrolytes. *Disc. Faraday Soc.*, **42**, 248–254

Matthews, B. A. and Rhodes, C. T. (1970) Some observations on the use of the Coulter Counter Model B in coagulation studies. *J. Colloid Interface Sci.*, **32**, 339–348

Miller, B. V. and Lines, R. W. (1988). Recent advances in particle size measurements: a critical review. *CRC Crit. Rev. Anal. Chem.*, **20**, 75–116

Mysels, K. J. (1959) *Introduction to Colloid Chemistry*, Interscience, New York

Pefferkorn, E., Pichot, C. and Varoqui, R. (1988) Size distribution of latex aggregates in flocculating dispersions. *J. Phys. France*, **49**, 983–989

Pelssers, E. (1988) Single particle optical sizing. Aggregation of polystyrene latices by salt and polymer. Thesis, Agricultural University, Wageningen

Pluta, M. (1989) *Advanced Light Microscopy*, Vol. 2, Chapter 12, Elsevier, Amsterdam

Pusey, P. N. (1981) Light scattering. In *Colloidal Dispersions*, (ed. Goodwin, J. W.) Royal Society of Chemistry, London, 129–142

Reerink, H. and Overbeek, J. Th. G. (1954). The rate of coagulation as a measure of the stability of silver iodide sols. *Disc. Faraday Soc.*, **18**, 74–84

Schmidt, P. W. (1989) Use of scattering to determine the fractal dimension. In *The Fractal Approach to Heterogeneous Chemistry* (ed. Avrir, D.) John Wiley, New York, 67–79

Slater, R. W. and Kitchener, J. A. (1966) Characteristics of flocculation of mineral suspensions by polymers. *Disc. Faraday Soc.*, **42**, 267–275

Sonntag, H. (1993) Coagulation kinetics. In *Coagulation and Flocculation, Theory and Applications.* (ed. Dobiáš, B.) Marcel Dekker, New York, 57–99

Sonntag, H., Schilov, V., Gedan, H., Lichtenfeld, H. and Dürr, C. (1986) Slow coagulation of polystyrene dispersions investigated by single particle light scattering. *Colloids Surfaces*, **20**, 303–317

Starchev, K., Peikov, V., Stoylov, S. P., Petkanchin, I. B., Streb, K. D. and Sonntag, H. (1993) Kinetics of reaction limited cluster aggregation studied by electric light scattering. *Colloids Surfaces*, **76**, 95–100

Stewart, R. F. and Sutton, D. (1986) Characterization of the structure of concentrated flocculated suspensions. *Part. Sci. Technol.*, **4**, 251–264

Stoylov, S. P. (1991) *Colloid Electrooptics*, Academic Press, London

Tiller, F. M., Shen, Y. L. and Adin, A. (1990) Capillary suction theory for rectangular cells. *Res. J. Water Poll. Control Fed.*, **62**, 130–136

Trimm, H. H., Parslow, K. and Jennings, B. R. (1984) Electric birefringence. A simple apparatus for determining physical parameters of macromolecules and colloids. *J. Chem. Educ.*, **61**, 1114–1118

van de Hulst, H. C. (1957) *Light Scattering by Small Particles*, Wiley, New York (Republished by Dover Publications Inc., New York, 1981)

van Zanten, J. H. and Elimelech, M. (1992) Determination of absolute coagulation rate constants by multi-angle light scattering. *J. Colloid Interface Sci.*, **154**, 1–7

Weiner, B. B. (1992) Twenty seven years of QELS: a review of the advantages and disadvantages of particle sizing with QELS. In *Particle Size Analysis* (eds. Stanley-Wood, N. G. and Lines, R. W.) Royal Society of Chemistry, Cambridge, 173–185

Whittington, P. N. and George, N. (1992) The use of laminar tube flow in the study of hydrodynamic and chemical influences on polymer flocculation of *Escherischia coli*. *Biotechnol. Bioeng.*, **40**, 451–458

Williams, R. A., Amarasinghe, W. B. K., Simons, S. J. R. and Xie, C. G. (1991) Sedimentation behaviour of complex polydisperse systems. *Powder Technol.*, **65**, 411–432

Wong, K., Cabane, B. and Duplessix, R. (1988) Interparticle distances in flocs. *J. Colloid Interface Sci.*, **123**, 466–481

Young, W. D. and Prieve, D. C. (1991) An improved light-scattering technique for measuring the flocculation rate of colloids. *Langmuir*, **7**, 2887–2892

Zeichner, G. R. and Schowalter, W. R. (1979) Effects of hydrodynamic and colloid forces on the coagulation of dispersions. *J. Colloid Interface Sci.*, **71**, 237–253

Zhang, H-J. and Xu, G-D. (1992) The effect of particle refractive index on size measurement. *Powder Technol.*, **70**, 189–192

Zhou, Z., Wu, P. and Chu, B. (1991) Cationic surfactant induced fractal silica aggregates: a light scattering study. *J. Colloid Interface Sci.*, **146**, 541–555

Zollars, R. L. (1980) Turbidimetric method for on-line determination of particle number and particle size distribution. *J. Colloid Interface Sci.*, **74**, 163–172

Zrınyi, M., Kabai-Faix, M. and Horkay, F. (1988) On the sediment volume of colloidal aggregates 1. A fractal approach to the problem. *Prog. Colloid Polymer Sci.*, **77**, 165–170

Chapter 10

Experimental techniques in particle deposition kinetics

Nomenclature

A_c	cross-section area
a_c	radius of a spherical collector
a_p	particle radius
C_s	scattering cross-section
C	particle concentration
C_0	influent (bulk) particle concentration
C_e	effluent concentration of particles
f	porosity of a porous medium
I	light intensity
I_0	intensity of incident light
L	bed depth
l	path length of light in a suspension
N	number of particles per unit volume
N_c	number of collectors in a differential volume element defined by eqn (10.5)
O_D	optical density
Q	volumetric flow rate
r_p	dimensionless interparticle spacing defined by eqn (10.1)
U	approach (superficial) velocity of fluid
z	axial coordinate
α_{exp}	experimental collision efficiency
η	actual single collector efficiency
η_0	theoretical single collector efficiency for favourable condition (i.e. with no double layer interaction)
η_{exp}	experimental single collector efficiency
κ	reciprocal Debye length
τ	turbidity

This chapter describes experimental techniques for measuring particle deposition rates from flowing suspensions in a variety of deposition systems. In addition to their practical application in technological and natural processes, these methods can be used to test fundamental theories of colloid deposition and interaction. The collector geometries that will be considered are classified as planar and curved collector surfaces. The most common planar surfaces in particle deposition studies include parallel-plate channel, rotating disc and stagnation-point flow cell. Collectors with curved surfaces include isolated spherical collectors, isolated cylindrical collectors and packed-bed columns of these collectors.

Experimental methods for studying particle deposition onto collectors are divided into direct and indirect methods. In direct methods, the number of particles deposited on the collector surface as a function of time is counted directly by microscopic observation. Direct methods are very useful for studying the dynamics of particle deposition. They can also provide information about the distribution of deposited particles and the shape and structure of retained particles. Indirect methods, on the other hand, are based on the measurement of bulk concentration changes before and after contact with the collector surface (as in the packed-bed technique). The amount of deposited particles and particle deposition rates can be calculated from the change in particle concentration.

10.1 System requirements

Particle deposition systems are designed in such a way that the physical and chemical conditions during the deposition experiment are well controlled. The selection of experimental materials, methods and chemical and physical conditions can be greatly influenced by the desire to rigorously test quantitative theories of particle deposition. The usual features and requirements in deposition systems are described below. These are also discussed later in this chapter, where each of the experimental systems is described separately.

10.1.1 Hydrodynamic conditions

The flow of colloidal suspensions through the collector (or collectors) of interest can be controlled by a variety of methods. The most common method involves the use of a pump that transfers the suspension from a feeding reservoir to the collectors. Peristaltic (roller) pumps with no less than three rollers are recommended in order to provide smooth flow. Using a dumping (surge) bottle after the pump can further ensure pulse-free flow. The flow rates generated by peristaltic pumps can be varied over a wide range. For very low flow rates, syringe pumps are most suitable for providing pulse-free flow.

Another method for controlling the flow through a deposition system is to use a constant-head overflow tank. In this method, a suspension which is continuously stirred in a feeding reservoir is pumped to an overflow tank located above the deposition system. The purpose of this tank is to maintain a constant head. An overflow line returns the overflow suspension from this tank to the feeding reservoir, and the suspension from the constant-head tank flows through the deposition system by gravity. The flow rate is controlled by an adjustable valve that can be located at the end of the line leaving the deposition system. Alternatively, one can use a Mariott bottle containing the particle suspension, above the deposition system (instead of the overflow tank and feeding reservoir), in order to maintain a constant head.

Flow rate can also be controlled by using a reservoir which is pressurized by a gas (such as N_2). In this method, the reservoir with the suspension of interest is connected to the deposition system. Pressurized gas at the top drives the suspension through the deposition system. The flow rate is controlled by regulating the applied pressure.

10.1.2 Solution chemistry

Solution chemistry has a marked effect on the electrokinetic properties of particles and collectors, and hence on the rate of particle deposition. In particle deposition studies, deposition rates can be controlled by altering the solution chemistry. It is desirable that particle deposition experiments be conducted under carefully controlled chemical conditions. Solution chemistries can be controlled by varying the: (1) ionic strength, (2) pH, (3) counterion type, and (4) solute type.

The ionic strength of the solution is usually controlled by the addition of a 1–1 inorganic electrolyte (such as KCl, NaCl or $NaNO_3$). It is commonly accepted that these electrolytes do not specifically interact with the surfaces of particles and collectors (i.e., indifferent electrolytes). The pH of the solution can be controlled by addition of small amounts of a strong acid or base. The ions OH^- and H^+ are potential determining ions for many particles and collectors, and therefore they significantly affect their electrokinetic properties. Furthermore, a change in the counterion type (i.e., monovalent, divalent or trivalent) or the solute type (e.g., by adding a surfactant or a polyelectrolyte) can also have a dramatic effect on the electrokinetic properties of particles and collectors, and hence on particle deposition rates.

10.1.3 Colloidal suspensions

In order to rigorously test fundamental theories of particle deposition and colloidal interaction, it is essential that monodisperse colloidal suspensions with well-defined chemical-colloidal properties be used. Synthetic latex particles are most commonly used to meet these criteria. Latex particles are spherical, monodisperse and contain well-defined chemical functional groups. The preparation and characterization of these particles is well established (e.g. Homola and James, 1977; Kamel *et al.*, 1981; Buscall *et al.*, 1985). Particles with a variety of chemical functional groups can be prepared (e.g. carboxylate, sulphate or amidine groups); these are also available commercially from a variety of sources. Disadvantages in the use of latex particles for deposition studies are associated with their anomalous electrokinetic behaviour in the presence of simple electrolytes (e.g. van den Hoven and Bijsterbosch, 1987; Elimelech and O'Melia, 1990a) and the possible presence of a 'hairy layer' on the particles that may affect deposition behaviour (Varennes and van de Ven, 1987).

Other types of particles that are used in deposition kinetics studies include metal oxides, such as iron and aluminium oxide colloids. Various methods to prepare such model particles are described in the literature (e.g. Matijevic and Scheiner, 1978; Penners and Koopal 1986). A major drawback with the use of these particles is the difficulty in producing spherical and smooth particles.

10.1.4 Clean and smooth collectors

Collectors of planar and curved surfaces are usually made of glass. Most glass surfaces are negatively charged in aqueous solutions of pH higher than 4, but they can be coated with polymers with certain functional groups to modify their surface charge (e.g. Marshal and Kitchener, 1966; Adamczyk *et al.*, 1989). Coating of glass surfaces with a cationic polymer will produce a positively charged collector. Spherical collectors used in deposition studies can also be made of steel (Kuo and Matijevic, 1980) and polymethyl methacrylate (Rajagopalan and Tien, 1977). Granular materials for deposition studies in packed beds include glass or steel beads (e.g. Elimelech and O'Melia, 1990bc; Vaidyanathan and Tien, 1991), sand grains (e.g. Ghosh *et al.*, 1975), and fibres of different materials (e.g. Gregory and Wishart, 1980).

Collector surfaces can be coated with a conducting film, such as indium tin oxide, so that the surface becomes an electrode (Albery *et al.*, 1990). The potential of the coated surface relative to a reference electrode can be adjusted, thus enabling the control of particle deposition rate at a fixed solution chemistry. This method was used to study the rate of deposition of carbon black colloids in a wall-jet (stagnation point-flow) system (Albery *et al.*, 1990). More details about this method can be found in that reference.

In particle deposition studies, it is imperative that collector surfaces be smooth. The roughness of collector surfaces can be inspected by scanning electron microscopy. It is also desirable that the collector surface be free of impurities. A common practice is to wash the collector surface with a strong acid solution followed by a thorough rinse with distilled water. Acid solutions for cleaning glass generally consist of nitric, chromic or hydrochloric acid. Hydrofluoric acid should not be used to clean glass surfaces because it attacks the glass surface vigorously.

The method of cleaning can significantly affect the measured deposition rates in deposition experiments where repulsive double layer interactions predominate. It was reported that particle deposition rates onto collectors cleaned with chromic acid are much lower than deposition rates onto similar collectors cleaned with hydrochloric or nitric acids (Dabros and van de Ven, 1983; Litton and Olson, 1993). It is speculated that chromic acid oxidizes organic impurities which are present on the surface of collectors. These organic impurities may provide sites for preferential deposition.

10.1.5 Non-aggregated suspension

When studying particle deposition onto collector surfaces from flowing suspensions, it is necessary to minimize particle–particle interactions prior to deposition. This can be achieved by using a dilute particle suspension, by minimizing the time in which the electrolyte solution is in contact with the suspended particles, and by shortening the travel time of the suspended particles through the deposition system.

The theories of particle deposition presented in this book are for dilute colloidal suspensions in which many-body interactions in the bulk solution and the interfacial region are negligible. A criterion to neglect many-body interactions in the bulk can be established based on a dimensionless average interparticle spacing (Chari and Rajagopalan, 1985):

$$r_p = \left(\frac{3}{4\pi C_0}\right)^{1/3} \frac{1}{a_p} \tag{10.1}$$

Here, C_0 is the number concentration of particles in the bulk, and a_p is the particle radius. For systems in which $r_p \gg (1 + 1/\kappa a_p)$, it may be assumed that the suspension is dilute enough to neglect many-body interactions (κ is the Debye reciprocal length described in Chapter 2).

In order to minimize coagulation of particles prior to deposition, the electrolytes should be applied to the colloidal suspension shortly before the suspension enters the deposition system. This is a common practice in particle deposition in packed beds, and will be discussed later in this chapter. However, in some systems, such as the rotating disc, this procedure cannot be applied, and electrolytes are added to the suspension before the deposition experiment starts. In these systems, the study of particle deposition at high electrolyte concentrations needs to be carefully analysed, since particles will probably aggregate prior to deposition. The extent of coagulation can be estimated from Smoluchowski theory for coagulation kinetics described earlier in Chapter 6. In designing deposition experiments, it is desirable that the coagulation half-life time of the suspension be sufficiently large compared to the time the particles are in contact with the electrolyte solution.

10.2 Particle-counting methods

10.2.1 Microscopic means

In direct methods, particle deposition rates are determined by counting the number of deposited particles at different time intervals. Some of the earliest works using this method are the studies of Marshal and Kitchener (1966) and Hull and Kitchener (1969) with the rotating disc system. In this method, the collector surface, usually a microscope coverslip, is taken and dried in a desiccator. The surface of the collector is then illuminated under a microscope and the particles are detected as points of scattered light. The counting can be facilitated by an automatic counter receiving an image from a video camera on the microscope trinocular. The surface density of the deposited particles is determined from counting of many squares covering different areas of the collector. In the rotating disc system, a radial dependence of particle deposition rates is observed. As a result, it is imperative that particle counting be conducted at concentric ring surfaces so that the radial dependence can be determined. Such a procedure is described in detail by Martin (1990).

Direct particle counting by microscopic means during the deposition process can be done with the use of stagnation-point flow and parallel-plate channel deposition systems (Adamczyk, 1989; Sjollema and Busscher, 1989, 1990). In these systems, the collector surface is made of a transparent material, such as glass or mica. Observations of the deposition process are carried out with an optical microscope under dark-field illumination. With the use of a TV camera, micrographs of the collector surface at various time intervals can be taken, from which particle deposition rates can be calculated. This experimental set-up will be described in more detail later in this chapter.

10.2.2 Particle counters

In indirect methods, a particle counter analyser can be used to determine the concentration of particles in the samples taken from the inlet and outlet of the deposition system. Particle counters are available commercially. The most common counters are those operating with the Coulter principle and the light blockage (obscuration) method. Particle counting was applied successfully in particle deposition studies in packed-bed columns (e.g. FitzPatrick and Spielman, 1973; Tobiason and O'Melia, 1988). Particle counters of the type mentioned above are effective for suspensions with particle diameter larger than about 0.5 μm.

Particle counting with the Coulter principle is conducted as follows. A dilute suspension of the sample to be analysed is made in an electrolyte solution. This suspension is then stirred and drawn through a small aperture by means of a vacuum source. A current, passing through the aperture between two electrodes, enables the particles to be sensed by the momentary changes in the electrical resistance as they pass through the aperture. These changes in resistance may be detected as a series of voltage pulses, where the height of each pulse is proportional to the volume of the particle which produced it. These pulses are then amplified, counted and allocated to the correct size class. The Coulter principle is unaffected by changes in material composition, surface texture, refractive index or light interaction effects. The disadvantage of this method is the necessity to use an electrolyte as a suspending medium; this electrolyte can induce aggregation of particles in the suspension. A detailed description of the principles, operation and calibration of the Coulter counter is given in Hunter (1989) and Allen (1990). Some of the features and limitations of this particle counter in relation to particle aggregation were discussed in Chapter 6.

Particle counting can also be carried out with the light blockage technique (Letterman, 1991). In light blockage instruments, particles in the sample being analysed flow through a narrow constriction in a tube past a collimated light beam. As a particle passes through the beam, it casts a shadow on a photodetector which sends a voltage pulse to a pulse-height analyser. The particles are counted as they block the light and their size is calculated from the height of the voltage pulse. The height of the pulse is proportional to the projected area of the particle. More details about this technique were given in Chapter 6.

10.2.3 Light scattering

There are two principal ways of monitoring light scattering from colloidal suspensions. The first method is to measure the scattered light directly at some fixed angle to the incident beam. The second method is to measure the turbidity, i.e. the reduction in intensity of the transmitted light due to the scattering of colloidal particles. Turbidity measurements are much more simple than light-scattering measurements, and they can be applied to monitor the ratio of the concentration of particles entering the deposition system to that of particles leaving the system.

The turbidity (τ), also known as the extinction coefficient, is related to the transmitted light by (van der Hulst, 1957; Kerker, 1969):

$$I = I_0 \exp(-\tau l) \tag{10.2}$$

Here I is the intensity of the light after passing through a length l of suspension, and I_0 is the intensity of the incident light. For a monodisperse suspension containing N particles per unit volume, the turbidity can be expressed as:

$$\tau = NC_s \tag{10.3}$$

where C_s is the scattering cross-section of a particle, which depends on the size and refractive index of the particle and on the wavelength of the incident light (Kerker, 1969). The turbidity is related to the optical density O_D by:

$$\tau = 2.3\left(\frac{O_D}{l}\right) \tag{10.4}$$

Hence, it is shown that, for a monodisperse colloidal suspension, the turbidity and optical density are linearly proportional to the number concentration of particles. This means that capture of particles in the deposition system will cause a decrease in the turbidity (or optical density) of the suspension.

In deposition systems, the reduced concentration of particles, C_e/C_0, can be determined from turbidity or optical density measurements. This can easily be done with the use of a conventional spectrophotometer. The optical densities of the colloidal suspension entering the deposition system and that leaving the deposition system at various time intervals are determined. It is most convenient to employ a spectrophotometer with a flow-through cell because it allows for continuous monitoring of the particle concentration in the effluent of the deposition system. The ratio of the effluent to the influent optical densities yields C_e/C_0. This ratio can be used to calculate the particle deposition rate in the deposition system, as will be shown later in this chapter.

10.3 Model deposition systems

10.3.1 Stagnation-point flow

A novel experimental system for directly observing the deposition of collodial particles in the vicinity of a stagnation-point flow was developed by Dabros and

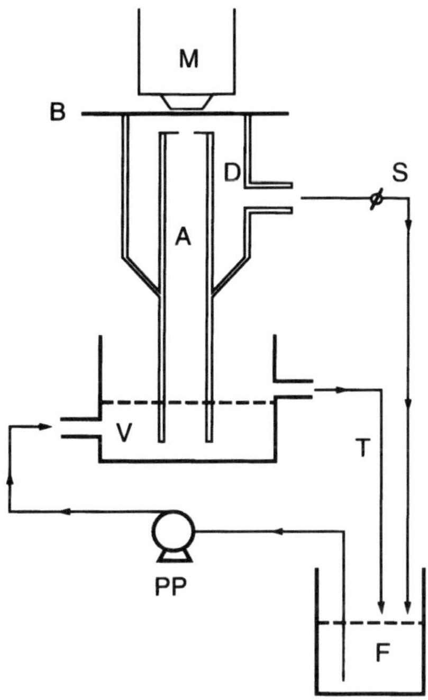

Figure 10.1 *A schematic description of the experimental set-up of a stagnation-point flow system (adapted from Dabros and van de Ven, 1987). The various components of the system are described in the text*

van de Ven (1983, 1987). A schematic description of the experimental system is shown in Figure 10.1. In this system, the stagnation flow is formed when the colloidal suspension in the tube **A** is forced to flow through a circular hole in a thin plate towards a flat transparent collector **B**. This is a modification of the original system of Dabros and van de Ven (1983), in which the liquid flowed out of a capillary. A close look at the region surrounding the stagnation point is shown in Figure 10.2.

The colloidal suspension flows under gravity from vessel **V** to vessel **F** through tubes **A** and **D**. Flowing out of tube **A** through a circular hole, the fluid impinges on a collector (a coverglass) and a radially symmetrical flow distribution around the stagnation point is formed. The intensity of the flow is controlled using precision valve **S**. A peristaltic pump **PP** returns the fluid from vessel **F** to **V** and excess fluid overflows from **V** to **F** through tube **T**. The deposition of particles onto the collector is observed from above by an optical microscope **M**. Dark-field illumination is necessary to observe small particles. The microscope image can be recorded by a TV camera for further analysis.

In the system used by Dabros and van de Ven (1987), the area of observation that was analysed had dimensions of 203×284 or 235×455 μm; the stagnation point was at the centre of this area. When particles deposited uniformly on the

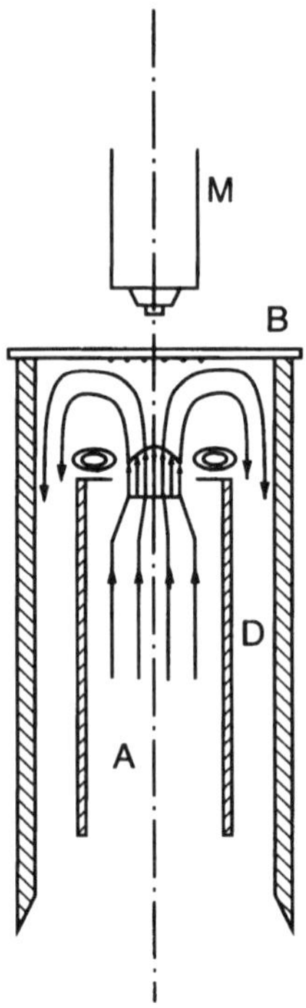

Figure 10.2 *A schematic view of the stagnation point in the system described in Figure 10.1*

surface, the particle deposition rate was determined by counting the particles deposited on the total area at various time intervals. For non-uniform deposition, the total observation region was divided into 96 equal rectangles and the number of particles in each rectangle was determined as a function of time. An alternative approach is to divide the total area into rings concentric about the stagnation point, as described previously.

In the set-up used by Dabros and van de Ven (1987), the separation between the collector coverglass and the outflow plate was equal to 1.81 times the area of the hole; the radius of the hole was 1.0 mm. Thus, the area of observation was less than 20% of the radius of the hole, ensuring uniform accessibility of particles in this region. The Corning coverglass slides which were used as surface collectors had dimensions of 25×50 mm. Each deposition experiment was carried out with a new coverglass.

A major disadvantage of the set-up described in Figure 10.1 is the difficulty in conducting deposition experiments at high ionic strengths. At high ionic strengths, the colloidal suspension in the reservoirs **V** and **F** will undergo aggregation. This can be minimized by modifying the system. In a modified system, the particles should be suspended in distilled water and pumped directly to a tube leading to the stagnation point. The electrolyte solution can be added by a low-flow-rate pump directly into the tube carrying the suspension from the particle reservoir. After the suspension leaves the deposition system, it should be wasted rather than recirculated. This approach has been used successfully in particle deposition in packed-bed columns, as will be described later.

Another method to monitor the deposition rate in a stagnation-point flow was developed by Albery *et al.* (1985). As described below, this method is based on the measurement of the light scattered by deposited particles from the evanescent wave (Schumacher and van de Ven, 1991). With this technique, the deposition rate of small, submicrometre particles (that cannot be seen by a light microscope) can be monitored.

A schematic description of the stagnation-point flow (wall-jet) apparatus using this technique is given in Figure 10.3. The colloid suspension is pumped by a peristaltic pump from below and the jet impinges on a microscopic slide. Light is focused onto the edge of the microscopic slide, and is trapped in the slide by total internal reflection. The light, however, penetrates about half a wavelength into the solution. As a result, particles deposited onto the surface of the slide scatter the light. The scattered light is measured by an optical microscope fitted with a photomultiplier. The wall-jet cell and the microscope table can be driven by an electric motor so that the microscope scans the scattered light along the diameter of the disc surface. This system was recently modified (Albery *et al.*, 1990) to include a mixing cell of the type used in stopped flow kinetics on the upstream side of the jet. The particles and solution are mixed at this point to minimize problems of colloid coagulation prior to deposition. In addition, the system employs a microscope slide covered with a conducting film so that its potential can be controlled. As described previously in Section 10.1.4, the rate of

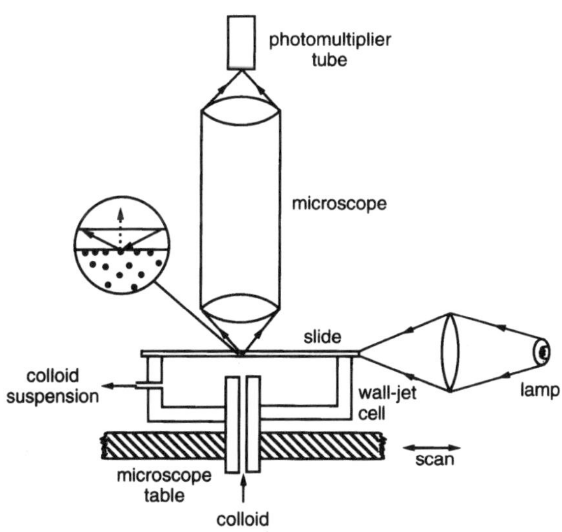

Figure 10.3 *A schematic description of the wall-jet apparatus with the evanescent wave scattering technique (adapted from Albery* et al., *1985)*

particle deposition can be controlled by adjusting the potential of the coated collector.

10.3.2 Rotating disc

A detailed description of the design and operation of a rotating disc system for studying mass transfer in electrochemical reactions was reported by Riddiford (1966). This apparatus was first applied to particle deposition by Marshal and Kitchener (1966) and Hull and Kitchener (1969). A schematic description of this apparatus is shown in Figure 10.4. The drive and shaft are of precision drill press to prevent possible vibration of the disc. The shaft is pressed into a plastic end piece (the disc), as shown in Figure 10.4. The collector surface is usually a 22-mm circular microscope coverglass that can be held to the plastic disc surface by a vacuum grease. The surface chemistry of the glass can be modified by polymer coatings as described by Martin (1990) and Marshal and Kitchener (1966).

In the rotating disc system, the particle suspension is contained in a glass beaker supported by an adjustable tray or a lab jack. These are used to change the elevation of the beaker, and, thus, the level of the disc surface. Following the

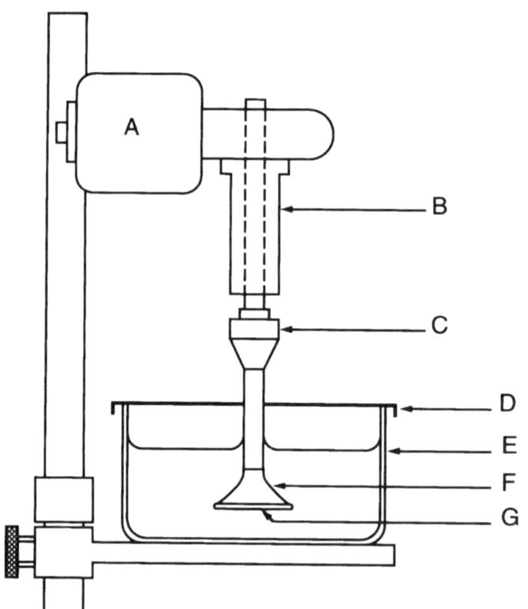

Figure 10.4 *A schematic description of a rotating disc apparatus (adapted from Marshal and Kitchener, 1966). The various components of the system are: (A) constant-speed motor; (B) roller bearing; (C) collet chuck; (D) cover; (E) glass container with particle suspension; (F) plastic disc; and (G) deposition surface*

design considerations suggested by Levich (1962) and Riddiford (1966), the Reynolds number of the disc should not exceed 2×10^5 in order to ensure laminar-flow conditions.

The basic steps required for obtaining data from the rotating disc system are (1) the preparation of particle suspension, (2) the exposure of the coverglass to the suspension with the desired solution chemistry and hydrodynamics, and (3) the enumeration of deposited particles. The final suspension is prepared in the beaker of the rotating disc apparatus by diluting a small amount of concentrated particles in a solution of desired chemistry while the disc is rotating. Martin (1990) found that direct immersion of the disc in a particle suspension could lead to high levels of deposition that were not eliminated by subsequent rotation in a particle-free solution. Thus, to avoid particle deposition during immersion, it was suggested that a proper amount of concentrated suspension be added by a pipette close to the shaft above the disc as the disc is rotating. With this procedure, complete mixing can be achieved within less than 30 s after adding the particles.

After exposure of the disc to the suspension for a desired time, the disc is stopped and removed from the suspension by slowly lowering the beaker with the adjustable tray or the jack. The resultant drop hanging from the disc surface is drawn off (e.g. with the corner of a tissue paper). The disc is then rinsed by rotation at the same speed for a short period of time (less than a minute) in a particle-free solution of the same chemical composition as the test suspension. Following the rinse, the disc is again removed from the beaker as described above. The coverglass is then gently pushed off the disc onto a holder which contacts the glass surface face only at the rim. The disc is then drained and dried in a desiccator before particle enumeration.

10.3.3 Parallel-plate channel

The deposition of particles on flat surfaces from flowing suspensions can also be studied experimentally with the use of a parallel-plate channel apparatus. In this method, a colloidal suspension is forced to flow in between two parallel plates, on which deposition takes place. A detailed description of this method was given by Bowen and Epstein (1979), Sjollema and Busscher (1989, 1990) and Sjollema *et al.* (1989).

A schematic description of the parallel-plate flow cell used by Sjollema and Busscher (1989, 1990) is shown in Figure 10.5. Deposition of particles is directly observed on either the bottom or top plate, 30 mm from the inlet of the cell, with channel dimensions of $0.6 \times 38 \times 55$ mm. The particle suspension is recirculated through the cell by a peristaltic pump or a combination of a pump and hydrostatic pressure. The hydrodynamics in the channel can be controlled by adjusting the flow rate of the recirculation pump.

The deposition of particles on the collector surface of the cell as a function of time can be observed through a microscope as illustrated in Figure 10.5. In the system used by Sjollema and Busscher, the number of particles deposited on the surface was counted from images obtained by a camera mounted on a phase-

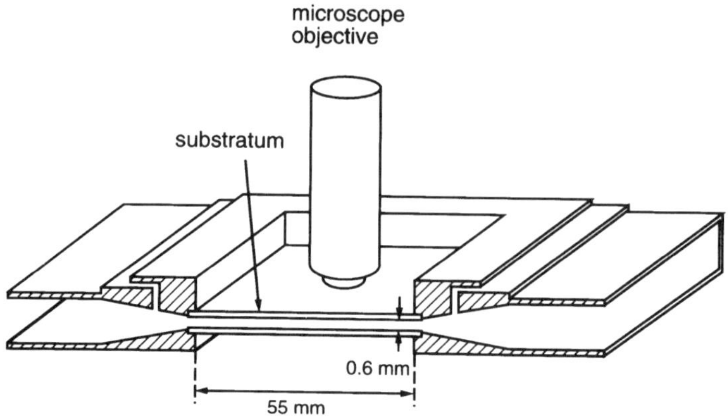

Figure 10.5 *A cross-section of the parallel-plate flow cell used by Sjollema and Busscher (1989)*

contrast microscope. The camera was coupled to an image analyser that calculated the number of deposited particles. A detailed description of the real-time enumeration of particles by automated image analysis is given by Sjollema *et al.* (1989). The deposition rate of particles is calculated from the slope of a plot of the number of attached particles as a function of time.

A different particle-counting method for determining the rate of particle deposition in a parallel-plate channel was employed by Bowen and Epstein (1979). In their study, particle deposition rates were measured using a radioactive tracer technique. The particles were tagged with a γ-emitter, and their accumulation was monitored by a collimated detector mounted on the channel wall. This method offers a means of obtaining the rate of particle accumulation on the surface in real time.

The measurement of particle accumulation in the radioactive tracer technique is accomplished by circulating the tagged particle suspension through the test cell. Under conditions where particle deposition is favourable (i.e. no significant repulsive forces between the particles and the flat surface), the counting system is left to operate in continuous recycle throughout the run. A shorter counting period is usually employed initially in order to capture more data points at the initial stage of the deposition. When deposition is unfavourable (i.e. low deposition rates due to repulsive forces), it is usually necessary to periodically stop the flow and flush out the test cell before counting. In this case, the temperature and electrolyte conditions of the flush solution should be similar to those of the suspension. It is also necessary that the flush rate be smaller than the recirculating flow rate to prevent particle detachment. Based on this description of the tracer method, it is clearly shown that there are major limitations associated with it. The major limitations are: (1) it is not possible to obtain continuous data of particle deposition when chemical conditions are not favourable for deposition; (2) the design and operation of this system are complex; and (3) the use of radioactive materials in the preparation of suspensions and during the experiment may be harmful.

10.3.4 Packed-bed technique

In particle deposition with the packed-bed technique, a well-defined granular material is packed in a column to form a porous bed with a fixed porosity. The granular material is usually composed of spherical beads, washed sand or fibres. The columns are usually made of an inert, transparent material such as glass or Plexiglas. The packed-bed column technique has been used extensively in numerous particle deposition studies (e.g. Kallay *et al.*, 1983; Tien, 1989; Elimelech, 1991).

A schematic description of a typical packed-bed system is presented in Figure 10.6. The system includes a feeding reservoir for the particle suspension, an electrolyte stock solution, peristaltic pumps to deliver the suspension and electrolyte solution, and a column packed with the granular material. As mentioned previously in this chapter, a dumping (surge) bottle can be added after the suspension-feeding pump in order to obtain a smooth flow. In a typical particle deposition experiment, the concentrations of particles entering and leaving the packed column are measured by one of the various methods described previously. The removal of particles by the granular medium is found from the difference between the inlet and outlet concentration of particles. A particle deposition run is composed of the following steps: (1) packing the column with clean granular material; (2) flushing the packed column with an acidic solution; (3) flushing the packed column with distilled (or deionized) water, followed by a flushing with particle-free electrolyte solution; (4) adjusting the flow rate to

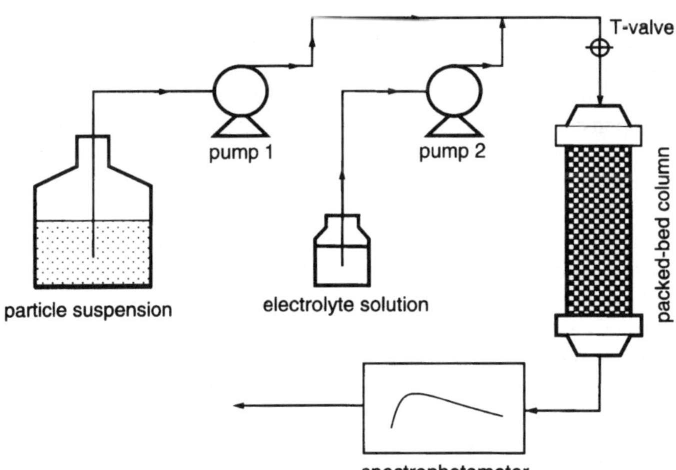

Figure 10.6 *A schematic description of a packed-bed column system used for particle deposition studies*

produce a desired approach velocity; (5) measuring the influent concentration of particles; and (6) starting the flow of the particle suspension through the column and monitoring the breakthrough concentration of particles. These steps are described in detail below.

The required weight of a clean, dry granular medium for a specified bed depth and porosity is added gradually to the clean column. Prior to adding the granular medium, the column is partially filled with deionized water to allow wet packing. After packing, the column is flushed with an acidic solution (pH in the range of 2–3) followed by passing deionized water through the packed bed at a high flow rate to stabilize the pH of the solution. It is well known that granular materials made of glass can take up protons from solution and thereby increase the pH (Tobiason, 1987; Elimelech, 1989; Paul, 1982). The flushing process to stabilize the pH may last for a long period when the size of the granular medium is small, because of the large surface area exposed to water. Flushing of the packed bed with an acidic solution is essential to achieve a near constant-pH effluent.

Before the initiation of the deposition experiment, the packed bed is flushed with a particle-free solution to equilibrate the bed to the deposition experiment conditions. The solution chemistry at this stage should be similar to that to be used in the deposition experiment. This step is required for obtaining reproducible results in the deposition experiments. It is also recommended that the flow be adjusted to produce the desired approach velocity in this step. The approach (superficial) velocity is defined as the volumetric flow rate divided by the inner cross-section area of the column.

After flushing with the particle-free solution, the flow is diverted from flowing through the column by switching a T-valve connected just ahead of the column entrance, and the solution is replaced by a suspension with the desired particle concentration. A stock electrolyte solution with a predetermined concentration is pumped by a low-flow-rate peristaltic pump and is mixed with the particle suspension, ahead of the T-valve, at a desired dilution ratio. In-line mixing can be achieved by serial expansions and contractions in the tubing ahead of the column. The particle concentration in the flowing suspension is measured several times. Then, the suspension is diverted to flow through the column by switching the T-valve; this time is defined as $t=0$ for the deposition run. The particle concentration in the effluent of the column is measured either continuously or at predetermined time intervals during the deposition run. This allows a breakthrough curve of particles to be established, from which the particle deposition rate can be calculated. The concentration of particles is measured by one of the methods described previously.

The deposition rate can be calculated from the removal of particles by the granular material in the packed column. Accumulation of particles in the packed bed is equal to the removal of particles by one collector multiplied by the number of collectors in the bed. Based on the definition of a single collector efficiency (eqn (5.93)), the particle removal by one collector of radius a_c is $\eta\pi a_c^2UC$, η being the actual single collector removal efficiency. Considering a differential slice of a packed bed, with a height dz and a cross-section area A_c, the number of collectors in the differential volume, N_c, is given by:

$$N_c = \frac{3(1-f)A_c\,dz}{4\pi a_c^3} \tag{10.5}$$

Hence, the accumulation of particles in the volume element can be expressed as:

$$\frac{3}{4}\,\eta(1-f)\,\frac{U}{a_c}\,CA_c\,dz = -\,Q\,dC \tag{10.6}$$

where Q is the volumetric flow rate. Considering the entire bed depth for which $C=C_0$ at $z=0$ and $C=C_e$ at $z=L$ (L is the bed depth), and noting that $U=Q/A_c$, one can write:

$$\int_{C_0}^{C_e} \frac{dC}{C} = -\frac{3}{4a_c}\,\eta(1-f)\int_0^L dz \tag{10.7}$$

The actual (experimental) single collector efficiency, η_{exp}, is derived from this equation; it is given by:

$$\eta_{exp} = -\frac{4}{3}\,\frac{a_c}{(1-f)L}\,\ln\,(C_e/C_0) \tag{10.8}$$

Here C_e/C_0 is the value taken from the particle breakthrough curve in the initial stage of deposition. At the early stage of the deposition process, η_{exp} can be viewed as the 'clean-bed' single collector removal efficiency.

10.4 Determination of experimental collision efficiencies

The collision efficiency in particle deposition (also referred to as sticking probability or attachment efficiency) is defined as the ratio of the rate at which particles attach to a surface (collector) to the rate at which they collide with the surface. The physical interpretation of the collision efficiency and the procedures to calculate theoretical collision efficiencies were described previously in Chapter 5. Methods to calculate collision efficiencies from experimental data of particle deposition are described below.

Experimental collision efficiencies can be determined by two approaches. In the first approach, collision efficiency is calculated as the ratio of the actual (measured) deposition rate (expressed as Sh or η) to the theoretical deposition rate under favourable chemical conditions (Sh_0 or η_0). The latter is calculated from the convective-diffusion equation with no double layer interaction. In the second approach, the collision efficiency can be calculated from the ratio of the actual deposition rate to the experimental deposition rate under chemical conditions favourable for deposition. Chemical conditions favourable for deposition can be achieved when deposition experiments are conducted at high electrolyte concentration or with solution chemistries that eliminate repulsive double layer forces between particles and collectors (e.g. by adjusting the pH or adding specifically interacting solutes).

Another method to achieve favourable chemical conditions for deposition is by reversing the charge of the collectors, so that the particles and collectors become oppositely charged. This can be achieved by coating the collector with polymers with functional groups bearing a charge opposite to the original charge of the collector. However, when conducting deposition experiments under this condition to obtain the favourable deposition rate, it is imperative that experiments be conducted at ionic strengths larger than 10^{-3} M. At ionic strengths smaller than 10^{-3} M, significant enhancement in particle deposition rates can be achieved because of the long range of the attractive double layer interactions (see discussion of this subject in Chapters 5 and 11).

Some difficulties in the determination of collision efficiencies from the second approach do exist. At high electrolyte concentrations, particles can aggregate prior to deposition. This results in calculated deposition rates which are probably different from the true value for a non-aggregated suspension. In addition, at high ionic strength, particles approaching the collector can experience additional repulsive interactions due to the existence of hydration forces (Israelachvili, 1992) and friction of liquid against the counterions of the interacting double layers (Muller, 1987; van de Ven, 1988).

A disadvantage of the first approach for determining experimental collision efficiencies is the difficulty of calculating the theoretical deposition rate under favourable chemical conditions (i.e. in the absence of double layer interaction). Approximate analytical expressions are available, but these should be used carefully. For particles smaller than about 0.1 μm in diameter, the use of the Smoluchowski–Levich approximation may be adequate. In this approximation, the deposition rate is calculated without the inclusion of van der Waals, electrical double layer and hydrodynamic interactions. Expressions based on the Smoluchowski–Levich approximation for various deposition systems were presented in Chapter 5. However, for Brownian particles larger than about 0.1 μm, expressions based on the Smoluchowski–Levich approximation overestimate the rate of particle deposition because hydrodynamic interaction becomes important. Hence, favourable collision efficiencies smaller than unity can be obtained with the use of this approximation. For non-Brownian particles, the correlation equation of Rajagopalan and Tien (1976) may be used to calculate η_0. However, under certain physical and chemical conditions, this correlation equation may produce values which are different from exact calculations (Tobiason, 1987). In order to obtain more accurate values for the favourable deposition rates of Brownian as well as non-Brownian particles, it is necessary to calculate them from a numerical solution of the complete convective-diffusion equation as described in Chapter 5.

For particle deposition in packed beds, the experimental collision efficiency based on the first approach is given by:

$$\alpha_{exp} = -\frac{4}{3}\frac{a_c}{(1-f)\eta_0 L}\ln(C_e/C_0) \tag{10.9}$$

This expression was derived from eqn (10.8) using $\eta_{exp}=\alpha_{exp}\eta_0$. The single collector efficiency in this equation is defined by eqn (5.93); it is obtained from the ratio of the overall deposition rate to the convective flux of particles towards the projected area of the solid collector. When the definition given by eqn (5.92) is used for the single collector efficiency, a simple mass balance analysis shows that the experimental collision efficiency is calculated from (Tobiason, 1987):

$$\alpha_{\exp} = -\frac{4}{3}\frac{a_c}{(1-f)^{1/3}\eta_0 L}\ln(C_e/C_0) \qquad (10.10)$$

Based on eqn (10.9) (or 10.10), it is evident that, in order to measure low values of the collision efficiency, it is necessary to employ deeper beds (increase L) or smaller collectors (decrease a_c).

Bibliography

Allen, T. (1990) *Particle Size Measurement*, Chapman and Hall, London

Buscall, R., Corner, T. and Stagman, J. F. (eds) (1985) *Polymer Colloids*, Elsevier Applied Science Publishers, London and New York

Hunter, R. J. (1989) *Foundations of Colloid Science*, Vol. I, Oxford University Press, Oxford

Kerker, M. (1969) *The Scattering of Light and other Electromagnetic Radiation*, Academic Press, New York

References

Adamczyk, Z. (1989) Particle deposition from flowing suspensions. *Colloids Surfaces*, **39**, 1–37

Adamczyk, Z., Siwek, B., Zembala, M. and Warszynski, P. (1989) Enhanced deposition of particles under attractive double-layer forces. *J. Colloid Interface Sci.*, **130**, 578–587

Albery, W. J., Kneebone, G. R. and Foulds, A. W. (1985) Kinetics of colloidal deposition studies onto a wall-jet cell. *J. Colloid Interface Sci.*, **108**, 193–198

Albery, W. J., Fredlein, R. A., Kneebone, G. R., O'Shea, G. J. and Smith, A. L. (1990) The kinetics of colloidal deposition under conditions of controlled potential. *Colloids Surfaces*, **44**, 337–356

Allen, T. (1990) *Particle Size Measurement*, Chapman and Hall, London

Bowen, B. D. and Epstein, N. (1979) Fine particle deposition in smooth parallel-plate channels. *J. Colloid Interface Sci.*, **72**, 81–97

Buscall, R., Corner, T. and Stagman, J. F. (eds) (1985) *Polymer Colloids*, Elsevier Applied Science Publishers, London and New York

Chari, K. and Rajagopalan, R. (1985) Deposition of colloidal particles in stagnation-point flow. *J. Chem. Soc. Faraday Trans 2*, **81**, 1354–1366

Dabros, T. and van de Ven, T. G. M. (1983) A direct method for studying particle deposition onto solid surfaces. *Colloid Polymer Sci.*, **261**, 694–704

Dabros, T. and van de Ven, T. G. M. (1987) Deposition of latex particles on glass surfaces in an impinging jet. *Phys. Chem. Hydrodynam.*, **8**, 161–172

Elimelech, M. (1989) The effect of particle size on the kinetics of deposition of Brownian particles in porous media. PhD. Dissertation, The Johns Hopkins University

Elimelech, M. (1991) Kinetics of capture of colloidal particles in packed beds under attractive double layer interactions. *J. Colloid Interface Sci.*, **146**, 337–352

Elimelech, M. and O'Melia, C. R. (1990a) Effect of electrolyte type on the electrophoretic mobility of polystyrene latex colloids. *Colloids Surfaces*, **44**, 165–178

Elimelech, M. and O'Melia C. R. (1990b) Effect of particle size on collision efficiency in the deposition of Brownian particles with electrostatic energy barriers. *Langmuir*, **6**, 1153–1163

Elimelech, M. and O'Melia, C. R. (1990c) Kinetics of deposition of colloidal particles in porous media. *Environ. Sci. Technol.*, **24**, 1528–1536

FitzPatrick, J. A. and Spielman, L. A. (1973) Filtration of aqueous latex suspensions through beds of glass spheres. *J. Colloid Interface Sci.*, **43**, 350–369

Gohsh, M. M., Jordan, T. A. and Porter, R. L. (1975) Physicochemical approach to water and wastewater filtration. *J. Environ. Eng. Division, ASCE*, **101**, 71–86

Gregory, J. and Wishart, A. J. (1980) Deposition of latex particles on alumina fibers. *Colloids Surfaces*, **1**, 313–334

Homola, A. and James, R. O. (1977) Preparation and characterization of amphoteric polystyrene latices. *J. Colloid Interface Sci.*, **59**(1), 123–134

Hull, M. and Kitchener, J. A. (1969) Interaction of spherical colloidal particles with planar surfaces. *Trans Faraday Soc.*, **65**, 3093–3104

Hunter, R. J. (1989) *Foundations of Colloid Science*, Vol. I, Oxford University Press, Oxford

Israelachvili, J. N. (1992) *Intermolecular and Surface Forces*, Academic Press, New York and London

Kallay, N., Nelligan, J. D. and Matijevic, E. (1983) Particle adhesion and removal in model systems. Part 6 – Kinetics of deposition of hematite particles on steel. *J. Chem. Soc. Faraday Trans. I*, **79**, 65–74

Kamel, A. A., El-Aasser, M. S. and Vanderhoff, J. W. (1981) The preparation and characterization of an ideal model colloid. *J. Dispersion Sci. Technol.*, **2**, 183–214

Kerker, M. (1969) *The Scattering of Light and other Electromagnetic Radiation*, Academic Press, New York

Kuo, R. J. and Matijevic, E. (1980) Particle adhesion and removal in model systems. III. Monodispersed ferric oxide on steel. *J. Colloid Interface Sci.*, **78**, 407–421

Letterman, R. D. (1991) *Filtration Strategies to Meet the Surface Water Treatment Rule*, American Water Works Association, Denver, Colorado

Levich, V. G. (1962) *Physicochemical Hydrodynamics*, Prentice Hall, New Jersey

Litton, G. M. and Olson, T. M. (1993) Colloid deposition rates on silica bed media and artifacts related to collector surface preparation methods. *Environ. Sci. Technol.*, **27**, 185–193

Marshal, J. K. and Kitchener, J. A. (1966) The deposition of colloidal particles on smooth solids. *J. Colloid Interface Sci.*, **22**, 342–351

Martin, R. E. (1990) Quantitative description of bacterial deposition and initial biofilm development in porous media. PhD Dissertation, The Johns Hopkins University, Baltimore, Maryland

Matijevic, E. and Scheiner, P. (1978) Ferric hydrous oxide sols. *J. Colloid Interface Sci.*, **63**, 509

Muller, V. W. (1987) Viscous resistance in electrolyte solutions when charged surfaces approach each other. *Langmuir*, **3**, 621–625

Paul, A. (1982) *Chemistry of Glasses*, Chapman and Hall, London

Penners, N. H. C. and Koopal, L. K. (1986) Preparation and optical properties of homodisperse hematite hydrosols. *Colloids Surfaces*, **19**, 337–349

Rajagopalan, R. and Tien, C. (1976) Trajectory analysis of deep bed filtration with the sphere-in-cell porous media model. *Am. Inst. Chem. Eng. J.*, **22**, 523–533

Rajagopalan, R. and Tien, C. (1977) Single collector analysis of collection mechanisms in water filtration. *Can. J. Chem. Eng.*, **55**, 246–255

Riddiford, A. C. (1966) The rotating disk system. *Adv. Electrochem. Electrochem. Eng.*, **4**, 47–116

Schumacher, G. A. and van de Ven, T. G. M. (1991) Evanescent wave scattering studies on latex–glass interactions. *Langmuir*, **7**, 2028–2033

Sjollema, J. and Busscher, H. J. (1989) Deposition of polystyrene latex particles toward polymethylmethacrylate in a parallel plate flow cell. *J. Colloid Interface Sci.*, **132**, 382–394

Sjollema, J. and Busscher, H. J. (1990) Deposition of polystyrene particles in a parallel plate flow cell. 1. The influence of collector surface properties on the experimental deposition rate. *Colloids Surfaces*, **47**, 323–336

Sjollema, J., Busscher, H. J. and Weerkamp, A. H. (1989) Real-time enumeration of adhering microorganisms in a parallel-plate flow cell using automated image analysis. *J. Microbiol. Methods*, **9**, 73–78

Tien, C. (1989) *Granular Filtration of Aerosols and Hydrosols*, Butterworth Publishers, Stoneham, Massachusetts

Tobiason, J. E. (1987) Physicochemical aspects of particle deposition in porous media. PhD Dissertation, The Johns Hopkins University, Baltimore, Maryland

Tobiason, J. E. and O'Melia, C. R. (1988) Physicochemical aspects of particle removal in depth filtration. *J. Am. Water Works Assoc.*, **80**, 54–64

Vaidyanathan, R. and Tien, C. (1991) Hydrosol deposition in granular media under unfavorable surface conditions. *Chem. Eng. Sci.*, **46**, 967–983

Van de Hulst, H. C. (1957) *Light Scattering by Small Particles*, Dover Publications, Inc., New York

Van de Ven, T. G. M. (1988) On the role of ion size in coagulation. *J. Colloid Interface Sci.*, **124**, 138–145

Van den Hoven, Th. J. J. and Bijsterbosch, B. H. (1987) Streaming currents, streaming potentials and conductances of concentrated dispersions of negatively-charged, monodisperse polystyrene particles. Effect of adsorbed tetraalkylammonium. *Colloids Surfaces*, **22**, 187–205

Varennes, S. and van de Ven, T. G. M. (1987) Deposition and detachment of latex particles at glass surfaces exposed to flow. *Phys. Chem. Hydrodynamics*, **9**, 537–559

Chapter 11

Theoretical predictions compared to experimental observations in particle deposition kinetics

Nomenclature

A	Hamaker constant
a_p	particle radius
a_R	radius of curvature of a protrusion
B	constant used in eqns (11.9)–(11.11)
C_e	effluent particle concentration
C_0	influent particle concentration
e	elementary charge of an electron, 1.60×10^{-19} C
$f(a)$	geometrical factor used in eqn (11.4)
$f_1(\chi_1)$	probability density function of random variable χ_1
$f_2(\chi_2)$	probability density function of random variable χ_2
h	separation distance (surface to surface) between a particle and collector
I	ionic strength
k	Boltzmann's constant, 1.38×10^{-23} J K^{-1}
m	empirical constant used in eqn (11.9)
N_A	Avogadro's number, 6.02×10^{23} mol^{-1}
n	empirical constant used in eqn (11.9)
R	radius of a surface patch
Sh	Sherwood number
s	empirical constant used in eqn (11.9)
T	absolute temperature
α	collision efficiency
α_m	mean collision efficiency defined by eqn (11.1)
Γ	function used in eqn (11.5)
ε_0	dialectic permittivity in vacuum, 8.854×10^{-12} C^{2}J^{-1}m^{-1} (SI units)
ε_r	relative dialectic permittivity of water, 78.5 at 25°C
ζ	electrokinetic (zeta) potential
η	single collector removal efficiency
κ	inverse Debye length
ϕ_1	interaction energy between a protrusion and a particle
ϕ_2	interaction energy between a particle and a smooth surface

ϕ_i	interaction energy of type i defined by eqn (11.4)
ϕ_{max}	energy barrier (based on the DLVO theory)
ϕ_T	total interaction energy defined by eqn (11.3)
χ_1	physical or chemical property of particles
χ_2	physical or chemical property of collectors
ψ_c	surface potentials of collectors
ψ_p	surface potentials of particles

Theories of particle deposition can be tested by conducting deposition experiments with model particles and collectors under controlled chemical and physical conditions. Such studies are useful in investigating the role of colloidal and hydrodynamic interactions in particle deposition. In order to investigate the role of such interactions, it is essential that the surface properties of the particles and collectors be well defined. In this chapter, experimental results from well-controlled deposition experiments are compared with theoretical predictions. Two distinct cases are discussed. The first case deals with deposition in the presence of repulsive double layer interactions, the so-called 'unfavourable' deposition. In the second case, we will deal with deposition in the presence of attractive double layer interactions. The latter case applies to situations where particles and collectors are oppositely charged. Discrepancies between theory and experiments are discussed and various explanations for these discrepancies are analysed. Finally, a semi-empirical approach for predicting collision efficiencies in salt-induced colloid deposition (or aggregation) is described. Although this chapter deals with particle deposition kinetics, most of the analysis and discussions related to the discrepancies between theory and observations also apply to coagulation kinetics.

11.1 Deposition with repulsive double layers

Several well-defined studies dealing with deposition of Brownian and non-Brownian particles on various collector geometries are available in the literature. In the following sections, a few of these studies will be described in order to illustrate the behaviour of observed particle deposition rates compared to those predicted by theory. These studies, among others, are summarized in Tables 11.1 and 11.2.

In the first study, Gregory and Wishart (1980) investigated the deposition of 0.172-μm latex particles on aluminium fibres packed in a laboratory-scale column. In their study, the effect of pH on the rate of particle deposition at a fixed ionic strength was examined. It was found that, at pH values where the latex particles and aluminium fibres were similarly charged (i.e. at pH values higher than the isoelectric point of the aluminium fibres), particle deposition rates were reduced. However, they were not reduced to the large extent predicted by theory. Experimental particle deposition rates were several orders of magnitude larger than those predicted from theories which incorporate the DLVO theory in the convective-diffusion equation. Gregory and Wishart stated: 'In spite of the near

Table 11.1 *Summary of deposition experiments with Brownian colloids in the presence of repulsive double layers*

Reference	*Deposition system*	*Colloidal particles (suspensions)*	*Collector*	*Solution chemistry*	*Main findings and conclusions*
Marshal and Kitchener (1966)	Rotating disc	Carbon black particles Diameter: 0.45 μm (some experiments with 0.18-μm particles)	Glass surfaces Glass surfaces coated with polymers	Cationic surfactant Anionic surfactant KCl electrolyte	Large discrepancy between theory (DLVO combined with mass transfer) and observations when repulsive double layer interactions predominate Discrepancy with theory was attributed to charge heterogeneity and roughness of the collectors
Hull and Kitchener (1969)	Rotating disc	Polystyrene latex particles Diameter: 0.308 μm	Glass surfaces Glass surfaces covered with cationic polymer Glass surfaces covered with anionic polymer	Anionic surfactant NaCl electrolyte	Large discrepancy between theory and observations when particles and collectors are similarly charged Good agreement with mass transfer theories when particles and collectors are oppositely charged Discrepancy was attributed to collector surface charge heterogeneity and roughness
Bowen and Epstein (1979)	Parallel-plate channel	Uniform silica spheres Diameters: 0.402, 0.600 and 0.649 μm	Glass surfaces Glass surfaces covered with cationic or anionic polymers	NaCl electrolyte	The same as above Discrepancy with theory was attributed mainly to surface charge heterogeneity
Gregory and Wishart (1980)	Packed-bed column	Polystyrene latex particles Diameter: 0.172 μm	Alumina fibres	pH was controlled by acid or base Ionic strength was controlled by a buffer solution	The same as Bowen and Epstein (1979)

Table 11.1 *Continued*

Reference	*Deposition system*	*Colloidal particles (suspensions)*	*Collector*	*Solution chemistry*	*Main findings and conclusions*
Kallay *et al.* (1983)	Packed-bed column	Haematite particles Diameter: 0.150 µm	Stainless steel spheres	$NaNO_3$ electrolyte	Large discrepancy between theory and observations when repulsive double layer interactions are significant
Sjollema and Busscher (1989)	Parallel-plate channel	Polystyrene latex particles Diameters: 0.736 and 0.820 µm	Polymethylmethacrylate surface	Ionic strength was controlled by a PBS buffer	Poor agreement between theory and observations when repulsive double layer interactions are important Discrepancy was attributed to surface charge heterogeneity and to protruding chains on the surface of the latex particles
Elimelech and O'Melia (1990a,b)	Packed-bed column	Polystyrene latex particles Diameters: 0.046, 0.121, 0.378 and 0.753 µm	Glass beads	KCl electrolyte $CaCl_2$ electrolyte	Poor agreement between theory and observations when repulsive double layer interactions predominate Observed collision efficiencies are insensitive to particle size Several explanations for observed discrepancies are suggested and discussed.
Litton and Olson (1993)	Packed-bed column	Polystyrene latex particles Diameter: 0.245 µm	Quartz sand Glass beads	pH was controlled by acid or base NaCl electrolyte	Poor agreement between theory and experiments Discrepancy was attributed to surface heterogeneity

Table 11.2 *Summary of particle deposition experiments with non-Brownian particles in the presence of repulsive double layer interactions*

Reference	*Deposition system*	*Colloidal particles (suspensions)*	*Collector*	*Solution chemistry*	*Main findings and conclusions*
FitzPatrick and Spielman (1973)	Packed-bed column	Latex particles Diameters: nine suspensions in the range of 0.71–21.0 μm	Glass beads	HNO_3 to vary the pH NaCl electrolyte $CaCl_2$ electrolyte	No discussion is given for agreement between theory and observations when repulsive double layer interactions predominate
Yoshimura *et al.* (1980)	Packed-bed column	Latex particles Diameters: 2.0, 3.7, 5.7 and 11.3 μm	Glass beads	NaCl electrolyte	Poor agreement between theory and observations when repulsive double layer interactions predominate No explanations for this discrepancy are provided
Tobiason and O'Melia (1988)	Packed-bed column	Polystyrene latex particles Diameters: 4 and 12 μm	Glass beads	$NaNO_3$ electrolyte $NaNO_3$ and $Ca(NO_3)_2$ Acid or base to vary the pH	Poor agreement between theory and observations Several explanations for this discrepancy are suggested and discussed
Vaidyanathan and Tien (1991)	Packed-bed column	Latex particles Diameters: five suspensions in the range of 4.2 and 25.7 μm	Glass beads	NaCl electrolyte Acid or base to vary the pH	Poor agreement between theory and observations when repulsive double layer forces predominate A model for surface charge heterogeneity was proposed; this model narrows the gap between theory and observations

ideality of the experimental system, there are serious discrepancies between observed capture efficiencies and those calculated from modern theories incorporating colloidal and hydrodynamic interactions.'

In the second study, Elimelech and O'Melia (1990a) investigated the kinetics of deposition of several submicrometre latex particles on glass beads using the packed-bed column technique. Experimental collision efficiencies of these suspensions at various ionic strengths were calculated from the measured deposition rates. The curves of the experimental and theoretical collision efficiencies as a function of electrolyte (KCl) concentration are shown in Figure 11.1. The theoretical collision efficiencies were calculated using the interaction-force boundary-layer approximation, as described in Chapter 5. A large discrepancy between theoretical and experimental results is observed under unfavourable chemical conditions. Theoretical collision efficiencies are many orders of magnitude smaller than those determined experimentally. It is also observed that, in contrast to theoretical predictions, particle size has no effect on experimental collision efficiencies.

In another study of particle deposition (Tobiason, 1987; Tobiason and O'Melia, 1988), collision efficiencies of non-Brownian latex particles (4 μm in diameter) were measured as a function of $Ca(NO_3)_2$ concentration. The ionic strength in those experiments was kept constant by adjusting the $NaNO_3$ concentration of the solution. The theoretical and experimental collision efficiencies as reported in those studies are presented in Figure 11.2. The results are similar to those obtained with submicrometre particles in the studies described previously. It was found that theoretical collision efficiencies calculated from trajectory analysis are many orders of magnitude smaller than

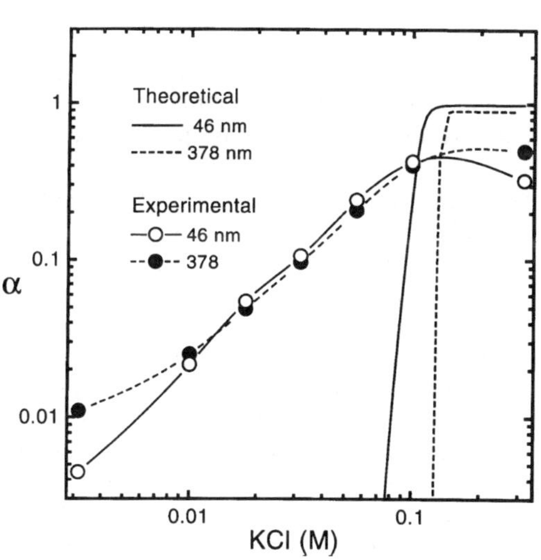

Figure 11.1 *Comparison of theoretical and experimental collision efficiencies of two different suspensions of Brownian (submicrometre) latex particles (after Elimelech and O'Melia, 1990a). The particle diameter of the suspensions is indicated*

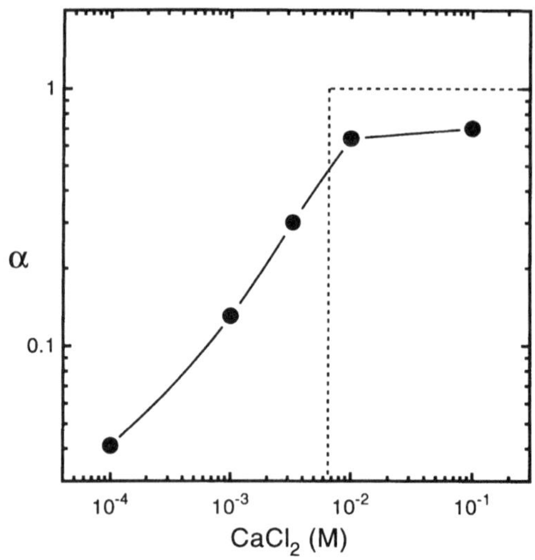

Figure 11.2 *Comparison of theoretical and experimental collision efficiencies of non-Brownian latex particles (4 μm in diameter) in packed-bed columns. The graphs were constructed from the data reported by Tobiason (1987) and Tobiason and O'Melia (1988). Theoretical predictions are in dashed lines and experimental data are in filled circles*

those determined experimentally. From the results described in Figures 11.1 and 11.2, it is demonstrated that the discrepancy between theory and experiments is more pronounced with non-Brownian particles. This is because of the higher energy barriers predicted by the DLVO theory for non-Brownian particles. Theories of deposition of non-Brownian particles predict that the transition from favourable to unfavourable deposition results in an abrupt drop of the collision efficiency from one to zero.

A common feature of the three investigations discussed above, the studies summarized in Tables 11.1 and 11.2, and many others (e.g. Rajagopalan and Tien, 1977; Matijevic and Kallay, 1983; Thompson *et al.*, 1983), is the large discrepancy between theory and observations when repulsive double layer interactions predominate. Particle deposition rates and collision efficiencies calculated from current particle deposition models (which incorporate the DLVO theory) are many orders of magnitude smaller than those observed experimentally. Experimentally determined collision efficiencies are sensitive to the ionic strength of the solution and to the electrokinetic potentials of particles and collectors but not to the large extent predicted by theory. These discrepancies are universal since they are obtained from studies with: (1) Brownian and non-Brownian particles, (2) particles of various surface chemistries (e.g. synthetic latex particles and metal oxides), (3) collectors of various geometries (planar and curved surfaces), and (4) collectors of various materials (e.g. glass, mica and steel).

Another serious disparity exists with respect to the effect of particle size on particle deposition rates. Particle deposition theories incorporating the DLVO theory predict that collision efficiencies and critical deposition concentrations are dependent on particle size (see model predictions in Chapter 5). Experimentally determined collision efficiencies and critical deposition concentrations, on the other hand, are virtually independent of particle size (as shown in Figure 11.1). Both discrepancies are possibly related to the inadequacy of the DLVO theory to describe quantitatively the actual interaction energy during particle–collector interaction when double layer forces are repulsive. This will be discussed in detail later in this chapter. When repulsive double layer forces are eliminated, however, particle deposition rates are generally in good agreement with those predicted by theory.

11.2 Deposition in the presence of attractive double layers

Particle deposition in the presence of attractive double layer interactions can provide additional insight into the role of colloidal and hydrodynamic interactions in particle deposition. Systematic studies of particle deposition in the presence of attractive double layer interactions have only recently begun to appear in the literature. These studies are described below.

In the first study, Adamczyk *et al.* (1989) investigated the deposition of negatively charged 0.94-μm latex particles on oppositely charged collectors in a stagnation-point flow cell. The collector surfaces were thin mica sheets covered with 3-amino-triethoxysilane polymer. The coating layer converted the natural negative charge of the mica surfaces to a positive charge due to the positively charged amino functional groups.

The experimental normalized particle fluxes (particle flux divided by the bulk number concentration of particles) determined at various ionic strengths and flow velocities are compared to those predicted by theory in Figure 11.3. A relatively good agreement between theoretical and experimental results is observed in this figure. Particle deposition rates increase with a decrease of ionic strength. This effect is much more pronounced at higher flow rates (or higher Peclet numbers). The inverse relation between deposition rates and ionic strength was discussed previously in Chapter 5. The mechanisms which are involved in the enhancement of particle deposition rates in the presence of attractive double layer interactions are discussed later in this chapter.

In the second study (Elimelech, 1991), the deposition of 0.48-μm amine latex particles onto glass beads was investigated using the packed-bed column technique. These experiments were conducted in aqueous medium at pH 6. Under these conditions, the particles are positively charged while the glass beads are negatively charged.

The particle breakthrough curves (i.e. the concentration of particles in the outlet of the column relative to the influent particle concentration as a function of time) at different ionic strengths are presented in Figure 11.4. It is observed that the removal of particles by the clean collectors increases (C_e/C_0 decreases)

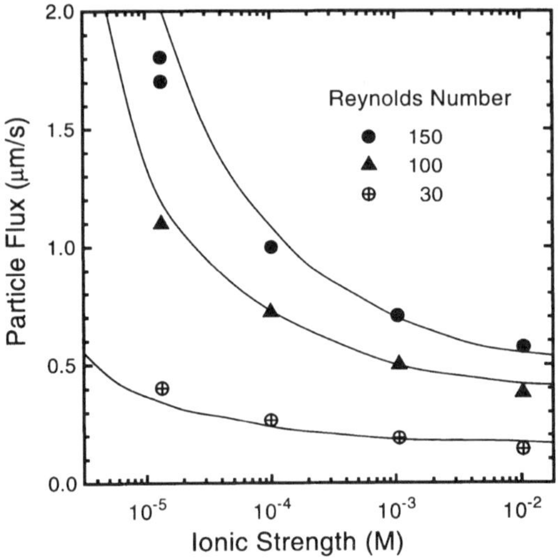

Figure 11.3 *Normalized initial particle flux as a function of ionic strength (1–1 electrolyte) of 0.94-μm latex particles deposited in a stagnation-point cell. The Reynolds number for each curve is indicated in the figure. The symbol points are experimental results, while the continuous lines are theoretical predictions (after Adamczyk* et al., *1989)*

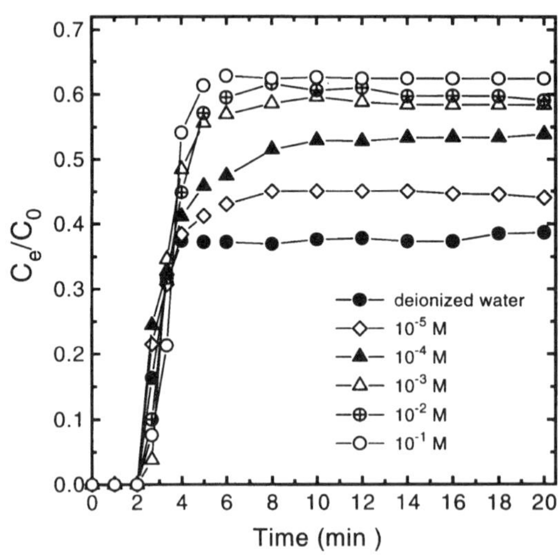

Figure 11.4 *Particle breakthrough curves at different KCl concentrations. In all experiments the approach velocity was 0.14 cm s^{-1}, the collector diameter was 0.46 mm, and the bed (media) depth was 16 cm (after Elimelech, 1991)*

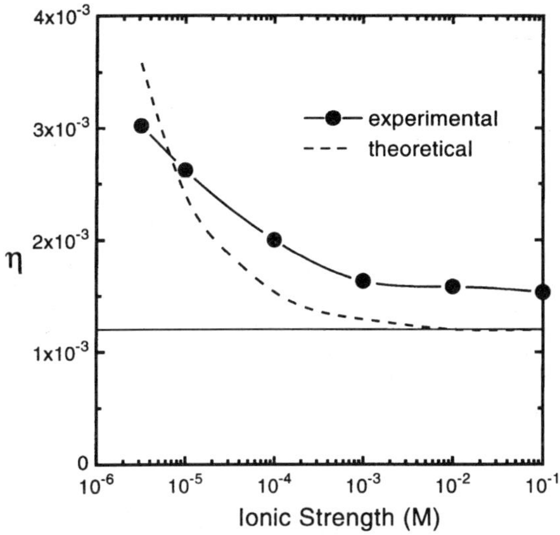

Figure 11.5 *Comparison of theoretical and experimental particle deposition rates as a function of ionic strength (M). The horizontal line is of theoretical η in the absence of electrical double layer interaction (after Elimelech, 1991)*

as the ionic strength decreases. Figure 11.5 compares the calculated dimensionless deposition rates, η, to theoretical predictions. The experimental single collector efficiencies (η) were calculated from the C_e/C_0 values of the breakthrough curves shown in Figure 11.4. The theoretical values were calculated from a numerical solution of the convective-diffusion equation with colloidal and hydrodynamic interactions fully incorporated. It is observed that theoretical predictions are in good agreement with the experimental results. The experimental deposition rates are higher than the theoretical predictions at most by 30% (at high ionic strength). Furthermore, it is shown that the shape of the theoretical and experimental curves is comparable.

In a more recent study, Elimelech (1994) investigated the effect of particle size on the kinetics of particle deposition under attractive double layer interactions. Experimental deposition rates of four particle suspensions with diameters ranging from 0.08 μm to 2.51 μm were compared to theoretical predictions based on a numerical solution of the convective-diffusion equation with colloidal, hydrodynamic and gravitational forces fully incorporated. Theoretical and experimental results revealed that the enhancement in particle deposition rate is dependent on particle size. As with the studies discussed previously, the agreement between experimental results and theoretical predictions was relatively good.

The agreement between theory and observations in particle deposition with attractive double layer interactions contrasts with the serious discrepancies in deposition with repulsive double layer interactions. The results discussed above demonstrate that the convective-diffusion equation incorporating the DLVO theory is adequate to describe the deposition of colloidal particles when double layer interactions are attractive.

11.2.1 Mechanisms of particle deposition with attractive double layers

The increase in colloid deposition rates at low ionic strength, when particles and collectors are oppositely charged, is attributed to the increase in the range and magnitude of the attractive double layer interactions as the ionic strength decreases (Adamczyk *et al.*, 1989; Elimelech, 1991; Elimelech and Song, 1992). At low ionic strengths, double layer interactions extend to large distances (several hundreds of nanometres). This results in an increase in the transport rate of particles towards the collector surface, and, consequently, an increase in the rate of particle deposition. At high ionic strengths, on the other hand, the range of the attractive double layer interactions becomes relatively small. As a result, particle deposition rates are not enhanced.

Calculated double layer interaction energy profiles between the colloidal particles and glass beads used in the experiments described in Figures 11.4 and 11.5 are presented in Figures 11.6 and 11.7. In these calculations, the experimentally determined electrokinetic (zeta) potentials were used as surface potentials. The calculated profile for van der Waals attraction is included in Figure 11.7. It is illustrated that the range of double layer interactions extends to

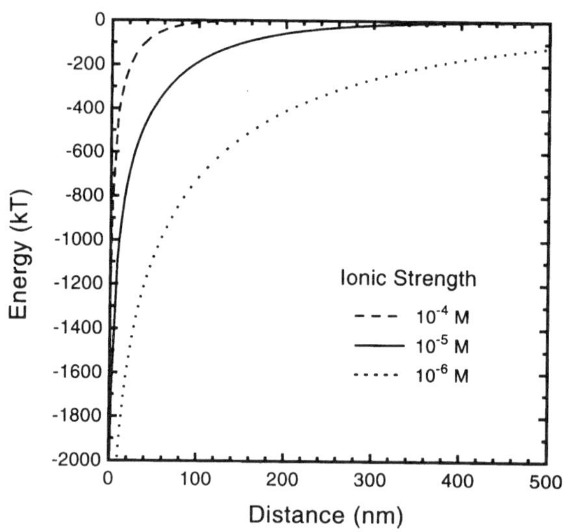

Figure 11.6 *Electrical double layer interaction energy profiles at low ionic strengths (10^{-6}, 10^{-5} and 10^{-4} M of a 1–1 electrolyte) when particles and collectors are oppositely charged. The following parameters were used in the calculations: temperature 25°C, particle diameter 477 nm, and surface potentials of particles and collectors as the zeta potentials described in Elimelech (1991)*

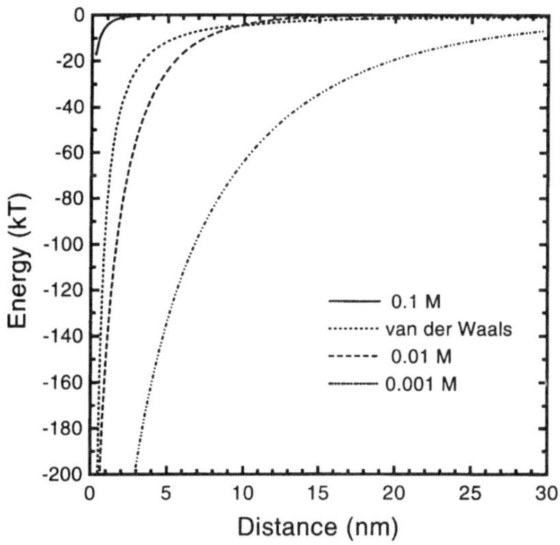

Figure 11.7 *Electrical double layer interaction energy at high ionic strengths (10^{-3}, 10^{-2} and 10^{-1} M of a 1–1 electrolyte) when particles and collectors are oppositely charged. The van der Waals attraction energy profile is also included. The following parameters were used in the calculations: temperature 25°C, particle diameter 477 nm, Hamaker constant 1×10^{-20} J, and surface potentials of particles and collectors as the zeta potentials described in Elimelech (1991)*

very large distances at low ionic strengths. The long-range attractive energies enhance the transport of colloids towards the collector, which results in higher deposition rates. At ionic strengths larger than 10^{-3} M, on the other hand, the range of electrical double layer attraction is reduced and becomes comparable to that of van der Waals attraction.

A schematic description of the enhancement of particle deposition onto a spherical collector in the presence of attractive double layer interactions is given in Figure 11.8. When the oppositely charged diffuse double layers of the particle and collector overlap during particle deposition, the particle will change its path and will be transported towards the collector due to the attractive forces. We can identify a particle path, such as that described in Figure 11.8, and view it as a limiting trajectory. At the given ionic strength, all particles with trajectories closer to the surface will be removed, since their diffuse double layers will overlap with that of the collector. For larger particles, the limiting trajectory is located further from the surface, resulting in greater enhancement of deposition rates relative to that of smaller particles. The enhancement in colloid deposition rates in the presence of attractive double layer interactions can also be explained as an 'interception effect'. The interception of particles will increase with an increase in the thickness of the diffuse double layers (i.e. lower ionic strengths) and with an increase in particle size.

Further discussion on the role of particle size in particle deposition in the presence of attractive double layer interactions was given by Elimelech (1994).

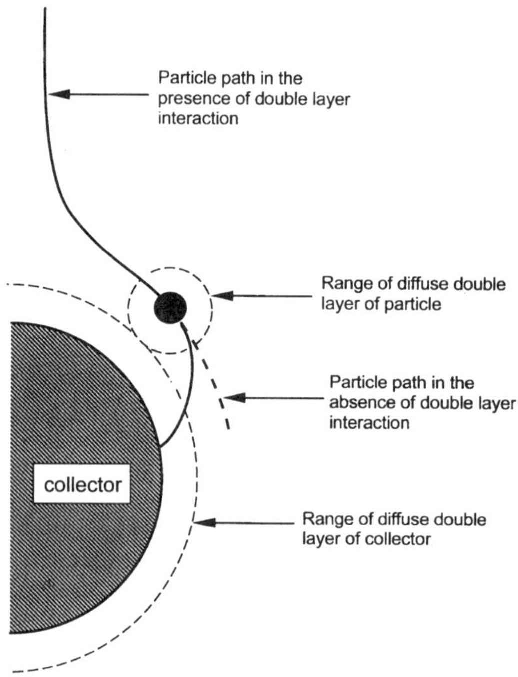

Figure 11.8 *Schematic representation of the enhancement of particle deposition onto a spherical collector in the presence of attractive electrical double layer (EDL) interactions at low ionic strength (after Elimelech and Song, 1992)*

It was found in that study that the enhancement in particle deposition at low ionic strength passes through a maximum which corresponds to a size around 1–2 μm. This maximum is determined by the interplay between the size of the particles and the range of the attractive double layer interactions.

11.3 Possible explanations for observed discrepancies in unfavourable deposition

Two major discrepancies between theory and experiment in particle deposition in the presence of repulsive electrical double layer interactions were discussed previously in this chapter. These are as follows: (1) experimental collision efficiencies are many orders of magnitude larger than the predicted values; and (2) experimental collision efficiencies and critical deposition concentrations are independent of particle size. Similar discrepancies were observed in studies of coagulation kinetics (e.g. Overbeek, 1982; Penners and Koopal, 1987). Several explanations for these discrepancies are discussed below.

11.3.1 Distribution of surface properties

Particles and collectors usually exhibit a distribution in measured physical and chemical properties such as electrophoretic mobility and particle size (e.g. Rajagopalan and Chu, 1982; Prieve and Lin, 1982; Jia *et al.*, 1990). It is well known from DLVO theory that particle deposition rate and collision efficiency are sensitive to the surface potentials of particles and collectors and to the particle size of the suspension. Hence, a distribution in these properties may have a notable effect on predicted collision efficiencies.

The effect of a distribution of any property of particles and collectors, which is used in the theoretical calculation of the collision efficiency, can be evaluated by calculating the mean value of the collision efficiency. In this approach, the property to be studied is assumed to be a random variable. Thus, if χ_1 and χ_2 denote a property χ of the particles and collectors, respectively, having jointly independent probability density functions $f_1(\chi_1)$ and $f_2(\chi_2)$, then the mean value of the collision efficiency can be calculated from:

$$\alpha_m = \int_{-\infty}^{\infty} \int_{-\infty}^{\infty} f_1(\chi_1) f_2(\chi_2)\, \alpha(\chi_1, \chi_2)\, d\chi_1 d\chi_2 \tag{11.1}$$

where $\alpha(\chi_1, \chi_2)$ is the collision efficiency, calculated based on the principles described in Chapter 5, with χ_1 and χ_2 as random variables.

The double integration of eqn (11.1) can be carried out numerically by discretizing the expression as follows:

$$\alpha_m = \sum_{i=1}^{N} \sum_{j=1}^{N} f_1(\chi_{1i}) f_2(\chi_{2j})\, \alpha(\chi_{1i}, \chi_{2j})\, \Delta\chi_{1i}\, \Delta\chi_{2j} \tag{11.2}$$

Here N is the number of the subintervals in which the random variables are divided between the lower and the upper limit, and $\Delta\chi_{1i}$, $\Delta\chi_{2i}$ are the corresponding increments in χ.

It was shown earlier in this book (Chapter 5) that particle deposition rates and collision efficiencies are very sensitive to surface potentials of particles and collectors when repulsive double layer interactions predominate. In current models, it is assumed that at a given solution chemistry, all particles and collectors have constant surface potentials. Electrophoretic mobility measurements of various colloidal particles indicate that there exists a wide distribution of mobility values (e.g. Rajagopalan and Chu, 1982; Jia *et al.*, 1990). Thus, particles with surface potentials smaller than the mean value can deposit and, therefore, increase the overall rate of particle deposition.

The effect of a distribution in surface potentials on collision efficiency in particle deposition can be assessed by the framework described earlier (eqns (11.1) and (11.2)). Such an approach was used to investigate the effect of a distribution in surface potentials on coagulation kinetics (Prieve and Ruckenstein, 1980), and kinetics of deposition of non-Brownian (Vaidyanathan, 1986; Tobiason, 1987) and Brownian (Elimelech and O'Melia, 1990a) particles in porous media. It is commonly assumed that the surface potentials of particles and

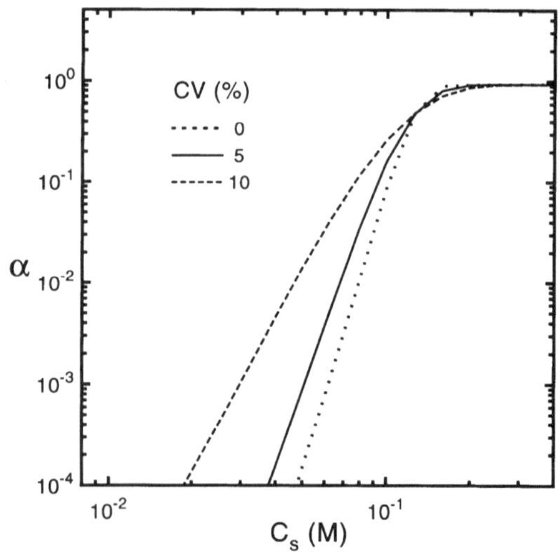

Figure 11.9 *Theoretical stability curves assuming a normal distribution of surface potentials of particles and collectors. The collision efficiencies were calculated from eqn (11.1) for various coefficients of variation (CV). The following parameters were used in the calculations: Hamaker constant = 10^{-20} J; particle diameter = 0.1* μm*; collector diameter = 0.2 mm; temperature = 25°C;* C_s is the molar concentration of 1–1 electrolyte

collectors are normally distributed; normal distribution is usually observed in electrophoretic mobility measurements (Jia *et al.*, 1990).

Representative results of theoretical stability curves for various coefficients of variation (standard deviation divided by the mean in percentage) in the surface potentials are shown in Figure 11.9. This range of coefficients of variation agrees with measurements of electrophoretic mobilities (Tobiason, 1987; Elimelech, 1989). The collision efficiencies were calculated using the interaction-force boundary-layer approximation; the principles of this approach were described in detail in Chapter 5. The results demonstrate that a distribution in surface potentials tends to decrease the slope of the stability curves. This is a result of contributions from combinations of particle and collector surface potentials in the lower part of the distribution curve (i.e. below the mean values). Although some improvement in the agreement between theory and observations is achieved with this modelling, experimental results are still many orders of magnitude larger than those predicted with the inclusion of a distribution in surface potentials (Elimelech and O'Melia, 1990a). In addition, a distribution in surface potential does not resolve the discrepancies with respect to the effect of particle size on collision efficiency and critical deposition concentration.

A much better agreement with theory can be obtained by using higher coefficients of variation in the surface potential of particles and collectors, but such large distributions are probably not applicable to real colloidal suspensions or collector surfaces. As will be discussed in the following subsection, however, it may be more reasonable to assume large distributions in *local* surface properties, such as the local surface potential of collectors.

11.3.2 Heterogeneity of surface charge

Background

Most solid surfaces in aqueous media are heterogeneously charged at both the microscopic (molecular) and macroscopic levels. Surface charge heterogeneity is attributed to the complexity of the crystalline structure of solids and to their complex chemical composition (e.g. Sposito, 1984; Jaroniec and Madey, 1988; Hiemstra *et al.*, 1989a,b). Surface-bound impurities may be an additional source of charge heterogeneity. For instance, it has been reported that cleaning methods of glass surfaces which are used as collectors in deposition experiments can significantly influence the rate of particle deposition under unfavourable chemical conditions (Dabros and van de Ven, 1983; Litton and Olson, 1993).

The electrostatic interaction energy between two ideal surfaces having uniform charge distribution can be calculated using the classic electrical double layer theory, as described in Chapter 3. Based on this assumption of ideality, several analytical expressions have been developed for double layer interaction energy between two surfaces, including constant surface potential, constant surface charge, and intermediate cases (see Table 3.2 for analytical expressions). These expressions, however, are not directly applicable to interactions between real surfaces which have non-homogeneous distributions of surface charge.

Several mathematical models have been developed to describe surface heterogeneities which are periodic in nature (Richmond, 1974, 1975; Schuhmann, 1984; Schuhmann and Depenoux, 1987; Kuin, 1990). Richmond (1974, 1975) modelled the interaction potential between two particles having either smeared out (random) charge distribution or localized charges. Richmond demonstrated that the double layer interaction energy between two plates depends on the relative configuration of the charge distribution on the surfaces. Schuhmann (1984) and Schuhmann and Depenoux (1987) modelled a surface containing a network of charged spots superimposed on a uniform background of opposite charge by utilizing a double Fourier expansion to represent the potential distribution. Using Schuhmann's model, Kuin (1990) was able to show that the interaction potential between a particle and a surface depends largely on the periodicity of the surface potential. As periodicity decreases from macroscopic to atomic dimensions, its effect on the distribution of surface potential diminishes to zero.

Studies of adsorption of potential-determining ions onto heterogeneous surfaces have been helpful in clarifying the effect of surface charge heterogeneity on double layer interaction energy. Koopal and Van Riemsdijk (1989) examined differences in adsorption to heterogeneous surfaces by utilizing a Stern layer capacitance model. With their model, which distinguished between random and patchwise heterogeneity, they demonstrated that a random heterogeneous surface (at a molecular scale) has double layer characteristics equivalent to those of a homogeneous surface, whereas with patchwise heterogeneity and large patches, the surface has double layer properties resembling a mosaic of the different patches. They also identified the transition from small to large patches as $\kappa R = 1$, where κ is the inverse Debye length (eqn (2.13)) and R is the characteristic dimension of the patch. Koopal and Dukhin (1993) examined the adsorption of charge-determining ions onto a patchwise heterogeneous surface containing two types of patches. They demonstrated that the electrochemical behaviour of such a heterogeneous surface can be described on the basis of a homogeneous analogue and a parameter which characterizes the degree of heterogeneity.

The important role of surface charge heterogeneity in particle deposition kinetics under unfavourable chemical conditions has been noted for a long time. Hull and Kitchener (1969), Bowen and Epstein (1979) and Gregory and Wishart (1980) reported the results of controlled colloid deposition experiments under unfavourable chemical conditions. They observed that experimental deposition rates were many orders of magnitude higher than theoretical predictions based on the classic DLVO theory. The discrepancy was attributed to inherent physical and electrochemical heterogeneities of collector surfaces. These investigators further suggested that particle deposition occurs preferentially onto favourable sites, resulting in initial deposition rates much higher than those predicted based on the average collector surface potential. Bowen and Epstein (1979) stated:

> As a consequence, the interaction of real surfaces may be represented by a wide distribution of possible interaction energy curves. Thus, for negative particles and negative walls, because the process is reaction-rate controlled, deposition occurs onto local, energetically favorable area, resulting in an initial deposition rate which is much more rapid than predicted by the interaction energy based on the measured ζ potentials.

Additional studies relating the anomalous particle deposition rates observed in experiments under unfavourable chemical conditions to surface charge heterogeneity have been reported for deposition in packed beds (Elimelech and O'Melia, 1990a; Vaidyanathan and Tien, 1991), stagnation-point flow (Adamczyk *et al.*, 1983; Dabros and van de Ven, 1983) and parallel-plate channels (Sjollema and Busscher, 1989, 1990). Kihira *et al.* (1992) have also attributed the discrepancies between theory and experiments (in heterocoagulation kinetics) to surface charge heterogeneities which result from the discreteness of surface charge.

Quantitative approaches to investigate the effect of surface charge heterogeneity on initial particle deposition rates, under unfavourable chemical conditions, have been recently proposed by Vaidyanathan and Tien (1991) and Song *et al.* (1994) for non-Brownian and Brownian particles, respectively. The results of these studies are summarized below.

The model of Vaidyanathan and Tien (1991)

Vaidyanathan and Tien (1991) and Vaidyanathan (1992) assumed that collector surfaces, such as glass surfaces commonly used in deposition studies, have two kinds of OH functional groups. The more acidic sites are assumed to dissociate completely and have a negative charge, while the less acidic sites are assumed not to contribute to the negative charge of the collector due to their negligible dissociation. The charged regions on the surface were modelled as uniform circular patches. The patches were assumed to be much smaller than a typical spherical collector in the packed bed, but much larger than a single OH surface site.

The model of Vaidyanathan and Tien involved the determination of the surface potential of the patch and its relation to the electrokinetic (ζ) potential of the collector, the calculation of the double layer force between a particle and the patched surface, and the determination of particle deposition rate on the patched surface (collector). Numerical calculations with arbitrarily selected patch sizes and collector fractions with patched regions showed that theoretically predicted collision efficiencies become less sensitive to ionic strength. Closer agreement between experimental results and theoretical predictions was obtained with the use of the patched surface model. These model calculations clearly demonstrated the importance of surface charge heterogeneity in particle deposition in the presence of repulsive double layer interactions. The model, however, cannot resolve the discrepancies between theory and observations because of its limited predictive capabilities and the arbitrary choice of the parameters characterizing the patched surfaces (i.e. size and surface potential of patches, distribution of patches, and fraction of the collector having undissociated functional groups).

The model of Song *et al.* (1994)

In spite of the modelling efforts described above, no general theory has been presented for which double layer interactions between heterogeneously charged surfaces may be determined with relative ease. Song *et al.* (1994) have recently introduced the concept of nominal surface potential as a means of calculating the double layer interaction energy between a particle and heterogeneous collector surfaces with the use of the current DLVO theory.

The nominal surface potential of a heterogeneous surface is equivalent to the potential of a homogeneous surface which would produce the same double layer interaction with a particle as the heterogeneous surface, at an identical separation distance. A value for the nominal potential of a surface should not be taken as an attempt at describing the actual potential of a heterogeneously charged surface; instead, it is only intended to represent the *homogeneous analogue* of a heterogeneously charged surface. By replacing the actual surface potential with a nominal surface potential, double layer interaction energies are easily calculated between heterogeneous surfaces by using the general expressions developed for interaction between homogeneously charged surfaces (Table 3.2).

Song *et al.* (1994) have developed a general theoretical approach for the calculation of colloid deposition rate onto heterogeneously charged surfaces. Patchwise and random (continuous) distribution models were used to quantitatively describe surface charge heterogeneity and its effect on the kinetics of colloid deposition. It was shown that with the concept of nominal surface potential, the general equation for calculating collision efficiencies (or deposition rates) with a distribution of surface properties (eqn (11.1)) can be used to calculate deposition rates onto heterogeneously charged surfaces. In the case of surface charge heterogeneity, a distribution in local surface potential of collector surfaces is considered rather than a distribution of mean (overall) surface potentials of collectors or particles.

Theoretical predictions showed that surfaces with only minor amounts of charge heterogeneity have particle deposition rates that are orders of magnitude larger than similar surfaces having no charge heterogeneity. Hence, a consideration of a distribution in local surface potential is of paramount importance, since minor surface charge heterogeneities will probably not change the mean surface potential of a collector surface but will dramatically change the particle deposition rate. It was further shown by Song *et al.* (1994) that the sensitivity of particle deposition rate to solution ionic strength decreases as the degree of surface charge heterogeneity increases.

Parameters characterizing the surface charge heterogeneity of collectors can be identified from experimental data of colloid deposition by using the inverse procedure of parameter estimation. Song *et al.* (1994) have shown that these heterogeneity parameters can be used in conjunction with current theories of particle deposition to explain experimental results of colloid deposition rates under chemical conditions that are unfavourable for particle deposition. As with the previously discussed model of Vaidyanathan and Tien (1991), the major limitation of this model is the inability to determine directly the heterogeneity parameters for particles and collectors at a given solution chemistry.

11.3.3 Surface roughness of particles and collectors

In the classical DLVO theory of colloidal stability, it is assumed that the interacting bodies are ideally shaped and perfectly smooth (i.e. spherical particles and perfectly smooth collectors). In real systems, however, surface irregularities always exist. The poor agreement between observations and theory in studies concerned with the kinetics of particle deposition has been attributed to surface roughness by several investigators (e.g. Marshal and Kitchener, 1966; Hull and Kitchener, 1969).

The disparity with respect to the effect of particle size on the kinetics of particle aggregation was originally attributed to the effect of surface roughness (Reerink and Overbeek, 1954; Overbeek, 1982). Overbeek (1982), one of the architects of the DLVO theory, discussed the strong and weak points of the DLVO theory. It was pointed out that one of the weakest points of the theory is the lack of influence of particle size in coagulation kinetics. Overbeek (1982) stated: 'It is not clear at this moment whether this discrepancy is due to the experiments (particles are not spherical, particle surface not smooth enough) or whether serious flaw of the theory shows here.'

Reerink and Overbeek (1954) suggested that the total interaction energy may be determined by the radii of curvature of the protrusions rather than by the curvature of the interacting particles. This idea can be applied in a simple and semiquantitative way to particle deposition (Elimelech and O'Melia, 1990a) as described below.

In this approach, roughness is assigned to particles and collectors as small half-spheres protruding from the surface. A schematic description of a spherical

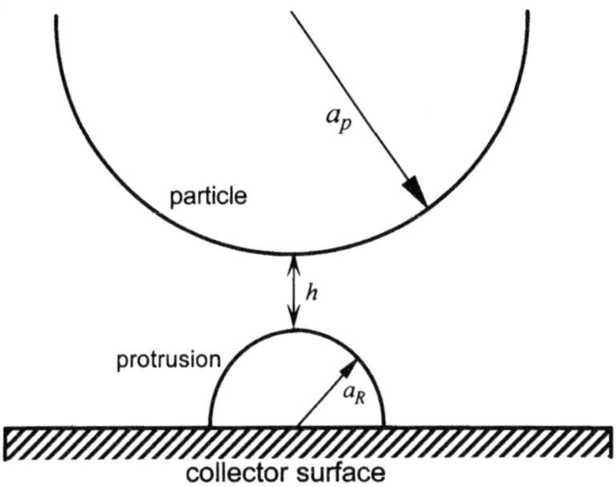

Figure 11.10 *Modelling of surface roughness in particle deposition. A rough collector with a protrusion having a radius of curvature* a_R *interacting with a smooth, spherical particle of radius* a_p

particle interacting with a rough planar collector (or a spherical collector which is much larger than the particle) is presented in Figure 11.10. As a first approximation, the total interaction energy is calculated as a superposition of interactions. Thus, for a case of a smooth spherical particle interacting with a rough collector, the interaction energy is considered as the sum of: (1) particle–half-sphere (protrusion) interaction, ϕ_1; and (2) particle–smooth collector interaction, ϕ_2. The distance of closest approach in calculating ϕ_2 is determined by the size (radius) of the protrusion on the surface of the collector. Based on this model, the total interaction energy as a function of the separation distance, h, between a spherical particle and a surface with a half-sphere protrusion of radius a_R is:

$$\phi_T(h) = \phi_1(h) + \phi_2(h + a_R) \qquad (11.3)$$

Model calculations for the total interaction energies (ϕ_1, ϕ_2, and ϕ_T) of a smooth particle interacting with a rough collector, based on the model described by eqn (11.3), are shown in Figure 11.11. For protrusions much smaller than the size of the interacting particle, surface roughness causes a decrease in the height of the energy barrier. This is because of the smaller contribution from the particle–smooth collector repulsion, owing to the presence of a distance of closest approach. For instance, the height of the energy barrier of the colloidal system described in Figure 11.11 is $23kT$ as compared to $212kT$ when roughness is not considered. As the radius of curvature of the protrusion increases, the energy barrier starts to increase at a certain protrusion size, as a result of the increase in repulsion caused by the particle–protrusion interaction (ϕ_1). This repulsion increases with the increase in protrusion size more significantly than the decrease of the repulsion due to the increase in the distance of closest approach.

Theoretical calculations of collision efficiencies as a function of electrolyte concentration for a sphere interacting with a rough collector, with this surface roughness model, were carried out by Elimelech and O'Melia (1990a).

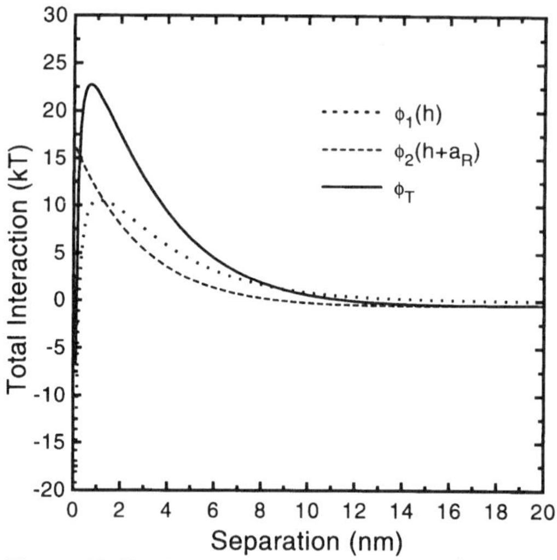

Figure 11.11 *Interaction energy curves when a rough collector with a protrusion having a radius of curvature of 10 nm interacts with smooth particles. The following parameters were used in the calculations: particle diameter = 0.4 μm; surface potentials of particles and collectors = –45 and –30 mV, respectively; Hamaker constant = 10^{-20} J; ionic strength = 0.01 M; and temperature = 24°C (after Elimelech and O'Melia, 1990a)*

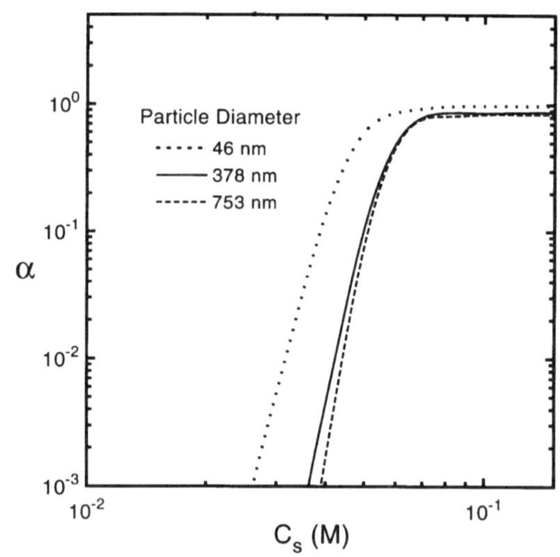

Figure 11.12 *Theoretical stability curves of three particle suspensions interacting with rough collectors. The radius of curvature of protrusions on collectors has been taken as 10 nm. The following parameters were used: surface potentials of particles and collectors were replaced by the mean measured zeta potentials, collector diameter = 0.2 mm, approach velocity = 0.136 cm s^{-1}, porosity = 0.4, temperature = 24°C (after Elimelech and O'Melia, 1990a)*

Interaction energies were calculated from eqn (11.3), and collision efficiencies from the interaction-force boundary-layer approximation. The stability curves for three colloidal suspensions of different size are presented in Figure 11.12. The results show that the slopes of the stability curves are much less sensitive to particle size (compared to predictions with smooth collectors). However, the slopes are still much steeper than the experimental results, and the critical deposition concentrations are still dependent on particle size. Hence, this simple and idealized model for surface roughness cannot resolve the discrepancies between theory and observations in particle deposition kinetics. Furthermore, for rough surfaces, distributions in protrusion size and location on collectors should be considered, as well as the interaction of a particle with more than one protrusion. However, these important aspects of surface roughness are very difficult to model and their effect of collision efficiency cannot be assessed at the present time.

The effect of surface roughness on van der Waals attraction energy was studied by Czarnecki and Dabros (1980) and Czarnecki (1986). According to their model, rough particles have solid cores and rough shells with given radial density distributions. It was found that at all small separations, surface roughness reduces the van der Waals attraction energy by a factor of approximately H/S, where H is the separation between the outermost peaks of the surface irregularities and S is the distance between the mean profiles of particles. The influence of surface roughness was found to be significant at small separations only. At larger separations, the effect of surface roughness becomes insignificant and classical expressions for smooth bodies may be adequately used. To determine whether it is necessary to consider surface roughness, the ratio of the thickness of surface rough layers to the separation between the bodies may be used. In the cases where the thicknesses of the rough layers are of the same order of magnitude as the surface–surface separation, the effect of surface roughness should be considered. The calculations of Czarnecki (1986) showed that for small separations, surface roughness may cause a decrease in the van der Waals attraction by a factor of 10 or more. As a result, it can be concluded that consideration of surface roughness in calculating van der Waals attraction energies cannot resolve the discrepancy between theory and experiments in particle deposition or coagulation in the presence of repulsive double layer interactions. Moreover, this can result in poorer agreement between theoretical predictions and experimental results since higher repulsive energies are obtained.

11.3.4 Dynamics of interaction

Several types of double layer interaction are possible during the interaction of a particle with a surface in particle deposition. In general, these depend on the relaxation (equilibration) time of ions in the double layer relative to the encounter (collision) time of a particle with a surface. Current theories for calculating electrical double layer interaction energies are for interaction at constant potential (Hogg *et al.*, 1966), interaction at constant charge (Wiese and Healy,

1970; Gregory, 1975) and intermediate interactions (Gregory, 1975; Chan *et al.*, 1976; Healy *et al.*, 1980). These theories are considered 'static' since the various dynamic interfacial processes taking place during particle collision in deposition are not considered. Constant-charge interactions produce the largest interaction energies, and constant-potential interactions result in the lowest calculated repulsive energies. Intermediate interaction modes produce intermediate repulsive energies.

The application of all types of 'static' interaction theories cannot resolve the discrepancies with respect to the magnitude of the collision efficiencies. Theoretical collision efficiencies are still extremely sensitive to the surface potentials of particles and collectors with the use of all of these electrical double layer interaction energy expressions. These also do not resolve the discrepancies discussed previously with respect to particle size. All static methods to calculate double layer interaction energies show a linear dependence of the interaction energy on particle size, and, consequently, a strong dependence of collision efficiency on particle size.

Studies of interfacial electrodynamics (Dukhin and Lyklema, 1987; van Leeuwen and Lyklema, 1987) suggest that, by considering the dynamics of interacting double layers, some basis for the apparent insensitivity of colloidal stability to particle size may be found. In those studies, it was found that the extent of lateral adjustment of charge during interaction is dominated by hydrodynamic drag. It was postulated that since the drag force depends on particle size, it may counteract the effect of particle size on the collision efficiency in particle coagulation and deposition.

More recently, Dukhin and Lyklema (1990) treated semiquantitatively the dynamics of double layer interaction by considering transient double layer disequilibration of interacting spherical particles. In their derivation, they considered desorption–adsorption disequilibrium of charge-determining ions as the origin of relaxation-determined retardation. An expression for the stability ratio of coagulating spheres, which includes an additional resistance term accounting for relaxation, has been derived. Their calculations showed that, under conditions where the retardation dominates, stability ratios become independent of particle size. Hence, this analysis can resolve the discrepancy with respect to particle size in particle–particle interaction phenomena. However, the resistance due to relaxation increases the stability ratios in coagulation (or decreases collision efficiencies in deposition), therefore making agreement between theory and experiment poorer. In a more recent study, Kiljstra and van Leeuwen (1993) argued that double layer dynamics can explain the discrepancy with respect to particle size only if the time-scale for the encounter is strongly dependent on the height of the energy barrier. Their theoretical analysis, however, indicates that this condition is unlikely to be satisfied in particle coagulation and deposition phenomena.

The studies of Dukhin and Lyklema (1990) and Kijlstra and van Leeuwen (1993), although based on several simplifying assumptions, are preliminary steps toward a more rigorous treatment of interfacial electrodynamics in particle–particle and particle–collector interaction phenomena. However, one should note that the actual effect of double layer interaction dynamics on colloidal stability and collision efficiency may be further complicated by the coupling between surface roughness, surface charge heterogeneities and interfacial electrodynamics. These complex factors may have a profound effect on colloidal stability, but, at the present time, we do not have adequate theories to quantitatively assess their role.

11.3.5 Deposition in secondary minima

Secondary minima in the total interaction energy profile, with the framework of the DLVO theory, can be obtained at large to moderate separation distances (usually larger than several nanometres). Under constant chemical conditions, the depth of the secondary minimum increases with an increase in particle size and the Hamaker constant of the interacting media. Secondary minima result from attractive van der Waals forces, whose energies have longer range than repulsive double layer energies.

In principle, coagulation of colloidal particles in secondary minima is possible, if the resultant of the forces of the thermal energy of particles and the fluid drag is insufficient to drive the particles out of the secondary minima. Indirect evidence for this phenomenon in coagulation has been reported by several investigators (e.g. Kotera *et al.*, 1970; Takamura *et al.*, 1979; Ludwig and Peschel, 1988). Regarding coagulation experiments, Wiese and Healy (1970) argued that the apparent independence of stability ratios on particle size can be understood in terms of an energetic rather than a kinetic criterion for colloidal stability. By incorporating both primary and secondary minima in coagulation, the anomalous particle size effects were shown to be in qualitative agreement with their theoretical calculations. Wiese and Healy wrote:

> We therefore conclude that a general DLVO treatment incorporating primary and secondary minima does account for apparently anomalous particle size effects . . . The rate of secondary minima coagulation has never been treated quantitatively before; nevertheless, there is much evidence that such coagulation does occur in systems for which the interaction energy curves show large energy barrier and deep secondary minima.

Other theoretical and experimental studies relating anomalous particle size effects to coagulation in secondary minima have been reported (e.g. Hogg and Yang, 1976; Marmur, 1979; Penners and Koopal, 1987). These semiquantitative studies, however, did not consider the effects of fluid shear, which may prevent this type of coagulation.

While some evidence exists for coagulation in secondary minima, there is no clear evidence for particle deposition from flowing suspensions in secondary minima. Significant deposition of particles in secondary minima onto smooth surfaces is unlikely when the fluid in which the particles are suspended is moving. Theoretical analysis for the trajectory of a non-Brownian particle around a spherical collector in a creeping flow, based on a force balance, predicts that deposition in a secondary minimum is possible only in a small region at the rear stagnation point of the spherical collector (Spielman and Cukor, 1973;

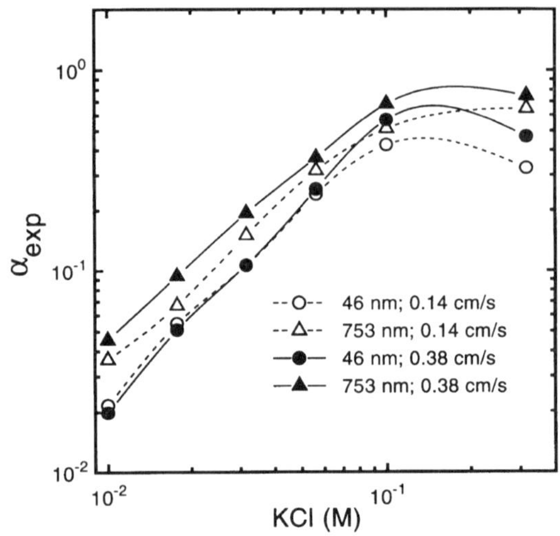

Figure 11.13 *Stability curves of two suspensions of latex particles as determined from particle deposition experiments at different approach velocities (KCl concentration in M, bed depth = 20 cm, porosity = 0.4, collector diameter = 0.2 mm, temperature = 24°C, pH = 6.7) (after Elimelech and O'Melia, 1990a)*

Rajagopalan and Tien, 1979; Tobiason, 1987). Similar predictions are expected for Brownian particles; particles trapped in a secondary minimum will translate and rotate due to fluid drag and shear until they reach the rear stagnation point where the net force (sum of van der Waals, electrical double layer and fluid drag forces) on the particle is zero. Furthermore, the width of the secondary-minimum region is on the order of a few nanometres, which implies that at most one particle can be collected by each collector.

Attractive forces for Brownian particles in secondary minima are small and can probably be overcome by fluid drag forces. The fluid drag force for a given particle can be varied systematically by changing the approach velocity of the fluid towards the collector. Elimelech and O'Melia (1990a) tested this effect by measuring collision efficiencies of colloidal particles in packed beds at two different approach velocities (0.14 cm s^{-1} and 0.38 cm s^{-1}). These results for two colloidal suspensions are presented in Figure 11.13. It is observed that the slope of the stability curves for the two colloidal suspensions at these approach velocities is comparable. If deposition of latex particles in secondary minima is feasible, it is expected that the almost three-fold increase of the approach velocity would have an effect on the particle concentration profile in the secondary minima. Therefore, experimental particle deposition rates and collision efficiencies would be different at these velocities. However, as observed, the stability curves are rather similar, indicating that significant deposition in secondary minima is not likely to occur under these conditions.

11.3.6 Possibility of additional forces

The large discrepancy between theoretical predictions and experimental results may be attributed to the fact that the DLVO theory considers only two types of interactions, the van der Waals and electrical double layer forces. This question was raised by Spielman (1977), who stated: 'When electrical double layer repulsion is significant, agreement between experiment and theory is so poor as to cast doubt upon the widely accepted view that van der Waals attraction and electrical double layer repulsion alone adequately characterize surface interactions.' Additional colloidal forces, not included in the classical DLVO theory, have been discovered by direct force measurements (Israelachvili, 1992; Christenson, 1988; Israelachvili and McGuiggan, 1988). These non-DLVO forces, which were discussed in some detail in Chapter 3, include hydration and hydrophobic forces.

Force measurements indicate that hydration forces are important only for ionic strengths larger than about 10^{-3} M (Israelachvili, 1992). These short-range forces are repulsive and, therefore, will increase the energy barrier for deposition or coagulation. No theoretical expressions that can be used to predict the hydration repulsive energy exist. In principle, inclusion of these repulsive forces in theoretical predictions of deposition or coagulation kinetics will result in smaller collision efficiencies and hence poorer agreement with observations. Hydrophobic forces, on the other hand, are attractive and could increase the agreement between observations and theoretical predictions. However, these forces only act between strongly hydrophobic surfaces. Discrepancies between theory and experiment in particle deposition are observed with hydrophobic as well as hydrophilic surfaces. Thus, the non-DLVO forces cannot resolve the large discrepancy between predicted and observed collision efficiencies.

Consideration of non-DLVO forces most probably would not resolve the discrepancy with respect to particle size effects in the kinetics of particle deposition or coagulation. All types of forces between interacting spherical particles or particles interacting with a smooth surface are dependent on the size of the particles. This is a result of the necessity to consider the radius of curvature of particles in the force calculations. The general form of an interaction energy of type i between two particles or a particle and a surface is:

$$\phi_i = f(a)\, g(\pi_1, \pi_2, \ldots, \pi_n) \tag{11.4}$$

where π_i are physical and chemical-colloidal variables, and $f(a)$ is a geometrical factor. The geometrical factor is given by $(a_1 a_2)/(a_1 + a_2)$ for the interaction of two particles with radii a_1 and a_2 (as in the case of coagulation), and by a_p for a particle of radius a_p interacting with a planar surface (as in the case of deposition). Thus, the dependence of the total interaction energy and hence the collision efficiency on particle size remains unchanged with the inclusion of non-DLVO forces.

11.4 A semi-empirical approach for predicting collision efficiencies

As shown in the previous sections, the classical DLVO theory, although elegant, certainly cannot be used directly to predict collision efficiencies in the initial stage of colloid deposition. An alternative approach is required to predict collision efficiencies in technological and natural colloidal processes. In this section, we present a semi-empirical approach to predict collision efficiencies in salt-induced colloid deposition (or aggregation). This approach was proposed by Elimelech (1992) and is described in more detail in that reference. The development and testing of this semi-empirical model are described below.

11.4.1 Theoretical formulation of a correlation equation

Approximate analytical expressions for the collision efficiency of small Brownian particles in the presence of repulsive electrical double layer interactions show that the collision efficiency in colloid deposition (and coagulation) can be presented in the *Arrhenius* form (Ruckenstein and Prieve, 1973; Spielman and Friedlander, 1974):

$$\alpha = \Gamma \exp\left(-\frac{\phi_{max}}{kT}\right) \tag{11.5}$$

Here ϕ_{max} is the height of the energy barrier in the interaction energy profile (absolute value), which, according to the DLVO theory, is the sum of van der Waals and electrical double layer interactions; k is the Boltzmann constant; and T is the absolute temperature. By analogy to chemical reaction kinetics, Γ is a function that can be viewed as a frequency factor, which depends on some physical and colloidal properties of the deposition system (e.g. Spielman and Friedlander, 1974). The energy barrier ϕ_{max}, located at a distance of a few nanometres from the surface of the collectors, can be considered as the 'activation energy' for collision. Particles that can overcome this activation energy will be captured by the collectors.

In unfavourable deposition or coagulation (i.e. when repulsive energy barriers exist), calculations show that the collision efficiency is dominated by the exponential term of eqn (11.5). A linear relationship between the logarithm of the theoretical collision efficiency and the logarithm of the energy barrier is obtained (Elimelech and O'Melia, 1990b). Predicted collision efficiencies are extremely sensitive to the height of the energy barrier. This dependence is also obtained from a numerical solution of the convective diffusion equation of Brownian particles (e.g. Adamczyk *et al.*, 1983). Furthermore, similar results are obtained for collision efficiencies of non-Brownian particles calculated by the trajectory analysis (e.g. Tien, 1989). In the trajectory analysis, the collision efficiency drops abruptly to a very small number above a critical value of the energy barrier. The drop in the collision efficiency in this case (i.e. for non-Brownian particles) is much more dramatic than in the case with Brownian particles.

Based on the DLVO theory, the height of the energy barrier (ϕ_{max}) between a spherical particle and a surface (in deposition) or between particles (in coagulation), in aqueous solutions at a given temperature, has the following form:

$$\phi_{max} = f(a_p, \psi_p, \psi_c, I, A) \tag{11.6}$$

where a_p is the particle radius; ψ_p and ψ_c are the surface potentials of particles and collectors, respectively; I is the ionic strength of the solution; and A is the Hamaker constant of the interacting media. Hence, in theory, the collision efficiency in colloid deposition depends predominantly on the variables controlling the height of the energy barrier (as listed in eqn (11.6)).

In our analysis, the inverse Debye length (κ) is used to depict the effect of ionic strength on the collision efficiency. The Debye length (κ^{-1}) is a characteristic length for the diffuse double layer thickness. At low ionic strengths, the value of the Debye length is high and repulsive double layer interactions extend to large distances. Since electrical double layer interactions arise from the overlap of diffuse double layers during the deposition process, the parameter κ will have a significant effect on the collision efficiency. The inverse Debye length, described previously in Chapter 2, is defined as (in SI units):

$$\kappa = \left(\frac{2 \times 10^3 \, e^2 N_A I}{\varepsilon_0 \varepsilon_r kT} \right)^{1/2} \tag{11.7}$$

Here e is the elementary charge of an electron; N_A is the Avogadro number; I is the ionic strength in mol l^{-1}, and ε_o and ε_r are the permittivity in vacuum and the relative permittivity of water, respectively.

It is a common observation in theoretical and experimental studies that colloid deposition and aggregation rates are highly dependent on the magnitudes of surface potentials of particles and collectors. Theoretical expressions for repulsive electrical double layer interactions (see Tables 3.1 and 3.2) show that the magnitude of these interactions at a given ionic strength is roughly proportional to the product of the surface potentials of the interacting bodies. Therefore, the effect of surface potentials of particles and collectors can be expressed by the parameter $\varepsilon_0 \varepsilon_r \psi_p \psi_c$. This parameter characterizes the repulsive electrostatic interactions between particles and collectors and has dimensions of force.

Theoretical predictions with the framework of the DLVO theory (like those presented in Figure 5.9) show that the particle size (diameter) has a marked effect on the collision efficiency (Elimelech and O'Melia, 1990a). This is a result of the linear dependence of van der Waals and electrical double layer interaction forces (and subsequently the energy barrier) on particle size. Experimental studies (Elimelech and O'Melia, 1990a), on the other hand, have found that experimental collision efficiencies in particle deposition are virtually independent of particle size (Figure 11.1). The lack of dependence of experimental collision efficiencies on particle size was also obtained in numerous well-controlled coagulation studies (e.g. Ottewill and Shaw, 1966; Penners and Koopal, 1987). Therefore, it can be assumed in this theoretical analysis that the particle radius (a_p) has no effect on the actual collision efficiency.

Based on the analysis discussed above, it is postulated that the actual collision efficiency in colloid deposition depends primarily on the following parameters: (1) $\varepsilon_o \varepsilon_r \psi_p \psi_c$, characterizing the electrical double layer repulsion forces; (2) κ, the inverse Debye length, characterizing the diffuse double layer thickness and the range of the repulsive double layer forces; and (3) A, the Hamaker constant of the interacting media, characterizing the universal van der Waals attraction forces. Hence, the collision efficiency should have the following general dependence:

$$\alpha = g(\varepsilon_0 \varepsilon_r \psi_p \psi_c, \kappa, A) \tag{11.8}$$

In order to perform dimensional analysis, eqn (11.8) is written in the form of:

$$\alpha = B(\varepsilon_0 \varepsilon_r \psi_p \psi_c)^m \kappa^n A^s \tag{11.9}$$

where B, m, n and s are constants. This equation must be dimensionally homogeneous, which results in $n = s$ and $m = -n$. Therefore, the collision efficiency reduces to:

$$\alpha = B \left(\frac{\kappa A}{\varepsilon_0 \varepsilon_r \psi_p \psi_c} \right)^n \tag{11.10}$$

or in the linearized form:

$$\log \alpha = \log B + n \log N_{col} \tag{11.11}$$

Here B and n are dimensionless constants that should be determined experimentally, and N_{col} is a dimensionless parameter characterizing the collision efficiency of the system. The latter is defined as:

$$N_{col} = \frac{\kappa A}{\varepsilon_0 \varepsilon_r \psi_p \psi_c} \tag{11.12}$$

11.4.2 Testing the correlation equation

There are a few studies of well-controlled colloid deposition experiments that can be used to test the semi-empirical correlation equation presented above. The requirements from such studies are: (1) that the deposition experiments be conducted under unfavourable chemical conditions, i.e. when repulsive double layer interactions predominate; (2) that the deposition experiments be conducted in the presence of indifferent electrolytes, so that particle deposition rates are not controlled by the effect of specific chemical interactions; (3) that the electrokinetic (ζ) potentials of particles and collectors at different electrolyte concentrations be known; and (4) that the reported deposition rates be those for the initial stage of deposition.

Three colloid deposition studies meeting the above conditions were used by Elimelech (1992) to test the relation described by eqn (11.11). Tabulated values of the collision efficiencies, the electrolyte concentrations, the electrokinetic potentials of particles and collectors, and the calculated $\log N_{col}$ of these studies can be found elsewhere (Elimelech, 1992). The electrokinetic (ζ) potentials of

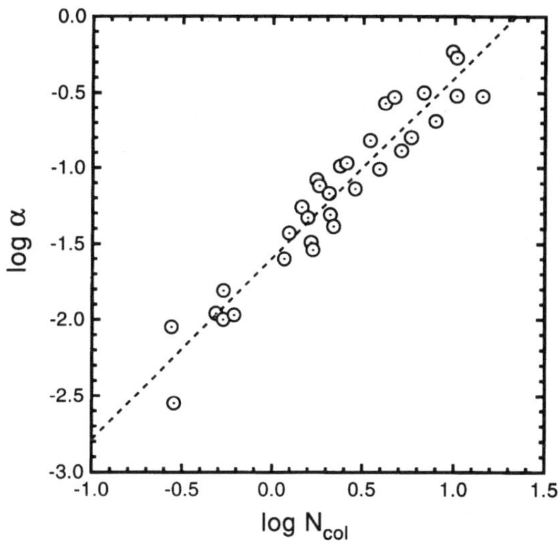

Figure 11.14 *Logarithm of the collision efficiency* (α) *as a function of the logarithm of the dimensionless parameter* (N_{col}) *for experimental data of latex particle deposition on glass beads. Linear regression analysis yields the following parameters:* $n = 1.19$, $\log B = -1.59$, *and* $r^2 = 0.91$

particles and collectors were used instead of the surface potentials in the calculation of $\varepsilon_o \varepsilon_r \psi_p \psi_c$. In addition, a value of 10^{-20} J was used for the Hamaker constant of the latex–water–glass media in the calculation of N_{col}.

A plot of $\log \alpha$ against $\log N_{col}$, using the data from the above-mentioned studies, is presented in Figure 11.14. It is clearly demonstrated that a fairly good linear correlation exists ($r^2 = 0.91$), in accord with eqn (11.11). The slope of the curve is close to unity ($n = 1.19$), suggesting that the actual collision efficiencies in these deposition experiments are inversely proportional to the product of the surface potentials of particles and collectors and linearly dependent on the Hamaker constant and the Debye reciprocal length (κ). Since κ is a function of $I^{1/2}$ (eqn (11.7)), the actual collision efficiency is proportional to the square root of the ionic strength. This finding indicates that the actual dependence of the collision efficiency on ionic strength and surface potentials of particles and collectors is much more moderate than that predicted by the current fundamental theories (see discussion in Chapter 5). As discussed previously, theoretical predictions within the framework of the DLVO theory indicate that collision efficiencies are extremely sensitive to these parameters.

In the analysis presented in this section, experimental collision efficiencies were correlated successfully to a dimensionless parameter that includes some measurable system properties. The correlation equation was developed to account for experimental results that cannot yet be explained quantitatively by the most fundamental theories. As discussed earlier in this chapter, other factors, such as chemical and physical heterogeneities, may have a significant effect on the actual collision efficiency in colloid deposition.

It should be emphasized that the results presented here are limited to collision efficiencies with indifferent electrolytes in the unfavourable deposition region. In practice, this region can be assigned for collision efficiency (α) smaller than about 0.5. In this case, the collision efficiencies are controlled by varying the electrolyte concentration of the solution. An increase in the electrolyte concentration results in a decrease in the absolute value of the surface (Stern) potentials of particles and collectors (Figure 2.2) and compression of the diffuse double layers. As a result, the range of the double layer forces and the height of the energy barrier are reduced so that the collision efficiency increases. The application of this approach to collision efficiencies in the presence of specifically interacting solutes is probably inappropriate. When specific interaction takes place, the collision efficiency will vary predominantly because of the change of the surface potentials caused by the specific interaction of solutes with the surfaces of colloids and collectors. The ionic strength and the Hamaker constant will make a small contribution to the variation of the collision efficiency in this case.

Bibliography

Hirtzel, C. S. and Rajagopalan, R. (1985) *Colloidal phenomena: Advanced Topics*, Noyes Publications, Park Ridge, New Jersey

Russel, W. B., Saville, D. A. and Schowalter, W. R. (1989) *Colloidal Dispersions* Cambridge University Press, Cambridge

Tien, C. (1989) *Granular Filtration of Aerosols and Hydrosols* Butterworth Publishers, Stoneham, Massachusetts

References

Adamczyk, Z., Czarnecki, J., Dabros, T. and van de Ven, T. G. M. (1983) Particle transfer to solid surfaces. *Adv. Colloid Interface Sci.*, **19**, 183–252

Adamczyk, Z., Siwek, B., Zembala, M. and Warszynski, P. (1989) Enhanced deposition of particles under attractive double-layer forces. *J. Colloid Interface Sci.*, **130**, 578–587

Bowen, B. D. and Epstein, N. (1979) Fine particle deposition in smooth parallel-plate channels. *J. Colloid Interface Sci.*, **72**, 81–97

Chan, D., Healy, T. W. and White, L. R. (1976) Electrical double layer interaction under regulation by surface ionization equilibria. *J. Chem. Soc. Faraday Trans I*, **72**, 2844–2865

Christenson, H. K. (1988) Non-DLVO forces between surfaces – solvation, hydration, and capillary effects. *J. Dispersion Sci. Technol.*, **9**, 171–206

Czarnecki, J. (1986) The effect of surface inhomogeneities on the interactions in colloidal systems and colloid stability. *Adv. Colloid Interface Sci.*, **24**, 283–319

Czarnecki, J. and Dabros, T. (1980) Attenuation of the van der Waals attraction energy in particle/semi-infinite medium system due to the roughness of particle surface. *J. Colloid Interface Sci.*, **78**, 25–30

Dabros, T. and van de Ven, T. G. M. (1983) A direct method for studying particle deposition onto solid surfaces. *Colloid Polymer Sci.*, **261**, 694–704

Dukhin, S. S. and Lyklema, J. (1987) Dynamics of colloid particle interaction. *Langmuir*, **3**, 94–98

Dukhin, S. S. and Lyklema, J. (1990) Dynamics of colloid particle-interaction: incomplete desorption relaxation. *Faraday Disc. Chem. Soc.*, **90**, 261–269

Elimelech, M. (1989) The effect of particle size on the kinetics of deposition of Brownian particles in porous media. PhD Dissertation, The Johns Hopkins University, Baltimore, Maryland

Elimelech, M. (1991) Kinetics of capture of colloidal particles in packed beds under attractive double layer interactions. *J. Colloid Interface Sci.*, **146**, 337–352

Elimelech, M. (1992) Predicting collision efficiencies of colloidal particles in porous media. *Water Res.*, **26**, 1–8

Elimelech, M. (1994) Effect of particle size on the kinetics of particle deposition under attractive double layer interactions. *J. Colloid Interface Sci.*, **164**, 190–199

Elimelech, M. and O'Melia, C. R. (1990a) Effect of particle size on collision efficiency in the deposition of Brownian particles with electrostatic energy barriers. *Langmuir*, **6**, 1153–1163

Elimelech, M. and O'Melia, C. R. (1990b) Kinetics of deposition of colloidal particles in porous media. *Environ. Sci. Technol.*, **24**, 1528–1536

Elimelech, M. and Song, L. (1992) Theoretical investigation of colloid separation from dilute aqueous suspensions by oppositely charged granular media. *Separat. Technol.*, **2**, 2–12

FitzPatrick, J. A. and Spielman, L. A. (1973) Filtration of aqueous latex suspensions through beds of glass spheres. *J. Colloid Interface Sci.*, **43**, 350–369

Gregory, J. (1975) Interaction of unequal double layers at constant charge. *J. Colloid Interface Sci.*, **61**, 44–51

Gregory, J. and Wishart, A. J. (1980) Deposition of latex particles on alumina fibers. *Colloids Surfaces*, **1**, 313–334

Healy, T. W., Chan, D. and White, L. R. (1980) Colloidal behavior of materials with ionizable group surfaces. *Pure Appl. Chem.*, **52**, 1207–1219

Hiemstra, T., Van Riemsdijk, W. H. and Bolt, G. H. (1989a) Multisite proton adsorption modeling at the solid/solution interface of (Hydr)oxides: a new approach. I. Model description and evaluation of intrinsic reaction constants. *J. Colloid Interface Sci.*, **133**, 91–104

Hiemstra, T., De Wit, J. C. M., and Van Riemsdijk, W. H. (1989b) Multisite proton adsorption modeling at the solid/solution interface of (Hydr)oxides: a new approach.
II. Application to various important (hydr)oxides. *J. Colloid Interface Sci.*, **133**, 105–117

Hogg, R. and Yang, G. C. (1976) Secondary coagulation. *J. Colloid Interface Sci.*, **56**, 573–576

Hogg, R., Healy, T. W. and Fuerstenau, D. W. (1966) Mutual coagulation of colloidal dispersions. *Trans Faraday Soci.*, **66**, 1638–1651

Hull, M. and Kitchener, J. A. (1969) Interaction of spherical colloidal particles with planar surfaces. *Trans Faraday Soc.*, **65**, 3093–3104

Israelachvili, J. N. (1992) *Intermolecular and Surface Forces*, Academic Press, New York and London

Israelachvili, J. N. and McGuiggan, P. M. (1988) Forces between surfaces in liquids. *Science*, **241**, 795–780

Jaroniec, M. and Madey, R. (1988) *Physical Adsorption on Heterogeneous Solids*, Elsevier, Amsterdam

Jia, X., Williams, R. A. and Green, M. (1990) The feasibility of selective separation of colloidal particles at a solid/liquid interface. *Chem. Eng. Commun.*, **91**, 199–223

Kallay, N., Nelligan, J. D. and Matijevic, E. (1983) Particle adhesion and removal in model systems. Part 6 – Kinetics of deposition of hematite particles on steel. *J. Chem. Soci., Faraday Trans I*, **79**, 65–74

Kihira, H. Ryde, N. and Matijevic, E. (1992) Kinetics of heterocoagulation. Part 2 – The effect of discreteness of surface charge. *J. Chem. Soc., Faraday Trans*, **88**, 2379–2386

Kiljstra, J. and van Leeuwen, H. P. (1993) Surface charge relaxation during coagulation. *J. Colloid*

Interface Sci., **160**, 424–434

Koopal, L. K. and Dukhin, S. S. (1993) Modeling of the double layer and electrosorption of patchwise heterogeneous surface on the basis of its homogeneous analogue: 1. Non-interacting patches. *Colloids Surfaces A*, **73**, 201–209

Koopal, L. K. and Van Riemsdijk, W. H. (1989) Electrosorption on random and patchwise heterogeneous surfaces: electrical double layer effects. *J. Colloid Interface Sci.*, **128**, 188–200

Kotera, A., Furusawa, K. and Kudo, K. (1970) Colloid chemical studies of polystyrene latices polymerized without any surface-active agents. **II**. Coagulation into secondary minimum. *Kolloid-Z. u. Z. Polymere*, **240**, 837–842

Kuin, A. J. (1990) Interaction potential of a dispersion of colloidal particles with a non-homogeneous surface potential. *Faraday Discuss. Chem. Soc.*, **90**, 235–244

Litton, G. M. and Olson, T. M. (1993) Colloid deposition rates on silica bed media and artifacts related to collector surface preparation methods. *Environ. Sci. Technol.*, **27**, 185–193

Ludwig, P. and Peschel, G. (1988) Evidence for secondary minimum coagulation in a silica hydrosol obtained by dynamic light scattering. *Prog. Colloid Polymer Sci.*, **77**, 146–151

Marmur, A. (1979) A kinetic approach to primary and secondary minimum coagulations and their combinations. *J. Colloid Interface Sci.*, **72**, 41–48

Marshal, J. K. and Kitchener, J. A. (1966) The deposition of colloidal particles on smooth solids. *J. Colloid Interface Sci.*, **22**, 342–351

Matijevic, E. and Kallay, N. (1983) Kinetics of deposition of colloidal metal oxides particles on a steel surface. *Croat. Chem. Acta*, **56**, 649–661

Ottewill, R. H. and Shaw, J. N. (1966) Stability of monodisperse polystyrene latex particles of various sizes. *Discuss. Faraday Soc.*, **42**, 154–163

Overbeek, J. Th. G. (1982) Strong and weak points in the interpretation of colloid stability. *Adv. Colloid Interface Sci.*, **58**, 408–422

Penners, N. H. G. and Koopal L. K. (1987) The effect of particle size on the stability of haematite (α–Fe_2O_3) hydrosols. *Colloids Surfaces*, **28**, 67–83

Prieve, D. C. and Lin, M. M. (1982) The effect of distribution in surface properties on colloid stability. *J. Colloid Interface Sci.*, **86**, 17–25

Prieve, D. C. and Ruckenstein, E. (1980) Role of surface chemistry in primary and secondary coagulation and heterocoagulation. *J. Colloid Interface Sci.*, **73**, 539–555

Rajagopalan, R. and Chu, R. Q. (1982) Dynamics of adsorption of particles in packed beds. *J. Colloid Interface Sci.*, **86**, 299–317

Rajagopalan, R. and Tien, C. (1977) Single collector analysis of collection mechanisms in water filtration. *Can. J. Chem. Eng.*, **55**, 246–255

Rajagopalan, R. and Tien, C. (1979) The theory of deep bed filtration. In *Progress in Filtration and Separation*, (ed. Wakeman, R. J.), Vol. 1, Elsevier, Amsterdam, 179–269

Reerink, H. and Overbeek, J. Th. G. (1954) The rate of coagulation as a measure of the stability of silver iodide sols. *Discuss. Faraday Soci.*, **18**, 74–84

Richmond, P. (1974) Electrical forces between particles with arbitrary fixed surface charge distributions in ionic solution. *J. Chem. Soci., Faraday Trans II*, **70**, 1066–1073

Richmond, P. (1975) Electrical forces between particles with discrete periodic surface charge distributions in ionic solution. *J. Chem. Soc., Faraday Trans II*, **71**, 1154–1163

Ruckenstein, E. and Prieve, D. C. (1973) Rate of deposition of Brownian particles under the action of London and double-layer forces. *J. Chem. Soc., Faraday Trans II*, **69**, 1522–1536

Schuhmann, D. (1984) Perturbations to the electrophoretic mobility due to electrical surface effects. *Phys. Chem. Hydrodynamics*, **5**, 383–388

Schuhmann, D. and Depenoux, B. (1987) The effect of electrical nonhomogeneities of particles on mean surface potentials. *J. Colloid Interface Sci.*, **116**, 159–167

Sjollema, J. and Busscher, H. J. (1989) Deposition of polystyrene latex particles toward polymethylmethacrylate in a parallel plate flow cell. *J. Colloid Interface Sci.*, **132**, 382–394

Sjollema, J. and Busscher, H. J. (1990) Deposition of polystyrene particles in a parallel plate flow cell. 1. The influence of collector surface properties on the experimental deposition rate. *Colloids Surfaces*, **47**, 323–336

Song, L., Johnson, P. R. and Elimelech, M. (1994) Kinetics of colloid deposition onto heterogeneously charged surfaces in porous media. *Environ. Sci. Technol.***28**, 1164–1171

Spielman, L. A. (1977) Particle capture from low speed laminar flows. *Annu. Rev. Fluid Mech.*, **9**, 297–319

Spielman, L. A. and Cukor, P. M. (1973) Deposition of non-Brownian particles under colloidal forces. *J. Colloid Interface Sci.*, **43**, 51–65

Spielman, L. A. and Friedlander, S. K. (1974) Role of electric double layer in particle deposition by convective diffusion. *J. Colloid Interface Sci.*, **46**, 22–31

Sposito, G. (1984) *The Surface Chemistry of Soils*, Oxford University Press, New York

Takamura, K., Goldsmith, H. L. and Mason, S. G. (1979) The microrheology of colloidal dispersions. IX. Effects of simple and polyelectrolytes on rotation of doublets of spheres. *J. Colloid Interface Sci.*, **72**, 385–400

Thompson, J. E., Kallay, N. and Matijevic, E. (1983) Particle adhesion and removal in model systems. *Chem. Eng. Sci.*, **38**, 1901–1908

Tien, C. (1989) *Granular Filtration of Aerosols and Hydrosols*, Butterworth Publishers, Stoneham, Massachusetts

Tobiason, J. E. (1987) Physicochemical aspects of particle deposition in porous media. PhD Dissertation, The Johns Hopkins University, Baltimore, Maryland

Tobiason, J. E. and O'Melia, C. R. (1988) Physiochemical aspects of particle removal in depth filtration. *J. Am. Water Works Assoc.*, **80**, 54–64

Vaidyanathan, R. (1986) Hydrosol filtration in granular beds. Master thesis, Syracuse University, Syracuse, NY

Vaidyanathan, R. (1992) Double layer calculations for the attachment of a colloidal particle with a charged surface patch onto a substrate. *Separat. Technol.*, **2**, 98–103

Vaidyanathan, R. and Tien, C. (1991) Hydrosol deposition in granular media under unfavorable surface conditions. *Chem. Eng. Sci.*, **46**, 967–983

Van Leeuwen, H. P. and Lyklema, J. (1987) Interfacial electrodynamics of interacting colloidal particles: geometrical aspects. *Bunsenges Phys. Chem.*, **91**, 288–291

Wiese, G. R. and Healy, T. W. (1970) Effect of particle size on colloid stability. *Trans Faraday Soci*, **66**, 490–499

Yoshimura, Y., Ueda, K., Mori, F. and Yoshioka, N. (1980) Initial particle collection mechanism in clean, deep-bed filtration. *Int. Chem. Eng.*, **20**, 600–608

Part IV

Applications and limitations of predictive modelling

Chapter 12

Performance of packed-bed filters

Nomenclature

A	Hamaker's constant
A_s	porosity-dependent parameter defined by eqn (12.14)
a_c	collector radius
b	radius of fluid envelope in the sphere-in-cell model
C	particle concentration
C_0	influent particle concentration
d_c	collector diameter
d_p	particle diameter
D_∞	bulk particle diffusion coefficient
f	filter medium porosity
g	gravitational acceleration constant
I	overall particle deposition rate onto a spherical collector
L	filter bed depth
N_G	gravitational force number used in eqn (12.15)
N_{LO}	van der Waals force number used in eqn (12.15)
p	parameter used in eqn (12.14); $p=(1-f)^{1/3}$
R	aspect ratio; $R=d_p/d_c$
T	absolute temperature
t	time
U	approach (superficial) velocity
z	axial coordinate
α	collision (attachment) efficiency
η	single collector removal efficiency
η_0	single collector efficiency of a clean-bed
η_D	single collector efficiency due to diffusion defined by eqn (12.11)
η_G	single collector efficiency due to gravity defined by eqn (12.13)
η_I	single collector efficiency due to interception defined by eqn (12.12)
θ	reduced time defined by eqn (12.3)
λ	filter coefficient
μ	fluid viscosity
ρ	density of fluid
ρ_p	density of suspended particles
σ	specific deposit

Granular packed-bed filtration is a process widely used in water and wastewater treatment for removal of suspended particulate matter. In this process, the water to be treated flows through a packed bed of granular material (the so-called filter medium) and particulate materials are collected by the granular medium by various physicochemical mechanisms. Granular filters have been found effective for removing particulates of all size ranges, including algae, protozoan cysts, bacteria, viruses, collodial clays and asbestos fibres. Increasingly stringent standards for drinking water and discharge requirements for wastewater have led to expanded installation of granular filter units in water and wastewater treatment plants. Some excellent reviews of the use of granular filters in water and wastewater treatment can be found in O'Melia (1985), Amirtharajah (1988) and Letterman (1991). In this chapter, we will illustrate the application of fundamental theories of particle deposition to the modelling of particulate removal in packed-bed filters.

12.1 Particle removal mechanisms

It is convenient to divide the overall removal (or capture) of suspended particles by a granular filter into two sequential steps: transport and attachment. In the first step, particles are transported from the bulk fluid to the vicinity of the filter grains. The rate of particle transport is determined primarily by physical forces such as those arising from fluid flow or gravity, and by interception due to the finite size of the particles (Yao *et al.*, 1971). The transport step in granular filters can be quantitatively described by the convective-diffusion and trajectory equations in a porous medium as described in detail in Chapter 5. The attachment step in granular filtration is controlled by the surface-chemical characteristics of the particles and granular media and by the solution chemistry. It is common in water treatment to add chemicals to the raw water in order to alter the surface properties of the suspended particles and, thus, to maximize the attachment of particles to filter grains. These chemicals, commonly referred to as coagulants, can be metal salts (such as aluminium sulphate or ferric chloride), polyelectrolytes, or a combination of a metal salt and a polyelectrolyte.

The transport mechanisms in granular filtration are illustrated in Figure 12.1. In this figure, a single grain from the filter medium is represented as an idealized spherical collector. Three predominant mechanisms are responsible for the transport of particles from the bulk fluid to the filter grain: interception, gravity (sedimentation) and Brownian diffusion. Transport of particles by interception occurs when the particle moving with the streamline comes into contact with the filter grain because of its finite size. Interception is accounted for as a boundary condition in quantitative descriptions of particle filtration. Large particles with densities greater than the suspending fluid can settle downwards, cross fluid streamlines, and contact the collector; this transport mechanism is denoted as gravity in Figure 12.1. Small (submicrometre) particles undergo Brownian motion in the water. This Brownian motion (diffusion) can result in contact with the filter grains. The efficiency of transport by interception and gravity increases with an increase in the size of the suspended particles, while transport by

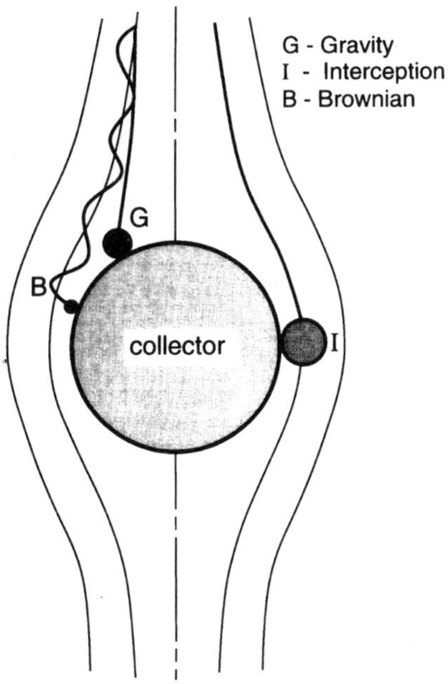

Figure 12.1 *Particle transport mechanisms towards a spherical collector in water filtration. The thick lines describe the particle path (trajectory) while the thin lines describe the fluid streamlines*

diffusion increases with a decrease in particle size. In general, under physical conditions typical for water and wastewater filtration, the transport of submicrometre particles is dominated by diffusion while that of larger (non-Brownian) particles is dominated by interception.

In addition to the three transport mechanisms discussed above, particles can contact filter grains due to physical screening. If a suspended particle is larger than the smallest pore through which fluid is flowing, the particle can be retained by the medium; this is referred to as straining. Straining is important in granular-bed filtration, when the ratio of the suspended particle diameter to the grain diameter is greater than about 0.2 (Herzig *et al.*, 1970). As the size of the pores is reduced due to the capture of particulate materials during filtration, the size of particles removed by straining becomes smaller.

12.2 Modelling of particle removal in granular filtration

Filtration theories that describe the capture (or removal) of particles in packed-bed filters can be classified into two types: phenomenological (macroscopic) and fundamental (microscopic). The ultimate aim of filtration models is to predict removal of particles as a function of suspension characteristics, physical and chemical properties of the filter and flow conditions. Important suspension characteristics are density, size, size distribution and surface chemistry of the particles; and temperature, viscosity and solution chemistry of the suspending medium. The parameters which describe the properties of packed-bed filters are bed depth, grain size, grain shape, grain surface chemistry and porosity. Another important parameter in packed-bed filtration is the approach (superficial) velocity. The approach velocity is defined as the ratio of the volumetric flow rate to the filter cross-section area. In the present chapter, we will focus on the microscopic (fundamental) approach for modelling particle removal in filtration. This will be preceded by a short description of the phenomenological approach.

12.2.1 Phenomenological theories

The phenomenological approach to modelling the process of filtration has developed from attempts to depict changes in the concentration of particles flowing through the filter. This type of formulation was first proposed by Iwasaki (1935) and has been developed extensively since that time. Phenomenological theories of particle removal in granular filters are derived from a conservation of mass expression and a rate expression to describe the rate of particle collection by filter grains. These are described by (Tien, 1989):

$$\frac{\partial \sigma}{\partial \theta} + U \frac{\partial C}{\partial z} = 0 \tag{12.1}$$

$$\frac{\partial \sigma}{\partial \theta} = - \lambda U C \tag{12.2}$$

and the corrected time, θ, is given as:

$$\theta = t - \int_0^z \frac{dz}{(U/f)} \tag{12.3}$$

Here σ is the specific deposit (volume of solid deposit per volume of bed), C is the concentration of particles in the fluid, z is an axial coordinate, U is the approach velocity, f is the bed porosity, and λ is the so-called filter coefficient. In these equations, the axial and radial dispersions were neglected; their effect is usually negligible in particle filtration (Tien, 1989). Since filtration begins with a clean filter bed, the initial and boundary conditions are:

$$C = 0, \sigma = 0 \quad \text{for } z \geq 0, \theta \leq 0 \tag{12.4}$$

$$C = C_0, \qquad \text{at } z = 0, \theta > 0 \tag{12.5}$$

where C_0 is the particle concentration of the suspension before entering the filter (i.e., the influent concentration of particles).

In the phenomenological approach, the physical and chemical properties of the suspension and the flow field in the granular bed are not explicitly accounted for.

These are implicitly incorporated in the filter coefficient λ. The filter coefficient cannot be determined from the physical and chemical properties of the suspension and filter bed; rather, it must be determined from experiments with the given suspension and filter bed. Furthermore, the experimentally determined λ cannot be applied directly to other systems. In general, λ changes with time as particles accumulate within the filter bed. A detailed description of phenomenological expressions for λ and for the application of this approach to granular filtration can be found in Tien (1989).

12.2.2 Fundamental theories

The aim of the fundamental approach is to describe the removal of particles in the initial stage of filtration (the so-called 'clean-bed removal') based on fundamental theories, such as mass transfer, hydrodynamics, and colloid and surface chemistry. In this approach, the filter bed is modelled as an assemblage of single or unit collectors which have a certain known geometry. The fluid flow field around or through this geometry has to be described analytically based on theories of low Reynolds number hydrodynamics. Such model collectors include the isolated sphere or cylinder, the sphere-in-cell model and the constricted tube model (Tien, 1989). Happel's flow field for the sphere-in-cell porous medium model (Happel, 1958) was described previously (Chapter 5).

In the fundamental approach, the removal of particles is represented by a single or a unit collector removal efficiency, usually denoted as η. The single collector removal efficiency is defined as the ratio of the overall particle deposition rate onto the collector to the convective transport of upstream particles towards the projected area of the collector. For an isolated spherical collector, the single collector removal efficiency is:

$$\eta = \frac{I}{UC_0 \pi a_c^2} \tag{12.6}$$

where I is the actual deposition rate on the collector with a radius a_c. The single collector removal efficiency is then related to the removal efficiency of the entire packed-bed filter through a mass balance (see eqn (10.8)). For a packed-bed filter composed of uniform spheres, this results in:

$$\ln (C/C_0) = -\frac{3}{4}\frac{(1-f)\eta L}{a_c} \tag{12.7}$$

where L is the depth of the granular medium in the filter. Thus, the overall removal efficiency of particles in the filter is $(1-C/C_0)$. It can be shown (e.g. Tien, 1989) that the single collector removal efficiency is related to the filter coefficient, λ, by the expression:

$$\lambda = \frac{3}{4}\frac{(1-f)}{a_c}\eta \tag{12.8}$$

The term $3(1-f)/4a_c$ can be viewed as the effective number of spherical collectors per unit depth of the filter medium.

The goal of the fundamental theories is to predict η (or I) from consideration of microscopic (fundamental) theories. Such an approach for describing particle deposition in porous media was described in detail in Chapter 5. However, as discussed in Chapter 11, the current theories fail to predict η when repulsive double layer interactions predominate. As a result, it is necessary to combine an empirical factor in predicting η. In this approach, we multiply the single collector efficiency, η_0, determined from physical considerations, by an empirical collision efficiency factor, α, which describes the fraction of collisions with filter grains that results in attachment. Thus, the overall single collector removal efficiency can be described as:

$$\eta = \alpha\eta_0 \tag{12.9}$$

Here η_0 can be calculated using one of the approaches described in Chapter 5, without the inclusion of electrical double layer interaction in the calculations. The collision efficiency, α, can be found experimentally. However, in well-operated granular filtration units, it is usually assumed that $\alpha = 1$, since repulsive double layer forces are eliminated due to addition of coagulants. In other words, if coagulants are added, all particle collisions with the filter grains will result in attachment.

The first fundamental expression for η_0 in water filtration was developed by Yao *et al.* (1971). Their work was based on earlier developments in aerosol filtration (Friedlander, 1958; Fuchs, 1964). In their theoretical derivation, they assumed that the overall single collector efficiency is the sum of three transport mechanisms:

$$\eta_0 = \eta_D + \eta_I + \eta_G \tag{12.10}$$

where η_D is the transport by diffusion, η_I is the transport by interception, and η_G is the transport by gravity. These transport efficiencies for a spherical collector in a porous medium are defined as:

$$\eta_D = 4.0 A_s^{1/3} \left(\frac{D_\infty}{U d_c}\right)^{2/3} \tag{12.11}$$

$$\eta_I = \frac{3}{2} A_s \left(\frac{d_p}{d_c}\right)^2 \tag{12.12}$$

$$\eta_G = \frac{(\rho_p - \rho)}{18\mu U} g d_p^2 \tag{12.13}$$

Here D_∞ is the bulk particle diffusion coefficient calculated from the Stokes–Einstein equation (eqn (6.3)); U is the approach velocity to the filter; d_p and d_c are the particle and collector diameters, respectively; μ is the viscosity of the suspending medium; ρ_p and ρ are the density of particles and the fluid, respectively; and A_s is a flow parameter to account for the effect of neighbouring collectors (grains) in the filter on the flow field around a single collector. Using

Happel's sphere-in-cell model (Happel, 1958), the porosity-dependent flow parameter is defined as:

$$A_s = 2(1 - p^5)/(2 - 3p + 3p^5 - 2p^6) \tag{12.14}$$

where $p = (1 - f)^{1/3}$. For a typical porosity of 0.4, A_s has a value of 38.

A more rigorous approach that includes corrections for the retardation of particle transport rate by hydrodynamic interaction was developed by Rajagopalan and Tien (1976). Based on numerical results from a solution of the trajectory equation under various physical conditions (in the absence of double layer interaction), they proposed a correlating equation for η_0. This equation, modified for use with eqns (12.7)–(12.9) (Rajagopalan *et al.*, 1982; Tobiason, 1987), is given by:

$$\eta_0 = 4.0A_s^{1/3}\left(\frac{D_\infty}{Ud_c}\right)^{2/3} + A_s N_{LO}^{1/8} R^{15/8} + 3.38 \times 10^{-3} A_s N_G^{1.2} R^{-0.4} \tag{12.15}$$

Here, N_{LO}, R and N_G are dimensionless numbers; N_{LO} is a van der Waals number given by $4A/9\pi\mu d_p^2 U$, A being the Hamaker constant of the interacting media; R is an aspect ratio given by d_p/d_c; and N_G is a gravitational force number given by $(\rho_p - \rho)gd_p^2/(18\mu U)$. It should be noted that only the second and third terms on the right-hand side of eqn (12.15) were obtained from Rajagopalan and Tien's correlation. The first term on the right-hand side, the transport by diffusion, eqn (12.11), was added to the equation with the assumption that the transport mechanisms are additive.

Another form of the correlation equation of Rajagopalan and Tien is obtained when the removal rate of particles by a unit collector is normalized by the convective flux of particles towards the projected area of the sphere-in-cell envelope (see eqn (5.92) and associated discussion). In this case, each one of the terms on the right-hand side of eqn (12.15) is modified by a factor of $(1 - f)^{2/3}$ to yield:

$$\eta_0 = (1 - f)^{2/3}\left\{4.0A_s^{1/3}\left(\frac{D_\infty}{Ud_c}\right)^{2/3} + A_s N_{LO}^{1/8} R^{15/8} + 3.38 \times 10^{-3} A_s N_G^{1.2} R^{-0.4}\right\} \tag{12.16}$$

where f is the porosity of the porous medium.

When eqn (12.16) is used to calculate the dimensionless removal rate by a unit collector of the filter, the following equation should be used to calculate the removal efficiency of particles (or C/C_0) in the packed-bed filter:

$$\ln (C/C_0) = -\frac{3}{4}\frac{(1 - f)^{1/3}\eta L}{a_c} \tag{12.17}$$

This equation is obtained from a mass balance over a packed bed composed of unit collectors with a radius $b = a_c(1 - f)^{-1/3}$. Similarly, the filter coefficient for this case is given by:

$$\lambda = \frac{3}{4}\frac{(1 - f)^{1/3}}{a_c}\eta \tag{12.18}$$

When used appropriately, identical particle removal efficiencies should be obtained from (1) eqn (12.7) along with eqns (12.9) and (12.15), and (2) eqn (12.17) along with eqns (12.9) and (12.16).

Finally, it should be noted that eqn (12.15) (or eqn (12.16)) provides an approximate value for η_0. The results obtained from this equation are adequate to illustrate the influence of various physicochemical parameters on particle removal by granular filters. When more accurate results are needed, the single collector efficiency, η_0, can be calculated from a numerical solution of the convective-diffusion equation as described in Section 5.4.

12.3 Predictions of filter performance

In this section, we will present predictions for the performance of granular filters as a function of several variables. These will be calculated using the equations presented earlier for the fundamental (microscopic) approach to model granular filtration. The variables that will be examined include the diameter of the suspended particle (d_p), the approach velocity in the filter (U), the diameter of the filter grains (d_c) and the depth of the filter medium (L). The reader should bear in mind that the predictions presented here are valid only for clean-bed removals. Under favourable chemical conditions, as in the case of properly operated filters in water treatment, removal efficiencies improve significantly with time due to 'filter ripening' (Tobiason and O'Melia, 1988; Tien, 1989). Thus, clean-bed predictions provide conservative estimates of filter efficiency.

12.3.1 Effect of particle size on the single collector removal efficiency

Calculated single collector removal efficiencies as a function of the particle size of the suspension are presented in Figure 12.2. The calculations were carried out using eqns (12.9) and (12.15) for two attachment (collision) efficiencies: $\alpha = 0.05$, representing filtration under unfavourable chemical conditions (poor or no chemical treatment), and $\alpha = 1$, representing filtration at optimal coagulant dose. The physical parameters used in these calculations (indicated in the figure) are typical of those used in water treatment. All pathogens of health concern, such as viruses, bacteria and protozoan cysts, fall in the size range shown in Figure 12.2.

As observed, the single collector removal efficiency of Brownian particles (i.e. particles smaller than about 1 μm) increases with a decrease in particle size. In this size range, transport of particles to the filter grains is dominated by diffusion. On the other hand, the single collector removal efficiency of non-Brownian particles (i.e. particles larger than about 1 μm) increases with an increase in particle size. In this size range, the dominant mechanisms by which particles are transported to the filter grains are interception and gravity. A minimum in the single collector removal efficiency is observed for particles having a diameter of about 2 μm.

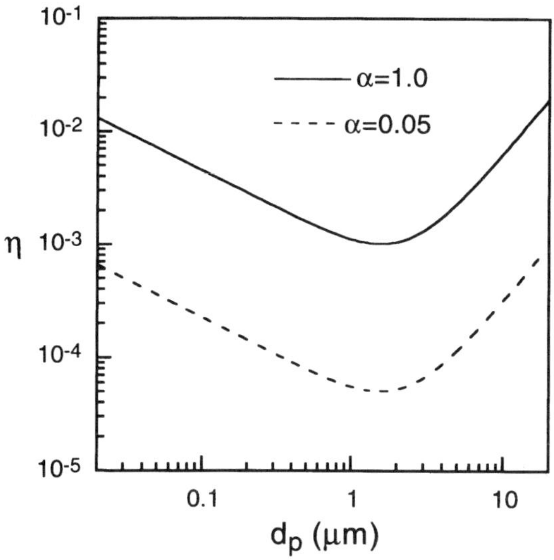

Figure 12.2 *Single collector removal efficiency as a function of particle diameter calculated for filtration at two different attachment (collision) efficiencies. The following parameters were used in the calculations:* $U = 0.14\ cm\ s^{-1}$, $L = 60\ cm$, $f = 0.4$, $d_c = 0.5\ mm$, $T = 293\ K$, $A = 1 \times 10^{-20}\ J$, *and* $\rho_p = 1.05\ g\ cm^{-3}$

12.3.2 Effect of particle size on clean-bed removal efficiency

The clean-bed removal efficiency of a packed-bed filter can be calculated from the single collector removal efficiency using eqn (12.7). The removal efficiency in percentage is defined as $100(1 - C/C_0)$. Calculated removal efficiencies as a function of particle size for the data shown above (Figure 12.2) are presented in Figure 12.3. It is demonstrated that, under optimal chemical treatment ($\alpha = 1$), a filter with a medium depth of 60 cm can significantly remove particles smaller than 100 nm (the size of many viruses) and larger than about 7 μm (the size of some protozoan cysts). A minimum in clean-bed removal is observed at a particle size around 1 μm (the size range of pathogenic bacteria). When there is no chemical treatment, or when the chemical treatment of the raw water is poor (represented by $\alpha = 0.05$), the removal efficiency of particles is poor at all size ranges. This is particularly noticeable in the size range of 0.1–10 μm.

An alternative way to present the removal efficiency curves is to use the concept of log removal. Current drinking water regulations express required removals of microbial contaminants by water treatment plants and filtration units in terms of '*x*-log removal' (Letterman, 1991). The *x*-log quantity is calculated by taking the negative (base 10) logarithm of the residual fraction of particles in the filter effluent (i.e. – log of C/C_0). For instance, with this concept 1-log removal corresponds to 90% removal efficiency and 2-log corresponds to 99% removal.

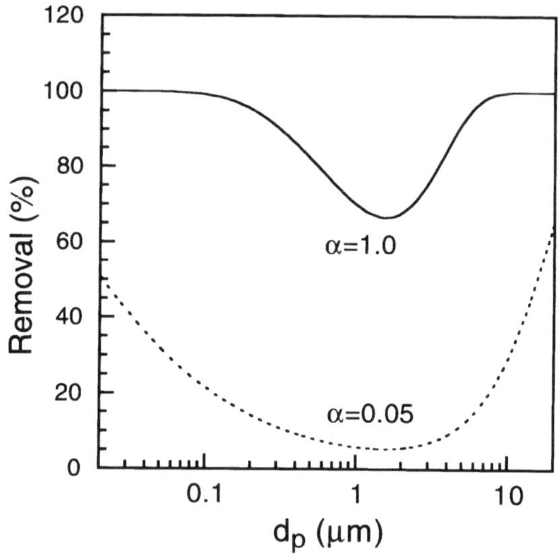

Figure 12.3 *Removal efficiency of particles as a function of particle size calculated for filtration at two different attachment (collision) efficiencies. The parameters used in the calculations are similar to those described in Figure 12.2*

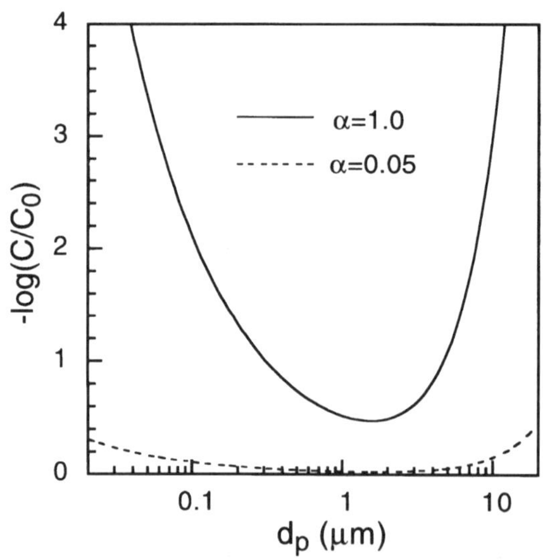

Figure 12.4 *Log removal of particles as a function of particle diameter calculated for the data shown in Figure 12.3*

Log-removal curves for the data presented earlier (Figures 12.2 and 12.3) are shown in Figure 12.4. The data demonstrate that well-operated filters can provide more than 2-log removal of particles in the size range of viruses (smaller than 100 nm) and protozoan cysts such as *Giardia* (7–10 μm). In the absence of coagulants ($\alpha = 0.05$), on the other hand, the removal of suspended particles by filters is very low.

12.3.3 Effect of filtration rate on the clean-bed removal

Predictions for the effect of fluid approach (superficial) velocity on the clean-bed removal are shown in Figure 12.5. In these predictions, it was assumed that the attachment efficiency is unity (i.e. favourable filtration). Calculations were carried out for three approach velocities, covering the range typically found in water treatment. Conventional water treatment plants (i.e. plants with coagulation, sedimentation and filtration units) usually operate at the lower part of this filtration rate range. On the other hand, direct filtration plants (i.e. plants that do not have sedimentation units) usually operate at higher filtration rates (Montgomery, 1985).

For particles larger than 1 μm, the predominant transport mechanism is interception. Transport by gravity is negligible since the density of the particles used in the calculations is not much larger than that of water. Because interception is independent of the approach velocity, the clean-bed removals of

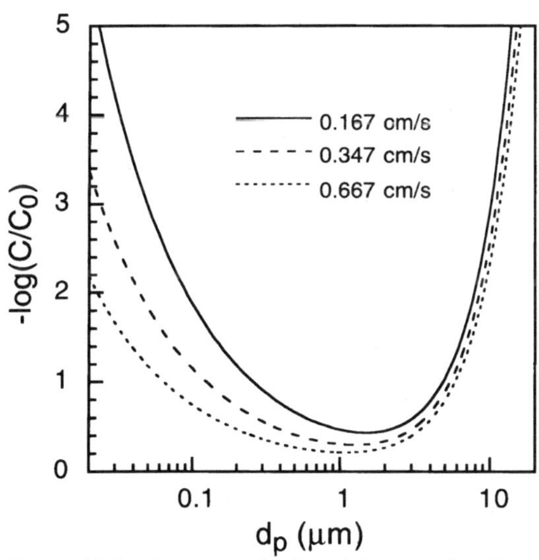

Figure 12.5 *Log removal of particles as a function of particle diameter calculated for three different filtration rates (approach velocities). The following parameters were used in the calculations:* $\alpha = 1.0$, $L = 60\ cm$, $f = 0.4$, $d_c = 0.5\ mm$, $T = 293\ K$, $A = 1 \times 10^{-20}\ J$, *and* $\rho_p = 1.05\ g\ cm^{-3}$

non-Brownian particles shown in Figure 12.5 are comparable. On the other hand, the removal efficiency of Brownian particles decreases with an increase in fluid velocity. At higher approach velocities, the removal of virus-size particles (i.e. particles smaller than 100 nm) can be smaller than 2-log. Removal efficiencies in general can be enhanced by increasing filter bed depth or decreasing the filter grain size, thus allowing the filter to operate at higher filtration rates.

12.3.4 Effect of filter grain size on clean bed removal

Model predictions for the effect of filter grain size on the removal of particles by clean beds are shown in Figure 12.6. In these predictions, it was assumed that $\alpha=1$ (i.e. favourable filtration). Calculations were carried out for three grain diameters covering the range typically used in water treatment plants. Conventional water treatment plants are usually designed with filter media having a smaller grain size compared to direct filtration plants (Montgomery, 1985).

As demonstrated in Figure 12.6, the filter grain size has a dramatic effect on the clean-bed removal for all particle sizes. Filters with small grains can result in high log removals of particles. However, the use of small grains is not recommended in water treatment since it results in higher head losses, clogging of filters and shorter filter runs. It is also observed that filters with large grains have poor removal efficiencies. However, as will be shown later, deeper beds can

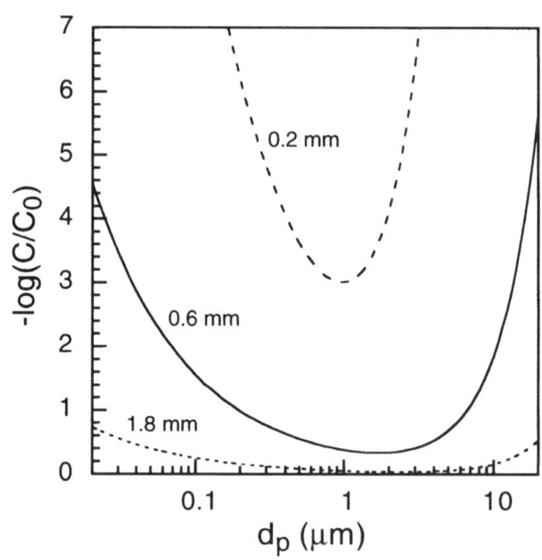

Figure 12.6 *Log removal of particles as a function of particle diameter calculated for three different filter grain sizes (diameters). The following parameters were used in the calculations:* $\alpha = 1.0$, $U = 0.14\ cm\ s^{-1}$, $L = 60\ cm$, $f = 0.4$, $T = 293\ K$, $A = 1 \times 10^{-20}\ J$, *and* $\rho_P = 1.05\ g\ cm^{-3}$

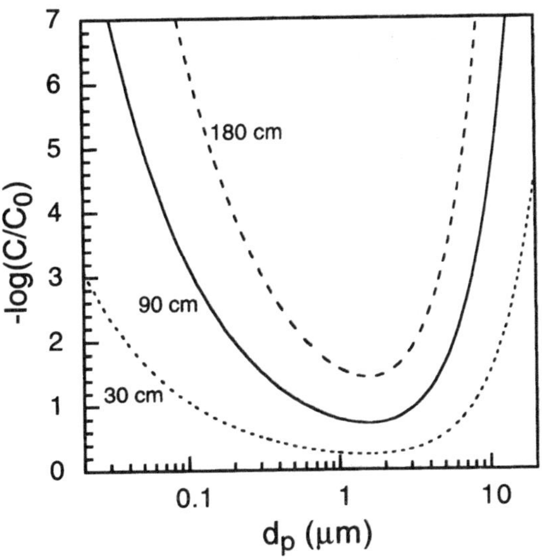

Figure 12.7 *Log removal of particles as a function of particle diameter calculated for three different filter bed depths. The following parameters were used in the calculations:* $\alpha = 1.0$, $U = 0.14\ cm\ s^{-1}$, $d_c = 0.5\ mm$, $f = 0.4$, $T = 293\ K$, $A = 1 \times 10^{-20}\ J$, *and* $\rho_P = 1.05\ g\ cm^{-3}$

be used to counteract this effect. Filters with bed depths larger than about 1 m are necessary in order to operate with high flow rates and large grain size (Montgomery, 1985).

12.3.5 Effect of filter bed depth on clean-bed removal

Model predictions for the effect of filter medium depth on the removal efficiency of particles by granular filters are shown in Figure 12.7. In these calculations, it was assumed that $\alpha = 1$ (i.e. filtration under favourable chemical conditions). As observed, removal efficiencies increase significantly as the filter medium depth increases. Deeper beds have more collectors and, therefore, can remove more particles from raw water. The predictions shown here, and in the previous subsections, demonstrate that, by a proper selection of filter medium depth, grain size and approach velocity, a desired removal efficiency can be achieved.

Bibliography

Letterman, R. D. (1991) *Filtration Strategies to Meet the Surface Water Treatment Rule*, American Water Works Association, Denver, Colorado

Montgomery, J. M. (1985) *Water Treatment Principles and Design*, John Wiley & Sons, New York, NY
Tien, C. (1989) *Granular Filtration of Aerosols and Hydrosols*, Butterworth Publishers, Stoneham, Massachusetts

References

Amirtharajah, A. (1988) Some theoretical and conceptual views of filtration. *J. Am. Water Works Assoc.*, **80**, 36–46
Friedlander, S. K. (1958) Theory of aerosol filtration. *Ind. Eng. Chem.*, **50**, 1161–1164
Fuchs, N. A. (1964) *The Mechanics of Aerosols*, Dover Publications, Inc., New York, NY
Happel, J. (1958) Viscous flow in multiparticle systems: slow motion of fluids relative to beds of spherical particles. *Am. Inst. Chem. Eng. J.*, **4**, 197–201
Herzig, J. P., LeClerc, D. M. and LeGoff, P. (1970) Flow of suspensions through porous media – applications to deep filtration. *Ind. Eng. Chem.*, **62**, 8–35
Iwasaki, T. (1935) Some notes on sand filtration. *J. Am. Water Works Assoc.*, **29**, 1591–1602
Letterman, R. D. (1991) *Filtration Strategies to Meet the Surface Water Treatment Rule*, American Water Works Association, Denver, Colorado
Montgomery, J. M. (1985) *Water Treatment Principles and Design*, John Wiley & Sons, New York, NY
O'Melia, C. R. (1985) Particles, pretreatment, and performance in water treatment. *J. Environ. Eng. ASCE*, **111**, 874–890
Rajagopalan, R. and Tien, C. (1976) Trajectory analysis of deep bed filtration with the sphere-in-cell porous media model. *Am. Inst. Chem. Eng. J.*, **22**, 523–533
Rajagopalan, R., Tien, C., Pfeffer, R. and Tardos, G. (1982) Letter to the Editor. *Am. Inst. Chem. Eng. J.*, **28**, 871–872
Tien, C. (1989) *Granular Filtration of Aerosols and Hydrosols*, Butterworth Publishers, Stoneham, Massachusetts
Tobiason, J. E. (1987) Physicochemical aspects of particle deposition in porous media. (PhD Dissertation), The John Hopkins University, Baltimore, Maryland
Tobiason, J. E. and O'Melia, C. R. (1988) Physicochemical aspects of particle removal in depth filtration. *J. Am. Water Works Assoc.*, **80**, 54–64
Yao, K. M., Habibian, M. T. and O'Melia, C. R. (1971) Water and wastewater filtration: concepts and applications. *Environ. Sci. Technol.*, **5**, 1105–1112

Chapter 13

Transport of colloidal materials in ground water

Nomenclature

a_c	collector (grain) radius
b	radius of a unit collector based on the sphere-in-cell porous medium model
C	particle concentration
C_L	particle concentration at $x = L_T$
C_0	initial particle concentration
f	porosity
L_T	travel distance
x	horizontal length coordinate
α	collision (attachment) efficiency
η_0	single collector removal efficiency under favourable chemical conditions

Inorganic and organic colloids are ubiquitous in ground water aquifers. Most contaminants in aquatic environments are readily associated with colloidal particles or are particles themselves. Examples include toxic metals and radionuclides adsorbed onto clays and metal oxides, synthetic organic compounds adsorbed onto organic colloids and macromolecules, and pathogenic microorganisms such as viruses and bacteria. In this chapter, we will review some of the current literature on the transport of colloidal contaminants in ground water and soils. More specifically, we will describe the transport of viruses in soils and ground water and the role of colloids in the transport of pollutants in ground water. The discussion given in this chapter is not intended to be exhaustive; rather, its purpose is to illustrate the important processes involved and the main factors that control the transport of colloidal materials. Following this discussion, we will present model predictions of travel distances of colloidal materials in saturated porous media. These predictions were calculated using theories of colloid deposition and filtration in porous media, presented earlier in this book.

13.1 Transport of viruses in soils and ground water

13.1.1 Introduction

Contamination of ground waters by pathogenic viruses is a major health concern. For instance, it was reported that almost half of the waterborne disease outbreaks every year in the United States are due to contaminated ground waters (Craun, 1979; Yates *et al.*, 1987; Snowdon and Cliver, 1989). Penetration of pathogenic viruses into ground waters is usually attributed to the following causes (Corapcioglu and Haridas, 1984; Snowdon and Cliver, 1989): (1) leaky sewer lines, (2) infiltration from septic tanks, (3) rain infiltration through sanitary landfills, (4) artificial recharge of ground waters by treated domestic sewage, (5) infiltration from lagoons and oxidizing ponds, and (6) land application of municipal wastewater. Although the number of viruses reaching the ground water from these sources can be significantly reduced while the water is flowing through the soil, the potential health hazards still exist since a single virus may be sufficient to cause a disease (Yates *et al.*, 1987). Field studies demonstrate that viruses can migrate considerable distances (e.g. Gerba and Goyal, 1985; Keswick *et al.*, 1982). In Sections 13.1.2 and 13.1.3, we will describe the colloidal properties of viruses and summarize previous studies on the transport of viruses in porous media.

13.1.2 Colloidal properties of viruses

Viruses are the smallest intracellular parasites known. They are colloidal in size and generally range from 20 nm to 200 nm (Pollard, 1953; Bitton, 1975). Since their nucleic acid is encased in a proteinaceous capsid, they behave as amphoteric colloids in which the surface charge is established by ionization of chemical functional groups. The surface chemical characteristics of viruses and their charge behaviour are very important in determining their transport in subsurface environments.

The surface charge of viruses is established by the protonation and deprotonation of functional groups of amino acids in the polypeptides comprising the virus capsid. The ionizable groups on the polypeptides are generally carboxyl, primary amine, phenolic hydroxyl, imidazole and guanidyl (Taylor, 1981). The majority of the functional groups are of the carboxyl and amino type. Surface charge of viruses develops by the following acid–base reactions (Taylor, 1981; Taylor and Bosmann, 1981):

$$\text{—COOH} \rightleftharpoons \text{—COO}^- + \text{H}^+ \tag{13.1}$$

$$\text{—NR}_2\text{H}^+ \rightleftharpoons \text{—NR}_2 + \text{H}^+ \tag{13.2}$$

$$\text{—Ph-OH} \rightleftharpoons \text{—Ph-O}^- + \text{H}^+ \tag{13.3}$$

The total charge at any pH depends on the surface concentration of the ionizable groups and on their dissociation constants. The pH at which a virus has no net charge, the isoelectric point, varies with capsid composition. At lower pH,

Table 13.1 *Representative isoelectric points (IEP) of different virus strains*

Virus	*IEP*	*Reference*
Reovirus	3.9	Floyd and Sharp (1978)
Rhinovirus	6.4	Kornat *et al.* (1975)
Polio 1 (LSc)	6.6	Gerba (1984)
Polio 2 (Sabin T2)	6.5, 4.5	Murray and Parks (1980)
Polio 1 (Chat)	7.5, 4.5	Ward (1978)
Vaccinia	3.8	Douglas *et al.* (1969)
Influenza A	4.5	Molodkina *et al.* (1986)
T4 Bacteriophage	4.5	Childs and Birboim (1975)
Reovirus	3.8	Taylor and Bosmann (1981)

net protonation of ionizable groups produces positive charge, whereas at high pH deprotonation dominates to give a net negative charge. Representative values of isoelectric points of several viruses are summarized in Table 13.1. A unique feature of some viruses, as shown in this table, is the presence of two conformational states with different isoelectric points (Gerba, 1984).

The electrokinetic properties of viruses can be determined by electrophoretic mobility measurements. Representative results for the electrophoretic mobility of reovirus as a function of pH and ionic strength are shown in Figure 13.1. The general behaviour shown in this figure is similar to that observed for various metal oxide colloids in which OH^- and H^+ act as potential-determining ions.

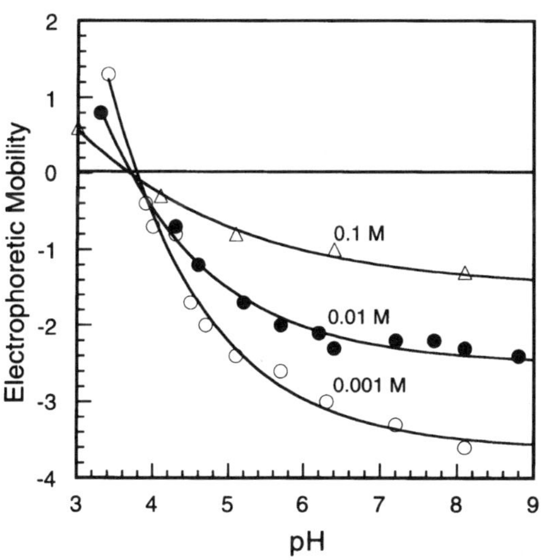

Figure 13.1 *Electrophoretic mobility (in $\mu m\, s^{-1}\, V^{-1}\, cm$) of reovirus type 3 as a function of pH and NaCl concentration (data from Taylor and Bosmann, 1981)*

Since simple electrolytes (such as NaCl) do not specifically interact with the surfaces of viruses, all the electrophoretic mobility curves in Figure 13.1 intersect at the isoelectric point (Stumm and Morgan, 1981; Stumm, 1992).

13.1.3 Studies on virus transport in soils and ground water

There are numerous field and laboratory investigations demonstrating the mobility of viruses in soils, ground waters and laboratory soil columns. Several summaries of such studies are available in the literature (e.g. Bitton, 1975; Corapcioglu and Haridas, 1984; Yates and Yates, 1988). These studies demonstrate that the transport of viruses is affected by the following factors: (1) type of soil and aquifer matrix, (2) pH of the solution, (3) ionic strength and electrolyte type of the solution, (4) virus type and characteristics, (5) presence of soluble organics, (6) flow rate, (7) degree of soil saturation, and (8) degree of virus aggregation (Gerba and Goyal, 1985; Yates and Yates, 1988).

The extent of virus transport in ground water aquifers was reported by several investigators. The horizontal travel distances of some representative viruses are summarized in Table 13.2. A variety of viruses, subsurface media and solution chemistries were involved in these studies. The results indicate that migration of viruses in subsurface environments ranges from extensive transport (several hundreds of metres) to very limited transport. The large differences in the extent of virus transport in these studies may be attributed to one or more of the following factors:

(1) transport of viruses through preferential flow paths, such as fissures and cracks
(2) the viruses, aquifer media and soils reported in the studies possessed different surface chemical characteristics
(3) the solution chemistry in the various ground waters was considerably different
(4) the texture of the aquifer solid matrix was fairly different.

The second and third factors outlined above are interrelated. Colloidal interactions between viruses and aquifer media are determined by their surface-chemical characteristics. These, in turn, are controlled by the solution chemistry

Table 13.2 *Representative horizontal travel distances of viruses in ground water aquifers*

Virus type	*Distance* (m)	*Reference*
Coliphage T4	1600	Keswick *et al.* (1982)
Bacteriophage	680	Martin and Thomas (1974)
Coxsackie B3	408	Gerba and Goyal (1985)
Poliovirus	180	Keswick *et al.* (1982)
Echo 12	3	Gerba and Goyal (1985)

of the ground water. The magnitude of the colloidal interactions determines the kinetics of virus capture, and, hence, the extent of virus transport in porous media.

Migration of viruses in subsurface environments can be enhanced by transport through preferential flow paths such as fissures and cracks typically found in karstic aquifers and other fractured media. These field-scale features are difficult to characterize and they cannot be simulated adequately in laboratory studies. In addition, there are no sound models to quantitatively assess or simulate the effect of preferential flow paths, in aquifers or soils, on virus transport. In addition to these physical heterogeneities, chemical heterogeneities of aquifer sediments may play an important role in virus transport in ground water. Finally, release (mobilization) of deposited viruses can enhance their mobility. Several investigators (e.g. Bales and Li, 1993) have reported release rates of viruses attached to granular media in laboratory columns. As with virus deposition rates, release rates of viruses are highly dependent on solution chemistry and can be significantly enhanced by perturbations in solution chemistry.

13.2 Transport of colloids and associated pollutants in ground water

13.2.1 Introduction

Colloids are found in virtually all aquatic environments, including ground water aquifers. Colloid genesis in ground water may occur by precipitation of supersaturated phases or by mobilization of existing colloidal phases (Nightingale and Bianchi, 1977; McDowell-Boyer *et al.*, 1986; Gschwend *et al.*, 1990; Ronen *et al.*, 1992). Many pollutants in aqueous media are readily adsorbed onto colloidal particles and mineral surfaces. Hence, the fate of pollutants in ground waters is determined in large part by the fate of the colloidal particles with which they are associated. Field studies demonstrate that colloids are mobile in subsurface environments. Mobile colloids can enhance the transport of organic and inorganic pollutants. The enhancement in the transport of pollutants sorbed to colloids is usually referred to as *facilitated transport* or *colloid-mediated transport* (McCarthy and Zachara, 1989).

A short description of the interactions between contaminants and colloidal particles is given below. In addition, representative studies dealing with the transport of colloids and colloid-associated pollutants in soils, ground waters and laboratory soil columns are summarized. Extensive reviews on colloidal transport in subsurface environments were given by McDowell-Boyer *et al.* (1986) and McCarthy and Zachara (1989).

13.2.2 Particle–pollutant interactions

A brief description of particle–pollutant interactions in aqueous systems is given in this section. Readers who are interested in this subject are referred to more comprehensive discussions, such as those given by Chiou *et al.* (1983), Karikhoff (1984), Dzombak and Morel (1987) and Stumm (1987, 1992).

Sorption of uncharged organic pollutants onto colloids is controlled by hydrophobic interactions. The degree of sorption depends primarily on the hydrophobicity of the contaminant, typically described by an octanol–water partition coefficient, and on the organic carbon content of the solid phase (Lyman *et al.*, 1982; Karickhoff, 1984; O'Melia, 1989). An estimate of the sorption of uncharged organic pollutants onto colloidal particles in ground waters and soils can be made on the basis of the octanol–water partition coefficient of the pollutant and the organic carbon content of the colloids and mineral surfaces (Karickhoff *et al.*, 1979; Chiou *et al.*, 1983; Karickhoff, 1984). The contribution of inorganic colloids to the uptake of uncharged organic pollutants is generally considered to be very small. For ionizable or ionic organic chemical species, such as organic acids and ionic surfactants, non-hydrophobic contributions to sorption are relatively important.

In contrast to the non-specific adsorption of hydrophobic organic pollutants, the adsorption of inorganic solutes such as toxic metals and radionuclides is conceived as a chemical coordination process (Stumm, 1992). This process involves specific interactions between solutes and reactive surface groups. The total energy of interaction is composed of an electrostatic and a chemical contribution. The interaction is dependent upon pollutant type and concentration, solution pH, ionic strength and surface characteristics of colloids (Dzombak and Morel, 1987). Both inorganic and organic colloids and surfaces are important in the adsorption and speciation of inorganic pollutants in ground water systems.

13.2.3 Previous work on colloid transport

There are numerous laboratory and field studies which demonstrate the mobility of colloids in porous media and their role in the transport of associated chemical compounds. These studies were conducted in ground water aquifers, unsaturated soils and packed-bed columns. Several of these studies will be described below. The results of these studies will be used to delineate the factors governing the transport of colloids in subsurface environments.

The mobility of radiolabelled clay colloids and bacteria in laboratory sand columns was studied by Chaplin and Eichholz (1976). They found that the clays and bacteria penetrated only 20 cm after 10 days. However, extensive transport of these colloids was observed after the addition of a detergent solution, which probably increased the colloidal stability of these particles. Robertson *et al.* (1984), in a field study of aquifer contamination by a waste pulp liquor, have found that the macromolecules tannin and lignin travelled horizontally up to 900 m from the discharge site over an 11-year period. The migration rate of these macromolecules was close to the average linear ground water velocity. Gschwend and Reynolds (1987) have reported extensive transport of ferrous phosphate colloids in an aquifer near a secondary sewage infiltration site. It was suggested that these colloids were formed by sewage-derived phosphate combining with ferrous iron released from the aquifer solids.

The enhancement in the transport of contaminants associated with colloids has been demonstrated in laboratory and field studies. Vinten *et al.* (1983) showed that pesticides can be transported into the soil by association with particles. Buddemeier and Hunt (1988) and Bates *et al.* (1992) have shown that radionuclides are associated with colloidal particles and that colloid-bound radionuclides can migrate considerable distances in a ground water aquifer. Results from laboratory batch studies, on the other hand, suggest that radionuclides are strongly bound to aquifer minerals and therefore will be immobile in ground waters in the absence of mobile colloids. Enhancement in the transport of trace organics associated with macromolecules was observed by Bengtsson *et al.* (1987) and Enfield and Bengtsson (1988). They have demonstrated that the transport of hexachlorobenzene in porous media is enhanced by about 25% when it is adsorbed onto mobile blue dextran macromolecules.

As with the case of virus transport, it can be seen from the studies discussed above, and many more, that the migration of colloids in subsurface environments ranges from extensive transport (several hundreds of metres) to very limited transport (a few centimetres). The large difference in the extent of colloid transport in these studies can be attributed to the factors discussed in the section on virus transport. The main factors are:

(1) transport of colloids through preferential flow paths, such as fissures, fractures and cracks in non-homogeneous media
(2) colloids, aquifer media and soils in the various experiments possessed different surface-chemical characteristics
(3) the solution chemistry in the various ground waters and laboratory columns investigated was considerably different
(4) the mineral textures of the aquifer solid matrix and column granular media were fairly different.

Chemical factors (second and third factors above) can have a dramatic influence on the rate of capture of colloids by the solid matrix, and, as a result, on the extent of their transport. Release of deposited colloids due to chemical perturbations can also increase the variability in the extent of transport reported in the literature.

The information on colloid transport is usually available from field and soil column studies, such as those described earlier, where mechanisms of colloid transport and interaction are not easily identified. The surface characteristics of the colloids and aquifer media as well as the solution chemistries in those field and column studies were not well defined. These may have a significant effect on the stability of the colloids, on particle–media interactions, and hence on the transport of the colloids and associated pollutants. For example, the presence of high concentrations of ionogenic organic compounds and macromolecules, from either natural or anthropogenic sources, will increase the stability and transport of the colloids. On the other hand, divalent metal ions (e.g. calcium and magnesium) can destabilize colloids and limit their transport.

13.3 Colloid travel distances in porous media

A simple approach for estimating the travel distance of colloidal particles in saturated porous media is presented in this section. This approach is based on theories of particle deposition and filtration described earlier in Chapters 5 and 12. We will begin with the derivation of the equation to calculate the travel distance of colloids and will follow with predictions for the effects of several chemical and physical variables on the extent of colloid transport.

13.3.1 Calculation of colloid travel distance

The kinetics of colloid capture (deposition) in porous media can be described by the following equation (Yao *et al.*, 1971; O'Melia, 1990):

$$\frac{\partial C}{\partial x} = -\frac{3}{4}\alpha\eta_0\frac{(1-f)}{a_c}C \tag{13.4}$$

Here C is the number concentration of colloids of a specific size; x is the horizontal distance (coordinate); a_c is the radius of the grains comprising the porous media – the grains are assumed to be spherical and uniform in size; f is the porosity of the porous media – the porous media are assumed to be homogeneous with a fixed porosity; α is the collision efficiency; and η_0 is the dimensionless capture rate of colloids under favourable chemical conditions. The last can be calculated rigorously from the approach described in Chapter 5 (eqn (5.93)) or it can be approximated by using the correlation equation of Rajagopalan and Tien (1976), eqn (12.15). The basic principles used in the derivation of eqn (13.4) were described in Chapter 10 (eqns (10.5) and (10.6)).

Integrating eqn (13.4) with the boundary conditions of $C=C_0$ at $x=0$ and $C=C_L$ at $x=L_T$ results in an expression for the travel distance of colloids in porous media:

$$L_T = -\ln(C_L/C_0)\left(\frac{4a_c}{3(1-f)\alpha\eta_0}\right) \tag{13.5}$$

The travel distance of colloids, L_T, can be defined arbitrarily as the distance where $C_L/C_0=0.001$, i.e. 99.9% of the colloids are captured by the porous media as they travel the distance L_T. Equations (13.4) and (13.5) provide a reliable estimate for colloid travel in saturated porous media since these equations were verified in numerous column experiments with various colloids, porous media and solution chemistries (e.g. Tobiason and O'Melia, 1988; Tien, 1989; Elimelech and O'Melia, 1990a,b). Since there are no adequate theories to predict the collision (attachment) efficiency α, it should be determined experimentally, using the packed-bed column technique described in Chapter 10. In colloid transport processes in aquatic environments (such as ground water), it is reasonable to assume that the collision efficiency ranges from about 0.001 to 1.

The single collector removal efficiency, η_0, can also be calculated based on the definition given by eqn (5.92) or from the correlation equation of Rajagopalan and Tien (Rajagopalan and Tien, 1976; Rajagopalan *et al.*, 1982) given by eqn (12.16) (see discussion in Section 12.2.2). In this case, a mass balance analysis over a packed bed composed of unit collectors of radius $b = a_c\,(1-f)^{-1/3}$ (sphere-in-cell flow model) shows that $(1-f)$ in eqn (13.5) should be replaced by $(1-f)^{1/3}$. In both approaches, however, identical L_T values should be obtained for given physical and chemical conditions.

13.3.2 Predictions of particle travel distance

Transport of particles in subsurface environments is determined in large part by the rate of their capture by stationary surfaces. The capture of particles by stationary surfaces is markedly affected by the chemical-colloidal interactions between particles and surfaces which, in turn, are controlled by solution chemistry and chemical characteristics of particles and surfaces. In addition, the overall capture of particles in soils and ground water aquifers is influenced by physical parameters, such as particle size, fluid velocity, grain size and water temperature. Predictions for the effect of these parameters on the travel distance of particles (L_T), calculated from eqn (13.5), are presented in this section. In all the simulations presented here, the travel distance will be presented as a function of particle size (diameter) for the various parameters outlined above.

Effect of chemical factors

As described previously in this book (e.g. in Sections 5.7 and 11.1), the role of chemical-colloidal factors in particle deposition kinetics can be represented by the collision (attachment) efficiency. Calculated travel distances for three collision efficiencies are presented in Figure 13.2. These collision efficiencies represent a wide range of chemical interactions. When $\alpha = 0.001$, the chemical conditions are unfavourable for deposition. In this case, repulsive double layer interactions predominate and particle deposition rates are negligible (only 1 collision out of 1000 results in attachment). When $\alpha = 1.0$, chemical interactions are favourable and particle deposition rates are transport limited. In this case, every collision of a particle with a solid surface results in attachment. The conditions represented by $\alpha = 0.03$ are intermediate; colloid deposition rates are small, yet, as will be shown later, they are still significant in influencing particle migration in porous media. The physical parameters used in these calculations (i.e. fluid velocity, grain size, porosity and water temperature) are typical for ground water aquifers. Comparable experimental conditions were reported by Harvey *et al.* (1989) in a field study of particle transport in a ground water aquifer.

As shown in Figure 13.2, under all chemical conditions, a maximum travel distance is observed at a particle size of about 2 μm. This size is comparable to that of many pathogenic bacteria. The maximum in the calculated travel distance of particles at this size range corresponds to the minimum in particle deposition rate (expressed as η_0 in eqn (13.5)). This was discussed in detail in the chapter on packed-bed filtration (Section 12.3).

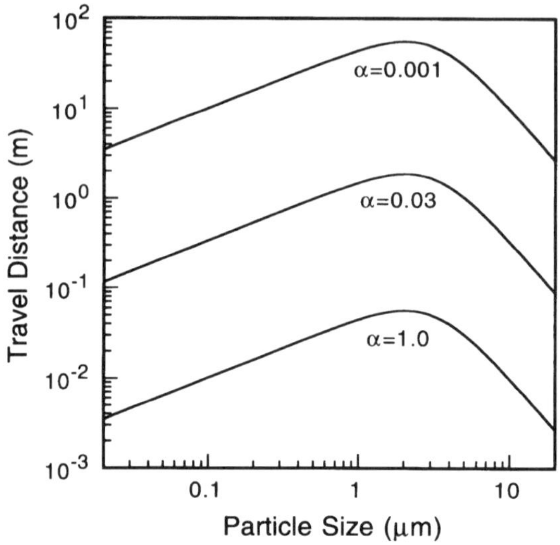

Figure 13.2 *Predicted colloid travel distances (L_T) as a function of particle diameter for various chemical conditions (represented by the collision efficiency α). The following parameters were used in the calculations: collector radius = 0.15 mm, approach velocity = 0.5 m d^{-1}, temperature = 10°C, porosity = 0.37, particle density = 1.05 g cm^{-3}, and Hamaker constant = 1×10^{-20} J*

The results in Figure 13.2 demonstrate that particle transport is highly dependent on chemical factors. When chemical conditions are favourable for deposition (i.e. $\alpha = 1.0$), the transport of particles of all sizes is extremely small (maximum travel distance of a few centimetres). In this case, we can assume that the particles are immobilized. Under unfavourable chemical conditions ($\alpha = 0.001$), on the other hand, the particles are mobile. Travel distances of 56 m are calculated for 2-μm particles. It is also demonstrated that very small particles (smaller than 0.1 μm) and large, non-Brownian particles (larger than 10 μm) can travel several metres in ground water aquifers, when chemical conditions are unfavourable for deposition. Intermediate travel distances are obtained in the case for $\alpha = 0.03$.

Effect of fluid approach velocity

The effect of fluid approach (superficial) velocity on the travel distance of particles is demonstrated in Figure 13.3. Calculations were carried out for chemical conditions unfavourable for deposition ($\alpha = 0.001$). The low fluid velocity (0.1 m d^{-1}) is typical of aquifers with low permeability. Higher velocities can be found in sandy aquifers, in geological formations with high permeability, or close to a pumping well.

The results demonstrate that particles are more mobile at higher fluid velocities. At higher fluid velocities, particle removal rate (or η_0 in eqn (13.5)) is smaller, resulting in greater travel distances. For instance, the travel distance

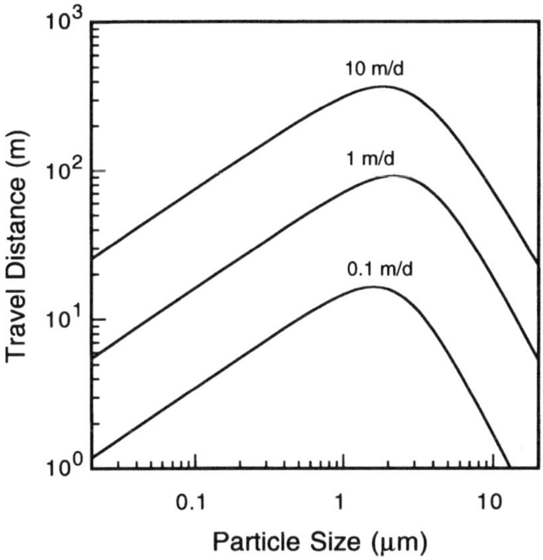

Figure 13.3 *Predicted colloid travel distances* (L_T) *as a function of particle diameter for various approach velocities (0.1, 1 and 10* $m\,d^{-1}$*) and for chemical conditions unfavourable for deposition (*$\alpha = 0.001$*). The following parameters were used in the calculations: collector radius = 0.15 mm, temperature = 10°C, porosity = 0.37, particle density = 1.05* $g\,cm^{-3}$*, and Hamaker constant =* 1×10^{-20} *J*

of particles with a size of about 2 μm can be greater than 350 m at a fluid velocity of 10 $m\,d^{-1}$. Although this velocity (i.e. 10 $m\,d^{-1}$) is higher than the average velocity of typical aquifers, such high velocities can develop in regions close to a pumping well or at certain locations in aquifers with structural heterogeneity.

Effect of aquifer grain size

Soils and aquifer media are granular and contain grains with a wide size range (Bear, 1972). The effect of media grain size on the travel distance of particles is illustrated in Figure 13.4. Calculations were carried out for chemical conditions unfavourable for deposition ($\alpha = 0.001$). As the grain size increases, the removal of suspended particles by the stationary media decreases, and, as a result, the travel distance of the particles increases. This can be seen directly from eqn (13.5), where a_c is the radius of the media grains. In addition, inspection of eqn (12.15) for η_0 shows that capture of particles by diffusional deposition (first term on the right-hand side of eqn (12.15)) and by interception (second term on the right-hand side of that equation) decreases as the grain size increases.

Effect of water temperature

The effect of water temperature on the travel distance of colloids is shown in Figure 13.5. As observed, the travel distance of colloids increases as the

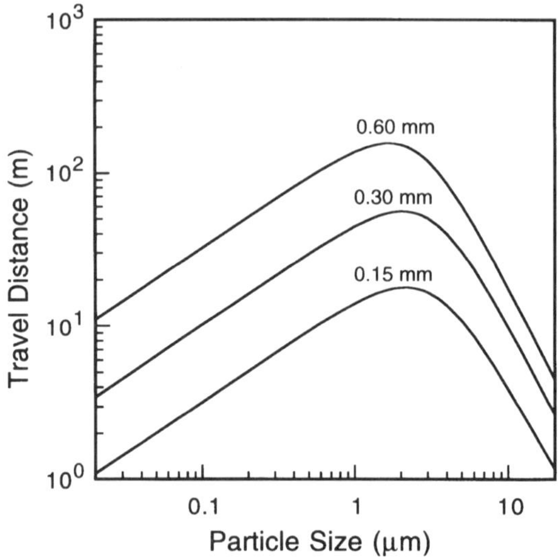

Figure 13.4 *Predicted colloid travel distances (L_T) as a function of particle diameter for various collector (grain) diameters (0.15, 0.30 and 0.60 mm) and for chemical conditions unfavourable for deposition ($\alpha = 0.001$). The following parameters were used in the calculations: approach velocity = 0.5 m d^{-1}, temperature = 10°C, porosity = 0.37, particle density = 1.05 g cm^{-3}, and Hamaker constant = 1×10^{-20} J*

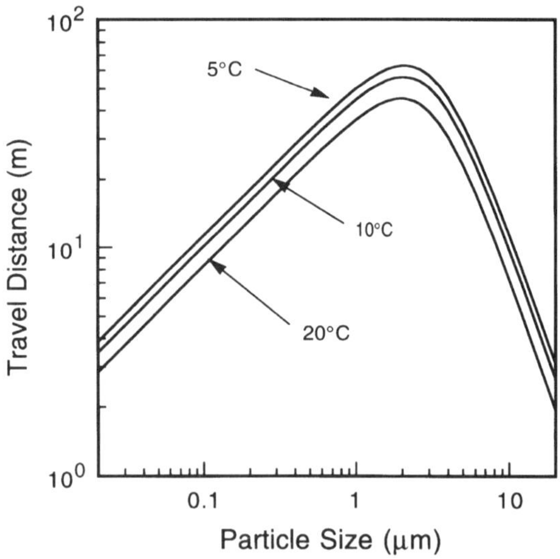

Figure 13.5 *Predicted colloid travel distances (L_T) as a function of particle diameter for various water temperatures (5, 10 and 20°C). The following parameters were used in the calculations: collector radius = 0.15 mm, approach velocity = 0.5 m d^{-1}, porosity = 0.37, particle density = 1.05 g cm^{-3}, and Hamaker constant = 1×10^{-20} J*

temperature of the water decreases. The effect of water temperature on the extent of particle migration in soils and aquifers is much smaller than the effect of the factors described earlier (i.e. chemical factors, fluid velocity and grain size).

The capture mechanisms of suspended particles are influenced by the temperature of the suspending medium. As the temperature of the water increases, capture of Brownian particles by soil and aquifer media increases because of the increase in the Brownian motion of particles. This can be seen from analysis of the first term on the right-hand side of eqn (12.15) or from a rigorous solution of the convective-diffusion equation as described in Chapter 5. The capture of non-Brownian particles also increases with temperature because of the decrease in the viscosity of the water (see the second and third terms on the right-hand side of eqn (12.15)).

Bibliography

Berkeley, R. C. W., Lynch, J. M., Melling, J., Rutter, P. and Vincent, B. (eds) (1981) *Microbial Adhesion to Surfaces*, Wiley Interscience, New York, NY

Bitton, G. and Marshal, K. C. (eds) (1980) *Adsorption of Microorganisms to Surfaces*, Wiley Interscience, New York, NY

Murray, J. P. (1978) Thermodynamics of poliovirus adsorption. PhD Dissertation, Stanford University, Stanford, CA, USA

Pollard, E. C. (1953) *The Physics of Viruses*, Academic Press Inc., New York, NY

Stumm, W. (ed.) (1987) *Aquatic Surface Chemistry: Chemical Processes at the Particle–Water Interface*, Wiley-Interscience, New York, NY

Stumm, W. (1992) *Chemistry of the Solid–Water Interface*, Wiley Interscience, New York, NY

Tadros, Th. F. and Gregory, J. (eds) (1993) *Colloids in the Aquatic Environment*, Elsevier Science Publishers, Amsterdam

Tien, C. (1989) *Granular Filtration of Aerosols and Hydrosols*, Butterworth Publishers, Stoneham, Massachusetts

References

Bales, R. C. and Li, L. (1993) MS-2 and poliovirus transport in porous media: hydrophobic effects and chemical perturbations. *Water Resources Res.*, **29**, 957–963

Bates, J. K., Bradley, J. P., Teetsov, A., Bradley, C. R. and Buchholtz ten Brink, M. (1992) Colloid formation during waste form reaction: implications for nuclear waste disposal. *Science*, **256**, 649–651

Bear, J. (1972) *Dynamics of Fluids in Porous Media*, Elsevier, Amsterdam

Bengtsson, G., Enfield, C. G. and Lindqvist, R. (1987) Macromolecules facilitate the transport of trace organics. *Sci. Total Environ.*, **67**, 159–164

Bitton, G. (1975) Adsorption of viruses onto surfaces in soil and water. *Water Res.*, **9**, 473–484

Buddemeier, R. W. and Hunt, J. R. (1988) Transport of colloidal contaminants in ground water: radionuclide migration at the Nevada test site. *Appl. Geochem.*, **3**, 535–548

Chaplin, J. B. F. and Eichholz, G. C. (1976) Fixation and remobilization of trace contaminants in simulated subsurface aquifer. *Health Phys.*, **30**, 215–219

Childs, J. O. and Birboim, H. C. (1975) Polyacrylamide gel electrophoresis of intact bacteriophage T4D particles. *J. Virol.*, **16**, 652–661

Chiou, C. T., Porter, P. E. and Schmedding, D. W. (1983) Partitioning equilibria between soil organic matter and water. *Environ. Sci. Technol.*, **17**, 227–231

Corapcioglu, M. Y. and Haridas, A. (1984) Transport and fate of microorganisms in porous media: a theoretical investigation. *J. Hydrol.*, **72**, 149–169

Craun, G. J. (1979) Waterborne disease – a status report emphasizing outbreaks in ground water. *Ground Water*, **17**, 183–191

Douglas, H. W., Williams, B. L. and Rondle, C. J. M. (1969) Micro-electrophoresis of pox virus in molar sucrose. *J. Gen. Virol.*, **5**, 391–396

Dzombak, D. A. and Morel, F. M. M. (1987) Adsorption of inorganic pollutants in aquatic systems. *J. Hydraul. Eng., ASCE*, **113**, 430–475

Elimelech, M. and O'Melia C. R. (1990a) Effect of particle size on collision efficiency in the deposition of Brownian particles with electrostatic energy barriers. *Langmuir*, **6**, 1153–1163

Elimelech, M. and O'Melia, C. R. (1990b) Kinetics of deposition of colloidal particles in porous media. *Environ. Sci. Technol.*, **24**, 1528–1536

Enfield, C. G. and Bengtsson, G. (1988) Macromolecular transport of hydrophobic contaminants in aqueous environments. *Ground Water*, **26**, 64–70

Floyd, R. and Sharp, D. G. (1978) Viral aggregation: effects of salts on the aggregation of poliovirus and reovirus at low pH. *Appl. Environ. Microbiol.*, **35**, 1084–1094

Gerba, C. P. (1984) Applied and theoretical aspects of virus adsorption to surfaces. *Adv. Appl. Microbiol.*, **30**, 133–168

Gerba, C. P. and Goyal S. M. (1985) Pathogen removal from wastewater during ground water recharge. In *Artificial Recharge of Ground Water* (ed. Asano, T.) Butterworth, Stoneham, Mass., 283–317

Gschwend, P. M. and Reynolds, M. D. (1987) Monodisperse ferrous phosphate colloids in an anoxic ground plume. *J. Contam. Hydrol.*, **1**, 309–327

Gschwend, P. M., Backhus, D. A., MacFarlane, J. K. and Page, A. L. (1990) Mobilization of colloids in groundwater due to infiltration of water at a coal ash disposal site. *J. Contam. Hydrol.*, **6**, 307–320

Harvey, R. W., George, L. H., Smith, R. L. and Leblanc, D. R. (1989) Transport of microspheres and indigenous bacteria through a sandy aquifer: results of natural and forced-gradient tracer experiments. *Environ. Sci. Technol.*, **23**, 51–56

Karickhoff, S. W. (1984) Organic pollutant sorption in aquatic systems. *J. Hydraul. Eng., ASCE*, **110**, 707–735

Karickhoff, S. W., Brown, D. S. and Scott, T. A. (1979) Sorption of hydrophobic pollutants on natural sediments. *Water Res.*, **13**, 241–248

Keswick, B. H., Wang, D. S. and Gerba, C. P. (1982) The use of microorganisms as ground-water tracers: a review. *Ground Water*, **20**(2), 142–149

Kornat, B. O., Longberg-Holm, K., Yin, F. H. and Noble-Harvey, J. (1975) Fractionation of biologically active and inactive populations of human rhinovirus type 2. *Virology*, **63**, 384–394

Lyman, W. J., Reehl, W. F. and Rosenblatt, D. H. (1982) *Handbook of Chemical Property Estimation Methods*, McGraw Hill Book Co., New York, NY

Martin, R. and Thomas, A. (1974) An example of the use of bacteriophage as a ground water tracer. *J. Hydrol.*, **23**, 73–80

McCarthy, J. F. and Zachara, J. M. (1989) Subsurface transport of contaminants. *Environ. Sci. Technol.*, **23**, 496–502

McDowell-Boyer, L. M., Hunt, J. R. and Sitar, N. (1986) Particle transport through porous media. *Water Resources Res.*, **22**, 1901–1921

Molodkina, L. M., Molodkin, V. M., Vostryukhina, O. A., Kolikov, V. M., Golikova, E. V. and Chernoberezhskii Y. M. (1986) Study of the electrophoretic mobility of A1 and A3 influenza viruses. *Colloid J. USSR*, **48**(1), 66–70

Murray, J. P. and Parks, G. A. (1980) Poliovirus adsorption on oxide surfaces. In *Particulates in Water* (eds. Kavanaugh, M. C. and Leckie, J. O.) American Chemical Society, Washington, DC, 97–133

Nightingale, H. I. and Bianchi, W. C. (1977) Ground-water turbidity resulting from artificial recharge. *Ground Water*, **15**, 146–152

O'Melia, C. R. (1989) Particle–particle interactions in aquatic systems. *Colloids Surfaces*, **39**, 255–271

O'Melia, C. R. (1990) Kinetics of colloidal chemical processes in aquatic systems. In *Aquatic Chemical Kinetics: Reaction Rates of Processes in Natural Water* (ed. Stumm, W.) Wiley-Interscience, New York, 447–474

Pollard, E. C. (1953) *The Physics of Viruses*, Academic Press Inc., New York

Rajagopalan, R. and Tien, C. (1976) Trajectory analysis of deep-bed filtration with the sphere-in-cell porous media model. *AIChE J.*, **22**, 523–533

Rajagopalan, R., Tien, C., Pfeffer, R. and Tadros, G. (1982). Letter to the Editor. *AIChE J.*, **28**, 871–872

Robertson, W. D., Barker, J. F., LeBeau, Y. and Marcoux, S. (1984) Contamination of an unconfined sand aquifer by waste pulp liquor: a case study. *Ground Water*, **22**, 191–197

Ronen, D., Magaritz, M., Weber, U., Amiel, A. J. and Klein, E. (1992) Characterization of suspended particles collected in groundwater under natural gradient flow conditions. *Water Resources Res.*, **28**, 1279–1291

Snowdon, J. A. and Cliver, D. O. (1989) Coliphages as indicators of human enteric viruses in ground water. *Crit. Rev. Environ. Control*, **19**, 231–249

Stumm, W. (ed.) (1987) *Aquatic Surface Chemistry: Chemical Processes at the Particle–Water Interface*, Wiley-Interscience, New York, NY

Stumm, W. (1992) *Chemistry of the Solid–Water Interface*, Wiley Interscience, New York, NY

Stumm, W. and Morgan, J. J. (1981) *Aquatic Chemistry*, 2nd edn, Wiley-Interscience, New York, NY

Taylor, D. H. (1981) Interpretation of the adsorption of viruses by clays from their electrokinetic properties. In *Chemistry in Water Reuse* (ed. Cooper, U. J.) Ann Arbor Sciences, Ann Arbor Michigan, 595–612

Taylor, D. H. and Bosmann, H. B. (1981) The electrokinetic properties of reovirus type 3: Electrophoretic mobility and zeta potential in dilute electrolytes. *J. Colloid Interface Sci.*, **83**, 153–162

Tien, C. (1989) *Granular Filtration of Aerosols and Hydrosols*, Butterworth Publishers, Stoneham, Massachusetts

Tobiason, J. E. and O'Melia, C. R. (1988) Physicochemical aspects of particle removal in depth filtration. *J. Am. Water Works Assoc.*, **80**, 54–64

Vinten, A. J. A., Yaron, B. and Nye, P. H. (1983) Vertical transport of pesticides into soil when adsorbed on suspended particles. *J. Agric. Fd Chem.*, **31**, 662–664

Ward, R. L. (1978) Mechanism of poliovirus inactivation by ammonia. *J. Virol.*, **26**, 299–305

Yates, M. V. and Yates, S. R. (1988) Modeling microbial fate in the subsurface environment. *Crit. Rev. Environ. Control*, **17**, 307–344

Yates, M. V., Yates, S. R., Wagner, J. and Gerba, C. P. (1987) Modeling virus survival and transport in the subsurface. *J. Contam. Hydrol.*, **1**, 329–345

Yao, K. M., Habibian, M. T. and O'Melia, C. R. (1971) Water and wastewater filtration: concepts and applications. *Environ. Sci. Technol.*, **5**, 1105–1112

Chapter 14

Advanced simulation of porous media and filtration processes

Nomenclature

a_H	hydraulic radius of pore
a_i	radius of the largest inscribed circle of a pore
a_p	particle radius
A_p	pore area
d_f	fibre diameter
l_i	length of tube *i*
L	depth of filter medium
N_f	number of fibres
N_p	number of pores
Δ_p	pressure differential
p_s	sticking probability
Q	flow rate
R_i	hydrodynamic conductance of tube *i*
x_1, y_1, z_1, a_1	position and radius of sphere 1
x_2, y_2, z_2, a_2	position and radius of sphere 2
ε	porosity or free area fraction
μ	viscosity

14.1 Classification of filtration processes and models

Of all the separation methods, filtration is the most widely used and extensively studied. All filtration processes rely on a porous medium (the filter) of one kind or another, which allows the fluid to pass through but retains the solids or dispersed phase(s) in the suspension. Such a physical separation is achieved either by deposition of suspended particles on the filter surface or by capture of particles in the porous medium.

14.1.1 Classification of filtration processes

When particles are larger than the surface pores of the medium, they will be retained on the surface of the medium. If the carrying fluid flows at right angles to the medium, the process is called *dead-end filtration*. In this case, the deposited particles will form a 'cake' on the filter surface. The formation of the cake generally gives rise to, and sometimes is essential for, an enhanced efficiency of particle capture. On the other hand, the growth of the cake also brings about an increasing pressure drop and/or a decreasing filtrate flux. When the pressure drop becomes too high or the flux too low, the filter must be cleaned (e.g. through back-flushing) to restore its permeability. This usually means disruption of the filtration operation. If, however, the feed flow is parallel to the filter surface, a process known as *cross-flow filtration*, a quasi-continuous operation may be achieved, since most of the particles retained on the filter surface will be swept away by the flow. Dead-end and cross-flow filtrations are complementary in that, whilst dead-end filtration is suitable for rather concentrated suspensions of relatively large particles, cross-flow filtration is most appropriate for dilute suspensions of very fine particles.

In a polydispersion, some suspended particles may be of a comparable size with the pores. These particles can be trapped in the porous medium and clog the medium internally. There are two distinguishable mechanisms. First, if the particle is larger than the pore opening, it will block the entrance of the pore and render the pore inaccessible to other particles. This is called pore blockage. When the flow is reversed (e.g. during back- flushing), the particle will be carried away and the pore entrance reopened. The second mechanism may take place when the particle is only slightly larger than, or about the same size as, the pore. In this case, the particle may enter the pore and stay inside permanently, i.e. it will not be carried out of the pore even if the flow is reversed. This irreversible process is termed pore blinding. Pore blinding is more likely to occur if the porous material (e.g. fibre filters) and/or the particles (e.g. oil droplets) are deformable under pressure. Pore blockage and blinding can take place during both dead-end and cross-flow filtrations of polydispersions.

When particles are much smaller than the surface pores and hence able to penetrate into the filter medium, retention of the particles takes place throughout the entire depth of the filter medium, hence the name deep-bed filtration. In this case, separation of particles from the suspension is achieved mainly through particle deposition on the interior surfaces of the porous medium. The deposited particles can often act as additional collectors to give an improved performance of the filter. This phenomenon is described as *filter ripening*, especially when dealing with granular filtration of hydrosols. As filtration proceeds, some of the pores will eventually be clogged by the deposited particles. This is also called pore blocking, but is different from the pore blockage, mentioned above, in the following aspects. First, in pore blockage, it is the *entrance* of the pore that is blocked – the pore itself may remain empty; whereas in pore blocking, the pore is literally *filled* with the deposited particles. Another distinction is that in pore blockage the particle is stopped by virtue of geometrical constraints of the pore opening, whereas in pore blocking the particles are immobilized mainly by physical attraction between particles and inner surfaces of the pore and/or morphological hindrance. Nevertheless, the practical result is the same, i.e. the pore is no longer available to other particles in the suspension.

14.1.2 Classification of filtration models

The long history of theoretical studies of filtration processes has resulted in a rich and diverse collection of filtration models. Several criteria for classifying the models have been put forward along with the theoretical development, each serving a particular purpose. To distinguish the simulation models described in this chapter from others, it is convenient to introduce yet another criterion, according to whether or not a realistic model structure is used to represent the filter medium. Those based on highly idealized model structures (e.g. pores represented as a bundle of uniform vertical channels) will be referred to as *conventional theories*, because they represent the mainstream development over the past decades. In contrast, those utilizing more realistic model structures (e.g. pores represented as voids between randomly packed spheres) are regarded as advanced models, since they should be capable of simulating filtration processes in a more rigorous and/or realistic manner.

Another relevant criterion for distinguishing filtration models is based on how the flow in a porous medium is visualized or treated. According to this criterion, filtration models may be classified into two categories as *external flow models* and *internal flow models* (Rajagopalan and Tien, 1979). In external flow models, the fluid is primarily visualized as flowing over collectors instead of pores, and the porous media are considered as an assembly of *orderly* arranged cylinders for fibre filters (Kirsch and Fuchs, 1968) or spheres for granular filters (Spielman and FitzPatrick, 1973). This approach is also exemplified earlier in Chapter 12. By contrast, in internal flow models, the emphasis is placed on the pores as opposed to collectors, and the porous media are viewed as a bundle of capillaries (Scheidegger, 1974) or constricted tubes (Payatakes *et al.*, 1973).

The use of a simplified geometrical model allows the conventional theories to concentrate on particle interactions with collector surfaces and the carrying fluid – the driving forces for colloidal particles to transport and deposit onto the collector surfaces. However, other important factors such as pore blockage and blinding caused by near-pore-sized particles are difficult to incorporate into the conventional theories, because details of the filter structure must be considered. Moreover, most of the conventional theories developed from the external flow models are based on analysis of a single, 'typical' collector and exploit the result to give predictions of the overall performance of the filter. A fundamental assumption implied by such a treatment is that flow conditions (e.g. velocity and direction) are *identical* for all the collectors comprising the porous medium. This assumption can hardly be justified with respect to the flow pattern in real filters.

Over recent years a different approach has been proposed and is being developed (Arnold *et al.*, 1991). The basic features of this approach are the adoption of much more realistic model structures for the filter media, and the use of the microscopic Langrangian-type descriptions, through computer simulations. Since filtration models based on this approach are built upon realistic model structures, the assumption of identical flow conditions throughout the filter medium and the implications therein can be avoided. In addition, the influence of the structure of the filter on its performance during filtration can readily be assessed.

Expressions for colloidal and hydrodynamic interactions used to formulate the trajectory equation for particles have been given in Chapters 3 and 4. The principles and basic techniques, together with the advantages and limitations, of computer simulations have been described in Chapters 7 and 8. For those considering simulation of granular (or deep-bed) filtration processes, Tien (1989) has provided some additional information on granular filtration. Hence, the focus of this chapter will be on how to obtain a realistic model structure for a particular type of filter media. Applications of these modelling strategies to filtration problems will be illustrated through a few case studies.

14.2 Random line network model

14.2.1 Generation of model structure

Non-woven fabrics (polymer fibres and glass fibres) are used extensively as filter or membrane media in filtration or coalescence of finely dispersed solids or liquids. In the manufacturing process, layers of fibres are laid down to build up a given thickness of the material. The layers are formed as a random network of filaments. The filaments may be of a different diameter, but in any case the fibre length is far greater than the diameter. The pores are formed from the spaces between adjacent and overlapping fibres. A photomicrograph of an actual fibre material is shown in Figure 14.1(a).

The features as observed from the photomicrograph provide the basis upon which the assumptions of a geometrical model, the *random line network* (*RLN*) model, for simulation of unneedled non-woven fibre mats are derived (Abdel-Ghani and Davies, 1985). The RLN model assumes that:

(1) the axes of the fibres within each layer are straight lines of random orientations. Therefore, the pores in each layer will be irregular external polygons
(2) the fibres are represented by long cylinders of a distribution of thickness, and span over the entire area considered by simulation
(3) where two fibres cross, one (the thinner) is allowed to bend over the other (the thicker) at the intersection point to maintain the layer structure.

Three different methods are available to generate the random line networks. Given a square section in which a random line network is to be formed, the line position representing the axis of a fibre within the simulation area, say a square, may be defined by:

- a random point *on* the square boundary and a random slope. This is called the surface (S) randomness.
- a random point *in* the square and a random slope. This is termed the interior (I) randomness
- a random perpendicular distance from a *fixed* point within the square and a random angular position of the perpendicular. This is known as μ randomness.

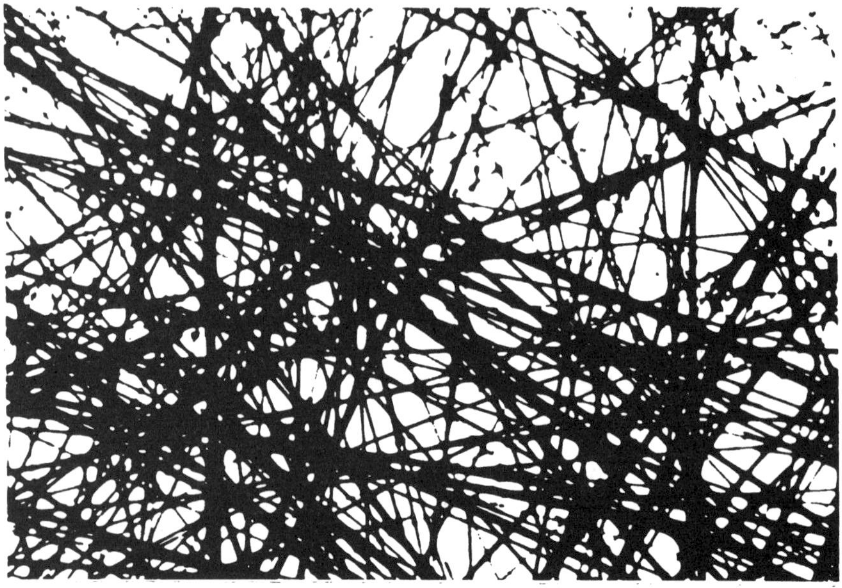

(a)

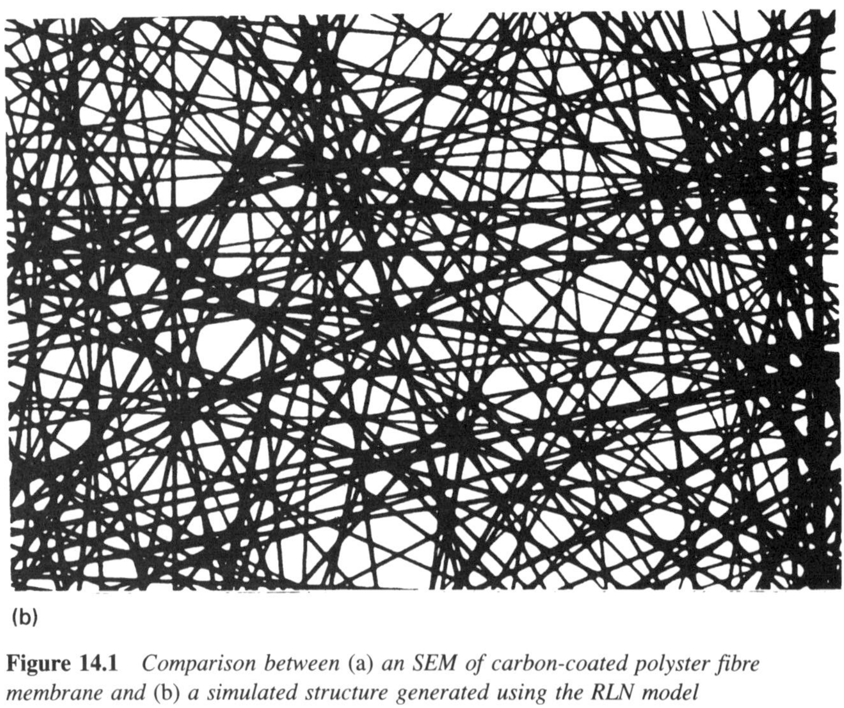

(b)

Figure 14.1 *Comparison between* (a) *an SEM of carbon-coated polyster fibre membrane and* (b) *a simulated structure generated using the RLN model*

The subtle differences between the three methods and their implications for the simulated structures have been discussed by Fairclough (1987).

During the simulation, as each fibre is added, the free area within the square is recalculated and compared with that of the actual filter material. The procedure is repeated until the calculated free area matches the desired value. A given thickness of material is then produced by stacking layers one above another.

An example of the simulated fibre mat is shown in Figure 14.1(b), for comparison with the actual material in Figure 14.1(a). Statistics of the generated random line networks have been compared with available theoretical predictions, and the agreement between the two is excellent. Good agreement has also been obtained between the simulated filters and a range of actual materials, including carbon fibres, glass fibres and polyesters, covering a mean fibre diameter from 3 μm to 32 μm and porosity from 85% to 98% (Abdel-Ghani and Davies, 1985).

The number of pores, N_p, and the number of fibres N_f, in each layer was found to be related by (Abdel-Ghani, 1983):

$$N_p = 0.1822N^{2.0458} \tag{14.1}$$

This relation is only accurate for thin fibre networks, and is likely to overestimate N_p for a given N_f if the edge effect and fibre thickness are considered.

Of more practical significance is the approximate relation between the mean pore area A_p, porosity ε, and mean fibre diameter d_f, given by Abdel-Ghani and Davies (1985):

$$A_p = \frac{\pi \varepsilon d_f^2}{(1 - \varepsilon)^2} \tag{14.2}$$

The RLN model described above is not the only one that has been put forward to represent fibrous structures. Back in the 1960s, Piekaar and Clarenburg (1967) had developed a geometrical model which is similar to the RLN model but for short and uniform fibres. Their conclusions concerning the structural properties of the filter were that the pore areas and their hydraulic radii could be described by log-normal distribution functions. These conclusions conflict with the simulation results of Abdel-Ghani and Davies (1985), who found that none of the pore areas, hydraulic radii and pore perimeters can be satisfactorily represented by log-normal distribution functions. Kaye (1989), in an attempt to explore the fractal structures of fibre filters, noted that fibre filters simulated by a RLN model appear to have two different fractal dimensions at different levels of inspection, but it is not clear whether this bimodal fractal is simply an artefact of the simulation or a fundamental property of the random line networks.

14.2.2 Simulation of coalescence of liquid droplets

To illustrate an application of the RLN model, consider the flow of a liquid dispersion through a fibre mat coalescer. Here the properties of interest are the capture efficiency of the media and the droplet penetration within the media. Both factors are difficult to quantify using conventional theories, especially when the mixture to be separated is a polydispersion. However, with the RLN model structure and a stochastic trajectory model, the conditions can be simulated using a Monte Carlo process, as described below.

The flow conditions here are restricted to the region where the capture mechanism is predominantly by *interception*. The dispersion arriving at the first layer is assumed to be homogeneous, i.e. there is an equal probability for a given droplet to be at any position in the cross-section of the filter medium. The simulation is commenced by selecting a droplet of diameter chosen randomly from a defined inlet distribution, at a random position above the first layer. The position of this droplet with respect to fibres in the first layer is then calculated. Direct interception of the droplet by a fibre is checked first. Then the drop–drop contact, which would result in coalescence and therefore drop rearrangement, is checked. In the event of coalescence, the resultant drop is assumed to be spherical, conserving the volume of the two droplets. The position of the new drop is calculated for point contact with the fibre on which the larger of the two droplets resided. The assumptions of coalesced drop being spherical and smaller droplets coalescing into larger ones can be justified by considering the size range of the droplets, which in the majority of practical cases is about 10 μm in diameter. After these checks (and rearrangements), if the original droplet is not captured in the first layer, it is passed through to the second layer, keeping its coordinates constant. The procedure is repeated until the droplet is captured or passes through the filter mat. Then a second droplet is introduced and its trajectory followed. By repeating the whole process for a large number of droplets, the capture efficiency and overall penetration can be calculated. Quantitative predictions in terms of droplet penetration show an exponential decay as the dispersion passes through the mat, and the capture efficiency is related to the particle diameter, fibre diameter and porosity area of the mat. Both are in line with experimental observations for deep-bed filtration.

The RLN model in conjunction with the stochastic trajectory model has also been applied to the separation of solids from dilute suspensions in deep-bed filtration to evaluate the effect of particle size distribution on filtration efficiency, pore blockage and blinding (Abdel-Ghani and Davies, 1984; Wilkinson and Davies, 1985). More recently, the model has been extended to include back-flushing (Chan, 1990), and can clearly demonstrate why flux performance of microfiltration membranes cannot be restored even with back-flushing and how back-flush frequency affects the on-stream time in filtration of the membranes. The results have revealed important characteristics which are likely to influence future design of filtration equipment.

14.3 Poisson point-line model

14.3.1 Generation of model structure

A large number of membrane filters used in ultra- and microfiltration are made from synthetic materials such as polymers, an example of which is the *stretched polymer* (*PTFE*) membrane. Membranes of this type are formed by heating and

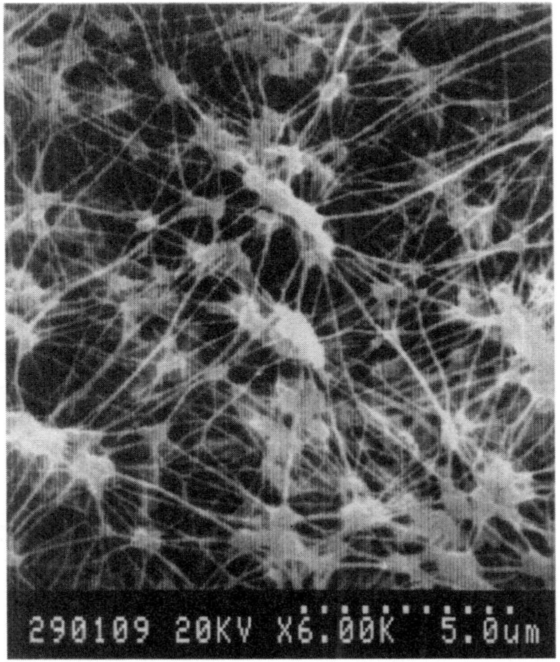

(a)

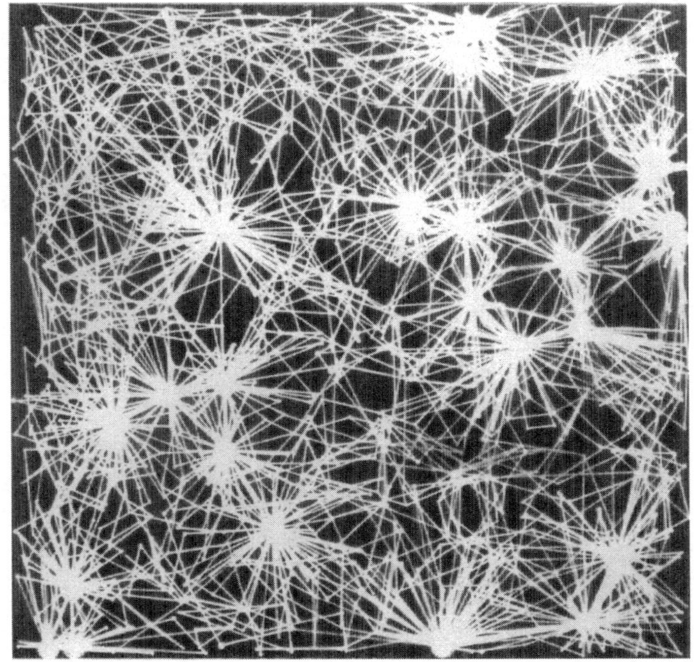

(b)

Figure 14.2 *Comparison between* (a) *an SEM of a Gore-Tex® PTFE stretched membrane and* (b) *a model structure generated using the PPL model*

stretching of dense polymer (PTFE) films. The stretching may take place randomly in all directions (almost as a dilation) or may be directional. The SEM photograph (Figure 14.2(a)) shows the formation of globules of polymers as a result of heating. The globules appear to be randomly distributed, and neighbouring globules are connected by thin and, mostly, straight filaments produced during the stretching process. When a section of the membrane parallel to the direction of permeate flow is examined, it is found that the filaments are formed at different levels but are parallel to the direction of stretching; and some can penetrate a certain depth or even the whole depth of the membrane.

These observations on the structure help to formulate the rules upon which a stochastic, geometrical model, the *Poisson point-line* (*PPL*) model, is built to simulate such membranes (Arnold and Davies, 1990):

(1) the globules are assumed to be randomly distributed in each layer, and hence they can be described by a Poisson point process. The number density and size distribution of the globules used in the simulation can be obtained by analysing photomicrographs of actual materials. The size distribution may be fitted by a normal distribution function
(2) penetration of globules into other layer(s) is determined by an overall probability (of globules in one layer extending down to the next layer). Alternatively, it can be done by assigning each globule a random variable representing the depth. In either case, the globules are, in effect, treated as cylindrical rods
(3) the network of filaments is produced by connecting nearest and second nearest globules with cylinders of a normal distribution thickness, for which the mean diameter and standard deviation are obtainable from SEM analysis of actual materials. The addition of filaments is terminated if the number density has reached the measured value. Alternatively, it may be determined by the porosity of the layer, which is computed after addition of each filament
(4) when the first layer is completed, the procedure is repeated for the second layer, whilst taking into account the globules inherited from the first one (due to penetration). Layers are then stacked to make up the thickness of the membrane. The minimum distance between two layers is defined as the sum of the maximum radius of filaments from each layer.

An example of the simulated structure is shown in Figure 14.2(b), which is made up of 45 layers, with 60 globules on each layer, a probability of 0.45 for overall penetration of globules and porosity of 0.961. For the sake of clarity, the diagram is plotted using a single line representation of filaments. The resemblance between the simulated and the actual membrane is remarkably good.

The type of membrane shown in Figure 14.3(a) is formed by stretching the original melt predominantly in one direction during the manufacturing process. More complex structures are produced if a shear deformation resulting from simultaneous elongation and compression in the orthogonal direction is carried out. This type of membrane appears to be very different from the first one. However, if the directional stretching is interpreted in terms of geometrical

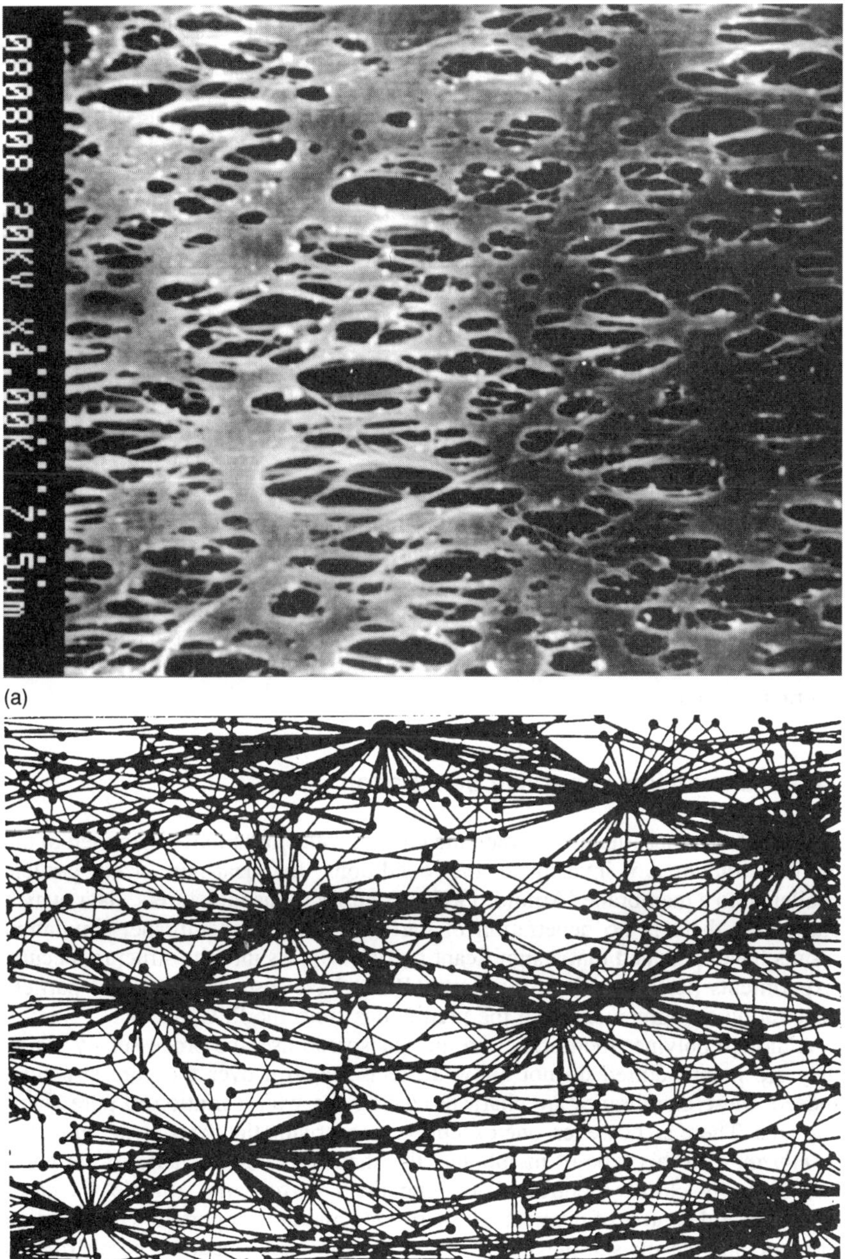

Figure 14.3 *Comparison between* (a) *an SEM of a Sartorius® PTFE membrane and* (b) *a PPL model structure after linear transformation*

transformation, the basic PPL model can also be used to simulate this type of membrane. Figure 14.3(b) shows one such an example, obtained by carrying out a linear affine transformation in one direction to each layer.

The structures produced by this model are much more complex than those generated by the RLN model. For example, the pores in each layer can be either convex or concave; the number of connections that a globule has is not necessarily the number of pores with which it associates. Furthermore, when techniques from graph theory are used to identify a non-isolated pore by following a path of connected vertices, the path does not necessarily return to the starting point and therefore a pore will not be identified. All these factors complicate the analysis, and characterization of such networks is by no means a trivial problem.

Pore statistics, including the pore order, pore area, perimeter and pore diameter, together with their frequency distribution, have been computed. Unlike a RLN network in which about 36% of the pores are triangular, in the PPL network the percentage of triangles is on average about 81%. The high frequency of small, low order pores means that such membranes are prone to blinding by entrained solid particles. Another conclusion arising from analysis of simulated structure, which was for a PTFE membrane of nominally 0.2 μm pore 'diameter', is that the definition of diameter adopted by manufacturers of this type of membrane is not entirely clear with respect to the distribution which exists. A pore rating derived from a bubble point or gas permeability test cannot properly relate to a distribution mean and hence is of little use in computing data on the internal structure.

14.3.2 Simulation of particle penetration

This can be further illustrated when the particle trajectory model is applied to the model structure to investigate the conditions for near-pore-sized particle challenge to a membrane (Arnold, 1992). Contrary to manufacturers' data, which claim that there is no penetration of particles into the membrane, simulation results have indicated that a significant particle ingress into the structure occurs. Experiments performed by Stoyanova (1991) have also demonstrated particle penetration, in agreement with the predictions from the PPL model.

Although polymer membranes are in widespread use, the processes by which they are manufactured do not allow a precise control over the pore structure. Indeed, these membranes are characterized by the wide variation in pore shape and area. There is, however, one exception, comprising the *track etched polymer membranes*. In this type of membrane, the shape, area and density of the pores can be fairly closely controlled, making them particularly suitable for cases where careful control of permeate or suspended matter is important (e.g. filtration of blood and plasma samples). This type of membrane is made by bombarding a polymer film (e.g. polycarbonate) with neutrons or α-particles. The high-energy beams effectively punch near-cylindrical holes in the film, which are then developed by chemical etching. The section is rather similar to that produced during the initial stage of a PPL simulation before globules are connected by

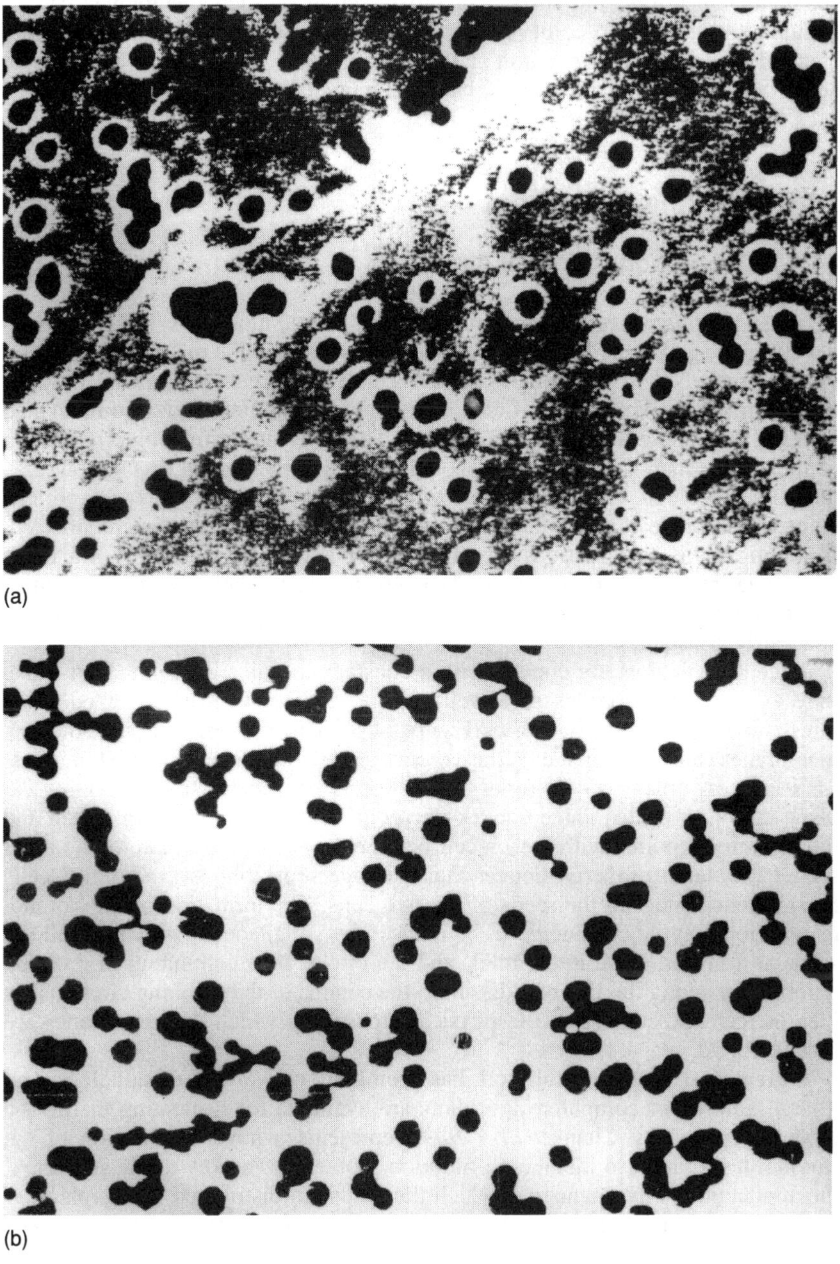

(a)

(b)

Figure 14.4 *Comparison of* (a) *an SEM of a track-etched polycarbonate membrane with* (b) *a simulated structure based on a constraint Poission point process*

filaments. Therefore, the surface or cross-section of such membranes can be simulated by a Poisson point process, with allowance for growth of points into areas of a controlled distribution of size (Chan, 1986). Comparison between the actual and simulated structure is given in Figure 14.4.

14.4 Tessellation models

14.4.1 Generation of model structures

Recently, *cellular ceramic membranes* based on alumina materials have been developed (e.g. Anotec Limited), for which it is claimed the mean pore diameter can be controlled in the range 0.02–2.0 μm. The membrane is made from an aluminium sheet. The sheet is first electropolished, and then anodized in an acid solution to obtain a thick layer of alumina. The pores are formed by propagation of sites where there is a higher current density and higher rate of dissolution (e.g., due to surface defects). In common with the track etched polymer membrane, Anotec membranes have a distinct feature that the pores are not interconnected. In addition, the pore sections are all irregular external polygons with a fairly narrow variation in shape and area. The coordination number of vertices is almost constant at three. The thickness of the ceramic walls separating adjacent pores is also fairly constant.

Since all polygons are convex with a constant coordination number of three, they exhibit the properties of a well-known geometrical pattern, called (two-dimensional) *Voronoi tessellation* (Figure 14.5). This suggests that the Anotec® membranes can be simulated by the Voronoi tessellation. For a given set of points (or seeds) on a plane, the Voronoi tessellation is constructed by subdividing the plane into a set of disjointed convex polygonal cells, each enclosing a seed and the constituent points within each cell being closer to its seed than to any other seeds. This latter property dictates that the edge separating two adjoining cells must be equidistant to the seeds of the two cells. The formation of the Voronoi tessellation may be considered to be induced by simultaneous growth of all the seeds at a uniform rate into circles, and the circles then continuing to grow by deformation along the line equidistant to the centres of the touching circles. This idea in fact also represents the physical process by which the membranes are manufactured.

A great deal of theoretical work has been done on Voronoi tessellations, and a large number of computer algorithms are available for generating them. The model proposed by Chan *et al.* (1989) represents a novel application of the Voronoi tessellation to membrane filtration problems. The key to the success of the application is the manner in which the seeds are distributed on the plane. If the position of each seed is considered independent of the presence of other seeds (i.e. a Poisson point process), the resulting pattern of the tessellation will be significantly different from that of the pore structure of the membrane: there is a much larger variation in pore shapes, with a significant number of triangular cells which do not appear in actual membranes.

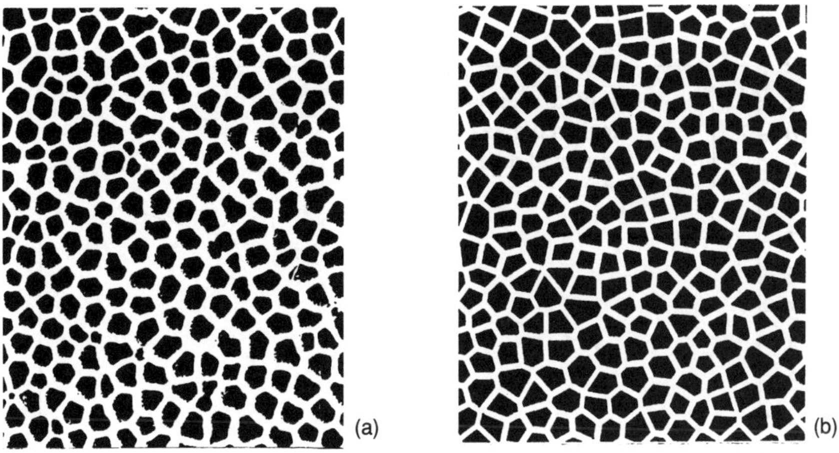

Figure 14.5 *Comparison between* (a) *an SEM of an Anotec® membrane and* (b) *a simulated structure based on the two-dimensional Voronoi tessellation*

To reduce the variation in pore shapes, two approaches can be adopted both having similar results on the resulting tessellation. In the first approach, a constraint is used along with the normal Poisson point process to control the minimum pair separation between seeds. In the second approach, a set of seeds is laid down using the normal Poisson point process and the plane is tessellated. Then the centroid of each cell is determined and taken as the new position for the seed, and the plane is tessellated again using the new set of seeds.

The simulation model based on the constraint Voronoi tessellation shows good qualitative and quantitative agreement between the predicted and the measured pore shape, area and perimeter distributions, as can be seen from Figures 14.6(a) and (b).

14.4.2 Prediction of solvent flux, pore blockage and blinding

The model can be used to predict solvent flux through the Anotec membrane. Assuming the flow is laminar, the relation between the flow rate Q and the pressure through a pore can be written as:

$$Q = \frac{\pi \Delta p}{8 \mu L} a_{\mathrm{H}}^4 \tag{14.3}$$

where Δp is the pressure drop across the membrane of depth L, μ is the viscosity of the fluid and a_{H} the hydraulic radius of the pore. Figure 14.6(b) compares the predicted and measured clean water flux at different pressures. Good agreement between the two is seen when the pressure is below 1.5×10^5 Pa. The difference at higher pressures may be attributed to errors in the experimental method (the time for emptying the vessel was too short for accurate flow measurements with the equipment used).

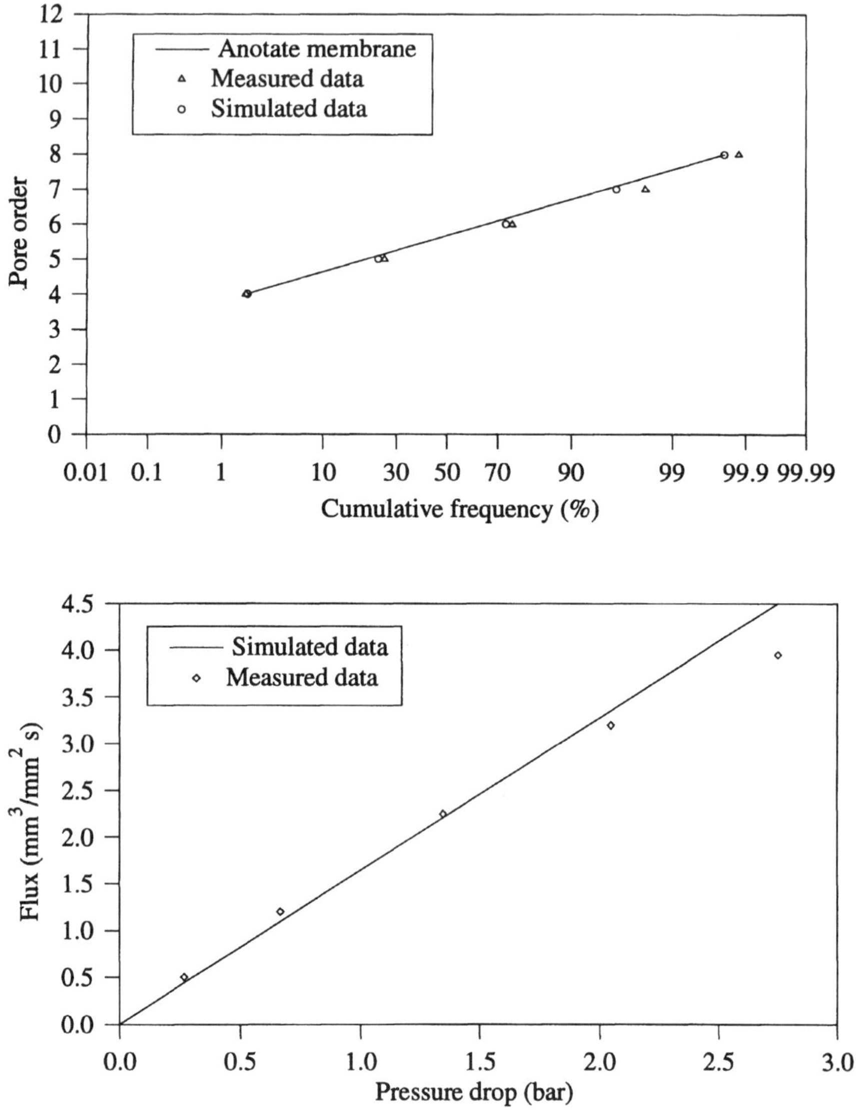

Figure 14.6 *Comparison between* (a) *the manufacturer's data (solid line), measured results of samples (hollow triangles) and simulation predictions (hollow circles) of the cumulative frequency distribution of pore order for Anotec membrane; and* (b) *the predicted (solid line) and measured (crossed squares) permeate flux through Anotec membrane samples at different pressure drops*

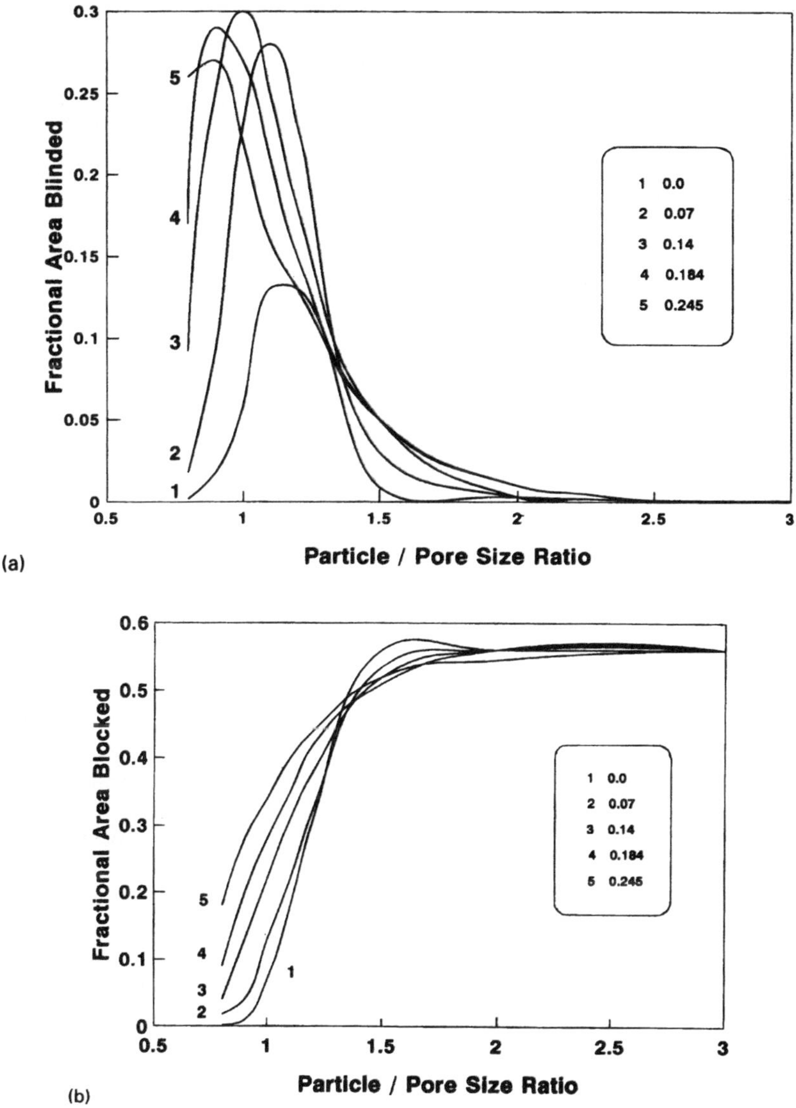

Figure 14.7 *The predicted variation of fractional area* (a) *blinded or* (b) *blocked, with particle size distributions*

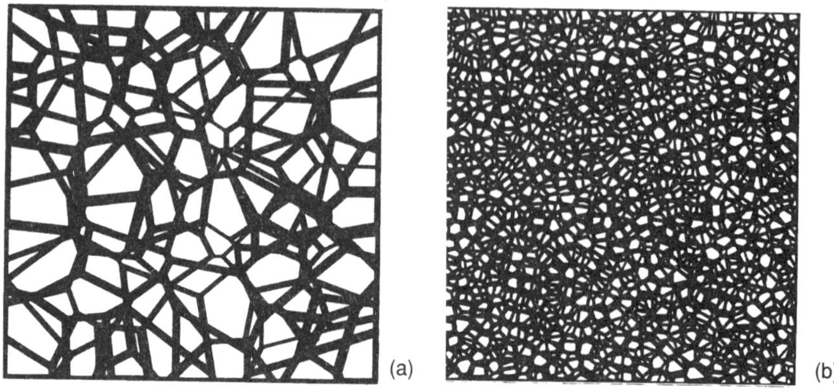

Figure 14.8 *Two-dimensional projection of the three-dimensional Voronoi tessellation used to simulate cellular foam membranes*

To assess the influence of particle size distribution on pore blinding and blockage during a filtration process, the following criteria, based on comparison between particle radius, a_p, and pore radius, a_i, were adopted in the simulation (Chan *et al.*, 1989):

$a_p < a_i$	entrainment of particle in the permeate, no effect on flux
$a_i \le a_p \le 1.1a_i$	blinding, irreversible loss of membrane flux and reduction of permeate flow
$a_p > 1.1a_i$	pore blockage, reversible by back-flushing and reduction in permeate flux

These were based on the experimental work of Rose and English (1973), and have also been used in conjunction with other models (e.g. RLN and PPL) to simulate filtration of solid particles. Figure 14.7 shows the results of the simulation for pore blinding and blockage. It can be seen that the size distribution of particles can have a profound effect on the performance of the membrane.

The concept introduced for the two-dimensional Voronoi tessellation applies equally well to three-dimensional space (or indeed space of any number of dimensions). Imagine that a set of points grow at uniform rate into spheres, and the spheres are then allowed to continue to expand along surfaces of contact until the whole space is tessellated; a three-dimensional Voronoi tessellation will be produced. In a three-dimensional Voronoi tessellation, cells become irregular convex polyhedra, and it is the face of the polyhedron that is equidistant from the seeds of two adjoining cells. If only the edges are retained, the resulting structure resembles a foam. This is evident from Figure 14.8. A reticulated polyurethene foam membrane is slightly different to this structure in that the cells are asymmetrical. Being formed from a bubble mixture, the hydrostatic pressure causes the cells to stretch in the vertical direction during manufacture. This, however, can easily be accommodated by applying a linear affine transformation in one direction.

14.5 Random packing of spheres

14.5.1 Generation of random packing of spheres

Thus far models of fibre filters, polymers and ceramic membranes have been described. This leaves yet another important type of filter media to be addressed: those which are themselves made from particles. Examples of this type of porous media include sintered membranes (e.g. used in cross-flow filtrations) and granular beds (e.g. used in deep-bed filtrations). Models for these filter media can be built on random packing of spheres.

The packing of particles has long been a subject of extensive study by both experimentalists and theoreticians from diverse disciplines because of its practical significance and academic interest. As a result, numerous algorithms exist for generating random packing of (mostly spherical) particles. The simple Monte-Carlo-type algorithm proposed by Vold (1963) for simulation of the structure of sediment can be used as the basis for models of porous media consisting of quasi-spherical particles, as well as the structure of deposit produced during filtration.

In this algorithm, spheres are dropped one by one in a straight line into a container. If the sphere touches the bottom of the container, it is allowed to stay. If it comes into contact with a previously dropped sphere before it reaches the bottom, it is allowed to roll down the surface of the sphere already in the sediment, until it either touches the bottom (or wall) of the container or a second sphere in the sediment. In the former case, it is stopped. In the second case, it begins to roll down the surfaces of the two dropped spheres while maintaining contact with both, until it either touches the bottom (or wall) of the container, or a third sphere. Then a test is performed to see if the sphere is now in a (mechanically) stable position. If it is, it will stay there; otherwise it will continue to roll down the surface of the sediment until a stable position is found. Sometimes a *sticking probability*, p_s , is used to control the number of contacts the sphere can make before it is stopped. The procedure followed is: at each contact, a random number x between 0 and 1 is generated and compared with p_s which is also between 0 and 1. If $x < p_s$, the sphere is stopped whether the current position is stable or not; if not, it will continue to roll down the surface of the sediment. Clearly, the sticking probability p_s will also dictate the overall packing density. If $p_s = 1$, an open, tree-like structure is produced (Figure 14.9), which resembles particle deposits on fibres frequently found in air filtration. If ps = 0, then a dense packing is produced and the structure is similar to that of a sintered metal membrane (Figure 14.10), and indeed granular beds as well.

In some sintered granular membranes, the distinguishable globules under the microscope cannot realistically be regarded as single spheres: rather they are themselves made of smaller particles fused together. An example of this type of membrane, a zirconium-coated carbon membrane, is shown in Figure 14.11(a). It is possible to simulate this kind of structure using the RPS model with the single spheres replaced by clusters of overlapping spheres. An example of the kind of structure generated by this method is given in Figure 14.11(b).

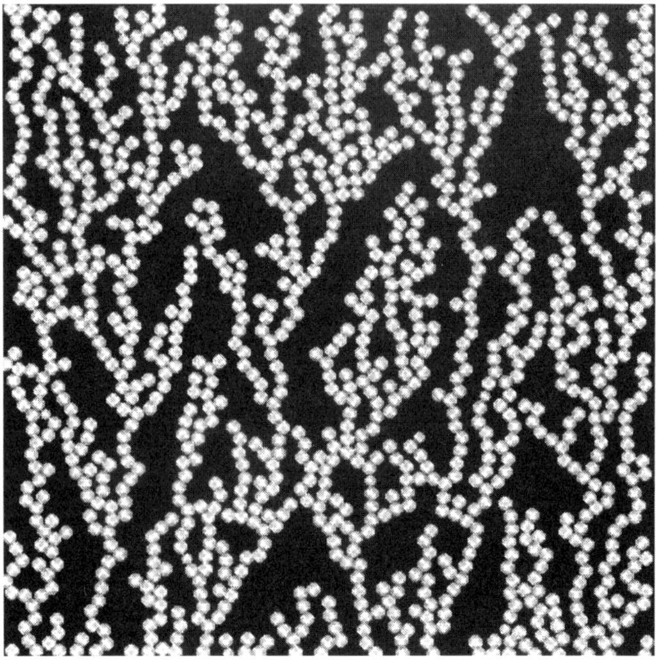

Figure 14.9 *A typical structure generated using the RPS model in two-dimensional space with the sticking probability equal to 1.0*

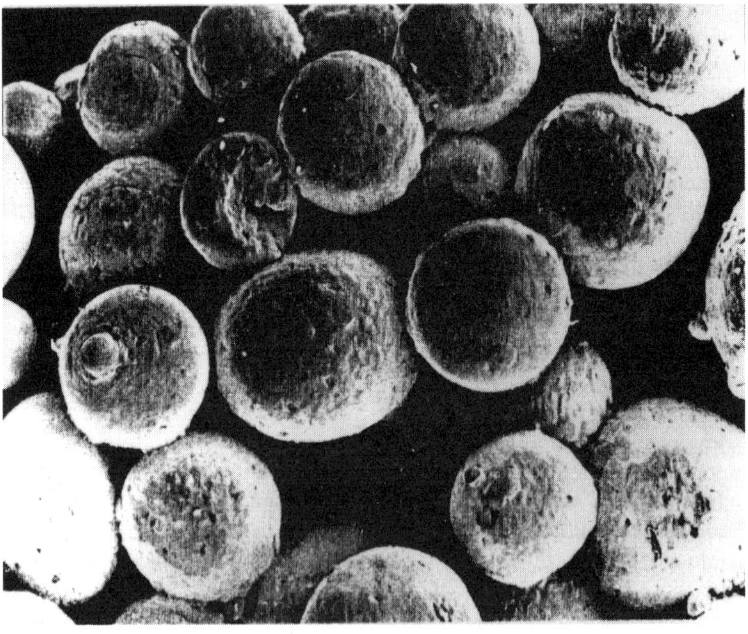

(a)

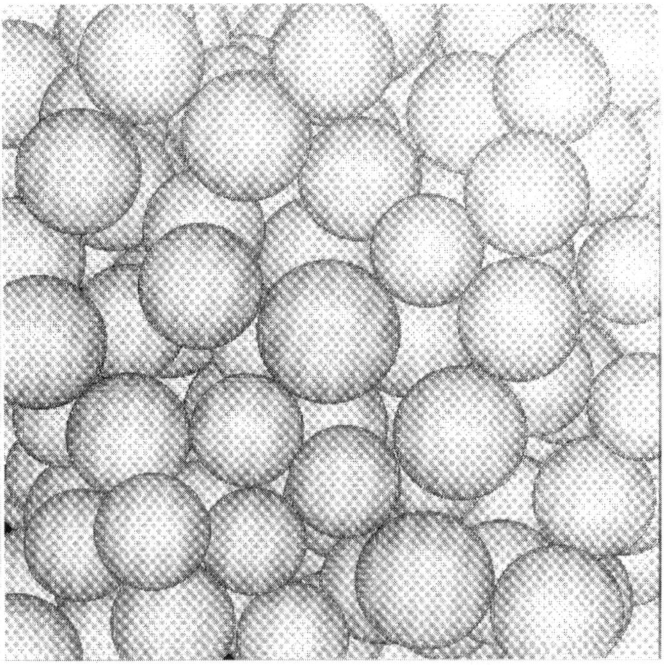

(b)

Figure 14.10 *Comparison between* (a) *an SEM of a sintered metal membrane and* (b) *the structure generated using the three-dimensional RPS model*

(a)

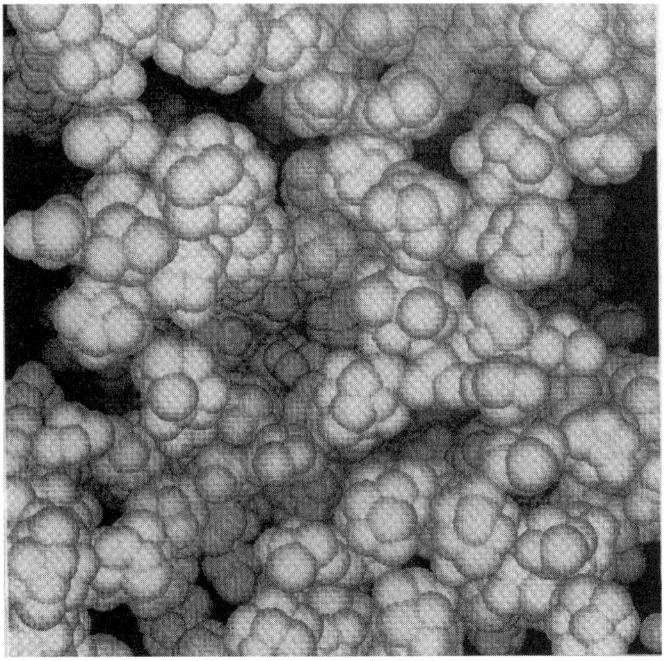
(b)

Figure 14.11 *Comparison between* (a) *an SEM of the cross-section of a zirconium-coated carbon membrane tube, and* (b) *the structure generated using the three-dimensional RPS model with subsequent replacement of each sphere with a cluster of overlapping small spheres*

14.5.2 Network models of granular porous media

In principle, given the size and position of each sphere in the random packing, the actual pore structure can be computed and from this calculations can be performed to predict the flow-pressure drop and particle collection efficiency. In practice, however, this presents an insurmountable problem both mathematically and computationally. Consequently, the problem has to be solved approximately. The approximation operates at three different levels or stages. First, the actual pores are modelled by some geometrically well-defined shapes, of which the simplest and oldest model is the cylindrical-tube representation. More sophisticated models use converging–diverging constricted tubes (Payatakes *et al.*, 1973), tetrahedrons (Chan and Ng, 1988), rectangular tubes (Ioannidis and Chatzis, 1993) or even fractals (Adler, 1991) to represent the pores. Second, the interconnected void spaces in the porous medium are modelled by a network of the tubes or shapes. Both two-dimensional and three-dimensional networks have been employed by various researchers (Dullien, 1991). However, with few exceptions (Rowley, 1988; Hollewand and Gladden, 1992), all the networks used to model porous media have been built upon a regular lattice, typically a square (two-dimensional) or a cubic (three-dimensional) lattice, although there are variations as to whether to consider explicitly the finite volume of nodes in the network (Ioannidis and Chatzis, 1993) or not (Imdakm and Sahimi, 1991). And finally, the flow through the porous medium is computed based on the established network, in a manner analogous to the calculation of electrical current in a resistive circuit using a so-called *nodal approach*. At each node in the network, the arithmetic sum of incoming and outgoing flows is zero:

$$\sum R_i \Delta p_i = 0 \tag{14.4}$$

where R_i is hydraulic conductance, and Δp_i the pressure drop, of tube i. If the flow through each tube is assumed to be laminar, then from eqn (14.3) it follows that the hydraulic conductance is given by:

$$R_i = \frac{\pi a_i^4}{8\mu l_i} \tag{14.5}$$

where a_i is the effective radius of tube i, and l_i its length. Writing down eqn (14.4) for each node of the network results in a set of linear algebraic equations for nodal pressures, and the pressure distribution and total pressure drop or flow rate can be calculated by solving these equations.

A disadvantage common to all network models mentioned above is that the size distribution of pores and the porosity of the porous medium must be prescribed before undertaking the simulation. Since this information has to be obtained elsewhere, those network models are not truly self-contained from simulation point of view. Moreover, it is well recognized that the topology as well as geometry of a network affects the fluid distribution in, and hence predictions of filter performance made from, the network. A key parameter describing the topology of a network is the average coordination number. In a randomly packed sand bed, the average coordination number is most likely to be 4, as indicated by results from the random packing of spheres, instead of 6 in a regular cubic (three-dimensional) lattice network. In this respect, a square lattice network, which has an average coordination number of 4, is a better representation of the granular porous media, and is simpler since it is a two-dimensional model. A more realistic network model is presented below which is a self-contained random network simulation model obtained from a tessellation of the random packing of spheres.

If the space occupied by a random packing of equal (or almost equal) spheres is divided using the Voronoi tessellation, with the centres of spheres being the seeds, then each node will lie in the middle of a unique pore body, and each edge will pass through a unique pore throat. In recognition of these features, Jia and Davies (1992) suggested that the Voronoi tessellation can be used as the skeleton upon which a random network of tubes can be built to represent the porous medium consisting of similarly sized spheres. The axes of the tubes coincide with (or are parallel to) the edges. The lengths of the tubes are equal to the lengths of the corresponding edges. The radius of each tube is calculated based on the average distance from the tube axis to each of the surrounding spheres. Since the tube is not an exact representation of the actual pore geometry, the radius of each tube is less well-defined and the exact value depends on which property the model should be made to preserve. For instance, to maintain the porosity of the packing, it should be adjusted so that the sum of the tube volume equals the total void space. As far as pore blockage/blinding is concerned, the tube radius should be assigned a value which corresponds to the narrowest part of the pore. When calculating the pressure drop or flow rate across the network, the tube radius should be adjusted so as to compensate for the errors caused by the replacement of an actual pore with a circular tube and the errors due to the lack of consideration of kinetic energy losses etc. in eqn (14.4).

If the packing consists of randomly-sized spheres, the Voronoi tessellation is no longer applicable, since it results in penetration of edges (and hence tubes) into spheres, and a more general tessellation method is needed. Such a general tessellation scheme has been devised, which, instead of a perpendicular bisector, uses an intersecting plane given by (Jia and Davies, 1992):

$$(x_2-x_1)x + (y_2-y_1)y + (z_2-z_1)z = (x_2^2+y_2^2+z_2^2-a_2^2)-(x_1^2+y_1^2+z_1^2-a_1^2) \tag{14.6}$$

to draw the boundary between one sphere (x_1, y_1, z_1, a_1) and another (x_2, y_2, z_2, a_2). The tessellation scheme has the following useful features:

- for a given set of spheres, the tessellation is uniquely defined. Each cell surrounds a unique sphere, and the edges pass through the pores and do not penetrate the spheres. If the spheres are of the same size, the structure is identical to the Voronoi tessellation
- it can be used for both two-dimensional and three-dimensional packings, with no restrictions on the packing density, whether the spheres overlap with each other, and whether the spheres are randomly packed or arranged in an orderly way
- analogous to the Voronoi tessellation, it may also be interpreted as corresponding to a growth process, only in this case the growth rate is different for each of the spheres.

It may be noted that most algorithms proposed to implement the Voronoi tessellation scheme cannot be adapted for use with this general scheme, because those algorithms rely heavily on the geometrical properties that are peculiar to the Voronoi tessellation. A rather primitive method has been used by Jia and Davies (1992) to implement the tessellation scheme, in which an initial box large enough to contain all the spheres in question is chopped step-by-step into the final shape, using the planes defined by eqn (14.6). The algorithm is not as slow as its primitive nature might suggest, as long as the packing is not very large.

Figure 14.12 shows an example of flow distribution in a horizontal cross-sectional plane of a random packing of spheres, calculated using the method

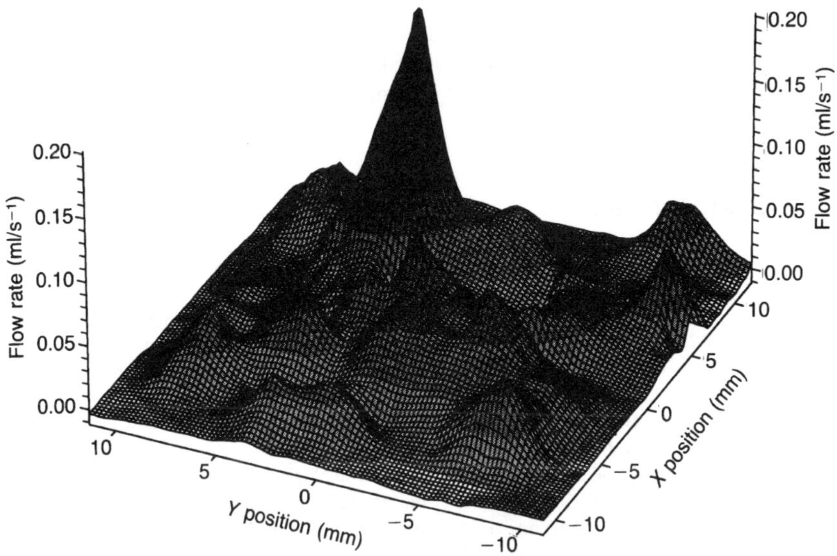

Figure 14.12 *Flow distribution in the cross-sectional plane at the top of a three-dimensional random packing of 940 spheres of 1 mm mean radius with 0.1 mm standard deviation, within a 20 × 20 × 20 mm simulation box. Porosity is 0.49. The corresponding random network contains 13 280 tubes, 6765 nodes, 171 inlets and 210 outlets. The total volume flow rate is 4 ml s^{-1}, or 10 mm s^{-1} in terms of superficial velocity. The pressure drop is 18.9 Pa for water at room temperature*

described above. It can be seen that the flow varies dramatically from point to point, with some pathways being more preferred than others by the flow.

14.5.3 Simulation of filtration processes

The model structure, and the random network of tubes derived from it, can easily be incorporated into existing simulation models or be used as the basis to develop other simulation models. Recent examples of the existing simulation models that can be used, directly or after some modifications, with the model structure include MC simulations by Imdakm and Sahimi (1991) and Adamczyk *et al.* (1992, 1993), and BD simulations by Peters and Gupta (1989) and Adamczyk *et al.* (1993).

The intricate details of these simulation models will not be described further here. However, it is worth noting that all these models can take full account of colloidal interactions involved in the process (e.g. electrical double layer and van der Waals forces) and, to some extent, hydrodynamic interactions (e.g. drag force and wall effect) as well. Evidently, this is an area of significant industrial interest, the ultimate long-term goal being process-realistic three-dimensional simulations which incorporate physicochemical as well as hydrodynamic factors. In practice, the complexities of quantitative predictions incorporating explicit descriptions of microhydrodynamics in porous media may be limited (see Chapter 4). However, qualitative predictions based on models described in conjunction with some experimental measurements are feasible. Such information will allow sensitivity analyses to be undertaken which are of importance in design and operation of colloidal filtration processes.

14.6 References

Abdel-Ghani, M. S. (1983) Coalescence and flow of secondary dispersions in fibre beds, PhD Thesis, University of Manchester Institute of Science and Technology

Abdel-Ghani, M. S. and Davies, G. A. (1984) Filtration of dilute suspension using non-woven wire cloths. In *Solid–Liquid Separation* (ed. Gregory, J.) Ellis-Horwood, Chichester, 342–357

Abdel-Ghani, M. S. and Davies, G. A. (1985) Simulation of non-woven fibre mats and the application to coalescers. *Chem. Eng. Sci.*, **40**, 117–129

Adamczyk, Z., Siwek, B. and Zembala, M. (1992) Kinetics of localized adsorption of particles on homogeneous surfaces. *J. Colloid Interface Sci.*, **151**, 351–369

Adamczyk, Z., Siwek, B. and Zembala, M. (1993) Influence of transport mechanism on adsorption of interacting colloid particles. *Colloid Surfaces*, **76**, 115–124

Adler, P. M. (1991) Fractal porous media. In *Transport Processes in Porous Media* (ed. Bear, J. and Coropcioglu, M. Y.) Kluwer Academic Publishers, Netherlands, 725–743

Arnold, S. T. (1992) Simulation of the structure and applications of streched polymer membranes, PhD Thesis, University of Manchester Institute of Science and Technology

Arnold, S. T. and Davies, G. A. (1990) A stochastic model to simulate the structure of stretched polymer membranes. *Proc. Vth World Filt. Cong.*, **1**, 541–547

Arnold, S. T., Chan, S. C., Jia, X., Lim, J. H. P. and Davies, G. A. (1991) Statistical models to describe the structure of porous membranes Presented at the Triennual Meeting of Institution of Chemical Engineers South Africa, Durban, South Africa

Chan, K. L. (1986) Modelling of crossflow micro- and ultra-filtrations. MSc Thesis, University of Manchester Institute of Science and Technology

Chan, K. L., Lim, J. H. and Davies, G. A. (1989) Prediction of the structure and blinding of an inorganic cellular membrane in dead-end filtration. *Int. Tech. Conf. Membrane Separation Processes*, Brighton, UK 121–141

Chan, S. C. (1990) Studies of separation of dilute dispersions. PhD Thesis, University of Manchester Institute of Science and Technology

Chan, S. K. and Ng, K. M. (1988) Geometrical characteristics of the pores space in a random packing of equal spheres. *Powder Tech.*, **54**, 147–155

Davies, C. N. (1973) *Air Filtration*, Academic Press, London

Dullien, F. A. L. (1991) Structure of porous media. In *Transport Processes in Porous Media*, (ed. Bear, J. and Coropcioglu, M. Y.) Kluwer Academic Publishers, Netherlands, 3–41

Fairclough, A. R. N. (1987) A simulation of non-woven fibre mats and the application to dilute particle separation processes, PhD Thesis, University of Manchester Institute of Science and Technology

Fairclough, A. R. N., Chan, K. L. and Davies, G. A. (1987) An analysis of fibre filters for filtration of dilute suspensions. *Chem. Eng. Res. Des.*, **65**, 396–406

Hollewand, M. P. and Gladden, L. F. (1992) Modelling of diffusion and reaction in porous catalysts using a random three-dimensional network model. *Chem. Eng. Sci.*, **47**, 1761–1770

Imdakm, A. O. and Sahimi, M. (1991) Computer simulation of particle transport processes in flow through porous media. *Chem. Eng. Sci.*, **46**, 1977–1993

Ioannidis, M. A. and Chatzis, I. (1993) Network modelling of pore structure and transport properties of porous media. *Chem. Eng. Sci.*, **48**, 951–972

Jia, X. and Davies, G. A. (1992) A model to describe the structure of sintered metal membrane. *IChE Instn Chemical Engineer Research Event* (Rugby), University of Manchester Institute of Science and Technology

Kaye, B. H. (1989) *A Random Walk Through Fractal Dimensions*, VCH Publishers, New York

Kirsch, A. A. and Fuchs, N. A. (1968) Studies on fibrous aerosol filters and diffusional deposition of aerosols in fibrous filters. *Ann. Occup. Hyg.*, **11**, 299

Lim, J. H. F. and Davies, G. A. (1990) A stochastic model to simulate the growth of anchorage dependent cells on flat surfaces. *Biotech. and Bioeng.*, **36**, 547–562

Payatakes, A. C., Tien, C. and Turian, R. M. (1973) A new model for granular porous media. *AIChE J.*, **19**, 58–76

Peters, M. H. and Gupta, D. (1989) Brownian dynamics simulation of convective transport and adsorption of hydrodynamically interacting particles onto surfaces. In *Studies in Physical and Theoretical Chemistry*, Vol. 63, *Math/Chem/Comp 1988* (ed. Graovac, A.) Elsevier, Amsterdam, 387–410

Piekaar, H. W. and Clarenburg, L. A. (1967) Aerosol filters – Pore size distribution in fibrous filters. *Chem. Eng. Sci.*, **22**, 1399–1408

Rajagopalan, R. and Tien, C. (1979) The theory of deep bed filtration. In *Progress in Filtration and Separation* (ed. Wakeman. R. J.) Elsevier, Amsterdam, 179–269

Rose, H. E. and English, J. E. (1973) The influence of blinding material on the results of test sieving. *Trans Instn Chem. Engrs*, **51**, 14–21

Rowley, M. E. (1988) Flow and separation of dispersions in media beds and tilted plate separators. PhD Thesis, University of Manchester Institute of Science and Technology

Scheidegger, A. E. (1974) *The Physics of Flow Through Porous Media*, University of Toronto Press, Toronto

Spielman, L. A. and FitzPatrick, J. A. (1973) Theory for particle collection under London and gravity forces. *J. Colloid Interface Sci.*, **42**, 607–623

Stoyanova, M. G. (1991) Structure and fouling of stretched polymer membranes. MSc Dissertation, University of Manchester Institute of Science and Technology

Tien, C. (1989) *Granular Filtration of Aerosols and Hydrosols*, Butterworth, Boston

Vold, M. J. (1963) Computer simulation of floc formation in a colloidal suspension. *J. Colloid Sci.*, **18**, 684–695

Wilkinson, E. T. and Davies, G. A. (1985) A stochastic model for the filtration of dilute suspensions using non-woven cloths. *Can. J. Chem. Eng.*, **63**, 891–902

Chapter 15

Application of simulation techniques to colloidal dispersion systems

Nomenclature

a	particle radius
a_c	radius of spherical collector
d	particle diameter
f_c	constraining force
$g(r)$	pair distribution function
L_{SB}	linear dimension (or length) of the simulation box
M_n	Mason number
n	floc size expressed as the number of primary particles
N	number of particles
N_{run}	number of simulation runs
Q	particle interaction strength
Pe	Peclet number
r	interparticle centre-to-centre separation distance
T_c	program execution time
Δt	time step
W_{SB}	width of simulation cell
δ_D	thickness of diffusion boundary layer
δ_H	effective range of hydrodynamic interactions
η_r	relative viscosity
τ	specified arrival time
τ^*	reduced stress
τ_y	yield stress
ϕ	volume fraction
ϕ_A	area fraction
γ	shear rate

The principles of simulating the behaviour of colloidal systems described in preceding chapters will be illustrated in this chapter using practical examples showing the application of selected simulation methods. The examples are arranged in increasing order of complexity, starting with a Monte Carlo simulation of the aggregation of a simple three-particle system and concluding with a Stokesian dynamics simulation of concentrated suspension in shear flow involving many-body hydrodynamics. Comparison of the simulation results with experimental data or results from other independent theoretical models is given where this information is available.

15.1 MC simulation of triplet formation

The first case concerns the simulation of the structure of aggregates formed by three particles (i.e. triplets or trimers), namely their *bond angle distribution*, for which some experimental measurements are available (Ito *et al.*, 1990). The bond angle of a triplet is defined as the largest angle in the triangle formed by the centres of the three primary particles. Clearly, for a triplet of equal spheres, the bond angle must be between 60° and 180°.

The experimental measurements were carried out by directly observing the collision process in a dilute polymer latex suspension at different electrolyte concentrations, using an ultramicroscope connected to an image processing system. The experimental conditions are summarized in Table 15.1.

The total colloidal interaction between the primary particles is evaluated as the sum of the electrical double-layer repulsion, van der Waals attraction and Born repulsion. The Born repulsion is included here so as to create a finite depth for the primary minimum. Alternatively, a minimum surface separation may be used instead, to achieve the same effect. Values of the parameters used by the expressions are listed in Table 15.2. The size of the simulation box was selected on the basis of the particle number density and the particle zeta potential on the surface charge density given in Table 15.1. Further details and discussion on the

Table 15.1 *Conditions for experimental measurements of triplet structure*

Parameter	*Value*	*Unit*
Mean diameter of particles	0.95	μm
Particle number density	1.42×10^8	cm^{-3}
Electrolyte (NaCl) concentration	1.0, 0.1 and 0.01	$mol\,dm^{-3}$
Surface charge density	–6.4	$\mu C\,cm^{-2}$

Table 15.2 *Conditions used in the MC simulation*

Parameter	*Value*	*Unit*
Mean particle radius	0.475	μm
Particle zeta potential	26.8	mV
Electrolyte concentration	1.0, 0.1 and 0.01	$mol\,dm^{-3}$
Effective Hamaker constant	1.0×10^{-20}	J
Dimension of the simulation box	26.8	μm
Collision diameter	0.5	nm

choice of these and other values listed in Table 15.2 have been given elsewhere (Jia, 1992).

By neglecting the kinetic aspects and concentrating on the structural properties alone, the problem is well suited for the MC simulation. To simulate the formation of triplets, a cubic simulation box is employed with periodical boundary conditions on all the sides and containing only three particles. To avoid possible bias caused by the initial configuration on the final structure of the formed triplet, the three particles are well-separated from one another at the start of the simulation.

The total interaction potential versus separation distance curves at the three different electrolyte concentrations shown in Table 15.2 are quite different. At high (1.0 M) salt concentration, the interaction profile only exhibits a primary minimum, and particles are expected to *coagulate* irreversibly. At low (0.01 M) salt concentration, the energy barrier is considered to be insurmountable, and particles are expected to *associate* with one another and the process is reversible. Different simulation procedures were used to allow for the two different modes of aggregation. The simulation procedure for coagulating particles is as follows. Initially, all three particles are moved one by one during each step, until a doublet has formed. Then only the vagrant particle is moved, since trial moves for the two irreversibly coagulated particles are deemed to be rejected. The reference acceptance ratio (Section 8.4) at the initial stage is set to be 0.5, and is lowered to 0.3 once the doublet has formed, in order to make the vagrant particle more 'energetic' and thus enhance its chances of colliding with the doublet within a given number of trial moves. The simulation procedure for associating particles is to move all three particles during each step, irrespective of their degree of aggregation, in order to accommodate the reversibility of the association. In both cases, the simulation stops at the formation of the triplet, and the final configuration is recorded. The bond angle distribution is obtained by analysing a large number of such final configurations. Alternatively, the simulation can be allowed to continue after the formation of the triplet and sample configurations are recorded at regular intervals.

Figure 15.1 compares the simulation predictions and the experimental measurements. It can be seen that the simulation results agree qualitatively with experimental data in the three cases, and are quantitatively comparable in the first two cases. It is interesting to note that the distribution for the first case resembles half the wave of a sinusoidal curve. This is in accordance with a theoretical model (Ito *et al.*, 1990) based on considerations of steric hindrance in the case of rapid coagulation. Possible reasons for the lack of exact quantitative agreement between simulation and experiment, especially for the third (reversible association) case, have been given and discussed in detail elsewhere (Jia, 1992).

15.2 MC simulation of magnetic flocculation

Compared with ordinary colloids, magnetic particle dispersions have distinctive properties and behave quite differently. This is because, in addition to the usual

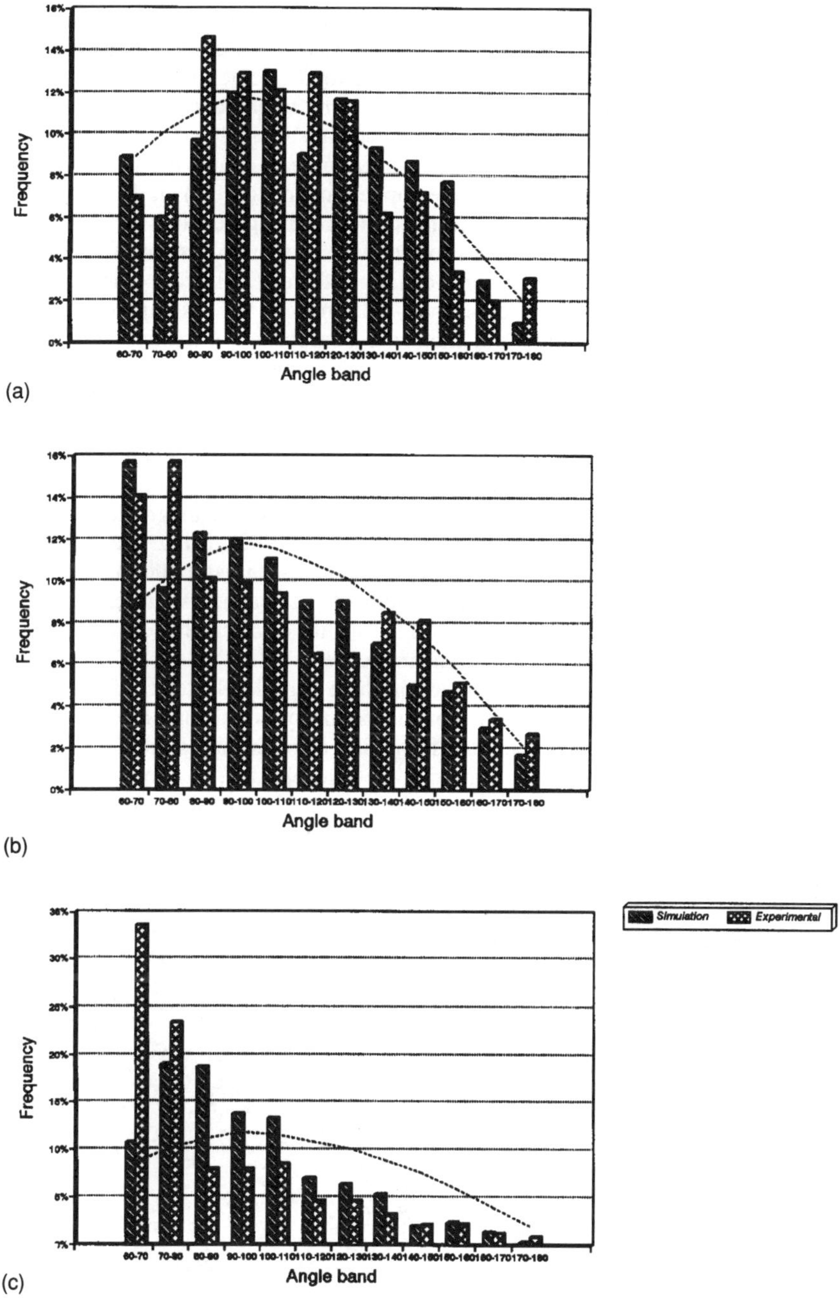

Figure 15.1 *Comparison between simulation, experiment and theoretical data for bond angle distributions. (a) 1.0 M; (b) 0.1M; (c) 0.01M*

isotropic colloidal forces such as electrostatic and van der Waals forces, magnetic particles are also subject to magnetic interactions, which are uniaxial by nature. Since the magnetic interactions, and hence the total interaction energy, depend not only on the separation distance but also on the orientation of the particles, aggregates of magnetic particles tend to have an anisotropic chain-like structure. By applying MC simulation methods, it is possible to study the influence of process variables (e.g. the strength of the applied magnetic field) on the structural properties of magnetic particle dispersions. Hence, if aggregation through van der Waals force is dominant, the flocs formed will be more isotropic; if magnetic aggregation becomes more dominant, chains with relatively few side-arms or branches will form. In this section, we will give a few examples of such application to magnetic fluids and magnetic composites. *Magnetic fluids* are colloidal dispersions of small single-domain particles of ferri- or ferromagnetic materials in non-magnetic carrier fluids, whereas *magnetic composites* consist of micrometre-sized non-ferrous particles (metallic or non-metallic, spherical or acicular) dispersed in a non-magnetic fluid. Particles in a magnetic fluid are usually much smaller than those in a magnetic composite.

The first example is a two-dimensional MC simulation of magnetic fluids of cobalt particles, carried out by Chantrell *et al.* (1982) using the standard Metropolis MC procedure outlined in Chapter 8. In this simulation, the total energy of a particle is calculated as the sum of the magnetostatic energy, the energy due to entropic repulsion resulting from the surfactant coating and the magnetic dipole energy in an applied magnetic field. Other colloidal forces have been neglected. The MC simulation can produce particle cluster configurations appropriate to thermal equilibrium. For dispersion of cobalt particles 15 nm in diameter, the MC model predicted the formation of open loop structures in zero magnetic field and long chains in an applied field of 1.0 T, as shown in Figure 15.2. Bradbury *et al.* (1986) have carried out a similar simulation but for polydispersions of cobalt particles. The main effect of the polydispersity was

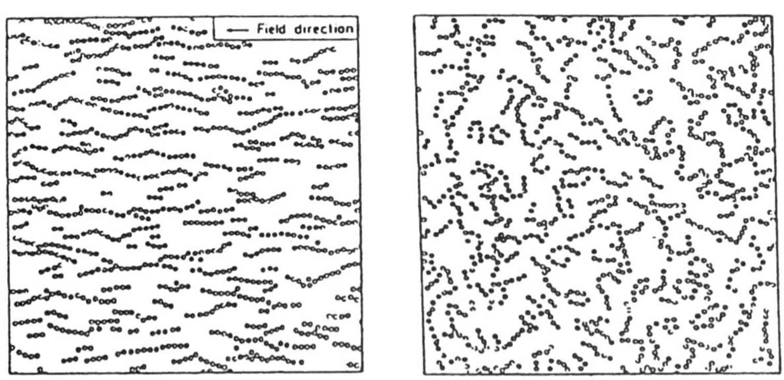

(a) (b)

Figure 15.2 *MC simulated configuration of magnetic fluid of 15 nm cobalt particles in* (a) *1-T applied field and* (b) *zero field*

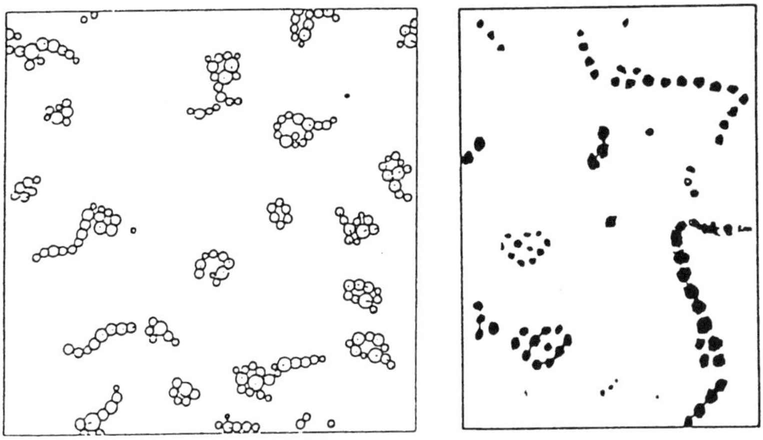

Figure 15.3 *Comparison between* (a) *simulated and* (b) *observed configurations for polydispersion of cobalt particles*

shown to be that the larger particles tend to aggregate and involve the smaller particles in the process. This can be seen in Figure 15.3. The similarity between the simulated configuration and the electronmicrograph, also shown in Figure 15.3, is good.

Magnetic composites are ideal systems to be simulated by two-dimensional MC methods, since the major forces involved in such systems are all potential forces and the systems are essentially two-dimensional. The suitability of the MC simulation methods for such systems can also be seen from the second example, which is again a standard two-dimensional Metropolis MC simulation, carried out by Davies *et al.* (1986). The simulation considers a magnetic composite consisting of polystyrene spheres of diameter 1 μm confined to a square simulation box of side 50 μm. Figure 15.4 shows an excellent agreement between simulation prediction and experimental observation under similar conditions of volume fraction and applied field. The variation of mean chain length (denoted by the number of the constituent primary particles n) with particle concentration (in terms of volume fraction ϕ) is given by an empirical relation as $n \propto \phi^{3/2}$. This relation can be reproduced by the simulation model as shown in Figure 15.4. The simulation also predicts that the chain length is proportional to the square of the mean particle diameter, i.e. $n \propto d^2$.

The third example is a MC simulation of magnetic flocculation in three-dimensional space (Jia, 1992). The simulation procedure remains essentially the same. Apart from the additional dimension, the difference between this simulation and the above simulations is the explicit inclusion of the electrostatic and van der Waals interactions and the omission of the magnetic dipole potential energy. The omission implies that all the particles in the magnetic dispersion have the same magnetic dipole moment and dipole orientation. Figure 15.5 gives

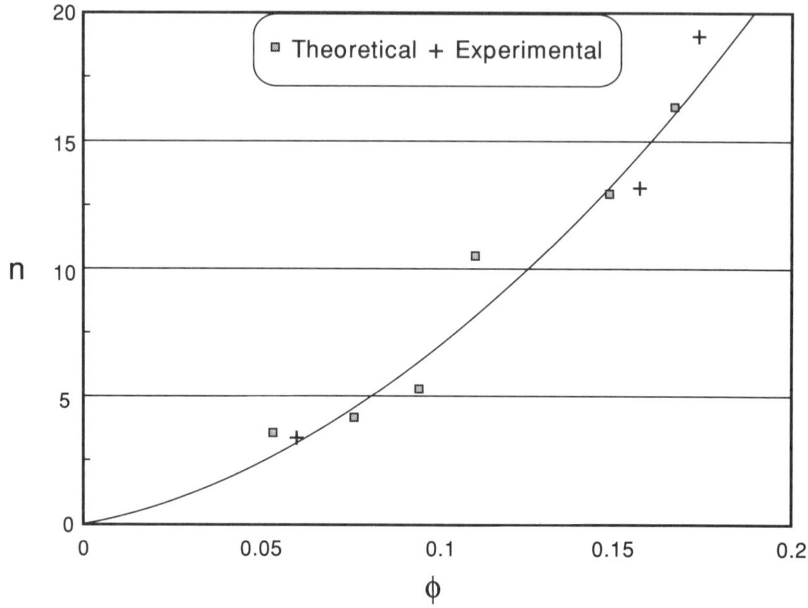

Figure 15.4 *Variation of mean chain length with particle concentration with a magnetic field of 10 mT applied parallel to the plane of the sample*

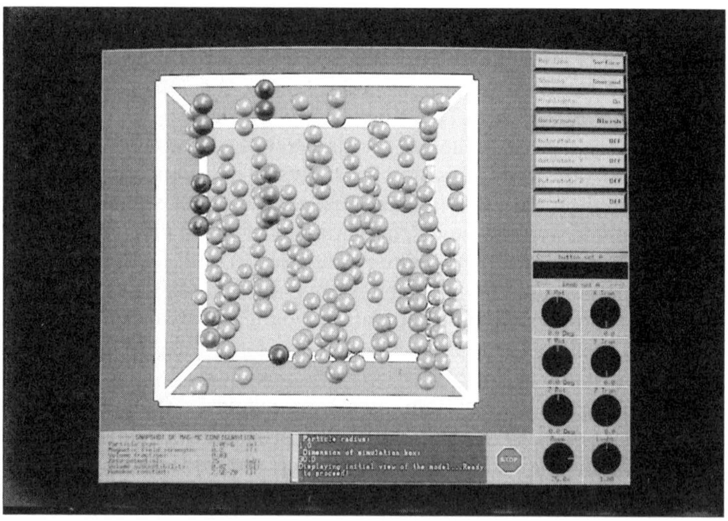

Figure 15.5 *Example of simulated magnetic flocs in three-dimensional space, with magnetic field lines in a vertical orientation*

an example of the simulated chain-like configuration in three-dimensional space. This method allows the three-dimensional form of the flocs (e.g. the degree of side-branching as a function of electrolyte conditions) to be predicted (Williams and Jia, 1991).

15.3 BD simulation of colloidal aggregation

The simplest type of BD simulation of colloidal aggregation is perhaps the case of the *dissociation* of a secondary-minimum flocculated doublet, where only two particles are involved and the hydrodynamic interaction is known exactly. Bacon *et al.* (1983) and Ansell *et al.* (1985) have carried out such simulations for doublets of spherical particles of 1 μm radius. In both studies, the simulation starts at a pair separation distance corresponding to the secondary minimum, and ends at a distance where the attraction potential is less than a few per cent of kT. The continuous DLVO-type interaction potentials are used and the time step is about 10 μs. The simulation results are in good agreement with the theoretical predictions of Chan and Halle (1984) using the first passage time method. Comparison between the simulation and the theory is given in Table 15.3, the simulation data are presented based on the N_{run} discrete dissociation events.

The complexity of the simulation increases when the flocculated doublet is also subject to a shear flow (Ansell *et al.*, 1985; Ansell and Dickinson, 1986a). The presence of shear entails the addition of two extra terms in the BD algorithm, one arises from the bulk motion of the undisturbed shear flow and the other accounts for the influence of interparticle hydrodynamics on the shear-induced motion of the two particles, as expressed in eqn (7.78). Using the BD simulation, the aggregate lifetime can be computed as a function of shear rate, depth of the secondary minimum and orientation of the doublet in relation to the flow field. The simulations show that if the secondary minimum is several kT deep, the doublet will rotate many revolutions before it dissociates.

For systems more complex than an isolated pair of particles, the many-body hydrodynamics take part in the processes or phenomena of interest. However, as has been mentioned in Chapter 4, no accurate analytical description is yet available for more than three-body hydrodynamic interactions. Besides, as more particles are involved hydrodynamically, evaluation of the diffusion tensor and

Table 15.3 *Results for doublet dissociation*

Depth of SM (kT)	*SM distance* (μm)	*Number of events*	*Dissociation time* (s) *(simulation)*	*Dissociation time* (s) *(theory)*	*Reference*
0.87	2.024	543	0.258 ± 0.015	0.253	Bacon *et al.* (1983)
4.7	2.008	185	1.49 ± 0.12	1.38	Bacon *et al.* (1983)
10.7	2.005	20	140 ± 40	164	Ansell *et al.* (1985)

generation of correlated random numbers becomes exceedingly time-consuming (see Chapter 8). Therefore, from both the theoretical and computational points of view, BD simulations of many-particle colloidal systems are only feasible if hydrodynamic interactions are considered in a less rigorous manner. This raises an important question as to how and to what extent the nature of the hydrodynamic approximations used in the BD algorithm affects the simulation results. Although the problem has not been investigated systematically, there is evidence (Ansell and Dickinson, 1987) that the structure and size distribution of small aggregates are not sensitive to the precise nature of the diffusion tensor used. Thus, as far as short-ranged structural properties are concerned, it seems justifiable to neglect multi-body hydrodynamics in the simulation of the aggregation of the DLVO particles. In the following example, the focus of the simulation is not the medium- to long-range disorder associated with the fractal-like characteristics of diffusion-limited aggregation process; the primary interest is confined to the short-range liquid-like structural features in relation to such factors as particle size and concentration, colloidal and hydrodynamic interactions, and the strength of the external force field.

Ansell and Dickinson (1986b) have performed a BD simulation of sediment formation. The DLVO potential is used for colloidal interactions and the Rotne–Prager diffusion tensor for hydrodynamic interactions. For the sake of efficiency, the process of a particle settling to become part of the sediment is broken into three stages. During the first stage, a spherical particle is descended, from a random lateral position well above the sediment, in a straight trajectory, to within $8a$ of the nearest particle already in the sediment. From then on, the Brownian is switched on, which signifies the onset of the second stage. At this stage, a large time step (50 μs) is used, and only the single particle diffusion tensor and the external force field (gravity) are considered. The third stage begins when the particle has reached a distance within $5a$ of the sediment. Now the time step is reduced to 20 μs. The DLVO force and the full Rotne–Prager tensor are included. Aggregation is deemed to take place if the colloidal interaction potential between the settling particle and its nearest neighbour in the sediment is more negative than $-10kT$. From the above brief description, it can be seen that the simulation procedure actually consists of three separate models: a ballistic trajectory model, a random walk model and the full BD model. Thus, it is interesting to compare the simulation results with those obtained from simulations using the pure ballistic model or pure Brownian model. Figure 15.6 shows the comparison of the simulation results graphically.

15.4 BD simulation of colloidal deposition

Although the dynamics of colloidal particles at a solid–liquid interface constitute an important fundamental process which underlies the kinetics of a variety of phenomena (such as particle adhesion and removal, sedimentation and filtration), studies of particle deposition by simulation techniques have lagged behind the theoretical and experimental developments, as well as their counterparts in other fields (such as colloidal aggregation and rheology). Here, in addition to the usual

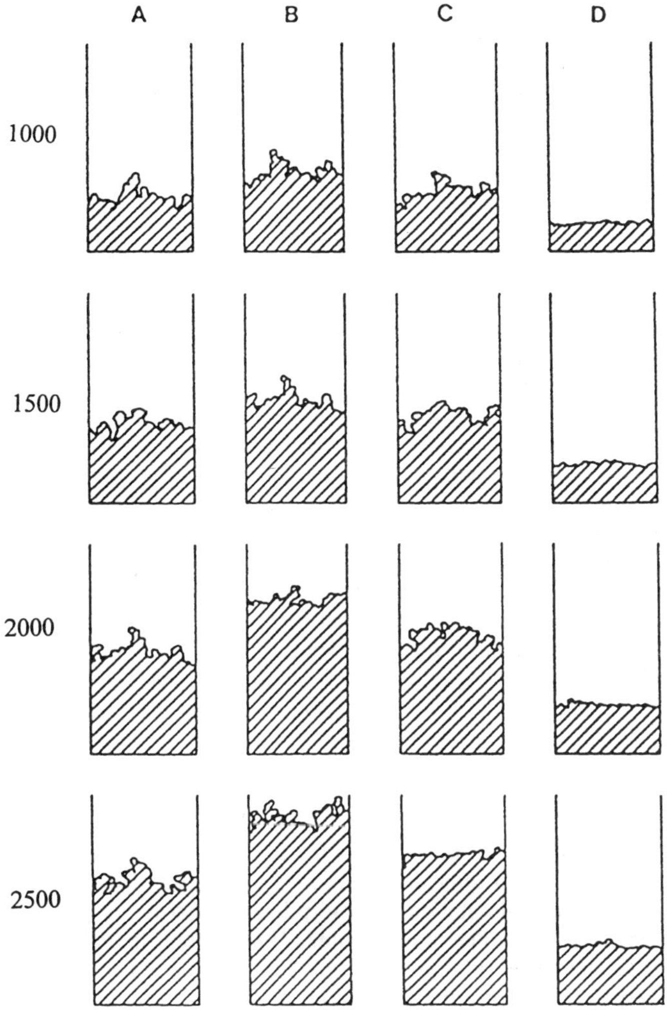

Figure 15.6 *Configurations generated using different simulation models for colloidal sediment formation after addition of 1000, 1500, 2000 and 2500 particles.* (A) *Pure Ballistic model.* (B) *Pure Brownian model.* (C) *BD model with 500g gravity strength.* (D) *BD model with 5000g gravity strength*

interparticle forces, the colloidal and hydrodynamic interactions between the particles and the collector surface must also be considered. Thus, it is perhaps not surprising that relatively fewer attempts have been reported that use the BD method to simulate the deposition process. Here, we shall give two examples of BD simulation of colloidal deposition, each representing an extreme case in terms of complexity and hence computational demand.

The first example is a BD simulation of adsorption of *non-interacting* particles onto a solid wall (Jia, 1992). Since the particles are non-interacting, study of a single-particle motion in the vicinity of the solid wall is sufficient to obtain the relevant information. The simulated system consists of a single spherical particle and two parallel planar walls. The particle is assumed to interact only with the nearer one of the two walls. As the three components of the Brownian (or random) term in the BD algorithm are generated independently in this case, and only the component normal to the wall actually contributes to the particle adsorption, the problem can be reduced to a one-dimensional simulation. The DLVO forces, Brownian motion and particle-wall hydrodynamics are considered by the simulation. The separation between the two parallel walls is chosen to be twice the effective range of the particle-wall interactions. By calculating the average time spent by the particle before it is adsorbed on either of the two walls, the rate of deposition can be estimated. Using this simple simulation procedure, deposition of non-interacting particles can be studied with respect to the particle size, ionic strength and surface charges. If surface roughness is included explicitly, the influence of shear flow, which in the vicinity of the solid wall is simplified to a simple shear, can also be studied. Despite the simplifications, the simulated deposition rates (Jia, 1992) have the same order of magnitude ($\sim 10^{-6}$ cm s^{-1}) as those measured by Sjollema and Busscher (1989) under similar conditions. Clark *et al.* (1987) have carried out a similar single-particle BD simulation, but they used a different arrangement for the simulation box. Their system has only one real solid wall, the other being replaced by a virtual wall which either absorbs or reflects the particle. Such an arrangement has only half the computational efficiency of the two-wall arrangement. Jia (1992) extended the single-particle BD simulation to include a second particle, in the form of a two-particle constraint BD simulation using Beenakker's tensor (eqn (4.67)), in order to study the blocking effect. It was found that on average the deposited particle can block an area about 3.5 times the projected area of the particle itself. If the second particle is allowed to move along the collector surface, then it is possible for aggregation to occur *after* deposition.

The next example is a BD simulation of adsorption of *interacting* particles onto a large spherical collector (Peters and Gupta, 1989). Here, the adsorbing sphere is located at the centre of the simulation box. Particles are generated at an upstream plane perpendicular to the flow, and moved according to their Brownian trajectories until they either adsorb on the collector surface or exit through the downstream boundary. Periodical boundary conditions and the minimum image convention are employed for the four sides parallel to the direction of the main flow, but not for the two sides normal to the flow direction. During the simulation, particles are continually generated at the upstream boundary in a spatially uniform random fashion according to a specified arrival time, τ. For very dilute dispersions ($\phi < 10^{-3}$), τ can be several times larger than the integration time step, Δt, so that a number of steps can be computed before a new particle is introduced. For concentrated dispersions ($\phi > 10^{-2}$), Δt may be greater than τ, and a number of particles need to be generated at each integration time step. The size of the simulation box is chosen such that $L_{SB} > 2(a_c + a + \delta_D)$, where a_c is the radius of the spherical collector, a the radius of the particle and

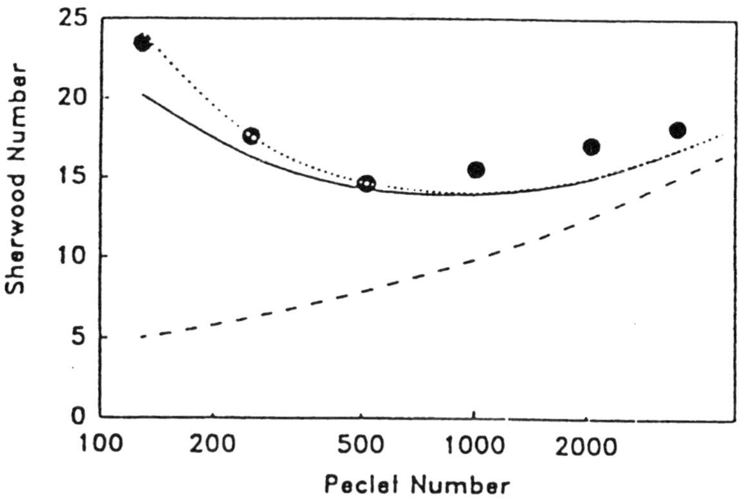

Figure 15.7 *Comparison between simulation and theory. Solid line: Peters' solution. Dotted line: Peters' solution with many-body colloidal interaction correction. Dashed line: Levich solution. Solid circle: Simulation results*

δ_D the thickness of the diffusion boundary layer. This is to ensure that particle interactions are not affected by the adsorbing surface, and that the infinite domain boundary condition for particle number density can be satisfied.

The simulation starts with particles uniformly distributed throughout the simulation box according to their free stream number density. New particle configurations are computed by simultaneously moving all the particles contained in the simulation box using the BD algorithm. Particles may either touch the fixed collector sphere and disappear from the system (which corresponds to *perfect-sink* deposition), or escape through the downstream boundary, or remain in the simulation box to be considered in the next integration step. The simulation results are in good agreement with numerical solutions of the convective-diffusion equations formulated and solved by Peters (1988), as shown in Figure 15.7.

A major difficulty in a BD simulation of this kind arises from the relatively long range of hydrodynamic interactions, which demand a large simulation box

Table 15.4 *Number of particles required for different volume fractions*

ϕ	$\delta_H/2a$	N	$W_{SB} \times L_{SB}$
1×10^{-2}	9.7	688	$60a \times 80a$
5×10^{-3}	18.7	537	$75a \times 80a$
2×10^{-3}	45.9	2959	$184a \times 184a$
1×10^{-3}	91.2	11595	$365a \times 365a$

Table 15.5 *Computation time for one time step for different numbers of particles*

N	*CPU time* (s)	N	*CPU time* (s)
68	0.79	341	79.51
137	5.56	410	136.39
205	17.81	500	210.59
274	41.64	1000	1466.65

and hence a large number of particles to be used. Table 15.4 shows the effective range of the hydrodynamic interactions, δ_H, as a function of particle volume fraction, ϕ. Also shown is the appropriate number of particles that would be required to perform the simulation. Table 15.5 gives the computation time for one time step on a Cyber 205 (a powerful vector computer) as a function of the number of particles. It can be seen that the required number of particles, N, increases sharply as the particle concentration (or volume fraction ϕ) decreases, and so does the execution time $T_c (T_c \propto N^{2.8}$, or $T_c \propto \phi^{-6})$. Assuming a particle volume fraction of 0.005%, the CPU time required to complete 7500 iterations would be about 438 hours on the Cyber 205 supercomputer!

15.5 Simulation of colloids under shear

The rheological behaviour of colloidal suspensions depends largely on the particle concentration and the extent of particle aggregation under shear. Computer simulation is a useful tool to relate the macroscopic rheological behaviour to the microstructure of colloidal suspensions, as can be seen from the following examples.

Melrose and Heyes (1993) have reported results of BD simulations of two model aggregated dispersions, namely, the electrorheological fluids (with anisotropic dipole interactions) and depletion flocculated systems (with isotropic interactions). In the absence of shear, both systems form an infinite gel-like aggregate network. Under shear, they shear thin and undergo shear-induced agglomeration to form more compact structures. The simulation model does not include many-body hydrodynamic interactions because of the computational constraints, as a relatively large number of particles need to be used. (Simulations which do include many-body hydrodynamics are discussed in Section 15.6.) This means that the structure in sheared suspensions of moderate to strongly interacting particles is assumed to be determined solely by the non-equilibrium statistical mechanics of the direct particle interactions, core-volume fraction and Brownian motion. Hence the simulated suspensions are idealized model systems. Since the interactions included can all be considered as thermodynamic interactions (in the sense that they determine the equilibrium structure), the simulation is able to capture the essential physics of structure formation in the model systems.

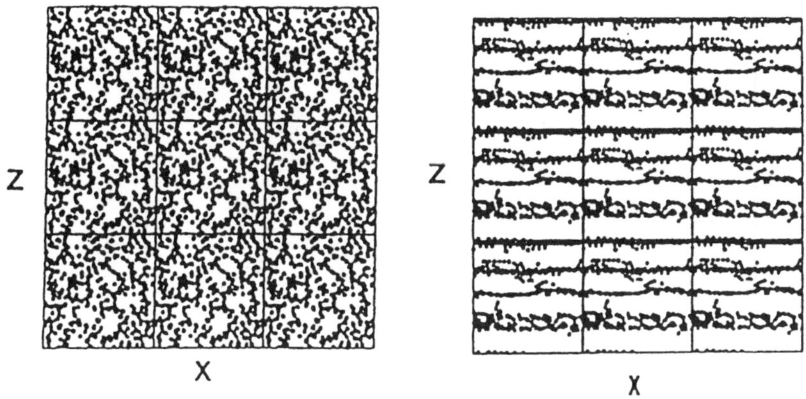

(a) (b)

Figure 15.8 *Transition of electrorheological suspension of 0.30 volume fraction gel-like network* (a) *to layered structure* (b)

Here we only illustrate a few sample results obtained for the first model system (the electrorheological fluids) using the simulation model. In the presence of a strong electric field, the apparent viscosity of an electrorheological suspension is appreciably higher than that manifested at zero field and the suspension is strongly shear thinning. This is because for a strong enough applied electric field and under shear, the initial gel-like network transforms into a phase of flowing layers. Figure 15.8 shows such a transition graphically. It was obtained using 500 particles at 30% volume concentration. Figure 15.8(a) is a snapshot of the structure of the suspension at rest without shear, generated after 12×10^4 time steps; Figure 8(b) is its counterpart generated after 10×10^4 time steps with the shear applied at a rate of 2×10^{-5} per time step. From the simulation results obtained over a wide range of Peclet number, *Pe*, and particle interaction strength, *Q*, at 30% solid volume concentration, a non-equilibrium phase diagram can be constructed, as shown in Figure 15.9. Four distinct phases exist. The ordered phases include a layered phase (LY) and the shear string phase (SS). A liquid phase (L2) intervenes between the shear string and layer phase. The Brownian line marks out the other liquid phase (L1) and is the extension of the equilibrium transition. The line of Stokesian transitions arises from the competition between the polarization and shear forces, and the line of Brownian transitions from that between Brownian and polarization forces. The cross-over (denoted by the dashed circle in Figure 15.9) from the Brownian to Stokesian transitions can be interpreted as a particle size effect.

The behaviour of the electrorheological fluids may be described by a power law in terms of the relative viscosity, η_r, against the Mason number, M_n (defined as the dimensionless ratio of the shear to dipole forces): $\eta_r \propto M_n^{-\alpha}$. If $\alpha = 1$, this corresponds to a Bingham model with a well-defined dynamic yield stress τ_y. For $\alpha < 1$, the model is pseudoplastic with no true yield stress. When $\alpha > 1$, it is shear

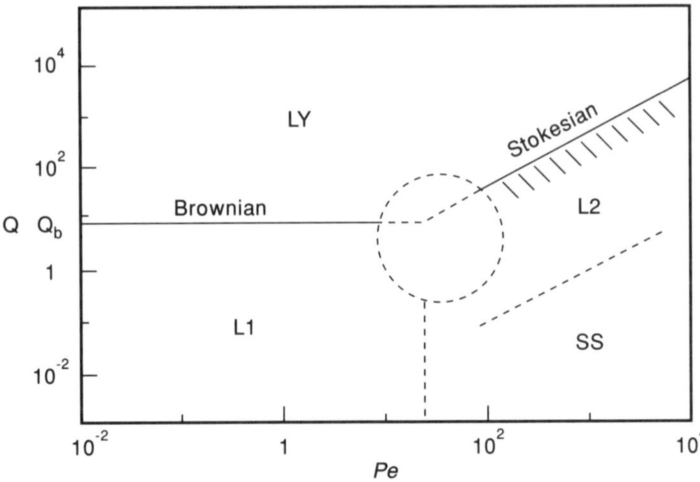

Figure 15.9 *Non-equilibrium phase diagram of the electrorheological model suspension at 30% volume concentration*

thickening. Figure 15.10(a) shows the relative viscosity versus Mason number plot obtained in conditions with minimal particle overlap and without Brownian motion (i.e. the Stokesian limit where the shear force by far exceeds Brownian forces). The data indicate that, over the range of M_n studied, the rheology is power law shear thinning. The insert to Figure 15.10(a) is a plot of the reduced stress τ^* against *Pe* for the low-M_n points which suggests that a dynamic yield stress exists. The data can be fitted quite well by the Casson equation, as can be seen from Figure 15.10(b). Since the simulation model uses aligned point dipoles

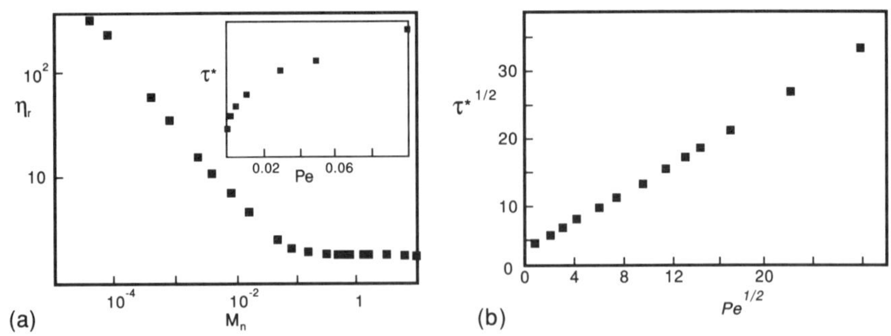

Figure 15.10 *Simulation prediction of the rheology of the model suspension.* (a) *Relative viscosity against Mason number in the Stokesian limit. The insert is a plot of reduced stress against shear rate for low-M_n points.* (b) *Casson plot of the square root of the reduced stress against the square root of the shear rate*

which underestimate the field-induced interactions, the viscosity and stress predicted by the simulation are understandably much lower than those of the real suspensions.

The interesting aspect of the next example is that the simulated particles are not spherical but plate-like. The simulation was conducted by Sherwood (1992) using a simplified version of the BD algorithm given by Ermark and McCammon (1978). The plate-like particles have dimensions of $2a \times 2a$ and zero thickness, and are placed in a monolayer within a unit cell with periodic boundary conditions. The forces considered by the simulation include shear force, electrical double layer and Born interactions, Stokes drag and couple on an isolated plate, and Brownian motion (translational as well as rotational). Hydrodynamic interactions between particles and inertia are neglected. The charges on the plates are represented by a series of point charges, each surrounded by an electrical double layer. This simplifies the calculation of electrical double force between the plates. If the charges are all identical, small liquid-crystal-like ordered domains are formed in the absence of shear (Figure 15.11(a)), and long-range ordering can be induced by shear (Figure 15.11(b)). If the charges on the edge of the plate are of a different sign from those on the plate surface, the particles can flocculate to form a gel spanning the whole unit cell in the absence of shear (Figure 15.12(a)). At low shear rates a single shear band appears in the gel (Figure 15.12(b)), and as the shear rate increases the typical size of flocs is found to decrease. Although the simulation results agree in general with one's intuition, some predictions (such as the long structure as shown in Figure 15.11 and the severe vulnerability of the gel to the imposed shear) are peculiar to the two-dimensional nature of the simulated system and may not be observable in the three-dimensional reality of flocculated suspensions.

(a) (b)

Figure 15.11 *Configurations of plates of equal charges* (a) *without shear and* (b) *under shear*

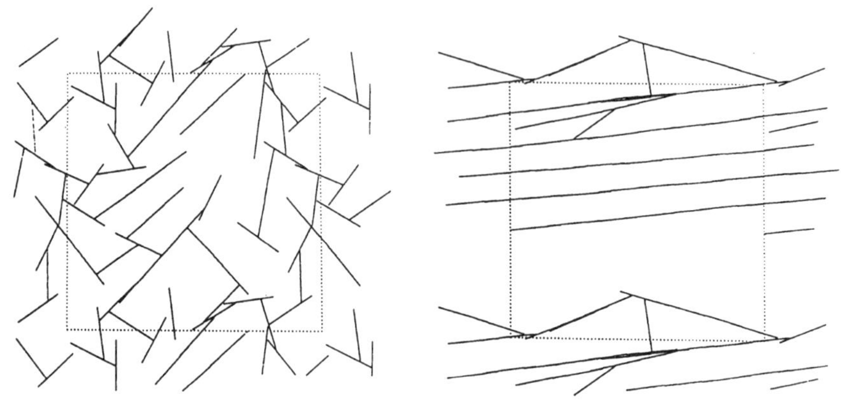

Figure 15.12 *Configurations of plate of opposite charges* (a) *without shear and* (b) *under shear*

Strictly speaking, the example given below does not fall neatly into the category of the BD simulation since Brownian motion is totally neglected. Nor does it belong to MD simulation as described in Chapter 7, since, first, it does not consider interactions between all the constituent particle species as a typical MD simulation would normally do, and, second, the equations of motion are not solved using the algorithms usually adopted by MD simulations. It has been chosen as an example here because this kind of simulation is not unusual when dealing with complex systems such as colloidal dispersions. The simulation methods described in Chapter 7 have been developed and matured in the course of studying simple liquids over the past three or four decades and they are now well established. The scenario in the area of colloids is quite different. Computer simulation of colloids has gained its popularity only fairly recently. There is no such thing as a 'standard' or well-established simulation method for colloidal systems in general; many simulation models are *ad hoc* models based on some basic ideas which have been originated from the simulation algorithms developed for simple liquids and were subsequently adapted for colloid simulations.

The example given here is called the sticky sphere model (Doi and Chen, 1989; Chen and Doi, 1989), and attempts to simulate the kinetics and rheological properties of aggregating colloids in shear flow. The colloidal particles are assumed to form a cluster when they touch. The cluster can deform, rupture and aggregate with other clusters. Apparently, the model acquires the name from one of its principal assumptions that upon collision the colloidal particles will simply stick together; the stuck spheres can roll over each other but are not allowed to slip at the contact point. For this to happen, whilst still allowing for the breakage of clusters, a maximum (or critical) value is assumed for the attractive constraining force, f_c. As such, the model can characterize all the interparticle (colloidal and hydrodynamic) forces using a single parameter, f_c, is assumed to have an infinitesimal effective range, which means particles will only interact with each upon contact. This greatly simplifies the problem, and since Brownian motion is also neglected, the dynamics of the system are deterministic, governed only by the hydrodynamics of each individual particle. Further details and mathematical treatments can be found in the original paper, and will not be given here. Suffice it to say that even with so many short-cuts, the computational requirements are still beyond most conventional computers, and the scale of the simulation system has to be kept small (up to 1000 particles) in order for one simulation run to be completed on a mainframe supercomputer within a reasonable time period (several hours of CPU time).

Despite the simplifications, the model still maintains the essential physics of shear-induced aggregation, and is thus able to give some qualitative understanding about the structure of aggregates, and the relation between structure and rheological properties; direct and quantitative comparison with experiments, though, is not to be expected. Using the sticky sphere model, the authors (Doi

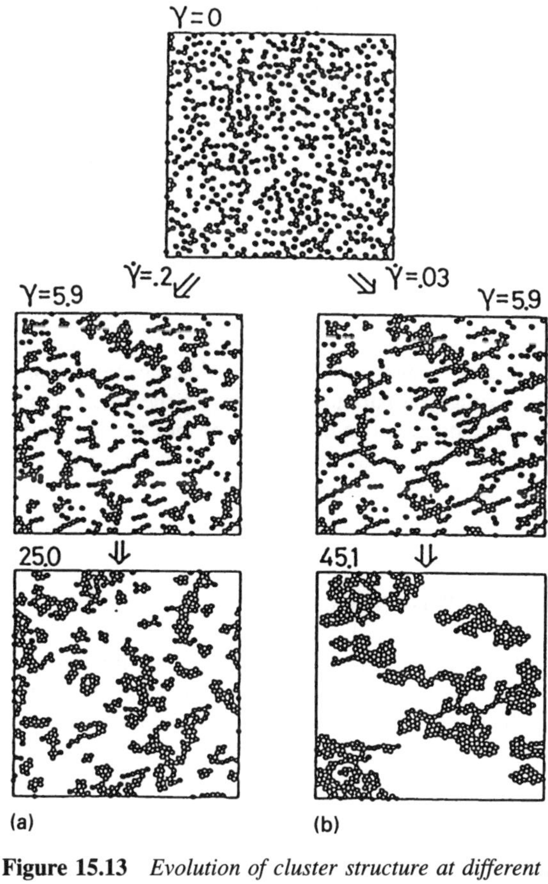

Figure 15.13 *Evolution of cluster structure at different shear rates*

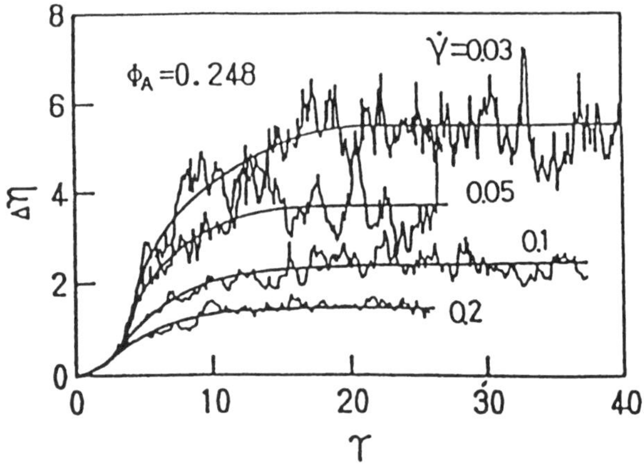

Figure 15.14 *Viscosity versus shear strain*

and Chen, 1989; Chen and Doi, 1989) have studied the time evolution of the structure of aggregates and macroscopic viscosity of the dispersion in shear flow. Figure 15.13 shows the simulated evolution of the structure of aggregates under different shear rates $\dot{\gamma}$. By comparing the two different cases shown in this figure, it can be seen that the shear-induced aggregates are chain-like initially but as the shear strain increases they curl to become more compact. In the steady state, the average cluster size is larger at low shear rates than at high shear rates.

In the transient state, the viscosity of the dispersion is found to increase in a sigmoidal manner. This can be seen in Figure 15.14. The results are qualitatively in agreement with experimental observations. The steady-state viscosity depends on the shear rate and obeys the same power law for all particle concentrations, and this suggests that the critical structural parameter governing the macroscopic viscosity of the dispersion is not the overall size of the cluster, but the size of the more rigid subclusters, the reason being that the bonds between these subclusters are flexible, making the whole cluster deformable under shear.

15.6 Stokesian dynamics simulations

The majority of BD simulations of colloidal suspensions reported in the literature make use of some simple far-field asymptotic expressions, such as the Oseen and Rotne–Prager tensors, for the diffusion tensor. Whilst these may be accurate for very dilute suspensions, they are not adequate to simulate the structure and behaviour of concentrated suspensions, where many-body interactions and lubrication play a predominant role. Bossis and Brady (1984) have developed an elegant method, called Stokesian dynamics, for numerically simulating the dynamics of interacting particles in Stokes flow. In its original form, Stokesian dynamics uses a MD-like approach to follow the time evolution of particles. Since then Brownian motion has also been incorporated (Brady and Bossis, 1988). The resulting algorithm provides a general and versatile framework, capable of predicting both microstructural and macroscopic properties of colloidal dispersions, and encompassing both high (Stokesian limit) and low (Brownian limit) *Pe* number limits as special cases with no change in approach or procedure. By comparison, the BD algorithm pioneered by Ermark and McCammon (1978) using Oseen or Rotne–Prager tensors is only applicable to dilute suspensions at the low *Pe* number limit. Details of the theoretical treatment of Stokesian dynamics are rather involved and will not be presented. Some simulation results will be given below to demonstrate the ability of the simulation algorithm, and to show the importance of many-body hydrodynamic and lubrication forces.

Using Stokesian dynamics with Brownian motion, Bossis and Brady (1987) have carried out simulations of a monolayer of identical Brownian hard spheres in a simple shear flow, and predicted the pair distribution function and self-diffusivity of the monolayer dispersion as a function of the Peclet number. In the purely Brownian case (i.e. $Pe=0$), the simulations reproduced the pair distribution function of equivalent hard disc system. This can be seen in Figure 15.15, which compares the radial distribution functions obtained from Stokesian dynamics simulation of a monolayer of 25 hard spheres by Bossis and Brady (1987) and Monte Carlo simulation of a hard disc fluid by Chae *et al.* (1969) at the same area fraction of 0.453. It is not a trivial achievement for a dynamic

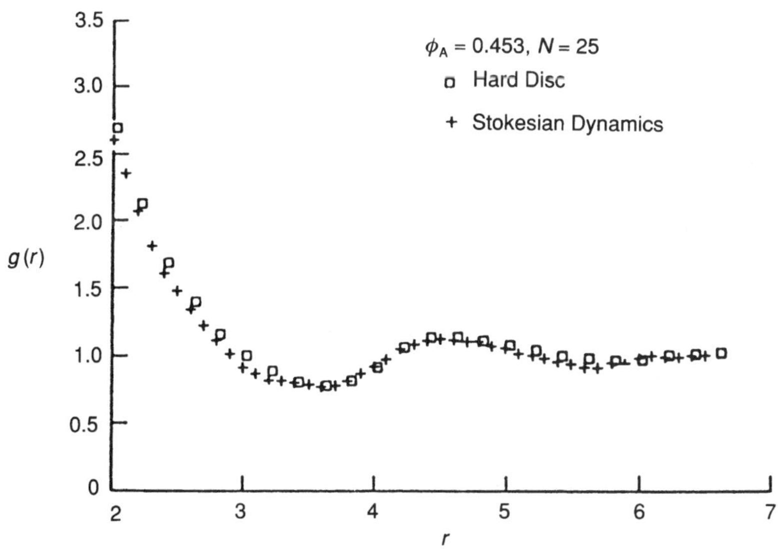

Figure 15.15 *Pair distribution function curves predicted by Stokesian dynamics simulation (Bossis and Brady, 1987) and Monte Carlo calculation (Chae* et al., *1969)*

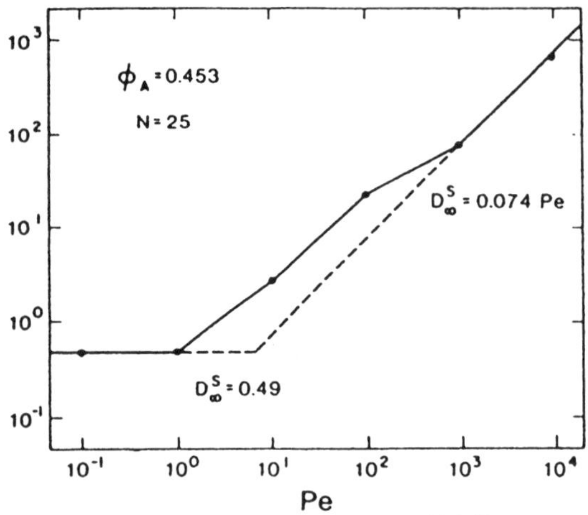

Figure 15.16 *The scaled long-time self-diffusivity as a function of* Pe

simulation involving no interparticle forces other than hydrodynamic to yield the correct pair distribution function, because only through a proper treatment of both near- and far-field hydrodynamic interactions will the correct $g(r)$ be obtained.

For self-diffusivity, the simulation predicted a clear transition from a Brownian motion dominated regime ($Pe < 1$) to a hydrodynamically dominated regime ($Pe > 10$). The most dramatic change is seen in the behaviour of the long-time self-diffusivity (Figure 15.16), which changes from being independent of Pe to growing linearly with Pe, with a cross-over around $Pe = 10$. Visual evidence of such a transition is shown in Figure 15.17. In the low Peclet number limit particles are well dispersed and the short-time self-diffusivity is larger than the long-time self-diffusivity. In the high Peclet number limit a large and cell-spanning cluster is formed along the compressive axis of the shear flow and the short-time self-diffusivity becomes smaller but the long-time self-diffusivity approaches infinity.

Although the Stokesian dynamics algorithm provides a rigorous and yet elegant method for incorporating many-body hydrodynamics into the BD-style simulations of colloidal suspensions and hence a general framework for undertaking dynamic simulation of colloidal systems, the increased complexity and, particularly, the demand for computational resources has limited its application to small-scale simulations. For example, the results shown in the Figures 15.15 and 15.16 were obtained using only 25 identical hard spheres in two-dimensional space. Only in the special cases, where $Pe = 0$ or ∞ and the governing equation can be further simplified, was the use of more particles feasible (Figure 15.17). However, with the increasing availability of high-performance computers and the advent of more powerful computers, this constraint can hopefully be lifted in the future.

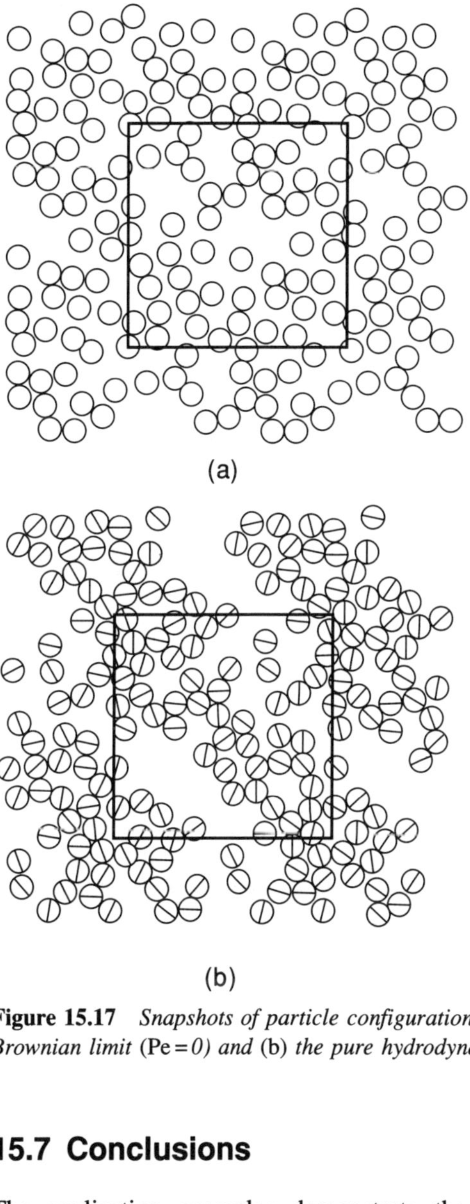

Figure 15.17 *Snapshots of particle configurations for 49 particles in the* (a) *pure Brownian limit* (Pe = *0) and* (b) *the pure hydrodynamic limit* (Pe^{-1} = *0)*

15.7 Conclusions

The application examples demonstrate the extent to which computational methods can be applied to predict the behaviour of colloidal dispersions. Most significantly, these methods provide the means to estimate the likely trends in dispersion behaviour when process and chemical conditions are changed. However, the principal limitations in the use of simulations based on microscopic models is that detailed knowledge of the dispersed phases, interaction forces and microhydrodynamics is required. Hence the simulation of complex colloidal dispersions, such as those containing a range of particle species with differing surface properties at high volume fraction in ill-defined hydrodynamic environments, is not possible. It is likely that computational limitations in performing calculations of many-body interactions will not, ultimately, be the limiting factor in allowing predictions to be made. The key technical limitation is the lack of knowledge concerning the fundamental physics, e.g. inadequate means of solving or describing hydrodynamic tensors, uncertainty of the surface physics and nature of non-DLVO interaction forces.

Despite these shortcomings there is a considerable incentive to facilitate the *engineering* and *visualization* of the dynamic behaviour of colloidal systems. The goal is similar to that already largely achieved in the use of molecular modelling for drug design, namely, to determine and optimize the formulation of colloidal dispersions and to describe their behaviour in process equipment and storage containers will have radical implications in manufacturing technology.

References

Ansell, G. C. and Dickinson, E. (1985) Aggregate structure and coagulation kinetics in a concentrated dispersion of interacting colloidal particles. *Chem. Phys. Lett.*, **122**, 594–598

Ansell, G. C. and Dickinson, E. (1986a) Sediment formation by Brownian dynamics simulation: effect of colloidal and hydrodynamic interactions on the sediment structure. *J. Chem. Phys.*, **85**, 4079–4086

Ansell, G. C. and Dickinson, E. (1986b) Brownian dynamics simulation of fragmentation of a large colloidal floc in simple shear flow. *J. Colloid Interface Sci.*, **110**, 73–81

Ansell, G. C. and Dickinson, E. (1987) Brownian dynamics simulation of the formation of colloidal aggregate and sediment structure. *Faraday Discuss. Chem. Soc.*, **83**, 167–177

Ansell, G. C., Dickinson, E. and Ludvigsen, M. (1985) Brownian dynamics of colloidal-aggregate rotation and dissociation in shear flow. *J. Chem. Soc. Faraday Trans 2*, **81**, 1269–1284

Bacon, J., Dickinson, E., Parker, R., Anastasiou, N. and Lal, M. (1983) Motion of flocs of two or three interacting colloidal particles in a hydrodynamic medium. *J. Chem. Soc. Faraday Trans 2*, **79**, 91–109

Bradbury, A., Menear, S. and Chantrell, R. W. (1986) A Monte Carlo calculation of the magnetic properties of a ferrofluid containing interacting polydispersed particles. *J. Magn. Magn. Mat.*, **54**–57, 745–746

Bossis, G. and Brady, J. F. (1984) Dynamic simulation of sheared suspensions. I. General method. *J. Chem. Phys.*, **80**, 5141–5154

Bossis, G. and Brady, J. F. (1987) Self-diffusion of Brownian particles in concentrated suspensions under shear. *J. Chem. Phys.*, **87**, 5437–5448

Brady, J. F. and Bossis, G. (1988) Stokesian dynamics. *Annu. Rev. Fluid Mech.*, **20**, 111–157

Chae, D. G., Ree, F. H. and Ree, J. (1969) Radial distribution functions and equation of state of the hard-disk fluid. *J. Chem. Phys*, **50**, 1581–1589

Chan, D. Y. C. and Halle, B. (1984) Dissociation kinetics of secondary-minimum flocculated colloidal particles. *J. Colloid Interface Sci.*, **102**, 400–409

Chantrell, R. W., Bradbury, A., Popplewell, J. and Charles, S. W. (1982) Agglomerate formation in a magnetic fluid. *J. Appl. Phys.*, **53**, 2742–2744

Chen, D. and Doi, M. (1989) Simulation of aggregating colloids in shear flow II. *J. Chem. Phys.*, **91**, 2656–2663

Clark, A. T., Lal, M. and Waston, G. M. (1987) Dynamics of colloidal particle in the vicinity of an interacting surface. *Faraday Discuss. Chem. Soc.*, **83**, 179–191

Davies, P., Popplewell, J., Martin, G., Bradbury, A. and Chantrell, R. W. (1986) Monte Carlo simulations of the structure of ferrofluid composites. *J. Phys. D*, **19**, 469–476

Doi, M. and Chen, D. (1989) Simulation of aggregating colloids in shear flow. *J. Chem. Phys.*, **90**, 5271–5279

Gupta, D. and Peters, M. H. (1985) A Brownian dynamics simulation of aerosol deposition on spherical collectors. *J. Colloid Interface Sci.*, **104**, 375–389

Ermark, D. L. and McCammon, J. A. (1978) Brownian dynamics with hydrodynamic interactions. *J. Chem. Phys.*, **69**, 1352–1360

Ito, K., Ono, T., Kitano, H. and Ise, N. (1990) Direct examination of association process of colloidal dispersion by visual imagery. *Int. Symp. Colloid Surf. Eng.*, San Diego, 21–25 August, 1990, 9–10

Jia, X. (1992) Computer simulation of colloidal aggregation and deposition. PhD Thesis, University of Manchester Institute of Science and Technology

Melrose, J. R. and Heyes, D. M. (1993) Simulations of electrorheological and particle mixture suspensions: agglomerate and large structures. *J. Chem. Phys.*, **98**, 5873–5886

Peters, M. H. (1988) Adsorption of interacting Brownian particles onto surfaces I. General formulation of the macroscopic diffusion equation and results for non-hydrodynamic force interactions. *J. Colloid Interface Sci.*, **121**, 179–192

Peters, M. H. and Gupta, D. (1989) Brownian dynamics simulation of convective transport and adsorption of hydrodynamically interacting particles onto surfaces. In *Studies in Physical and Theoretical Chemistry*, Vol 63, *Math/Chem/Comp 1988* (ed. Graovac, A.) Elsevier, Amsterdam, 387–410

Sherwood, J. D. (1992) Brownian dynamics simulation of a 2-D suspension of charged colloidal plates under shear. *J. Non-Newtonian Fluid Mech.*, **43**, 195–228

Sjollema, J. and Busscher, H. J. (1989) Deposition of polystyrene latex particles towards polymethylmethacrylate in a parallel flow cell. *J. Colloid Interface Sci.*, **132**, 382–394

Williams, R. A. and Jia, X. (1991) Simulation of magnetic flocculation behaviour of fine minerals. *Int. J. Min. Proc.*, **32**, 175–192

Author Index

Subject Index